METHANE CONVERSION BY OXIDATIVE PROCESSES

METHANE CONVERSION BY OXIDATIVE PROCESSES
Fundamental and Engineering Aspects

Edited by

E. E. Wolf

VAN NOSTRAND REINHOLD CATALYSIS SERIES

VNR SPRINGER SCIENCE+BUSINESS MEDIA, LLC

Originally published by Van Nostrand Reinhold in 1992

Library of Congress Catalog Card Number 91-28504

ISBN 978-94-015-7451-8

Manufactured in the United States of America

16 15 14 13 12 11 10 9 8 7 6 5 4 3 2 1

Library of Congress Cataloging-in-Publication Data

Methane conversion by oxidative processes: fundamental and engineering aspects/[edited by] Eduardo E. Wolf.

p. cm.

ISBN 978-94-015-7451-8 ISBN 978-94-015-7449-5 (eBook)

DOI 10.1007/978-94-015-7449-5

1. Synthetic fuels. 2. Methane. 3. Oxidation. I. Wolf, Eduardo E.

TP360.D58 1991

662′.669—dc20 91-28504

CIP

VAN NOSTRAND REINHOLD
CATALYSIS SERIES

Burtron Davis, Series Editor

Metal-Support Interactions in Catalysis, Sintering, and Redispersion, edited by Scott A. Stevenson, R.T.K. Baker, J.A. Dumesic, and Eli Ruckenstein

Molecular Sieves: Principles of Synthesis and Identification, R. Szostak

Raman Spectroscopy for Catalysis, John M. Stencel

Theoretical Aspects of Heterogeneous Catalysis, John B. Moffat

Biocatalysis, edited by Daniel A. Abramowicz

Methane Conversion By Oxidative Processes: Fundamental And Engineering Aspects, edited by E. E. Wolf

Contents

Series Introduction

A reasonable case could be made that the scientific interest in catalytic oxidation was the basis for the recognition of the phenomenon of catalysis. Davy, in his attempt in 1817 to understand the science associated with the safety lamp he had invented a few years earlier, undertook a series of studies that led him to make the observation that a jet of gas, primarily methane, would cause a platinum wire to continue to glow even though the flame was extinguished and there was no visible flame. Döbereiner reported in 1823 the results of a similar investigation and observed that spongy platina would cause the ignition of a stream of hydrogen in air. Based on this observation Döbereiner invented the first lighter. His lighter employed hydrogen (generated from zinc and sulfuric acid) which passed over finely divided platinum and which ignited the gas. Thousands of these lighters were used over a number of years. Döbereiner refused to file a patent for his lighter, commenting that "I love science more than money." Davy thought the action of platinum was the result of heat while Döbereiner believed the effect was a manifestation of electricity. Faraday became interested in the subject and published a paper on it in 1834; he concluded that the cause for this reaction was similar to other reactions. Berzelius viewed things differently from Faraday and introduced the concept of catalysis in 1835. Berzelius is credited with the recognition of catalysis even though his view that it was the result of a catalytic force was incorrect.

From the beginning, catalysis has been associated with practicality. Over the years, and especially in our century, industrial applications have been a strong driving force in determining the areas of intense catalytic research efforts. This is especially true in the case of natural gas, and of methane in particular. Within the past 20 years, real—or perceived—shortages have provided the basis for a period of intense research on methanation catalysis. During the same period, recognition of vast supplies of natural gas in locations where the demand is low or nonexistent has provided motivation for extensive research to convert methane to more valuable products, preferably liquid, by partial oxidation.

A science partly driven by economics will not always show an orderly progression of activity. An analogous situation applies to a series of books on catalysis: It is never possible to develop a series in orderly fashion. Potential authors find it increasingly difficult to justify the time needed to write a book. The current economics of book publishing in a specialized area such as catalysis requires the author to work primarily for personal satisfaction—the reward must be found in a paraphrase of the words of Döbereiner: "I love catalysis more than money."

This book provides another volume that surveys an emerging process—the partial oxidation of methane—and provides coverage of both basic and engineering aspects. This is entirely justified by the practical nature of catalysis.

Burtron H. Davis

Preface

The editor of the *Catalysis* series, Dr. Burtron Davis, approached me some time ago with a proposal to write a book on methane activation. This is an active field in the area of catalysis and energy conversion, and, although several review articles exist in the area, a comprehensive state-of-the-art book summarizing the latest research and economic status of methane conversion has not been put forward. Although I have been involved in research in this area since relatively early on, writing a book on the subject was quite beyond what my time commitments permitted. As I felt that a book in this subject was badly needed, I proposed instead to edit a book to which the main researchers in the area would contribute. Fortunately, the response from every coauthor that I approached was positive and, moreover, all but one delivered.

The objective of the book is not to summarize again what is already known from existing reviews, but to provide a comprehensive, in-depth report of the latest available research or engineering results on the conversion of methane via oxidative processes from each author's group. The book focuses primarily on the oxidative coupling of methane because this is the conversion route that has received the most attention, with one chapter devoted to partial oxidation because it has been less studied. Indirect conversion of methane is excluded because it is a mature subject that is discussed in many articles, books, and textbooks.

The book is organized into two parts. The first deals with fundamental aspects, consisting of chapters describing state-of-the-art work done in the leading laboratories around the world, whereas the second part deals with engineering aspects, including economic evaluations. The fundamental portion contains descriptions of studies concerning the reaction mechanism, catalyst characterization and evaluation, and the latest techniques used in such studies.

The engineering evaluation work is more restricted because it comes from leading industrial researchers subject to publication restrictions. It describes the engineering problems involved in the large-scale implementation of the process, including reaction and reactor engineering as well as economic evaluations of the process alternatives. The economic information is quite important because, due to proprietary restrictions, very little is known in the open literature of the latest economic assessment of the technology. Unfortunately, I was able to convince only a few authors from industry to write about the economic and engineering assessments and problems. Some who agreed at first had to withdraw due to legal restrictions imposed by their companies. Some obvious omissions are due to this reason and not due to the lack of willingness of this editor to find adequate sources. Nonetheless, I think that at least some perspective on the more applied, yet all-important, aspects of this process is placed alongside the more fundamental and basic information summarized in the first part, making this volume rather unique in the blend of scientific engineering and economic aspects. The summary (Chapter 16) attempts to present a critical assessment of the field based on the conclusions reached by the various authors in the open literature.

I wish to thank all the authors who took time from their busy schedules to put in print their views on the subject. Thanks to my friends G. Martin and C. Mirodatos for helping me during my stay at the Institut de Recherches sur la Catalyse, in Villeurbanne, France, where most of the reviewing took place. I also wish to thank my collaborator and graduate student Z. Kalenik for helping me with the details in handling the final manuscript. I also would like to explain that limitations on the size of the volume prevented me from including many other leading authors who otherwise should have been included.

Notre Dame

E. E. Wolf

I

Fundamentals

1

Formation and Reactions of Methyl Radicals over Metal Oxide Catalysts

J.H. Lunsford

Department of Chemistry, Texas A & M University, College Station, TX

TABLE OF CONTENTS

INTRODUCTION

Current interest in the oxidative coupling of methane has focused attention on the formation and reactions of methyl radicals. Although other intermediates may contribute to the formation of minor amounts of ethylene (Hutchings, Woodhouse, and Scurrell 1989), it is generally agreed that the reaction sequence involves the coupling of $CH_3\cdot$ radicals to form C_2H_6, followed by the dehydrogenation of C_2H_6 to form the desired product C_2H_4 (Nelson, Lukey, and Cant 1988; Mims et al. 1989). Within this rather simple mechanistic framework many ancillary issues remain to be reconciled, as will be evident in this and subsequent chapters of the book. Some of these issues include the site at which the initial $CH_3\cdot$ radicals are formed, the role of surface-initiated gas-phase radical reactions, and the origin of carbon oxides (CO and CO_2). The diversity of catalysts and operating conditions makes it difficult to generalize; however, we attempt to show that surface-generated gas-phase radicals are important intermediates in the catalytic process and that the fate of these radicals determines, in part, the selectivities that are obtained.

The conditions employed in the selective oxidation of CH_4 often approach those of combustion, which is dominated by free radical reactions. The high reaction temperatures, the presence of O_2 in the gas phase, and the inertness of the catalysts for reaction with $CH_3\cdot$ radicals make the oxidative coupling of CH_4 a particularly suitable reaction for a heterogeneous–homogeneous catalytic process; that is, a process in which surface-generated intermediates react further in the gas phase to form the desired products. Heterogeneous–homogeneous reaction schemes have been proposed for a number of oxidation reactions, but only recently have spectroscopic techniques been used to confirm the presence of reactive intermediates that emanate from the catalytic surface (Driscoll, Campbell, and Lunsford 1987; Garibyan and Margolis 1990). Even when radical intermediates are identified, it is a difficult task to prove that they are involved in the primary reaction channel. Yet with $CH_3\cdot$ radicals this seems to be the case.

Apart from some theoretical studies by Anderson and co-workers (Mehandru, Anderson, and Brazdil 1988), little is known concerning the adsorption properties of $CH_3\cdot$ radicals on metal oxides. On the predominant plane of MgO, methyl radicals are not strongly bonded, and at high temperatures entropy considerations would favor desorption of the radicals. As is shown later, the sticking coefficient of the radicals is small, even for relatively reactive oxides. Thus, once the radicals enter the gas phase they are able to collide many times with the surface before they undergo reaction. The low probability of reaction with the surface is important for obtaining high ethane plus ethylene (C_2) selectivities as the concentration of $CH_3\cdot$ radicals in the gas phase is small and collisions between two $CH_3\cdot$ radicals and a third body are infrequent, relative to collisions between the radicals and a surface.

In this chapter two spectroscopic methods for the detection of $CH_3\cdot$ radicals are discussed: the matrix-isolation electron spin resonance (MIESR) technique and resonance-enhanced multiphoton ionization (REMPI). Qualitative data showing the relationships between radical formation and catalytic properties are presented. In addition, quantitative evidence shows that the coupling of $CH_3\cdot$ radicals in the gas phase indeed

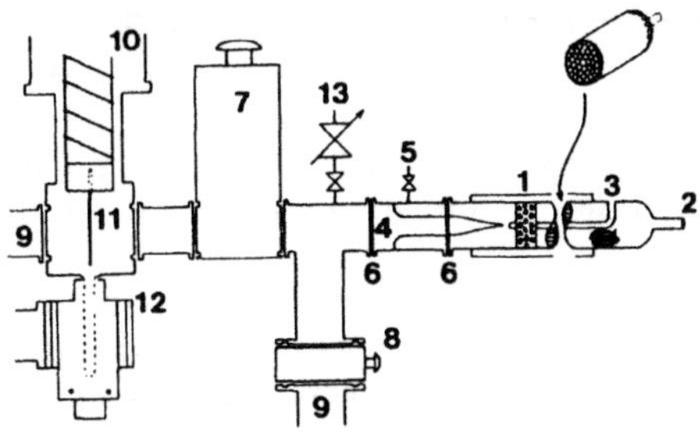

FIGURE 1-1. Schematic of ESR matrix-isolation system, coupled to the catalytic reactor: (1) catalysts; (2) gas inlet; (3) thermocouple well; (4) leak; (5) outlet to gas chromatograph; (6) O-ring joints; (7) gate valve; (8) butterfly valve, (9) to vacuum pumps; (10) vacuum shroud; (11) sapphire rod; (12) microwave cavity; (13) quadrupole mass spectrometer inlet. (Reprinted with permission from Martir and Lunsford 1981, *J. Am. Chem. Soc.* 103:3728–3782. Copyright 1981 American Chemical Society.)

provides a major pathway for the formation of C_2 products. Finally, secondary reactions of gas-phase $CH_3\cdot$ radicals with metal oxides are described in order to demonstrate the importance of this factor in determining C_2 selectivity.

EXPERIMENTAL METHODS

The MIESR system that has been used for most of the studies involving $CH_3\cdot$ radicals is shown schematically in Fig. 1.1. Several modifications of the system now exist. The system essentially consists of a hot catalyst bed that serves to generate the radicals, a differentially pumped region, and a cold finger on which the radicals are matrix-isolated for analysis by ESR spectroscopy. In the version developed in our laboratory (Martir and Lunsford 1981) the reactant gases, including Ar (which eventually forms the matrix), pass through a preheater and are then allowed to contact a thin bed of the analyst. Generally, a small quantity of catalyst chips is dispersed in several layers of quartz wool, although another variation includes a thin layer of catalyst deposited on the downstream side of a porous quartz frit. The leak into the low-pressure region is a hole in the end of a quartz tube. The pressure in the catalyst region will determine the size of the leak. The pressure at the matrix should be on the order of 10^{-4} to 10^{-5} torr, as higher pressures allow too much heat transfer to the matrix. The matrix is formed on a sapphire rod maintained at ca. 16 K. Cryogenic cooling is provided by a Displex model CSA 202 closed-cycle refrigerator.

Typically the matrix is formed over a 15- to 30-min period in order to build up an adequate concentration of radicals. If the catalyst is particularly active for radical formation, this time could be reduced to less than a minute. After the matrix is formed, the sapphire rod is lowered into a quartz shroud, and the microwave cavity is raised so that the shroud is in the cavity.

The bimolecular reaction between the $CH_3\cdot$ radicals requires a third body, therefore it is highly dependent upon the total pressure. As a consequence the radicals are stable once they enter the low-pressure region. If molecular oxygen is present in the product stream, $CH_3O_2\cdot$ radicals will be formed in regions where the temperature is low, and the spectrum of these radicals will be observed in addition to the spectrum of $CH_3\cdot$ (see below). Two pressure regions over the catalyst have been adopted for most of the experiments carried out in our laboratory: the "high-pressure" experiments were at a total pressure of 760 torr and the "low pressure" experiments were at ca. 1 torr. At the high pressure it is convenient to analyze the total product distribution by gas chromatography and to compare the rate of CH_4 conversion with the rate of $CH_3\cdot$ radical production. A disadvantage of the high-pressure experiment is the fact that a significant fraction of the $CH_3\cdot$ radicals will couple or otherwise react in the gas phase before they enter the leak. In the low-pressure mode the $CH_3\cdot$ radicals may travel several centimeters away from the catalyst bed without extensively reacting. Generally the partial pressure of oxygen is kept small so that the $CH_3\cdot$ radicals mainly couple rather than react with O_2, and the ESR spectrum is that of $CH_3\cdot$ rather than $CH_3O_2\cdot$ radicals.

For many experiments it is sufficient to know the relative rate of production of $CH_3\cdot$ radicals; however, the production of $CH_3\cdot$ radicals can be determined quantitatively by suitable calibration of the system. Three measurements must be made: (i) the number of radicals in the matrix; (ii) the collection–detection efficiency; and (iii) the fraction of the gas stream that enters the leak. The number of radicals is obtained by comparing the second integral of the $CH_3\cdot$ radical spectrum with that of a standard (Lunsford 1972). The collection–detection efficiency, that is, the fraction of the radicals entering the leak that are collected and detected, may be determined by introducing a known amount of a stable, gas-phase free radical into the system. Nitrogen dioxide has been used for this purpose (Campbell and Lunsford 1988; Driscoll, Campbell, and Lunsford 1987). Finally, the fraction of the gas stream entering the leak may be determined by measuring the rate of pressure drop from a fixed volume of gas that is attached to the leak. The overall error in the absolute rate of

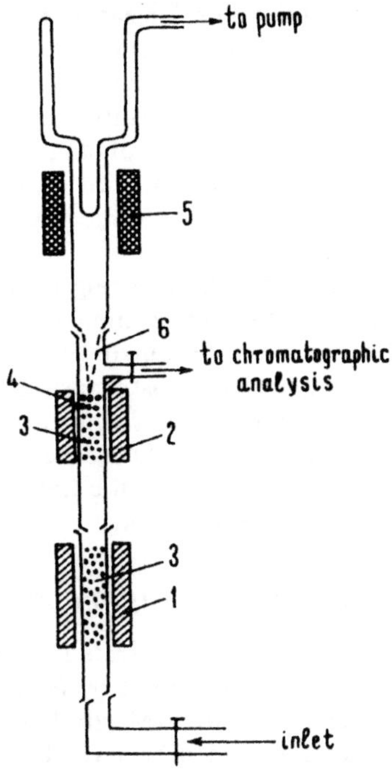

FIGURE 1-2. Schematic view of the apparatus: (1) pyrolysis unit for producing radicals; (2) catalytic quartz reactor; (3) quartz tube; (4) catalyst; (5) ESR spectrometer cavity; (6) diaphragm. (Reprinted from Garibyan and Margolis 1990 by courtesy of Marcel Dekker, Inc.)

radical production has been estimated to be ca. $\pm 30\%$, although the error in obtaining relative rates is considerably less.

A somewhat simpler matrix-isolation system, shown in Fig. 1-2, has been used by Soviet scientists to detect radicals formed during the partial oxidation of CH_4. The technique, first developed by Nalbandyan (Nalbandyan and Mantashyan 1975), is called the kinetic radical freezing method (KRFM). The essential elements are the same as those found in the MIESR system except that the radicals are frozen in a CO_2 matrix on a cold finger that is cooled with liquid nitrogen. Carbon dioxide may be added to the gas flow after the leak into the low-pressure region of the apparatus.

Resonance-enhanced multiphoton ionization has been used recently by two groups (Gulcicek, Colson, and Pfefferle 1990; Lee, Yu, and Lin 1990) to study gas-phase $CH_3\cdot$ radicals that emanate from metal oxide surfaces. The technique involves the use of photons from a laser to ionize the $CH_3\cdot$ radicals. In the REMPI experiments two photons simultaneously promote the radicals to an excited state, and a third photon excites the radical above its ionization potential. The ions formed by such a (2 + 1) REMPI process are then collected.

Perhaps the simplest method for detecting hydrocarbon radicals produced during the oxidative coupling reaction has been via the classical lead-mirror (Jones, Leonard, and Sofranko 1987). The disadvantages of the lead mirror are (i) that one cannot distinguish the types of radicals that are reacting with the mirror and (ii) that the mirror must be removed from the hot catalyst by several centimeters. Unless the linear velocity of the gas is very high or the pressure is low, most of the $CH_3\cdot$ radicals will couple in the gas phase before they reach the mirror.

HISTORICAL PERSPECTIVE

The earliest work to detect surface-generated gas-phase $CH_3\cdot$ radicals over metal oxide catalysts was carried out by Nersesyan and co-workers (Nersesyan et al. 1975) and by Martir (Martir 1981). In the latter study, CH_4 and O_2 or N_2O were passed over several oxides, including SiO_2, Al_2O_3, and MgO, and it was found that MgO was a superior catalyst for the formation of $CH_3\cdot$ radicals. The origin of this activity was thought to be either defect centers or impurities in the host oxide. Driscoll (Driscoll 1985) began a study of both possibilities and found that the addition of lithium, a common impurity in MgO, greatly enhanced the radical-forming ability of the oxide. Subsequently it was found, first by Liu (Liu 1983) and then by Ito (Ito and Lunsford 1985) that the same type of lithium-promoted MgO was an effective catalyst for the oxidative coupling of CH_4. Thus, what began as a fundamental study on radical formation resulted in the discovery of a new class of catalysts for the conversion of CH_4 to higher hydrocarbons.

METHYL RADICAL FORMATION AND REACTIONS OVER Li^+/ MgO AND RELATED CATALYSTS

The characteristic ESR derivative spectra of matrix-isolated $CH_3\cdot$ radicals and the corresponding $CH_3O_2\cdot$ radicals are shown in Fig. 1-3. These radicals were obtained over MgO, but, except for amplitude differences,

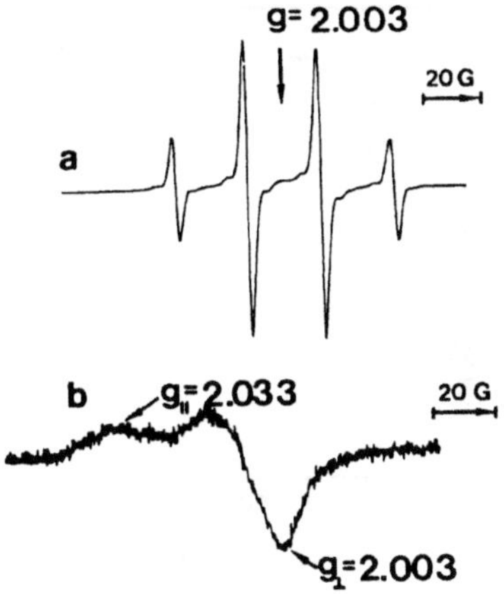

FIGURE 1-3. ESR spectra: (a) methyl radical ($CH_3\cdot$); (b) methyl peroxy radical ($CH_3O_2\cdot$). (Reprinted with permission from Driscoll et al. 1985, *J. Am. Chem. Soc.* 107:58–63. Copyright 1985 American Chemical Society.)

the same spectra have been obtained over many other oxide catalysts. If the line width does not change, radical concentrations are proportional to the amplitudes of the derivative spectra. By contrast, the shape of the spectrum of the $CH_3O_2\cdot$ radical is quite different from that of the $CH_3\cdot$ radical, and a comparison of concentrations would require one to obtain the second integral of both spectra. Even though the amplitude of the spectrum b is much less than spectrum a, as reflected in the signal-to-noise ratio, the number of spins that contribute to the two signals is similar. This is consistent with the fact that $CH_3\cdot$ radicals are being converted to $CH_3O_2\cdot$ radicals via the equilibrium

$$CH_3\cdot + O_2 \leftrightarrows CH_3O_2\cdot \qquad (1\text{-}1)$$

One of the means of demonstrating that gas-phase $CH_3\cdot$ radicals are important in the oxidative coupling reaction has been to correlate the relative radical formation rate with the catalytic conversion of CH_4. Although the data often have been obtained under different reaction conditions (e.g., different partial pressures of CH_4 and O_2), the correlations usually are very good for the simple oxides. For example, high-purity MgO is a rather poor $CH_3\cdot$ radical generator at $T \leq 720°C$, and it also is not effective as an oxidative coupling catalyst. But when Li^+ is added, both the radical generating ability and the activity for C_2 formation increase remarkably (Driscoll et al. 1985; Ito et al. 1985). For the analo-

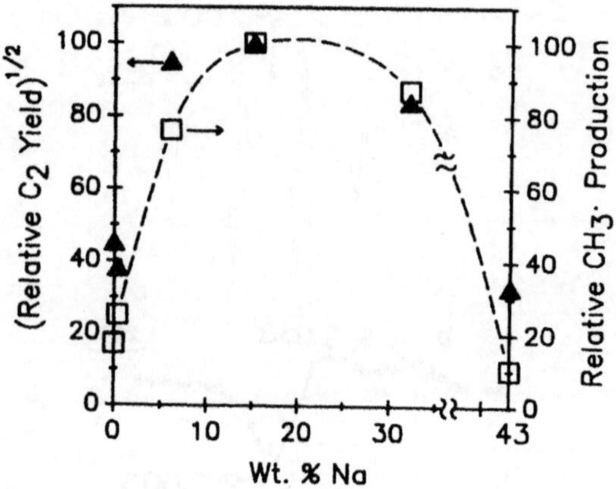

FIGURE 1-4. Relative $(C_2\ yield)^{1/2}$ (▲) and rate of methyl radical formation (□) as a function of Na^+ loading in a CaO catalyst; 43% Na^+ corresponds to pure Na_2CO_3. Reaction conditions for C_2 yield: $T = 675°C$; $He:CH_4:O_2 = 32:22:1.4$ ml min^{-1}; 760 torr. Reaction conditions for $CH_3\cdot$ production: $T = 700°C$; $Ar:CH_4:O_2 = 3.8:1.1:0.024$ ml min^{-1}; 1 torr. (Reprinted with permission from Campbell and Lunsford 1988, *J. Phys. Chem.* 92:5792–5796. Copyright 1988 American Chemical Society.)

gous Na^+/CaO catalyst this type of behavior is depicted in Fig. 1-4, where the square root of the relative C_2 yield and the relative $CH_3\cdot$ radical production are plotted as a function of the Na^+ ion content of the catalyst (Lin, Wang, and Lunsford 1988; Campbell and Lunsford 1988). The $(C_2\ yield)^{1/2}$ is used because the radical coupling reaction is bimolecular; that is, the probability of forming C_2H_6 is proportional to the square of the $CH_3\cdot$ radical concentration. The sample corresponding to 43 wt% Na was pure Na_2CO_3. The Na^+ ions in CaO give rise to $[Na^+O^-]$ centers, which are thought to be responsible for the activation of CH_4. Both the formation of $CH_3\cdot$ radicals and the production of C_2 compounds were proportional to the sodium loading, up to about 10 wt% Na, and at higher loadings the productivities decreased. The decrease presumably is a result of extensive coverage of the CaO by the less active Na_2CO_3 phase.

When the study was extended to include other combinations of group IA ions and group IIA oxides it became clear that several factors needed to be considered for there to be agreement between $CH_3\cdot$ formation and C_2 formation. One such factor is the presence of CO_2, which is a product of the oxidation reaction that inhibits both the CH_4 conversion and the formation of $CH_3\cdot$ radicals (Korf et al. 1987; Campbell and Lunsford 1988). When the MIESR experiments were carried out at low pressures, the partial pressure of CO_2 was small and the $CH_3\cdot$ radical production on

Table 1-1 Comparison of Atmospheric-Pressure MIESR Results for Alkali-Metal–Promoted MgO and CaO[a,b]

Catalyst	Relative Conversion	(Rel. C_2 Yield)$^{1/2}$	Relative $CH_3\cdot$ Production
7 wt% Li^+/MgO	100	100	100
20 wt% Na^+/MgO	75	71	74
30 wt% K^+/MgO	100	62	41
5 wt% Li^+/CaO	91	78	45
15 wt% Na^+/CaO	108	90	98
23 wt% K^+/CaO	120	80	35

[a]Relative conversion, (C_2 yield)$^{1/2}$, and $CH_3\cdot$ production are per gram, and their values are based on an assigned value of 100 for 7 wt% Li^+/MgO.
[b]Reaction conditions: $T = 700°C$, Ar : CH_4 : O_2 = 840 : 37 : 3.8 ml min^{-1}; 760 torr.

certain catalysts was abnormally high. When the results were obtained under the same conditions (i.e., high pressure) the agreement between (C_2 yield)$^{1/2}$ and $CH_3\cdot$ radical production was significantly improved. The results are shown in Table 1-1, from which it is evident that the K^+/CaO catalyst is an anomaly.

Although there are numerous correlations that suggest that $CH_3\cdot$ radicals couple in the gas phase to form C_2H_6, one could argue that the concentration of radicals in the gas phase is only proportional to some concentration on the surface, and the predominant coupling reaction occurs on the surface. In order to address this possibility, quantitative comparisons have been made between the production rates of gas-phase radicals and the C_2 products. Such experiments have now been carried out over Li^+/MgO (Campbell, Morales, and Lunsford 1987), Na^+/CaO (Lin, Wang, and Lunsford 1988), and La_2O_3 (Tong and Lunsford 1991). In each case the reaction was at atmospheric pressure and the flow rate was high (880 ml min^{-1}) so the $CH_3\cdot$ radicals would be rapidly swept out of the catalyst bed. It is important to note that the formation rates of $CH_3\cdot$ radicals and C_2 products were determined in a single experiment.

The variation in the $CH_3\cdot$ radical concentration as a function of the distance from the exit of the Li^+/MgO catalyst bed is shown in Fig. 1-5. At the CH_4 : O_2 ratios used in this experiment the $CH_3\cdot$ radicals are removed from the gas stream primarily via the coupling reaction. From the gas flow rate it is possible to determine the time required for the radicals to move from the exit of the catalyst bed to the leak. This is the time that the radicals react in the gas phase. A plot of $[CH_3\cdot]^{-1}$ versus time fits a straight line, as expected for a second-order reaction between $CH_3\cdot$ radicals. The slope of this line yields the rate constant for the radical

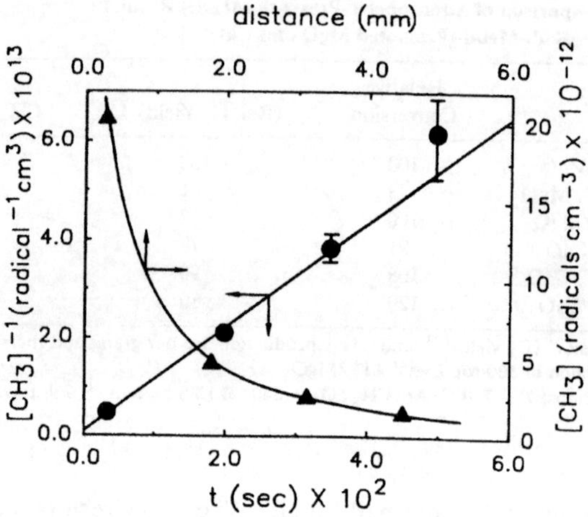

FIGURE 1-5. Plots of $[CH_3\cdot]$ versus probe distance and $[CH_3\cdot]^{-1}$ versus time (time for gas flow from catalyst bed to the probe). ● $[CH_3\cdot]$ concentration versus probe distance; ▲ inverse of $[CH_3\cdot]$ concentration versus time. (Reprinted with permission from Campbell, Morales, and Lunsford 1987, *J. Am. Chem. Soc.* 109:7900–7901. Copyright 1987 American Chemical Society.)

coupling reaction, and the intercept yields $[CH_3\cdot]$ at the exit of the catalyst bed. The rate constant for the gas-phase coupling of $CH_3\cdot$ radicals, obtained by this method, was in good agreement with values reported in the literature (Warnatz 1984).

Using the appropriate calibration methods as described in the experimental section, the absolute concentration of $CH_3\cdot$ was determined. Because this was a flow experiment the concentration may be transformed into a rate of production of $CH_3\cdot$ radicals. For the Li^+/MgO catalyst the rate of radical production was 1.8×10^{17} radicals per gram per second. Upon comparing this value with the rate of C_2 production, it is evident that more than 40% of the C_2 products were formed by the coupling of gas-phase $CH_3\cdot$ radicals. The value of 40% is a lower limit, as many radicals couple in the gas phase over the catalyst particles, that is, before they exit the catalyst zone and reach the leak. Some of these coupling reactions may occur in a boundary layer that is adjacent to the surface. In such a case the local $CH_3\cdot$ radical concentration would be greater and the oxygen concentration would be less than that found in the turbulent gas phase. Over Na^+/CaO and La_2O_3 catalysts it was found that $>28\%$ and $>71\%$, respectively, of the coupling reactions occurred in the gas phase. The La_2O_3 catalyst was painted on the surface of a fused-quartz frit. With this thin catalyst layer it was possible to reduce the number of coupling reactions that occurred prior to the gases entering the leak.

An apparent activation energy for the formation of $CH_3\cdot$ radicals may be obtained by varying the temperature of the catalyst and observing the change in the number of radicals collected. This activation energy, however, is affected by two reactions: the formation of $CH_3\cdot$ radicals via

$$CH_4 \xrightarrow[Li^+/MgO]{k_2} CH_3\cdot + (H\cdot)_s \tag{1-2}$$

and the radical coupling reaction

$$CH_3\cdot + CH_3\cdot + M \xrightarrow{k_3} C_2H_6 + M \tag{1-3}$$

Here $(H\cdot)_s$ means that hydrogen atoms are removed by the surface as OH^- ions, and M refers to a third body, which is required to remove excess energy from excited C_2H_6 molecules. The true activation energy for reaction (1-2) may be determined from a simple kinetic analysis that assumes that the two reactions are in steady state. Based on this approximation, one may show that (Campbell and Lunsford 1988)

$$0 = \frac{d[CH_3\cdot]}{dt} = k_2[CH_4] - k_3[CH_3\cdot]^2 \tag{1-4}$$

or

$$[CH_3\cdot] = \left(\frac{k_2}{k_3}\right)[CH_4]^{1/2} \tag{1-5}$$

Upon applying the Arrhenius relationship and noting that $E_3 \simeq 0$, it follows that

$$\frac{d \ln[CH_3\cdot]}{d(1/T)} \simeq -\frac{E_2}{2R} \tag{1-6}$$

The left-hand side of equation (1-6), divided by R, is the apparent activation energy E_a as determined by the number of $CH_3\cdot$ radicals collected. Thus, E_2, the true activation energy for reaction (1-2), is twice the measured apparent activation energy.

Over Li^+/MgO, E_a values of 24 kcal mol^{-1} in the low-pressure mode and 24–28 kcal mol^{-1} in the high-pressure mode have been observed using the MIESR system (Campbell and Lunsford 1988). Recently, using the REMPI technique, Lee and co-workers (Lee, Yu, and Lin 1990) have obtained a value of 22 kcal mol^{-1} for E_a over the same type of catalyst.

The agreement between the two methods obviously is good. Moreover, the corresponding value of E_2 = 48–56 kcal mol^{-1} is in agreement with the observed value of 55 kcal mol^{-1} for the catalytic conversion of CH_4 over a Li^+/MgO catalyst. Similarly, Gulcicek and co-workers (Gulcicek, Colson, and Pfefferle 1990) have used REMPI to detect $CH_3\cdot$ radicals produced on a Sr/La_2O_3 catalyst and have deduced a value for E_2 of 30.0 kcal mol^{-1}, which again is in agreement with the activation energy of 36.8 kcal mol^{-1} for ethane production over the same catalyst (DeBoy and Hicks 1988).

Although this section is mainly devoted to the formation of $CH_3\cdot$ radicals, it is interesting to compare these results with those obtained using other hydrocarbons. Of particular interest are the radicals formed when the products of the oxidative dimerization reaction, C_2H_6 and C_2H_4, react over various metal oxides. Results obtained during the oxidation of four hydrocarbons over six metal oxides are summarized in Table 1-2 (Driscoll and Lunsford 1985). It should be noted that the reaction temperature employed in this study was 475°C, which is well below the temperatures used in the oxidative coupling reactions. Nevertheless, it is evident that Li^+/MgO and MgO are far superior to the other oxides in generating $CH_3\cdot$ radicals from CH_4, and $C_2H_5\cdot$ radicals from C_2H_6. In view of the lower C—H bond strength in C_2H_6 it is surprising that the amount of $C_2H_5\cdot$ radicals collected was so small. Ethyl radicals undergo a unimolecular loss of H· atoms, and this reaction may lead to the results obtained. By contrast, the allylic C—H bond in propylene is weak and the allylic radical is stable, therefore large amounts of this radical were obtained, even over a catalyst such as Bi_2O_3 that is not effective in activating alkanes. At the other extreme, the C—H bond strength in C_2H_4 is large and no gas-phase vinyl radicals were detected, even over Li^+/MgO. When gas-phase radicals are not detected, one should keep in mind the possibility of subsequent surface reactions that might destroy the radical (see below).

If $CH_3\cdot$ radicals couple on a metal oxide surface, rather than in the gas phase, it should be possible to enhance this coupling reaction by increasing the surface area of the metal oxide. In order to explore this possibility a layer of sintered MgO, having a surface area of 12 m^2 g^{-1}, was placed downstream from a Li^+/MgO catalyst that served as a generator of $CH_3\cdot$ radicals (Campbell and Lunsford 1988). The objective was to determine whether the $CH_3\cdot$ radical concentration decreased as a result of the coupling on the MgO. The results of this experiment, together with the appropriate blank experiments, are given in Table 1-3. With the MIESR apparatus operating in the low-pressure mode, it was shown that quartz powder was not very effective as a source of $CH_3\cdot$ radicals and that MgO

Table 1-2 Radical Production over Various Oxide Catalysts[a]

Catalyst	Surface Area ($m^2\ g^{-1}$)	RADICAL AMOUNT ($nmol\ g^{-1}$)			
		$C_3H_5\cdot$ [C_3H_6(89)][b]	$C_2H_5\cdot$ [C_2H_6(98)][b]	$CH_3\cdot$ [CH_4(104)][b]	$C_2H_3\cdot$ [C_2H_4(108)][b]
Bi_2O_3	0.35	17.90 (51.1)[c]	Trace $CH_3\cdot$, $CH_3O_2\cdot$	0.056 (0.16)	Trace $CH_3\cdot$, $CH_3O_2\cdot$
γ-$Bi_2O_3\cdot MoO_3$	1.2	6.50 (5.4)	0	0	0
PbO(I)[d]	0.13	2.80 (21.5)	0	0.010 (0.08)	0
PbO(II)[e]	0.43	13.24 (30.8)	0	0.036 (0.08)	Weak unidentified signal
MgO	159.2	14.82 (0.1)	2.16 (0.01)	3.54 (0.02)	$CH_3\cdot$ (0.066)
14.6 wt% Li/MgO	29.8	65.52 (2.2)	10.78 (0.36)	38.18 (1.28)	$CH_3\cdot$ (0.184)

[a] Reactions carried out at 475°C, collection period = 30 min. Flows: Ar = 3.8 $cm^3\ min^{-1}$, O_2 = 0.024 $cm^3\ min^{-1}$, hydrocarbon = 1.08 $cm^3\ min^{-1}$.
[b] Reactant and C—H bond strength in kilocalories per mole.
[c] Numbers in parentheses are radical amounts based on surface area ($nmol\ m^{-2}$).
[d] Preconditioned at 450°C, 2.5 h, 300 cm^3 of O_2 per minute (chips).
[e] Preconditioned under vacuum (powder).

Table 1-3 Influence of MgO on Methyl Radical Recombination[a]

Solid Configuration	Relative Radical Concentration
QP_2	2.2
MgO	17
Li^+/MgO | QP	100
Li^+/MgO | QP | MgO	112

[a]Conditions: $T = 700°C$; 0.15 g of QP, 0.1 g of MgO, 0.08 g of 7 wt% Li^+/MgO; $Ar:CH_4:O_2 = 837:38:3.8$ ml min^{-1}; 760 torr.
[b]Quartz powder.

was only slightly more effective. As expected, Li^+/MgO produced a large flux of $CH_3\cdot$ radicals, and when the MgO was added to the system, after the Li^+/MgO, the production of radicals actually increased rather than decreased. In fact, the total was approximately the sum of the radical production on the pure MgO and the Li^+/MgO samples individually. Thus, it is clear that a metal oxide such as MgO does not adsorb or otherwise promote the surface coupling of $CH_3\cdot$ radicals at 700°C. In addition, MgO does not promote the total oxidation of $CH_3\cdot$ radicals to any significant extent.

The latter point has been addressed in additional detail because secondary reactions of $CH_3\cdot$ radicals with metal oxides may result in reactions other than the desired coupling reaction. If one estimates the partial pressure of $CH_3\cdot$ radicals, it is possible to show that on the average a given radical will collide with the surface approximately 10^5 times before it couples. Prior to these studies there was essentially no information available on the reactions of alkyl radicals with metal oxide surfaces.

The MIESR system was modified as shown in Fig. 1-6 so that a second metal oxide could be placed between the radical source and the leak into

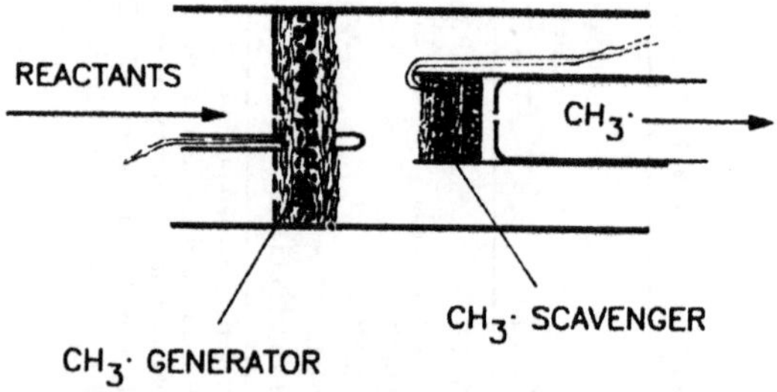

FIGURE 1-6. Apparatus for studying the reactivity of $CH_3\cdot$ radicals with metal oxides. (Reprinted with permission from Tong, Rosynek, and Lunsford 1989, *J. Phys. Chem.* 93:2896–2898. Copyright 1989 American Chemical Society.)

Table 1-4 Generation of Gas-Phase Methyl Radicals and their Reactions with Catalytic Surfaces[a]

Metal Oxide	Surface Area ($m^2\ g^{-1}$)	Relative Rate of Formation		Relative Reaction Efficiency	
		g^{-1}	m^{-2}	g^{-1}	m^{-2}
MgO(JM)	38	3	0.3	26	0.7
MgO(A)	92	9	0.5	100	1
Li^+/MgO(JM)	0.18	18	500	8	
Li^+/MgO(A)	0.45	20	220	5	
Li^+/MgO(Li_2CO_3)	2.7	100	180	3	

[a] $T = 700°C$, $P = 1$ torr for $CH_3\cdot$ radical formation; $T = 470°C$, $P = 1.7–1.9$ torr for $CH_3\cdot$ radicals reacting with a catalyst surface. Flow rates: Ar = 3.7 $cm^3\ min^{-1}$; CH_4 = 1.1 $cm^3\ min^{-1}$; O_2 = 0.025 $cm^3\ min^{-1}$.

the high-vacuum region of the system. A stable source of $CH_3\cdot$ radicals is Sm_2O_3. In the low-pressure mode, radical coupling between the source and the second oxide is not extensive. Reaction efficiencies of the oxides are reported relative to quartz chips by using the equation (Tong, Rosynek, and Lunsford 1989)

$$E_R = \frac{[CH_3\cdot]_Q - [CH_3\cdot]_s}{[CH_3\cdot]_Q} \times 100\% \qquad (1\text{-}7)$$

where $[CH_3\cdot]_Q$ and $[CH_3\cdot]_s$ are the radical amounts determined by ESR with quartz chips and the oxide of interest, respectively, as the scavenger. Quartz chips are almost inert both for generating and scavenging methyl radicals.

The relative rates of $CH_3\cdot$ radical formation and reaction with several MgO and Li^+/MgO oxides are given in Table 1-4 (Lunsford et al. 1989). Here, MgO(JM) and MgO(A) refer to Johnson Matthey Puratronic Grade and Aldrich Gold Label MgO, respectively. The former was chosen because it exists in the form of small cubes. With the Li^+/MgO(JM) and the Li^+/MgO(A) catalysts, the lithium was added as methyl lithium in diethyl ether and the material was subsequently oxidized in flowing molecular oxygen.

A number of interesting conclusions can be drawn from these results. First, the specific rate of formation of $CH_3\cdot$ radicals is essentially the same for the pure MgO catalysts; that is, the rate of radical formation is proportional to the surface area. Second, the addition of lithium had a dramatic effect on the specific rate of radical formation. Third, the specific rate of reaction of $CH_3\cdot$ radicals with the oxides was similar for the pure oxides. Fourth, on a per-gram basis, the Li^+/MgO catalysts are relatively unreactive with $CH_3\cdot$ radicals. The specific activity of these materials has not been listed because the fraction of the radicals removed from the gas stream was small and this introduced a large error into the reaction efficiency. The observation that the Li^+/MgO catalysts are efficient in generating gas-phase $CH_3\cdot$ radicals but not in reacting with these radicals is consistent with the good C_2 selectivities attained in the oxidative coupling reaction. It is interesting to note that the $Li^+/MgO(Li_2CO_3)$ catalyst was more active for CH_4 oxidation than the other two lithium-promoted catalysts, and essentially the same C_2 selectivity was obtained over the three catalysts, which is consistent with the relative rate of $CH_3\cdot$ radical formation per gram of catalyst.

Lithium-promoted MgO also is an effective catalyst for the oxidative dehydrogenation of C_2H_6 to C_2H_4, and $C_2H_5\cdot$ radicals were detected using the MIESR system. It was concluded that surface-generated gas-phase $C_2H_5\cdot$ radicals are intermediates in this process (Morales and Lunsford 1989). In the presence of O_2 the unimolecular decomposition of $C_2H_5\cdot$ radicals is replaced by the reaction

$$C_2H_5\cdot + O_2 \rightarrow C_2H_4 + HO_2\cdot \tag{1-8}$$

Because of this reaction channel the yield of C_2H_4 from C_2H_6 is considerably greater than the yield of C_2 products during the oxidative coupling of CH_4.

METHYL RADICAL FORMATION AND REACTIONS OVER THE LANTHANIDE OXIDES

Certain members of the lanthanide oxide series are among the most active catalysts for oxidative coupling, whereas others are complete oxidation catalysts. It was of interest in our laboratory to determine whether the formation and reactions of $CH_3\cdot$ radicals on the oxides reflected these variations in the overall catalytic behavior. In an early study it was demonstrated the La_2O_3 was ca. fourfold more active than Li^+/MgO in producing $CH_3\cdot$ radicals (Lin et al. 1986). A comparison of all of the oxides in the series revealed a complex pattern in activity for the generation of $CH_3\cdot$ radicals (Campbell, Zhang, and Lunsford 1988). With the exception of CeO_2, Pr_6O_{11}, and Tb_4O_7, the trend in radical-forming ability generally correlated with the basicity of the oxide. The three exceptions all have multiple oxidation states that are accessible, as is discussed in more detail later. There was good agreement between the relative specific

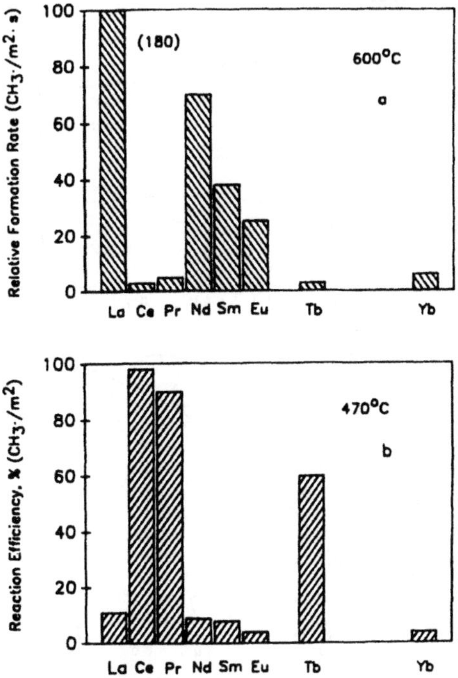

FIGURE 1-7. Methyl-radical generation and reaction with lanthanide oxides: (a) relative rate of formation normalized to unit time and surface area; (b) percentage of radicals that react with metal oxide. (Reprinted with permission from Tong, Rosynek, and Lunsford 1989, *J. Phys. Chem.* 93:2896–2898. Copyright 1989 American Chemical Society.)

activity of $CH_3\cdot$ radical formation and $(C_2 \text{ yield})^{1/2}$; the exception being Nd_2O_3, which exhibited a somewhat greater yield of C_2H_4 and C_2H_6 than anticipated on the basis of the $CH_3\cdot$ radical formation rate.

Selected members of the series were studied in more detail and the results are shown in Fig. 1-7 (Tong, Rosynek, and Lunsford 1989). Clearly, CeO_2, Pr_6O_{11}, and Tb_4O_7 are unable to produce gas-phase $CH_3\cdot$ radicals because they extensively react with these radicals. It is believed that the oxides react with the radical via a reductive addition reaction,

$$CH_3\cdot + M^{(n+1)+}O^{2-} \rightarrow M^{n+}(OCH_3)^- \tag{1-9}$$

through which CO_2 is formed at the elevated temperatures. The reactor shown in Fig. 1-6 has been modified so that the $CH_3\cdot$ radicals may react with a wafer of CeO_2 suitable for obtaining an infrared spectrum. Following the reaction with CeO_2 at room temperature, the infrared spectrum

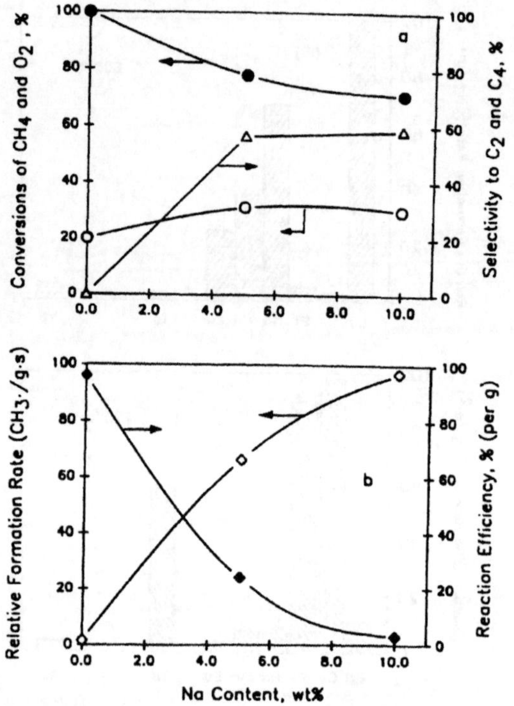

FIGURE 1-8. Effect of Na_2CO_3 addition to CeO_2 on (a) the catalytic oxidation of CH_4 and (b) the production of $CH_3\cdot$ radicals and their reaction with the catalyst: ○, CH_4 conversion; ●, O_2 conversion; △, combined C_2 and C_4 selectivity; ◇, relative formation rate of $CH_3\cdot$ radicals; ◆, reaction efficiency of $CH_3\cdot$ radicals with the catalysts. The catalytic reaction was carried out at 770°C, 1 atm, and a flow rate of 34 ml min^{-1}. The partial pressure of CH_4 and O_2 were 180 and 67 torr, respectively. The $CH_3\cdot$ formation rates and the reaction efficiencies were determined with the catalysts at 760 and 470°C, respectively. The gas pressures were 1 and 1.5 torr for the two cases. (Reprinted with permission from Tong, Rosynek, and Lunsford 1989, *J. Phys. Chem.* 93:2896–2898. Copyright 1989 American Chemical Society.)

revealed infrared bands in the 2924–1074 cm^{-1} region that are characteristic of a M—O—CH_3 surface species (Tong and Lunsford 1991). At 100°C this species was transformed into a surface formate ion.

The oxidation characteristics of CeO_2 may be modified significantly by the addition of Na_2CO_3 to the catalyst (Tong, Rosynek, and Lunsford 1989). As shown in Fig. 1-8, the modified CeO_2 catalyst was transformed from a totally nonselective catalyst to one that gave reasonably good yields

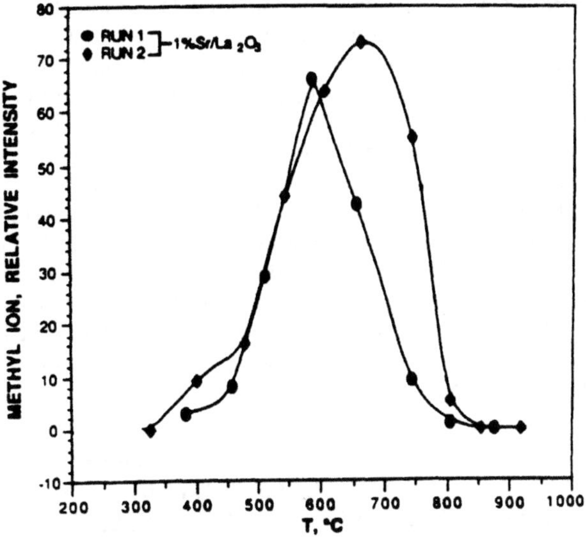

FIGURE 1-9. Plots of methyl radicals detected as a function of temperature, for 1% Sr/La_2O_3 catalyst: run 1, $p = 500$ torr; run 2, $p = 380$ torr. Run 1 results are for the fresh catalyst. Run 2 results are for catalysts aged under reaction conditions for 6–8 h. (Reprinted with permission from *J. Phys. Chem.* 94:7069–7074. Copyright 1990 American Chemical Society.)

of C_2 products. Likewise, the addition of Na_2CO_3 resulted in a decrease in the efficiency for reaction with $CH_3\cdot$ radicals and concomitantly the formation of $CH_3\cdot$ radicals. Results from ion scattering spectroscopy revealed that the surface was essentially covered with sodium compounds; that is, no CeO_2 was available for reaction with the gas phase (Tong, Rosynek, and Lunsford 1990). Subsequent work has shown that a similar phenomenon occurs with the other lanthanide oxides and that the resulting catalyst may be viewed as a Na_2CO_3–Na_2O–Na_2O_2 phase supported on the lanthanide oxide. The Na_2O_2 is believed to be responsible for the activation of CH_4.

The REMPI results of Gulcicek and co-workers (Gulcicek, Colson, and Pfefferle 1990) revealed a maximum in the production of $CH_3\cdot$ radicals over Sr/La_2O_3, as shown in Fig. 1-9. With the fresh catalyst the maximum was observed at 580°C, whereas with the used catalyst the maximum was shifted to ca. 650°C. The authors noted that with the aged catalyst the decrease in $CH_3\cdot$ radical production occurred when the system became oxygen-limited, but the reaction over the fresh Sr/La_2O_3 catalysts was not oxygen-limited. Using the same method for $CH_3\cdot$ radical detection, Lee and co-workers (Lee, Yu, and Lin 1990) reported that over MgO and Li^+/MgO catalysts the rate of $CH_3\cdot$ radical production became nearly constant at temperatures greater than 675°C. They attributed the leveling-off phenomenon to O_2 limitations or possibly to removal of $CH_3\cdot$ by both surface and gas-phase oxidation reactions. It seems unlikely that $CH_3\cdot$ radical coupling on the surface would increase with temperature as the entropy factor would favor desorption. Using the MIESR system, we have observed neither a maximum nor a leveling off in $CH_3\cdot$ radical production over La_2O_3 up to 600°C, over Li^+/MgO up to 730°C, over $LiNiO_2$ up to 751°C, and over $NaMnO_2/MgO$ up to 815°C. Clearly, more work is needed to determine the factors that result in the maxima or the leveling-off phenomena, which is perhaps not surprising as $CH_3\cdot$ radicals are very reactive intermediates.

METHYL RADICAL FORMATION AND REACTIONS OVER TRANSITION-METAL OXIDE CATALYSTS

Based on kinetic studies, Hatano and Otsuka (Hatano and Otsuka 1988) have suggested that oxidative coupling over the monophasic $LiNiO_2$ catalyst is primarily a surface phenomenon. In principle, it is possible that $CH_3\cdot$ radicals could be held more strongly to a surface that contains transition-metal ions. Such might also be the case for the $NaMnO_4/MgO$ catalyst (Sofranko et al. 1988). To determine the role of surface-generated gas-phase $CH_3\cdot$ radicals over these two catalysts the MIESR system was modified so that the leak was located within the catalyst bed (Tong and Lunsford 1990). In addition, the partial pressure of oxygen was increased so that it more closely corresponded to pressures used in conventional catalytic studies. With the usual $CH_4 : O_2$ ratios, which are typically > 10 so as to avoid the formation of $CH_3O_2\cdot$ radicals in the cooler regions of the system, it was observed that the $CH_3\cdot$ radical concentration was anomalously small for the $LiNiO_2$ and $NaMnO_4/MgO$ catalysts. Perhaps the greater oxygen partial pressures are required to produce the active form of oxygen on these transition-metal oxides or in the case of $LiNiO_2$ to prevent reduction of Ni^{2+} to Ni^0 (Pickering, Maddox, and Thomas 1989). But even at the higher partial pressures of oxygen it should be emphasized that the equilibrium represented by reaction (1-1) is shifted far to the left at 600–850°C. Thus, the dominant radical species above the hot catalyst is $CH_3\cdot$, not $CH_3O_2\cdot$.

In these experiments the conversion and selectivities were measured while the $CH_3\cdot(CH_3O_2\cdot)$ radicals were being collected. The Li^+/MgO

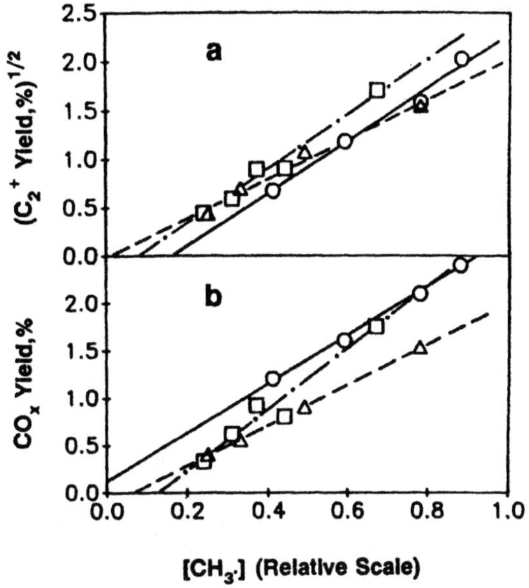

FIGURE 1-10. Relationship between Me· radical concentration in the gas phase above the catalyst and (a) C_2^+ yield or (b) CO_x yield: ○, $LiNiO_2$; △, $NaMnO_4/MgO$; □, Li/MgO. For each of the three catalysts the same relative scale was used for the Me· radical concentration. (Reprinted from Tong and Lunsford 1990).

catalyst was also studied for comparison with the two transition-metal oxides. It is evident from the results of Fig. 1-10a that for each of the three catalysts the $(C_2^+ \text{ yield})^{1/2}$ is linearly related to the concentration of gas-phase $CH_3\cdot$ radicals in the catalyst bed. These results indicate that all three catalysts behave similarly with respect to the formation of C_2H_6 via the coupling of gas-phase $CH_3\cdot$ radicals. This is particularly significant as the importance of gas-phase $CH_3\cdot$ radicals in the formation of C_2 products has been demonstrated quantitatively for Li^+/MgO. If the mechanism was exclusively a surface phenomenon, one would expect a positive intercept at the ordinate, that is, C_2^+ products would be formed without there being any gas-phase $CH_3\cdot$ radicals. In fact, the ordinate intercept was either zero or negative.

The CO_x yield (the combined yield of CO and CO_2) also increased in a linear manner with respect to the $CH_3\cdot$ radical concentration, as shown in Fig. 1-10b. The linear relation is consistent with the formation of CO_x via gas-phase $CH_3\cdot$ radicals; however, it does not exclude the formation of CO_x from C_2^+ products.

Similar experiments have been reported by Sinev and co-workers (Sinev, Korchak, and Krylov 1988) who found that the production of $CH_3\cdot(CH_3O_2\cdot)$ radicals was a linear function of CH_4 conversion over a K_2O/Al_2O_3 catalyst. A linear relationship was also found over θ-Al_2O_3, but at a given conversion the rate of $CH_3\cdot$ radical formation was considerably less over θ-Al_2O_3 and the yield of C_2 hydrocarbons was more than a factor of 10 less.

Secondary reactions of $CH_3\cdot$ radicals with transition-metal oxide catalysts may occur in the same manner as noted previously for certain members of the lanthanide oxide series. Additional evidence for the reducing ability of $CH_3\cdot$ radicals via reaction (1-9) has been demonstrated for a MoO_3/SiO_2 catalyst, which is a nonselective oxidation catalyst under the condition used for the oxidative coupling reaction. After exposure of the MoO_3/SiO_2 catalyst at 25°C to a gas stream containing $CH_3\cdot$ radicals, the solid phase was analyzed by ESR spectroscopy and a strong signal of Mo^{5+} was observed (Tong and Lunsford 1991). Presumably methoxide ions also are formed, and these lead to the formation of CO_x. At lower temperatures the methoxide ions can produce methanol or formaldehyde (Liu et al. 1984). It should be possible to develop a system in which $CH_3\cdot$ radicals, formed at elevated temperatures on one catalyst, are carried downstream where they react on another catalyst to form methanol and formaldehyde over MoO_3/SiO_2 at a lower temperature. To date, efforts to develop such a system in our laboratory have failed because at any reasonable pressure the radicals couple before they have an opportunity to react on the second catalyst.

Clearly the secondary reactions of $CH_3\cdot$ radicals with metal oxides are important in effecting selectivity, and any realistic model of a catalytic reaction must include them. In an effort to provide quantitative information on these reactions, reactive sticking coefficients s have been determined. The reactive sticking coefficient is defined as the fraction of molecules colliding with a surface that react either by chemisorption or by decomposition. The system described in Fig. 1-6 was modified such that the metal oxide of interest was placed after the leak into the high-vacuum region, rather than in front of it. The catalyst was loosely distributed between two layers of quartz wool. In a separate experiment, the total gas pressure in the zone containing the catalyst was measured. From a quantitative determination of the $CH_3\cdot$ radicals collected without the metal oxide present it was possible to calculate the partial pressure of radicals in the gas phase at the catalyst and thereby to calculate the total number of collisions in a given period. With the catalyst present the number of $CH_3\cdot$ radicals that reacted was determined and from this

Table 1-5 Reactive Sticking Coefficients and Activation Energies for $CH_3\cdot$ Radicals Reacting with Metal Oxides[a]

Oxide	S	E_a (kcal mol^{-1})
ZnO	1.8×10^{-5}	2.6
CeO_2	2.4×10^{-6}	2.2
MgO	1.2×10^{-7}	5.7

[a]The oxide was at 482°C.

information the sticking coefficient was calculated (Tong and Lunsford 1991).

Among the several metal oxides that we have investigated, ZnO and CeO_2 were found to be among the most reactive with respect to $CH_3\cdot$ radicals. Therefore, ZnO and CeO_2 were the primary oxides used to determine the reactive sticking coefficient. The sticking coefficient of other oxides are related to ZnO on the basis of their relative activities. The values obtained in this manner are listed in Table 1-5. It is somewhat surprising that the reactive sticking coefficient was as low as 1.8×10^{-5} on ZnO; however, as noted previously a $CH_3\cdot$ radical would make $>10^5$ collisions with the metal oxide before it reacted with another $CH_3\cdot$ radical. Thus, the observed sticking coefficient is consistent with the fact that unmodified ZnO is an active, but nonselective, catalyst for CH_4 oxidation (Zhang et al. 1988). By contrast, the sticking coefficient over MgO is more than two orders of magnitude less; therefore it is suitable as a catalyst or a support for the oxidative coupling reaction.

In a separate experiment the activation energies were determined over a wide range of temperatures (typically 200–500°C). The activation energies were found to be quite small on ZnO and CeO_2 (Table 1-5) but somewhat larger on MgO. Because of these low activation energies, the sticking coefficients at normal oxidative coupling conditions (e.g., 700°C) are not much greater than at 480°C, the temperature at which they were measured.

The low activation energies at first seem inconsistent with the low reactive sticking coefficients for $CH_3\cdot$ radicals reacting with metal oxides; however, the results can be understood in terms of a tunneling mechanism in which the electron on the $CH_3\cdot$ radical tunnels through the oxide ion and reduces the metal ion according to reaction (1-9). Even with very short residence times of the $CH_3\cdot$ radical on the surface, such a process would be reasonable. It is relatively easy to reduce Ce^{4+} to Ce^{3+} and Zn^{2+} to Zn^{+}, but the reduction of Mg^{2+} to Mg^{+} probably would not occur under these conditions. Rather it appears like the electron on the $CH_3\cdot$ radical would be transferred to oxide ion vacancies in MgO. An analogous process was observed by Smith and Tench (Smith and Tench 1968), who reacted MgO powder with H· atoms and observed the formation of trapped electrons and OH^- ions on the surface.

SUMMARY

Extensive research on the oxidative coupling of CH_4 has shown that $CH_3\cdot$ radicals, formed on the surface of a metal oxide catalyst, enter the gas phase, where they may either couple or initiate chain reactions. The number of radicals entering the gas phase is sufficient to account for the formation of ethane and ethylene, which are the desired stable products of the reaction. Those catalysts that are effective in producing gas-phase $CH_3\cdot$ radicals are active and selective for the oxidative coupling reaction. The activation energy for the formation of $CH_3\cdot$ radicals is approximately the same as the activation energy for CH_4 oxidation. If the metal ion in the catalyst has multiple oxidation states that are accessible, $CH_3\cdot$ radicals will react with the oxide more extensively than they couple to form C_2 products. Under the conditions of oxidative coupling, the reactions of $CH_3\cdot$ radicals with the metal oxide result in the formation of CO and CO_2. These carbon oxides also may be formed via gas-phase chain-branching reactions. The oxidative coupling of CH_4 is a complex interactive network of surface and gas-phase reactions that ultimately determines the conversion of CH_4 and selectivities for C_2H_4 and C_2H_6.

ACKNOWLEDGMENT

The author is indebted to Drs. W. Martir, D.J. Driscoll, K.D. Campbell and to Mr. Y. Tong who provided not only the technical assistance but also the scientific insight to make possible the research carried out in our laboratory. Financial support was provided by the Office of Basic Energy Sciences, Department of Energy.

References

Campbell, K.D., and J.H. Lunsford. 1988. Contribution of gas-phase radical coupling in the catalytic oxidation of methane. *J. Phys. Chem.* 92:5792–6.

Campbell, K.D., E. Morales, and J.H. Lunsford. 1987. Gas phase coupling of methyl radicals during the catalytic partial oxidation of methane. *J. Am. Chem. Soc.* 109:7900–901.

Campbell, K.D., H. Zhang, and J.H. Lunsford. 1988. Methane activation by the lanthanide oxides. *J. Phys. Chem.* 92:750–3.

DeBoy, J.M., and R.F. Hicks. 1988. Kinetics of the oxidative coupling of methane over 1 wt% Sr/La_2O_3. *J. Catal.* 113:517–24.

Driscoll, D.J. 1985. Ph.D. Dissertation, Texas A & M University.

Driscoll, D.J., K.D. Campbell, and J.H. Lunsford. 1987. Surface-generated gas-phase radicals: formation, detection, and role in catalysis. In *Advances in Catalysis*, ed. D.D. Eley, H. Pines, and P.B. Weisz, pp. 139–86. New York: Academic Press.

Driscoll, D.J., and J.H. Lunsford. 1985. Gas-phase radical formation during the reaction of methane, ethane, ethylene and propylene over selected oxide catalysts. *J. Phys. Chem.* 89:4415–18.

Driscoll, D.J., W. Martir, J.-X. Wang, and J. H. Lunsford. 1985. Formation of gas-phase methyl radicals over MgO. *J. Am. Chem. Soc.* 107:58–63.

Garibyan, T.A., and L.Ya. Margolis. 1990. Heterogeneous–homogeneous mechanism of catalytic oxidation. *Catal. Rev.—Sci. Eng.* 31:355–84.

Gulcicek, E.E., S.D., Colson, and L.D. Pfefferle. 1990. Multiphoton ionization detection of methyl radicals from catalytic oxidation of methane. *J. Phys. Chem.* 94:7069–74.

Hatano, M., and K. Otsuka. 1988. The oxidative coupling of methane on lithium nickelate (III). *J. Chem. Soc., Faraday Trans. 1* 85:199–206.

Hutchings, G.L., J.R. Woodhouse, and M.S. Scurrell. 1989. Partial oxidation of methane over oxide catalysts: Comments on the reaction mechanism. *J. Chem. Soc., Faraday Trans. 1* 85:2507–23.

Ito, T., and J.H. Lunsford. 1985. Synthesis of ethylene and ethane by partial oxidation of methane over lithium-doped magnesium oxide. *Nature (London)* 314:721–2.

Ito, T., J.-X. Wang, C.-H. Lin, and J.H. Lunsford. 1985. Oxidative dimerization of methane over a lithium-promoted magnesium oxide catalyst. *J. Am. Chem. Soc.* 107:5062–8.

Jones, C.A., J.J. Leonard, and J.A. Sofranko. 1987. The oxidative coupling of methane to higher hydrocarbons over alkali-promoted Mn/SiO_2. *J. Catal.* 103:311–9.

Korf, S.J., J.A. Roos, N.A. de Bruijn, J.G. van Ommen, and J.R.H. Ross. 1987. Influence of CO_2 on the oxidative coupling of methane over a lithium promoted magnesium oxide catalyst. *J. Chem. Soc., Chem. Commun.* 1433–4.

Lee, S.-P., T. Yu, and M.C. Lin. 1990. Effects of temperature and lithium on CH_3 radical formation in the CH_4 oxidative coupling reaction over MgO. *Int. J. Chem. Kinet.* 22:975–80.

Lin, C.-H., K.D. Campbell, J.-X. Wang, and J.H. Lunsford. 1986. Oxidative dimerization of methane over lanthanum oxide. *J. Phys. Chem.* 90:534–7.

Lin, C.-H., J.-X. Wang, and J.H. Lunsford. 1988. Oxidative dimerization of methane over sodium-promoted calcium oxide. *J. Catal.* 111:302–16.

Liu, H.-F. 1983. Unpublished results.

Liu, H.-F., R.-S. Liu, K.Y. Liew, R.E. Johnson, and J.H. Lunsford. 1984. Partial oxidation of methane by nitrous oxide over molybdenum on silica. *J. Am. Chem. Soc.* 106:4117–21.

Lunsford, J.H. 1972. Electron spin resonance in catalysis. In *Advances in Catalysis*, ed. D.D. Eley, H. Pines, and P.B. Weisz, pp. 265–344. New York: Academic Press.

Lunsford, J.H., M.D. Cisneros, P.G. Hinson, Y. Tong, and H. Zhang. 1989. Oxidative dimerization of methane over well defined lithium-promoted magnesium oxide catalysts. *Faraday Discuss. Chem. Soc.* 87:13–21.

Martir, W. 1981. Gas phase radicals formed on metal oxide surfaces. Ph.D. Dissertation, Texas A & M University.

Martir, W., and J.H. Lunsford. 1981. The formation of gas-phase π-allyl radicals from propylene over bismuth oxide and γ-bismuth molybdate catalysts. *J. Am. Chem. Soc.* 103:3728–32.

Mehandru, S.P., A.B. Anderson, and J.F. Brazdil. 1988. Methyl radical formation over Li-doped MgO. *J. Am. Chem. Soc.* 110:1715–19.

Mims, C.A., R.B. Hall, K.D. Rose, and G.R. Myers. 1989. Oxidative dimerization of CH_4/CD_4 mixtures: Evidence for methyl intermediate. *Catal. Lett.* 2:361–8.

Morales, E., and J.H. Lunsford. 1989. Oxidative dehydrogenation of ethane over a lithium-promoted magnesium oxide catalyst. *J. Catal.* 118:255–65.

Nalbandyan, A.B., and A.A. Mantashyan. 1975. *Elementarnye Protzessy v Medlennykh Gazofaznykh Realetzyakh*. p. 259. Erevan: Armenian Academy of Sciences.

Nelson, P.F., C.A. Lukey, and N.W. Cant. 1988. Isotopic evidence for direct methyl coupling and ethane and ethylene conversion during partial oxidation of methane over Li/MgO. *J. Phys. Chem.* 92:6176–9.

Nersesyan, L.A., I.A. Vardanyan, E.M. Kegeyan, L.Ya. Margolis, and A.B. Nalbandyan. 1975. Heterogeneous–homogeneous oxidation of methane. *Dokl. Akad. Nauk. SSSR* 220:605–7.

Pickering, I.J., P.J. Maddox, and J.M. Thomas. 1989. Probing changes in the structure and performance of a lithium nickel oxide catalyst by in situ X-ray diffraction during the high-temperature oxidative coupling of methane. *Agnew. Chem., Int. Ed. Engl., Adv. Mater.* 28:808–10.

Sinev, M.Yu., V.N. Korchak, and O.V. Krylov. 1988. Mechanism of oxidative condensation of methane into C_2 hydrocarbons over oxide catalysts. In *Proceedings of the 9th International Congress on Catalysis*, ed. M.J. Phillips and M. Ternan, pp. 968–73. Ottawa: The Chemical Institute of Canada.

Smith, D.R., and A.J. Tench. 1968. The reaction of hydrogen atoms at oxide surfaces. *Chem. Commun.* 1113–14.

Sofranko, J.A., J.J. Leonard, C.A. Jones, A.M. Gaffney, and H.P. Withers. 1987. Catalytic oxidative dimerization of methane over sodium-promoted Mn/SiO_2 and Mn/MgO. Preprints, Div. Petro. Chem., Am. Chem. Soc. 32:763.

Tong, Y., and J.H. Lunsford. 1990. Gas-phase coupling of methyl radicals during the partial oxidation of methane over transition metal oxide catalysts. *J. Chem. Soc., Chem. Commun.* 792–3.

Tong, Y., and J.H. Lunsford. 1991. Mechanistic and kinetic studies of the reactions of gas-phase methyl radicals with metal oxides. *J. Am. Chem. Soc.* 113:4741–46.

Tong, Y., M.P. Rosynek, and J.H. Lunsford. 1989. Secondary reactions of methyl radicals with lanthanide oxides: Their role in the selective oxidation of methane. *J. Phys. Chem.* 93:2896–8.

Tong, Y., M.P. Rosynek, and J.H. Lunsford. 1990. The role of sodium carbonate and oxides supported on lanthanide oxides in the oxidative dimerization of methane. *J. Catal. press.* 126:291.

Warnatz, J. 1984. Rate coefficients in the C/H/O system. In *Combustion Chemistry*, ed. W.C. Gardner, pp. 197–360. New York: Springer.

Zhang, H.-S., J.-X. Wang, D.J. Driscoll, and J.H. Lunsford. 1988. Activation and oxidative dimerization of methane over lithium-promoted zinc oxide. *J. Catal.* 112:366–74.

2

The Role of Gas-Phase Reactions during Methane Oxidative Coupling

Zbigniew Kalenik and Eduardo E. Wolf

University of Notre Dame, Notre Dame, IN

TABLE OF CONTENTS

INTRODUCTION

The objective of this chapter is to place in the proper perspective the effect of gas-phase reactions in the oxidative dimerization of methane. Obviously this concerns mainly the cofeed operation because gas-phase oxygen is not present in the cyclic mode in order to avoid extensive oxidation of the products. Furthermore, by gas-phase reactions we understand the homogeneous reactions that occur in the absence of a catalyst, because it has been proven convincingly by Lunsford's group (Ito et al. 1987; Campbell, Morales, and Lunsford 1987) and by many others, that the dimerization reaction indeed occurs in the gas phase via surface-initiated methyl radicals. Lane and Wolf (Lane and Wolf 1988) were among the first to study systematically the homogeneous gas-phase dimerization of methane. Prior to our work, most reports were concerned with catalytic effects and very little had been reported regarding homogeneous reactions in the absence of catalysts. At about the same time that our results were published, two other reports appeared in the literature confirming that, indeed, gas-phase reactions were relevant (Asami et al. 1987; Hutchings, Scurrell, and Woodhouse 1988). Thereafter, the gas-phase work has been repeated in many laboratories and the results of our earlier report have been confirmed (van der Wiele, Geerts, and van Kasteren 1990; Hatano et al. 1990). Unfortunately, the effect of the gas-phase reactions has been misinterpreted by many authors and is now seen as the limiting factor in the yield that can be attained in this process. This is only partially true because the real limiting factor is the lack of selective low-temperature catalysts that can operate under conditions in which gas-phase reactions are not significant.

The motivation of our work on the gas-phase reactions originated from our review of the literature, which as of 1988 listed catalytic results that had been obtained under a variety of experimental conditions, making a comparison of catalytic results rather difficult. Table 2-1 summarizes the main experimental conditions and catalysts used up to that time by various authors, illustrating the wide range of conditions used. A particularly bothersome condition utilized in many laboratories was the use of highly diluted streams, which distorted the yield reported in the early papers. In view of the rather broad spectra of experimental conditions used, we decided to design our experimental system so that the effect of the various parameters could be evaluated. Early in our program we found that under certain conditions the effect of various catalysts was not very clear; consequently, we tested the reactor without catalysts and to our surprise we found that under a specific set of experimental conditions, significant conversion and selectivities of C_2 product could be achieved in the gas phase. At this point, we set up an experimental program to test the various conditions under which the homogeneous uncatalyzed reactions become relevant. The results reported in this chapter summarize part of that work, along with results previously unpublished.

Although the emphasis of this chapter is on the gas-phase reactions, the effect of different catalysts on reaction selectivity is addressed in order to provide a balance between the selective catalytic routes and the unselective homogeneous reactions. The purpose is to reiterate the theme that we still do not have a catalyst that can make this process commercially feasible and to foster further work in the search for such catalyst, instead of the less-optimistic theme that the gas-phase reactions place a limit on this process. The last section of this chapter deals with the simulation of experimental results obtained in the preceding sections.

EXPERIMENTAL CONSIDERATIONS

Experimental Apparatus

Steady-State Studies

The experimental apparatus used during gas-phase experiments was a simple tubular reactor with a feed control system that permitted metering each gas independently via electronic flow controllers. The reactor was heated by an electric furnace controlled by a programmable temperature controller that monitored the temperature inside the reactor at its hottest point. The furnace was designed to provide preheating of the feed with a minimum heated volume. The temperature inside the reactor was measured with a K-type thermocouple shielded by a quartz thermocouple well, which permitted measurement of the axial temperature profile.

After testing several reactors of various types and shapes, the design adopted in the studies of methane dimerization consisted of a simple tubular flow reactor made of fused silica (referred as quartz) with an inside diameter of 0.95 cm and a heated length of 15 cm. The temperature

Table 2-1 A Summary of Literature Conditions for the Cofeeding of Methane and Oxygen[a]

REACTOR				FLOW RATES ($cm^3\ min^{-1}$)						
Material	Size (mm)	Temperature (°C)	Mass of Catalyst (g)	He	N_2	CH_4	O_2	$CH_4:O_2$[b]	P^*[c]	Reference
*[d]	8, o.d.	400–800	2.0	50.0	3.0	1.5	0.75	2.0	0.04	Aika et al. 1986
SS[e]	9, i.d.	550–700	0.035	50.0	40.0	100.0	10.0	10.0	0.55	Emesh Ali and Amenomiya 1986
Quartz	10, i.d.	700–800	1.0	0.0	344.0	56.2	6.4	8.75	0.15	Asami et al. 1986
Quartz	6, i.d.	740	0.05–0.10	0.0	9.2	28.0	2.8	10.0	0.77	Bytyn and Bearns 1986
Quartz	6, i.d.	650–750	0.05–0.10	0.0	9.2	28.0	2.8	10.0	0.77	Hinsen, Bytyn, and Bearns 1984
Quartz	14, i.d.	710	0.3 (ml)	0.0	16.0	20.0	4.0	5.0	0.60	Imai and Tagawa 1986
Quartz	23, o.d.	550–770	0.5–4.0	28.0	0.0	21.0	1.0	21.0	0.44	Ito et al. 1987
Quartz	23, o.d.	600–800	4.0	20.7	0.0	19.1	10.2	1.9	0.59	Ito and Lunsford 1985
*	8, o.d.	400–800	2.0	50.0	3.0	1.5	0.75	2.0	0.04	Iwamatsu et al. 1987
Quartz	*	600–800	*	0.0	16.6	20.8	4.2	5.0	0.60	Kimble and Kolts 1986
Quartz	25, i.d.	525–725	1.0	23.0	0.0	1.8	0.56	3.3	0.10	Lin et al. 1986
*	*	625–700	0.8	47.0	0.0	5.5	2.7	2.1	0.15	Lin et al. 1987
*	*	600–800	2.0	24.6	0.0	17.2	8.2	2.0	0.51	Matsuura et al. 1986
*	8, o.d.	400–800	2.0	50.0	3.0	1.5	0.75	2.0	0.04	Moriyama et al. 1986
*	*	750	1.0	33.3	0.0	1.8	0.90	2.0	0.08	Otsuka et al. 1986
Alumina	11, i.d.	750	2.0	47.0	0.0	9.6	4.3	2.3	0.23	Otsuka and Komatsu 1986
*	*	750	0.01–0.15	0.0	0.0	24.0	4.0	6.0	1.00	Otsuka, and Komatsu 1987a
*	*	750	1.0	33.3	0.0	1.8	0.90	2.0	0.08	Otsuka, Liu, and Morikawa 1986
Quartz	20, i.d.	600–800	0.5	30.6	0.0	6.7	0.15	45.5	0.18	Otsuka, Jinno, and Morikawa 1986
*	*	750	0.40	17.0	0.0	55.3	27.7	2.0	0.83	Otsuka and Komatsu 1987b
*	*	600–830	0.50	28.0	0.0	23.4	4.7	5.0	0.50	Yamagata et al. 1987

[a] In some reports, flow rates and temperatures were varied; this table summarizes the standard experimental conditions of each report. Some of the values in this table were determined by making some assumptions about the exact conditions when they were not specified by the authors.
[b] Mole ratio of methane to oxygen in the feed.
[c] Reactant partial pressure, $P^* = (P_{CH_4} + P_{O_2})/P_{total}$.
[d] *—Not enough information was provided by the author to determine this condition.
[e] SS—Stainless steel.

profile measured at the reactor center line was $\pm 1\%$ of the set point of the controller at 800°C in the middle 10 cm of the 15-cm heated length. Temperature end effects were present at the inlet and outlet of the reactor. In gas-phase studies, the reactor was operated without any packing material, except for comparative experiments to test the effects of quartz wool and quartz chips placed in the heated zone of the reactor. After realizing the effect of the gas-phase reactions and therefore of the empty heated volume, other reactor designs were used with the purpose of the minimization of the prereactor and postreactor volumes. For simplicity, these other reactor designs will be discussed briefly in the next section, which deals with the oxidation of the products of the methane dimerization reaction (i.e., ethane, ethylene, and carbon monoxide).

Product analysis was performed by gas chromatography (GC) using a thermal conductivity cell detector equipped with thermistors. The separation was carried out using two columns operated in parallel at isothermal conditions at 50°C. A carbosphere packed column ($\frac{1}{8}$ in. by 6 ft) was used for the separation of H_2, O_2, CH_4, and CO_2, and a HayeSep Q polymer packed column ($\frac{1}{8}$ in. by 6 ft) was used for the separation of carbon dioxide, ethane, ethylene, propylene, and propane. Water was condensed in a cold trap placed at the reactor exit before entering the gas chromatograph. Helium was used as the carrier gas in the GC columns as well as a diluent of the feed.

The closure of the carbon balances was always within 5% but in most cases was close to 2%; oxygen balances closed within 5% based on the overall mass balances. Further details on experimental procedures have been published elsewhere (Lane and Wolf 1988). Throughout this chapter, *conversion* is defined as the amount of carbon converted to products divided by the amount of carbon in the feed, *selectivity* is defined as the percentage of carbon converted to a particular product divided by the total amount of carbon converted, and *yield* is the product of selectivity and methane conversion.

Typical experimental conditions were as follows: total feed flow rate of 50 $cm^3\ min^{-1}$; mole ratio $CH_4 : O_2 = 2$; reactant partial pressure divided by total pressure was 0.7 for the gas-phase experiments and 0.4 for the catalytic experiments; and temperatures were in the range of 650–850°C. The total pressure was slightly higher than 1 atm. Specific conditions used in a given set of experiments are described in the corresponding figure caption.

Transient Studies

In order to investigate the role of the lattice oxygen as well as other mechanistic aspects of the oxidative coupling of methane a series of transient experiments was performed. During transient studies a combination of three-way solenoid valves and a 16-loop valve was used to generate either step or pulse functions in the methane feed. In the step mode, at time zero, the He feed to the reactor was switched to the selected stream (i.e., methane or methane–oxygen). The effluent products were collected at 5–20-s intervals in a multiloop sampling valve for later GC analysis. In the pulse mode of operation, a pulse containing one or more reactants was injected via a six-port valve into the reactor, and the effluent was either directed into the gas chromatograph for separation and detection via the TCD or FID detectors or was directly analyzed using a UTI 100 mass spectrometer.

The three catalysts systems used in these studies were 16% Li/TiO_2, 9% $Li/NiTiO_3$, and 17% $Li/La_2O_3 \cdot 2TiO_2$. Catalysts were prepared by wet impregnation followed by calcination at 650°C. These alkali-promoted titanates proved to be active and selective for hydrocarbon formation, with C_2 yields approaching the highest values reported in the literature. A detailed description of the experimental procedure can be found elsewhere (Miro et al. 1990), and thus only a brief description is given here.

Transient measurements were conducted in a new reaction apparatus that was especially designed and built for that purpose. The new flow system enabled operation with minimum influence of nonselective reactions: secondary combustion of products and the methane feed. In programing feed mode, methane and oxygen were fed to the reactor in an alternating fashion with helium purging between feed cycles, using three-way solenoid valves that could be activated by a timer. By installing a manifold with 16 loops downstream from the reactor exit, a convenient and simple way of analyzing effluent gases with respect to time was obtained. Gases were analyzed by gas chromatography and/or mass spectrometry (MS). Small changes in the gas feed that were registered by MS (within parts per million) allowed for very precise monitoring of all concentration changes of the reaction species interacting with the catalytic surface. Another interesting feature of the apparatus was the presence of two six-port sampling valves, which could be used to send pulses of oxygen or methane directly into the reactor.

METHANE DIMERIZATION

Gas-Phase Results

Reactor-Wall Effects

Our first concern was the use of a reactor made of a material that was sufficiently inert so that only the homogeneous gas-phase reaction was the

Table 2-2 A Comparison of Reactor Materials to Determine the Importance of Wall Effects

	CONVERSION (%)		PRODUCT SELECTIVITY (%)				
Reactor Material	CH_4	O_2	C_2H_6	C_2H_4	C_3s	CO_2	CO
Empty[a]	30.7	69.8	6.0	13.8	1.9	8.3	70.0
Quartz chips[b]	30.5	66.9	6.3	16.0	1.8	8.1	67.9
Quartz wool[a]	33.8	76.9	5.2	15.7	1.7	8.6	68.9
Alumina[a]	28.1	78.7	6.0	17.6	1.8	8.0	66.6
SS 304[a,c]	23.4	100.0	1.1	0.6	0.0	89.9	8.4

[a]Conditions: 800°C; $D = 0.75$; CH_4–O_2 mole ratio of 2; and $Q = 50$ cm^3 min^{-1} (at room temperature).
[b]Conditions: 800°C; $D = 0.75$; CH_4–O_2 mole ratio of 2; $Q = 33$ cm^3 min^{-1} (at room temperature).
[c]SS—stainless steel type 304.

predominant reaction pathway. Table 2-2 summarizes the results obtained in our first study of the gas-phase reactions (Lane and Wolf 1988). Due to this effect, reactors made of alumina and quartz gave essentially identical conversions; however, a reactor made of stainless steel yielded mainly CO_2. The role of quartz was less unequivocal because in our initial work we attained similar conversion in the empty reactor or when a section of the reactor was filled with an unspecified amount of quartz chips or quartz wool. Recently, these results were reevaluated under a different set of operating conditions, in light of results indicating that quartz was a scavenger of methyl radicals (Hatano et al. 1990; Kalenik and Wolf 1990a). In the later case a new reactor was specially designed to minimize the prereactor and postreactor volumes so that the heated volume was well-defined. As shown in Fig. 2-1, when the volume of quartz chips is comparable with that of the empty volume, quartz has a significant effect on methane conversion. In our original studies the empty volume was much larger than the volume occupied by the quartz chips; consequently, their effect was not detected.

Total Reactant Pressure or Dilution Effects

As evident from the results summarized in Table 2-1, most catalytic results have been obtained under conditions where, in order to avoid the explosive limits, He, N_2, or Ar was used as a gas diluent. This hid the effect of the total reactant partial pressure, which turned out to be one of the main variables in the gas-phase reaction; at the same time, unknown to many authors, it minimized the effect of the gas-phase reactions in catalytic studies conducted at high reactant partial pressures. Figure 2-2 shows the effect of P^*, which we initially, inadequately, called the dilution ratio and later, more appropriately, redefined as the reactant partial pressure. It can

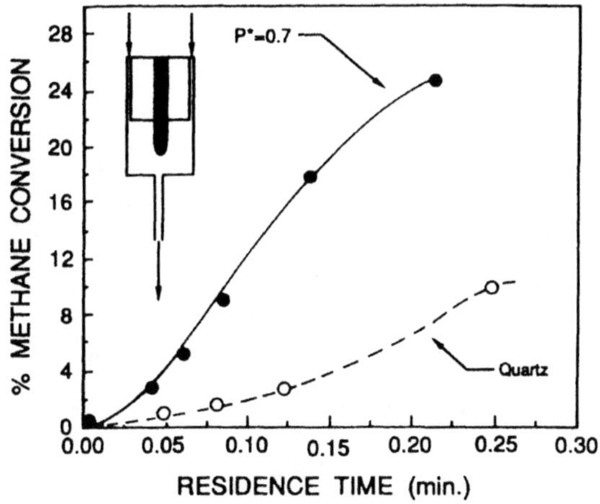

FIGURE 2-1. The effect of residence time on methane conversion at T = 750°C. The solid line corresponds to experiments with an empty reactor at P^* = 0.7. The broken line corresponds to results obtained with the reactor filled with quartz chips at P^* = 0.7.

be seen that methane conversion is a strong function of P^*, with negligible conversions below P^* = 0.4 but increasing with P^* and reaching a value as high as 21% when using no diluent in the feed (P^* = 1). C_2 selectivity reached about 30% under these conditions. The effect of the reactant partial pressures at other temperatures showed a similar trend although at a given value of P^*, methane conversion was lower at lower temperatures and higher at temperatures higher than 750°C. Selectivity to ethane decreased and CO selectivity increased with increasing methane conversion. In all gas-phase studies, CO_2 production was relatively low, with CO being the main carbon oxide.

Reactant Partial Pressure

As in the case of the total reactant pressure, many catalytic studies have been conducted under conditions of low oxygen partial pressure (i.e., $P_{O_2} < 0.2$), under which gas-phase reactions are not significant. However, as shown in Fig. 2-3, methane conversion increases as oxygen partial pressure increases. C_2 selectivity is strongly affected by oxygen partial pressure, with CO showing an ignition-type behavior as the oxygen partial pressure approaches 0.2. When methane was fed without oxygen, no reaction was observed at temperatures as high as 850°C. This clearly shows that the presence of oxygen is required to activate methane below this temperature.

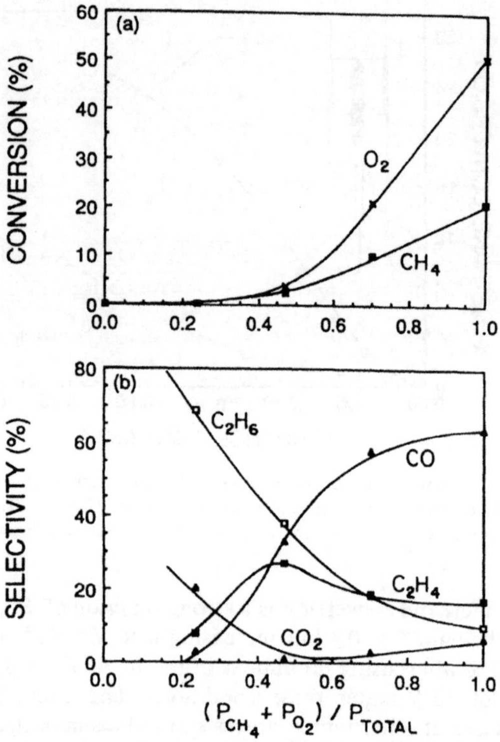

FIGURE 2-2. The effect of the reactant partial pressure on methane and oxygen conversions and on product selectivities at 750°C in an empty quartz reactor with a methane–oxygen feed mole ratio of 2 and with a total feed flow rate of 50 cm^3 min^{-1}: (a) ×, oxygen conversion; ■, methane conversion. (b) □, ethane selectivity; ■, ethylene selectivity; △, CO selectivity; ▲, CO_2 selectivity.

In order to investigate the effect of methane partial pressure, the oxygen partial pressure was kept low to avoid crossing into the explosive limits. As a result, in these experiments (reported in Fig. 2-4) the methane conversion was relatively low and decreased slightly as methane partial pressure increased. Due to the low conversion, C_2 selectivity remained relatively high and even increased slightly as methane conversion increased.

Kinetics

A better way to ascertain the effect of reactant partial pressure is to measure the reaction rate instead of the conversion. Kinetic studies were

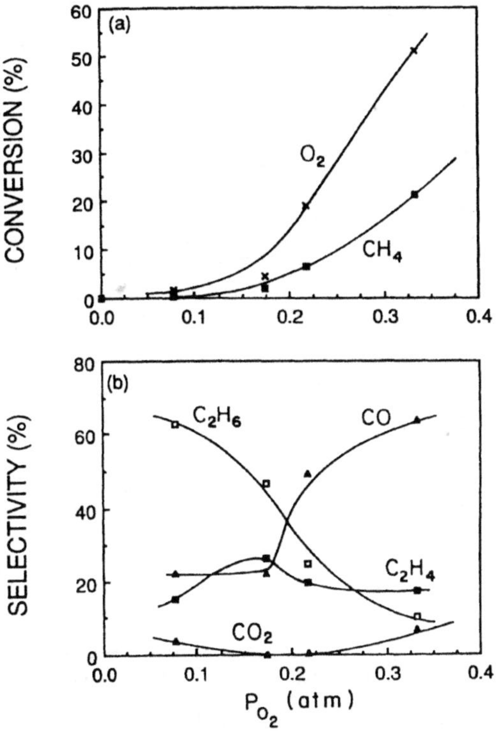

FIGURE 2-3. The effect of oxygen partial pressure on methane and oxygen conversions and on product selectivities at 750°C in an empty quartz reactor with a total feed flow rate of 50 cm^3 min^{-1} and a methane partial pressure of 0.66 atm: (a) ×, oxygen conversion; ■, methane conversion. (b) □, ethane selectivity; ■, ethylene selectivity; △, CO selectivity; ▲, CO_2 selectivity.

conducted under differential reaction conditions, that is, oxygen conversion below 10%. A power-law-type kinetic expression was written for each of the major products and a multivariable regression fit of these results was used to obtain an empirical estimate of the rate constants. The results summarized in Table 2-3 show that the rate of methane consumption (conversion) depended on a first power of the methane partial pressure and a second power of oxygen partial pressure. This agrees well with the conversion results presented previously, which show a strong dependence of methane conversion and selectivity on oxygen partial pressure and a slight effect on methane partial pressure.

As pointed out previously, many researchers have used diluted feeds in their catalytic studies (including the use of air) and, consequently, their

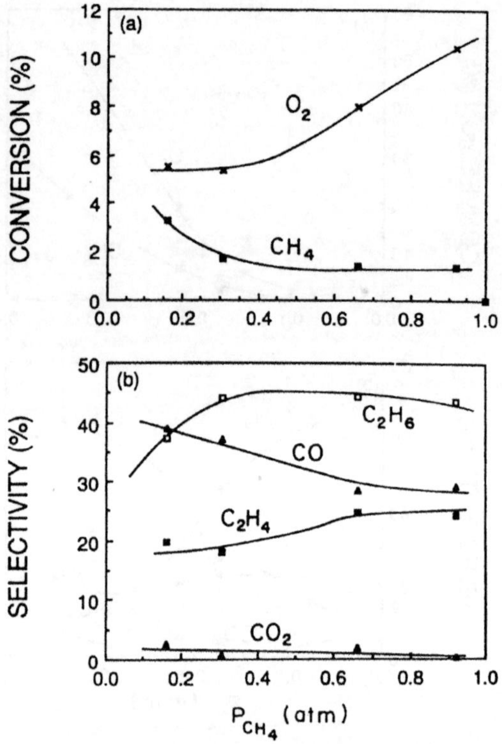

FIGURE 2-4. The effect of methane partial pressure on methane and oxygen conversions and on product selectivities at 750°C in an empty quartz reactor with a total feed flow rate of 50 cm^3 min^{-1} and an oxygen partial pressure of 0.08 atm: (a) ×, oxygen conversion; ■, methane conversion. (b) □, ethane selectivity; ■, ethylene selectivity; △, CO selectivity; ▲, CO_2 selectivity.

catalytic results were not affected by the gas-phase reactions. However, this fact has not been recognized clearly in the literature, as demonstrated by the apparent controversy on this issue that has appeared in the literature recently (Hatano et al. 1990; Kalenik and Wolf 1990a; Yates and Zlotin 1990).

Residence Time

Because the bed length and contact time have been reported to have a significant effect in the catalytic studies (Sofranko, Leonard, and Jones 1987; Otsuka and Komatsu 1986; Otsuka and Jinno 1986; Otsuka and

Table 2-3 Summary of Kinetic Parameters Determined from a Multivariable Linear Fit of the Low-Conversion Results[a]

	Preexponential Factor $\mathscr{A}_i$	Activation Energy E_i	Methane Dependence α_i	Oxygen Dependence β_i
CH_4 (conversion)	1.10×10^7	54.6	1.04	2.05
C_2H_6	3.95×10^5	51.8	1.04	1.78
C_2H_4	9.75×10^4	52.1	1.16	1.62
CO	3.23×10^{11}	71.6	0.53	3.70
CO_2	1.02×10^{-2}	29.5	−0.95	1.33

[a]The reaction rate is given be

$$\mathscr{R}_i = \mathscr{A}_i \exp(-E_i/RT)[P_{CH_4}]^{\alpha_i}[P_{O_2}]^{\beta_i}$$

where units are as follows: $\mathscr{R}_i[=]$ mol s^{-1}; $\mathscr{A}_i[=]$ mol s^{-1} $(atm)^{-(\alpha+\beta)}$; $E_i[=]$ kcal mol^{-1} s^{-1}; $P_i[=]$ atm.

Komatsu 1987), we decided to study its effect on the gas-phase reactions. Due to the temperature end effects in the open tubular reactor used in the first study (Lane and Wolf 1988), the exact volume of the reactor, which operated at the set-point temperature, was not exactly known; consequently, we used the reciprocal flow rate as a measure of residence time. The results shown in Fig. 2-5, which were taken in the reactor with well-defined volume at two different reactant partial pressures, indicate

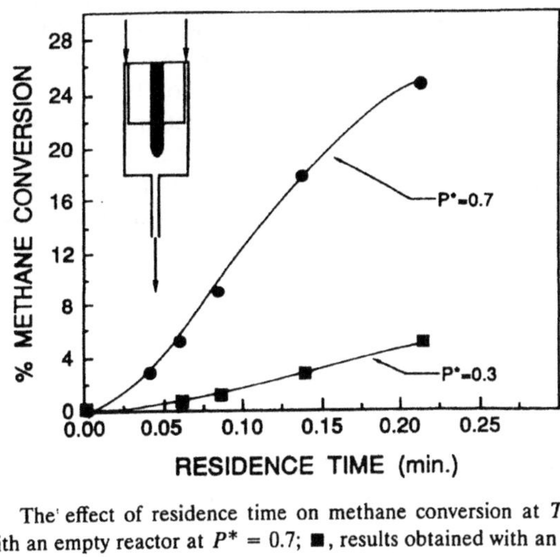

FIGURE 2-5. The effect of residence time on methane conversion at $T = 750°C$: ●, experiments with an empty reactor at $P^* = 0.7$; ■, results obtained with an empty reactor at $P^* = 0.3$.

that at $P^* = 0.7$ and at 750°C conversion in the gas-phase increased with increasing residence time and then leveled off as oxygen conversion reached high values. Ethane selectivity (not shown) decreased as methane conversion increased with a corresponding increase in the CO selectivity.

It should be noted, however, that in Fig. 2-5 the high value of reactant partial pressure, $P^* = 0.7$, was selected on purpose in order to enhance gas-phase reactions, and the results are different at low values of P^*, as also shown in Fig. 2-5. It can be seen that the effect of residence time depends on P^* and its effect is of less importance at $P^* = 0.3$. Again, the use of long residence time (or large-volume reactors) tends to enhance the gas-phase reactions, which otherwise might affect the catalytic results as in the case of the work reported by Yates and Zlotin (Yates and Zlotin 1988).

Temperature Effects

The effect of temperature is displayed in Fig. 2-6 for an oxygen partial pressure of 0.23 and $P^* = 0.7$, so that the gas-phase reactions are significant. It can be seen that the conversion of both methane and oxygen increases with increasing temperature. As in the other cases when conversion increases, C_2 selectivity decreases, and CO selectivity increases with increasing temperature. The effect is less pronounced at lower oxygen partial pressures but it follows the same trends as shown in Fig. 2-6.

The activation energy obtained in the kinetic studies gives a value of 54.6 kcal mol^{-1} for methane conversion, which in fact agrees fairly well with values reported for the gas-phase reaction of methane and molecular oxygen to yield methyl radicals and hydroperoxy radicals (Emesh Ali and Amenomiya 1986). The activation energies for the formation of ethane and ethylene were also similar, indicating that the rate-limiting step was the formation of methyl radicals in the gas phase. In the case of CO and CO_2 formation the fit was not as good, but this is not so surprising because these are secondary products and their partial pressures and the temperature dependence of reaction constants affects them indirectly via an intermediate.

Selectivity–Conversion Relationships

A common characteristic of the gas-phase results presented earlier is that regardless of the way that a given conversion was achieved, we observed that the C_2 (ethane plus ethylene) selectivity decreased as the conversion increased. This is shown in Fig. 2-7, which summarizes the results obtained in many different experiments. Figure 2-7 also shows that, in the *gas phase*, selectivity decreases as conversion increases. Very high selectivity can be attained at low conversion, a result that was common in the early

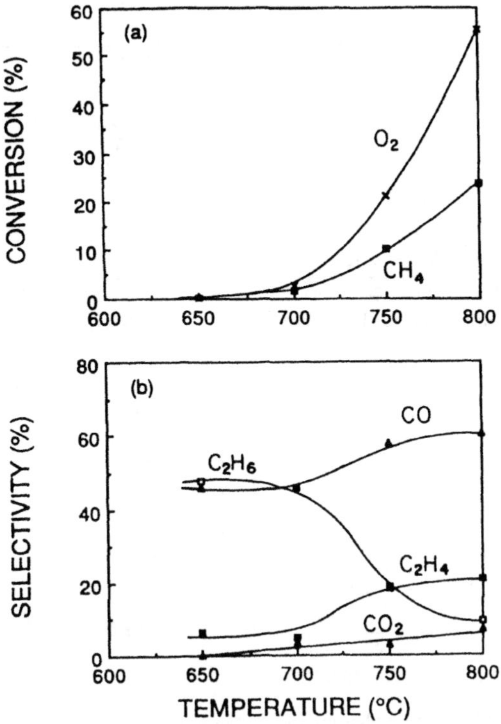

FIGURE 2-6. Influence of temperature effects on methane and oxygen conversions and on product selectivities in an empty quartz reactor with a total feed flow rate of 50 cm^3 min^{-1}, methane partial pressure of 0.66 atm, oxygen partial pressure of 0.23 atm, and dilution ratio $P^* = 0.7$: (a) ×, oxygen conversion; ■, methane conversion. (b) □, ethane selectivity; ■, ethylene selectivity; △, CO selectivity; ▲, CO_2 selectivity.

catalytic literature. This selectivity–conversion relationship *does not* imply that higher selectivities than those shown in Fig. 2-7 cannot be attained; in fact, as is demonstrated later, the role of the catalyst is to alter this relation and to provide higher selectivity at higher conversions than the ones attained in the gas phase at atmospheric pressure. Recent results at high pressure indicate that these trends are also significant at P higher than 1 atm.

Table 2-4 presents a summary of the conditions wherein the gas phase is important. It is clear that, for a given combination of reactant total pressure, reactant partial pressure, temperature, and residence time, the gas-phase reactions can be significant. Conversely, it is possible to select

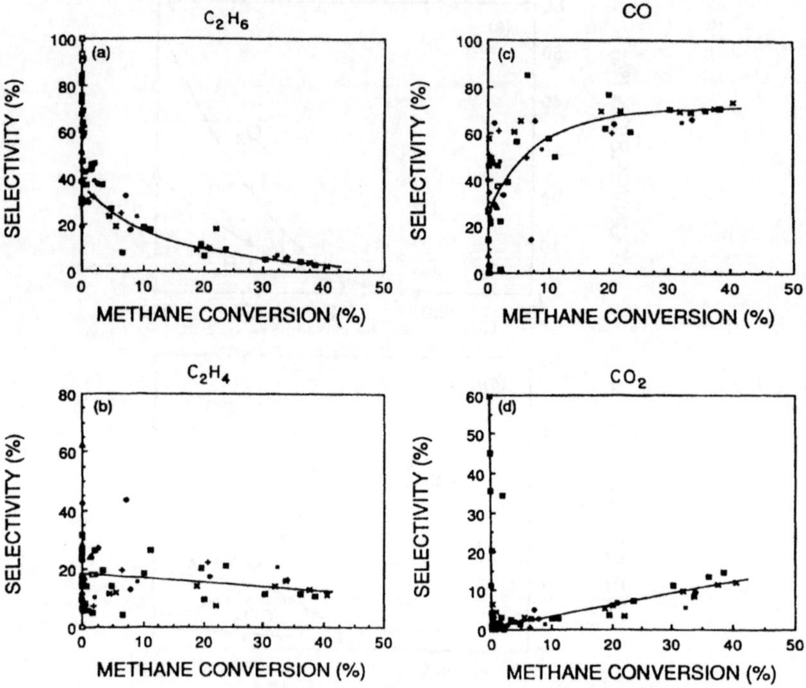

FIGURE 2-7. Master plots showing the overall selectivity that results at varying levels of methane conversion independent of the operating conditions. The resulting product selectivities are shown as a function of methane conversion for (a) C_2H_6, (b) C_2H_4, (c) CO, and (d) CO_2. The experimental conditions for each set of results are summarized in Table 2-4.

Table 2-4 Summary of Gas-Phase Studies

Parameter	CH_4 Conversion < 5%	CH_4 Conversion > 10%
Residence time (min)	< 0.1	> 0.1
Reaction partial pressure, P^*	< 0.4	> 0.4
Methane–oxygen ratio	> 3.0	< 3.0
Temperature	< 650°C	> 650°C

conditions in which the gas-phase reactions are not relevant and thus to study only the catalytic effects.

Catalytic Results

The main purpose of the studies previously reported concerning the homogeneous reaction was to define the conditions where the gas-phase reactions were important so we could study catalytic effects without their interference. Because this chapter is devoted to gas-phase studies, we do not review here all the catalytic studies by our group but only the main difference of the catalytic results with respect to the gas-phase results. It is necessary to compare gas-phase and catalytic results to emphasize the importance of the need for a more-selective catalyst that can activate methane at a lower temperature, so that gas-phase reactions of methane or its products are less important than the catalytic pathway. Although many catalysts have been studied for the dimerization of methane, our group has focused mainly on alkali-promoted titania-based catalysts. Here we present only the effect of temperature on the conversion and selectivity, and also some limited transient results to demonstrate the role of oxygen mobility on catalyst activity and selectivity.

Steady-State Results

The main promoter studied in our group has been Li, with Na studied to a lesser extent. The first support selected was TiO_2 because this material exhibited a peculiar activity for the CO oxidation (Lane, Miro, and Wolf 1989). Other supports selected have been a family of titanates (Lane, Kalenik, and Wolf 1989), $NiTiO_3$, which was inactive unless promoted with alkaline metal (Miro, Santamaria, and Wolf 1990a), followed by a $La_2O_3 \cdot 2TiO_2$ (Miro et al. 1990; Kalenik and Wolf 1991). For all the various catalysts studied it was found that the conversion increased with Li loading, reaching a maximum and then decreasing at higher loadings. On the other hand, selectivity increased with the degree of lithium loading. The results summarized here were obtained at conditions where the influence of gas-phase reactions on C_2 hydrocarbons yield *was not* important. The main variable controlling catalytic selectivity was the reactant partial pressure, which was maintained around 0.4 so that according to the previous studies gas-phase conversion was minimized. Figures 2-8–2-10 illustrate catalytic results in terms of conversion and selectivity versus temperature for 16% Li/TiO_2 (Lane, Miro, and Wolf 1989), 9% $Li/NiTiO_3$, and 1.6% $Na/NiTiO_3$ (Miro, Santamaria, and Wolf 1990a; Miro et al. 1990) and for the different loadings of the $Li/La_2O_3 \cdot 2TiO_2$ (Kalenik and Wolf 1991), respectively. The results reveal a general trend

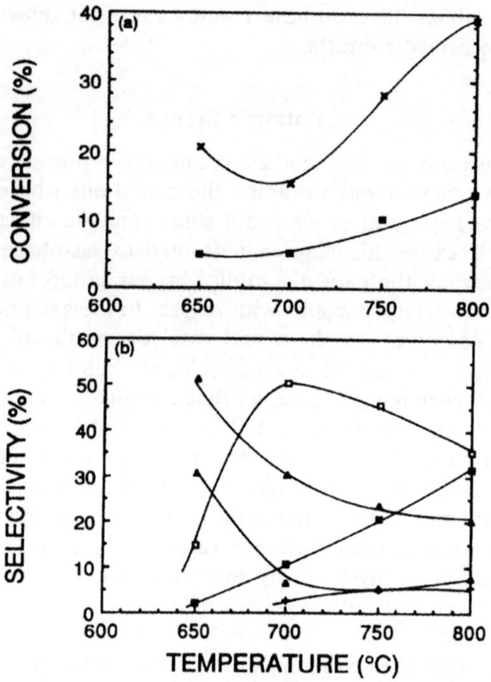

FIGURE 2-8. Temperature effects on CH_4 and O_2 conversions and product selectivities for the 16.2% lithium-titania catalyst with a CH_4–O_2 feed mole ratio of 4 and a partial pressure of reactants of 0.4: (a) ×, oxygen conversion; ■, methane conversion. (b) □, ethane selectivity; ■, ethylene selectivity; △, CO selectivity; ▲, CO_2 selectivity; +, C_{3+} selectivity.

that, whenever the oxygen conversion was below 100%, the characteristic selectivity–conversion relationship exhibited by the gas-phase reactions is no longer held in the presence of a catalyst. In fact, for the nickel titanate-based catalyst C_2 selectivity tends to increase as conversion (temperature) increases.

The effect of reactant partial pressure on the activation energy was also studied for the 16% Li/TiO_2 catalysts. Figure 2-11 shows that the activation energy for the ethane formation of the catalyzed reaction is a function of the reactant partial pressure P^*, varying from 25 kcal mol^{-1} at $P^* = 0.2$ to 38 kcal mol^{-1} at $P^* = 0.8$. Clearly, the increase in activation energy is due to the increased participation of the gas-phase reactions over the catalytic reactions when the reactant partial pressure increases. It should be noted, however, that the main carbon oxide in catalytic studies

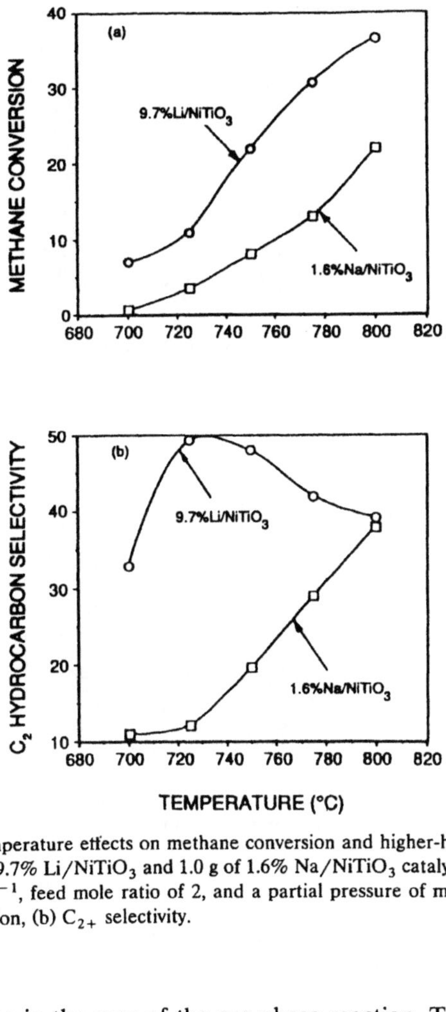

FIGURE 2-9. Temperature effects on methane conversion and higher-hydrocarbon selectivity using 0.25 g of 9.7% $Li/NiTiO_3$ and 1.0 g of 1.6% $Na/NiTiO_3$ catalyst with a feed flow rate of 100 $cm^3\ min^{-1}$, feed mole ratio of 2, and a partial pressure of methane of 0.32: (a) methane conversion, (b) C_{2+} selectivity.

is CO_2 not CO as in the case of the gas-phase reaction. The deviation of catalytic results from the gas-phase results is clearly seen when all published catalytic results (as of 1988, plus the results presented in Figs. 2-8–2-10) are presented in a master plot of selectivity versus conversion, as the one shown in Fig. 2-7. The catalytic results are shown as scattered points in the Fig. 2-12, wherein the gas-phase results shown in Fig. 2-7 are summarized as the solid line in the plot. It is clear that many catalysts yield higher amounts of C_2 products than the gas phase; however, in some

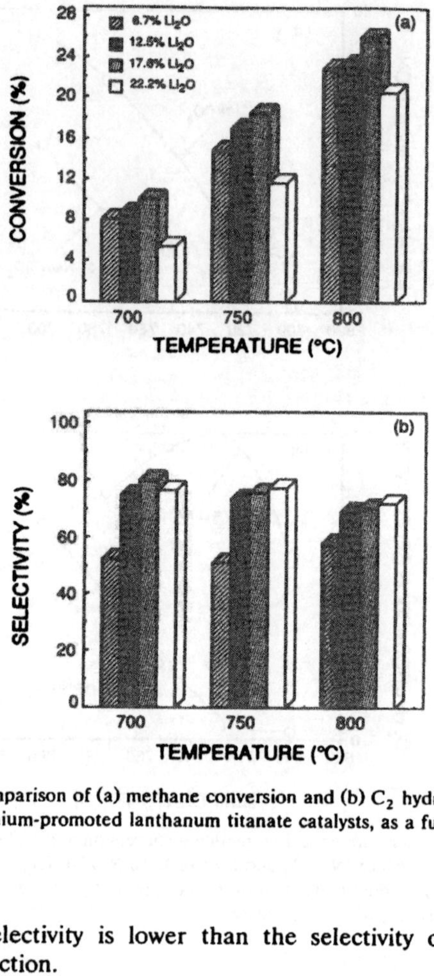

FIGURE 2-10. Comparison of (a) methane conversion and (b) C_2 hydrocarbons selectivities for a series of lithium-promoted lanthanum titanate catalysts, as a function of reaction temperature.

cases catalytic selectivity is lower than the selectivity obtained during homogeneous reaction.

Transient Results

Transient studies consisting of (i) step inputs of methane followed by a step of methane and oxygen mixture, (ii) oxygen pulses with different CH_4–O_2 ratios, (iii) continuous mass spectrometric monitoring of oxygen trace level in the gas phase, and (iv) steady-state isotopic switches show that the extend of participation of lattice oxygen on methane conversion and hydrocarbon production is a function of the promoter, the support

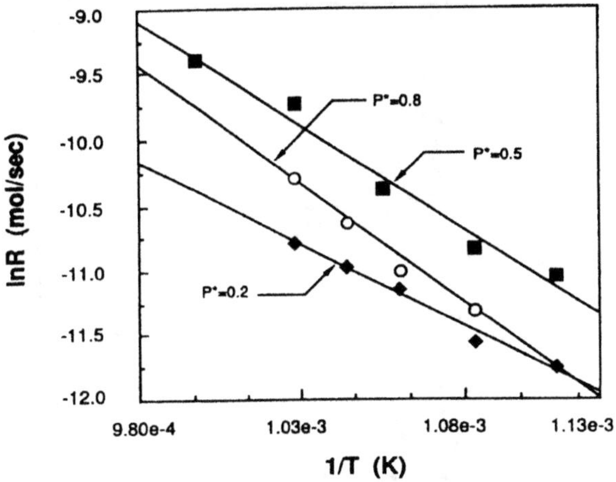

FIGURE 2-11. An Arrhenius plot for the rate of ethane production for the low-conversion data for three different reactant partial pressures. Activation energies for ethane formation calculated on the basis of these experiments were as follows: $E(P^* = 0.2) = 25.6$ kcal mol^{-1}; $E(0.5) = 32.4$ kcal mol^{-1}; $E(0.8) = 37.6$ kcal mol^{-1}.

used, and the reaction temperature. Methane steps over oxidized catalysts yielded CO and CO_2 for 16% Li/TiO_2, and considerable amounts of ethane and ethylene for Li/$NiTiO_3$ and Li/$La_2O_3 \cdot 2TiO_2$ catalysts. Since temperature programmed desorption (TPD) experiments did not reveal weakly absorbed oxygen on the catalysts surface, it is expected that lattice oxygen participates in the reaction and is selective for hydrocarbon formation in the case of nickel and lanthanum titanates. After introducing gas-phase oxygen into the system, a dramatic increase in C_2 production was observed for Li/TiO_2, and only a slight increase was observed for Li/$La_2O_3 \cdot 2TiO_2$ catalyst. On the other hand, the yield of C_2 products for the Li/$NiTiO_3$ phase remained constant after O_2 was allowed into the reactor. These data clearly indicate that lattice oxygen plays a very significant role during hydrocarbon formation in the case of lanthanum and nickel titanates. For the Li-promoted TiO_2, however, the species responsible for methane activation is the weakly adsorbed oxygen, which interacts with the surface of the catalysts. To investigate further the involvement of lattice oxygen in methane activation, mass spectrometric experiments were conducted. During these studies a trace of oxygen [~ 10 ppm] in helium gas was continuously monitored while introducing methane pulses over degassed samples of the catalysts at 750°C. These data show that for the Li/TiO_2 catalyst the gas-phase O_2 level remains

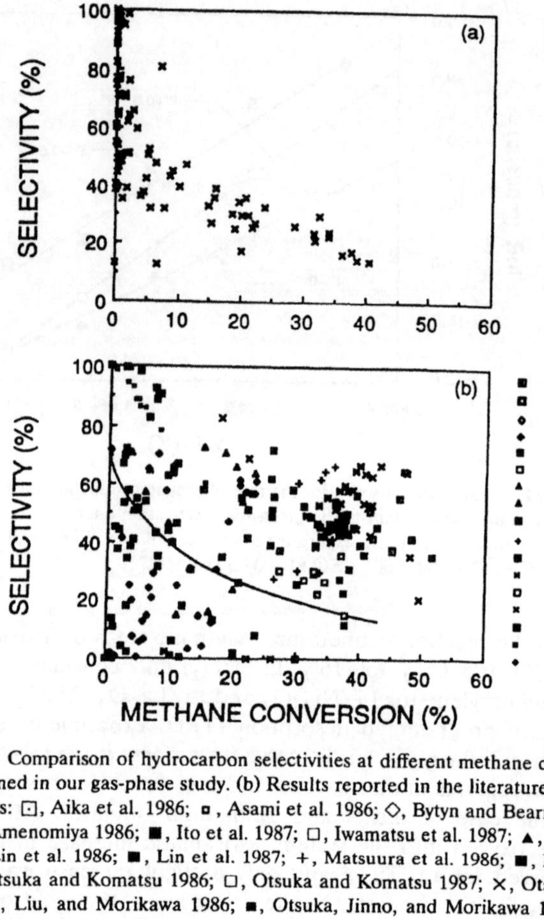

FIGURE 2-12. Comparison of hydrocarbon selectivities at different methane conversions. (a) Results obtained in our gas-phase study. (b) Results reported in the literature by various authors as follows: ⊡, Aika et al. 1986; ▫, Asami et al. 1986; ◇, Bytyn and Bearns 1986; ◆, Emesh Ali and Amenomiya 1986; ■, Ito et al. 1987; □, Iwamatsu et al. 1987; ▲, Kimble and Kolts 1986; ▲, Lin et al. 1986; ■, Lin et al. 1987; +, Matsuura et al. 1986; ■, Moriyama et al. 1986; × Otsuka and Komatsu 1986; □, Otsuka and Komatsu 1987; ×, Otsuka et al. 1986; ⧅, Otsuka, Liu, and Morikawa 1986; ▪, Otsuka, Jinno, and Morikawa 1986; ◆, Yamagata et al. 1987. The line represents a visual fit of the noncatalytic, gas-phase results shown in (a).

constant whereas for $Li/NiTiO_3$ and $Li/La_2O_3 \cdot 2TiO_2$ a significant decrease in gas-phase oxygen concentration occurs. The initial level of O_2 was regained only after additional pulses of oxygen were introduced into the reactor.

In order to investigate the interaction of methane with the catalyst surface, switches of $^{13}CH_4/^{12}CH_4$ under steady-state reaction conditions were used. These experiments which were carried out for the $Li/NiTiO_3$

and $Li/La_2O_3 \cdot 2TiO_2$ catalysts, indicated that no long-lived adsorbed methane is present in significant quantities on the surface of these catalysts under reaction conditions. On the other hand, the carbon dioxide signal showed long relaxation times, which was to be expected in view of the large pool of surface carbonates revealed by X-ray photoelectron spectroscopic analysis.

Summary

It is clearly shown that the gas-phase homogeneous reactions can play a significant role in the methane oxidative coupling and that the extent of their participation depends on the operating conditions applied. In the gas phase, ethane appears as the main product, followed by ethylene and carbon monoxide. The extent of the gas-phase reaction depends on the reactants' total and partial pressures, residence time (reactor volume), and temperature. The effect of oxygen partial pressure is more important than that of methane, and, in the homogeneous reaction, selectivity decreases with conversion. These results agree well with a mechanism involving the formation of methyl and hydroperoxy radicals, which is consistent with known results in methane combustion and with mechanisms proposed for the catalytic reaction. Mechanistic implications are discussed in the next section.

The results summarized in this section are now well-established and have been repeated by several authors in different laboratories and also have been presented by other co-authors of this book (van der Wiele, Geerts, and van Kasteren 1990). The merit of our work was to bring attention to the gas-phase reactions at a time when very little documentation of their effect existed. The role of the catalysts is to alter the reaction pathway so that the selectivity–conversion curve is modified to higher values than those predicted in the gas phase. As shown by our transient studies, as well as those of others, an important variable governing catalyst design appears to be the mobility of oxygen in the lattice of the oxide catalyst as well as the ability to control the active sites to prevent the deep oxidation of the reaction products.

ETHANE, ETHYLENE, AND CARBON MONOXIDE OXIDATION

After completing our work on the gas-phase reactions of methane, we realized that it was also important to study the gas-phase reactions of the products of methane oxidative coupling. This section presents additional evidence that the operating variables also affect gas-phase and catalytic oxidation of ethane, ethylene, and carbon monoxide. Thus, steady-state experiments on the oxidation of these reactants were conducted without a catalyst, as before, and also in the presence of the 16% Li_2O/TiO_2 (Kalenik et al. 1989) and other catalysts.

As shown in the preceding section, the reactor volume and its design are important factors in ascertaining the role of the gas-phase reactions during methane oxidative coupling experiments. Many catalytic experiments have been conducted in quartz tube reactors partially filled with a catalyst. Unless special designs are used, the heated empty volume of the reactor could significantly exceed or be equal to the volume taken by a catalyst. In order to investigate the influence of the reactor "empty" volume on ethane and ethylene conversions, gas-phase and catalytic experiments were performed in three different reactor configurations. In addition to the tubular reactor used previously, two other designs were applied to study the effect of the prereactor and postreactor volumes and thus, indirectly, the effect of residence time. The three different reactor designs used in the following experiments are shown schematically as the three inserts in Fig. 2-13: Reactor A consisted of the previously used clean quartz tube, whereas reactors B and C each consisted of an inner quartz tube (9.5 mm o.d.) with a thermowell, placed concentrically inside another quartz tube (10 mm i.d.). Gases were fed into the reactor volume through the annular space between the two cylinders (prereactor section for

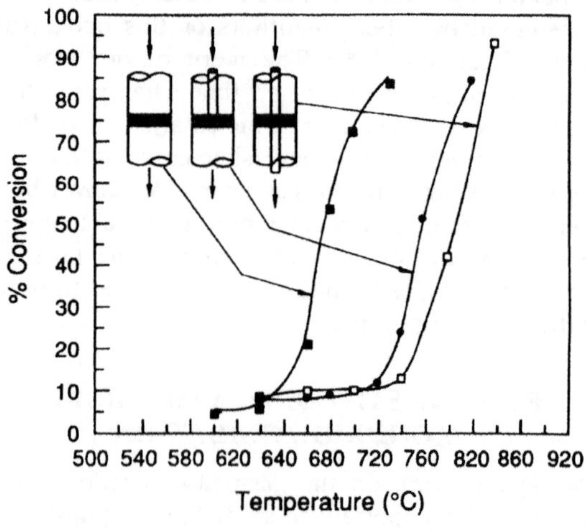

FIGURE 2-13. Gas-phase ethane conversion versus reaction temperature for different reactor configurations: ■, reactor A; ●, reactor B; □, reactor C.

designs B and C) and exited the reactor volumes through the 10-mm-i.d. tube (reactors A and B) or through the 3-mm-i.d. capillary tube (postreactor section for design C). Due to the fast flows in prereactor volumes (reactors B and C) and postreactor volume (reactor C), residence times in these two sections were very low and could be neglected. Because all experiments were conducted at the same gas-flow rates, the residence time of reactants in each reactor varied from 0.1 to 0.01 min, approximately. Another advantage of reactors B and C was the fact that inlet gases were not able to react before reaching the catalyst bed or the reactor volume. In the case of gas-phase experiments the reactor volume was filled with quartz chips so that the residence time will be similar to catalytic experiments. The quartz chips used in these experiments were broken fragments of fused silica placed in the reactor. The void fraction of the packing material was around 0.5 for both catalyst and quartz chips. Extreme care was taken in preparing the latter material without any exposure to a ceramic or any metal surface because it was found that quartz chips were susceptible to contamination when prepared in a clean ceramic mortar, which resulted in a significant catalytic effect during the oxidative coupling reaction. The catalytic effect of contaminated reactors also has been reported by Martin and co-workers (Martin et al. 1989), who found that gas-phase experiments conducted in the lithium-contaminated quartz reactor showed significantly increased methane conversion. The volume of quartz chips and/or catalyst applied in these experiments was 1 cm^3. Gases were diluted in helium, and a typical set of operating conditions for ethane, ethylene, and carbon monoxide oxidation was as follows: reactant–oxygen feed mole ratio of 2, reactant partial pressure $P^* = 0.06$, feed flow rate $Q = 100\ cm^3\ min^{-1}$ (measured at room temperature). In most experiments carbon balances closed within 3–5%.

Additional experiments were conducted to determine the influence of residence time and the effect of quartz chips on the conversion of ethylene. In this case measurements were taken in the reactor with a well-defined volume of 4.3 cm^3 with and without quartz chips. Experiments at different residence times were conducted at reactant partial pressures of 0.3 and 0.1, temperature 650°C, and at reactant-to-oxygen volume ratio of 2. Residence time (at STP) was changed by adjusting the volumetric flow rate of reactants.

Gas-Phase Results

Conversion–temperature curves for ethane oxidation are shown in Fig. 2-13 for the three different reactor designs shown in the insert. These curves exhibit ignition-type behaviour, with the ignition temperature

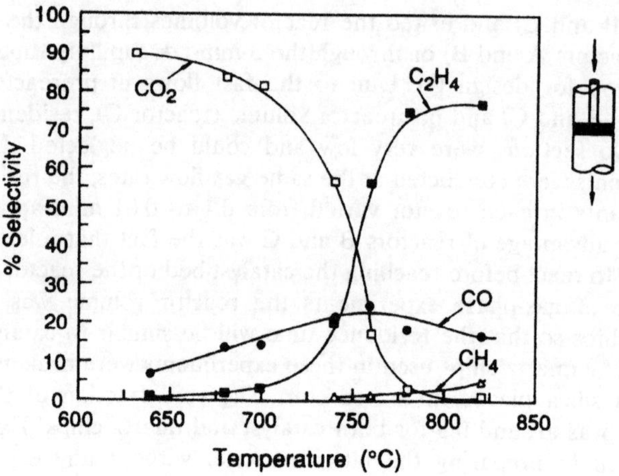

FIGURE 2-14. Product selectivities versus reaction temperature for the gas-phase conversion of ethane in the reactor with a minimum prereactor volume.

strongly depending on the reactor configuration. A good indicator of catalytic activity is the temperature at which 50% conversion of the limiting reactant occurs (i.e., the "light off" temperature, or LOT). It can be seen that the combustion of ethane starts at 700°C for an open tube reactor (i.e., higher residence time), at 780°C for the reactor with high postreactor volume, and at 815°C for the reactor with minimum volume (lower residence times). The main reaction products are CO_2, CO, and C_2H_4. It can be seen that when the gas-phase volume is kept to a minimum or the residence time is low (reactor C) ethane conversion in the gas phase occurs at temperatures that are similar to or higher than those used in the methane coupling reaction. However, when the gas-phase volume is large or the residence time is high, ethane conversion occurs at temperatures that are significantly lower than the methane coupling reaction. Selectivities of products obtained for the experiments using reactor B are shown in Fig. 2-14. Due to the depletion of gas-phase oxygen above the ignition temperature, selectivity of CO_2 decreases while CO selectivity increases as temperature increases, and at 800°C they both level off. Selectivity to ethylene reaches 80% and still increases due to the high-temperature dehydrogenation of ethane. Similar trends in selectivity–temperature curves were observed for the gas-phase experiments conducted at different reactor configurations.

Reactors B and C with minimized empty volumes were also used during gas-phase oxidation of ethylene. As expected, results of those experiments

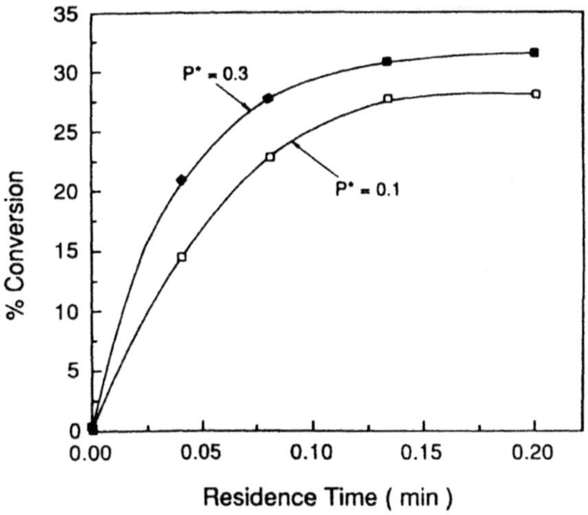

FIGURE 2-15. Conversion versus residence time for the gas-phase oxidation of ethylene at 650°C at two different reactant partial pressures.

(not shown) indicate that in the case of C_2H_4, LOT also depends on the reactor used. Oxidation of ethylene starts at 780°C for the reactor with the minimum prereactor volume (reactor B) and at 790°C for the reactor with the minimum prereactor and postreactor volumes (reactor C).

Because it was shown (Fig. 2-13) that the ethane conversion during oxidation experiments strongly depends on the reactor configuration, it was appropriate to conduct an additional set of measurements that would quantify the explicit dependance of the residence time and reaction conditions on the conversion of C_2 hydrocarbons. Ethylene oxidation experiments were conducted at two different reactant partial pressures, $P^* = 0.1, 0.3$, and at a temperature of 650°C in order to avoid oxygen-limited conditions. An additional experiment was conducted in the reactor filled with quartz chips in order to determine the influence of that material on the gas-phase conversion of ethylene. Figure 2-15 presents results of the oxidation experiments for two different reactant partial pressures as a function of residence time. It can be seen that the conversion of ethylene increases as residence time increases and later levels off due to the 100% oxygen conversion. For the smaller value of P^*, gas-phase conversion remains smaller, which is in good agreement with our previous reports on gas-phase oxidation of methane (Lane and Wolf 1988; Kalenik and Wolf 1990). Experiments conducted in the presence of quartz chips at reactant

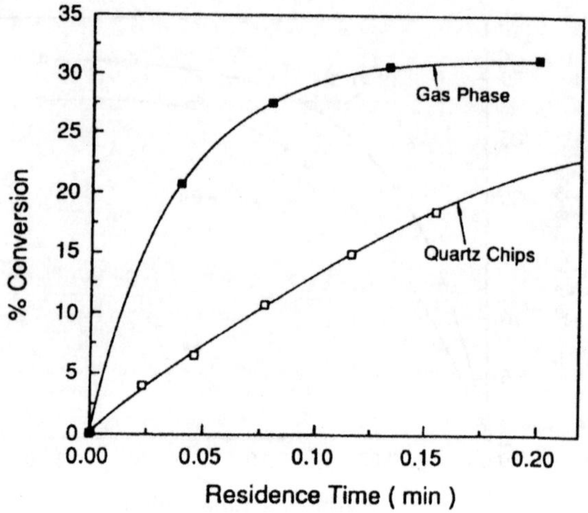

FIGURE 2-16. Conversion versus residence time for the gas-phase oxidation of ethylene with and without quartz chips at 650°C.

partial pressure of $P^* = 0.3$ (Fig. 2-16) indicate again that the presence of an inert material significantly inhibits gas-phase conversion of ethylene.

Catalytic Results

Comparison between gas-phase and catalytic oxidation of ethane was performed in the reactor with the minimum prereactor volume (reactor B). The need to use a new reactor after every catalytic experiment dictated this choice of reactor over the reactor with minimum prereactor and postreactor volumes (reactor C), which is more expensive. Figure 2-17a displays gas-phase and catalytic conversions of ethane as a function of reaction temperature. As expected, the presence of the 16% Li/TiO_2 catalyst significantly decreased ethane LOT from 780°C (no catalyst) to around 660°C for the 16% Li/TiO_2. The effect of catalytic versus homogeneous ethane oxidation, that is, the selectivity toward carbon oxides versus reaction temperature can be seen in Fig. 2-17b. At temperatures up to 750°C formation of carbon dioxide and carbon monoxide for the catalytic reaction approaches 13% and then slightly decreases due to the depletion of gas-phase oxygen, thermal dehydrogenation of ethane to ethylene and ethane oxidative dehydrogenation via secondary chain initiation reactions. In the case of catalytic oxidation of ethane, the main reaction products were C_2H_4, CO_2, and CO below ignition temperature and C_2H_4 and CO_2

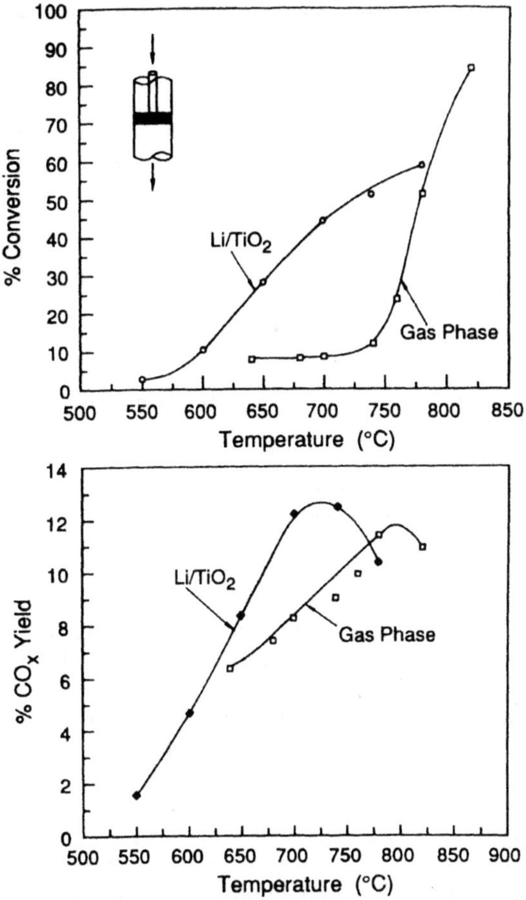

FIGURE 2-17. Comparison of gas-phase and catalytic oxidation of ethane in the reactor with minimum prereactor volume: (a) methane conversion versus reaction temperature for the homogeneous experiment and for the experiment on 16% Li/TiO_2 catalyst; (b) yield of CO_x product versus reaction temperature for the homogeneous experiment and for the experiment on 16% Li/TiO_2 catalyst.

above. The main reaction products for the gas-phase reaction were CO_2 and CO below the ignition point and C_2H_4 and CO above. It is clear that under the conditions used in these experiments, most of the ethane produced during catalytic oxidative coupling of methane say at 750°C, would have been converted to ethylene and CO. As this is not the case,

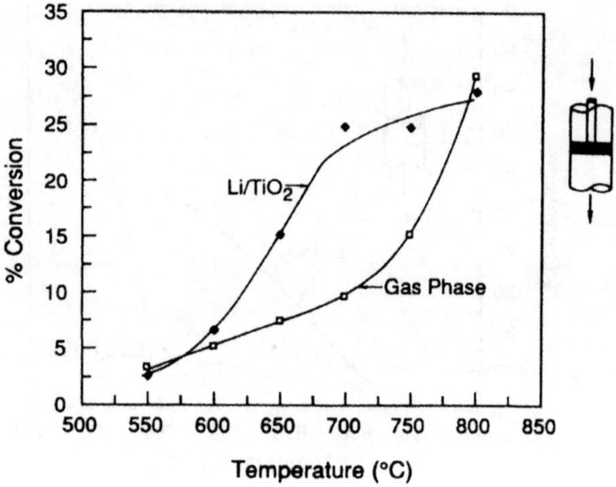

FIGURE 2-18. Comparison of ethylene conversion for the gas phase and catalytic oxidation of C_2H_4, in the reactor with minimum prereactor volume.

one should conclude that the presence of the co-reactants affects the oxidation sites responsible for the catalytic oxidation of methane to higher hydrocarbons.

Conversion and temperature results for the oxidation of ethylene in the gas phase and in the presence of the Li_2O/TiO_2 catalyst (Fig. 2-18) show similar trends as during ethane oxidation experiments. In the case of ethylene, however, the main reaction products are CO and CO_2, with small amounts of methane also observed. Again, it can be seen that the catalytic combustion of C_2H_4 starts at lower temperatures than the gas-phase reaction. Ignition temperatures are found to be ~ 660 and ~ 780°C for the catalytic and the gas-phase reactions, respectively. Because the depletion of gas-phase oxygen takes place for both reactions, the highest ethylene conversions obtained under the conditions used did not exceed 30%. Selectivity results indicate again that in the presence of the catalyst ethylene is converted mainly to CO_2, whereas for the gas-phase oxidation the main reaction product is CO. Comparison of ignition temperatures for ethane and ethylene oxidation in the gas phase in the same reactor suggests that ethane is converted at higher rates than ethylene. As in the case of ethane, these results would indicate that in the presence of the catalyst most of the ethylene formed during methane oxidative coupling would be converted to CO_2; because this is not the case we must assume that the presence of the co-reactants is decreasing this effect.

Carbon monoxide oxidation experiments indicated that no CO conversion was observed in the gas phase up to 800°C. The ignition temperature for a catalytic reaction was found to be around 730°C.

Summary

This section presented the following results: (i) Oxidation of ethane and ethylene occurs in the gas phase at temperatures that are lower than those typically used in methane oxidative coupling experiments. The LOT is a function of operating parameters, including the reactor volume or its equivalent—the residence time. (ii) The ignition temperatures for the catalytic conversion of ethane and ethylene are lower than those observed in the gas phase under identical conditions. The main product of the oxidation of C_2 hydrocarbons on the catalyst is CO_2. (iii) CO does not oxidize in the gas phase at temperatures below 850°C but its LOT decreases to 730°C in the presence of the catalyst. (iv) Small residence times, low reactant partial pressures, and the presence of an inert material (quartz chips) may significantly reduce gas-phase conversion of C_2 hydrocarbons during their oxidation.

Experiments similar to those described previously have been reported by others. Martin and co-workers (Martin et al. 1989), who conducted their experiments in the U-tube reactor and at reactant partial pressures P^* of 0.12, reported that the gas-phase oxidation of ethane and ethylene starts around 650°C. This temperature is substantially lower than the ignition temperatures observed in our research. The most probable reason for the discrepancies is the fact that the LOT is a strong function of the operating variables and reactor design. Ross and co-workers (Ross et al. 1989), who studied the reaction path of the oxidative coupling products over a Li/MgO catalyst in a well-mixed reactor concluded that most of the CO_x products are formed by catalytic oxidation of ethylene.

As pointed out previously, the ignition temperatures found for the oxidation of the C_2 hydrocarbons in the presence of a catalyst are significantly lower than the typical temperatures used in the methane coupling studies. Accordingly, ethane and ethylene should not be seen as products in the catalytic oxidative coupling of methane. Because this is not the case, as shown in the previous section and in many of the literature studies, it follows that the conditions for the oxidation of ethane and ethylene are different during methane oxidative coupling than those used in the catalytic studies reported in this section. This could occur due to the interaction of methane with the active sites that are also responsible for the oxidation of ethane and ethlene or because the operating conditions are vastly different than those that these reactants encounter during the methane coupling reaction. Nonetheless, in order to decrease oxidation of the reaction products the operating conditions should be adjusted to minimize the gas-phase reactions and more selective catalyst should be designed to minimize oxidation of these products.

REACTION PATHWAY CONSIDERATIONS

Mechanistic Considerations

Gas-Phase Reactions

The studies presented in this chapter regarding the gas-phase reactions provided evidence of their relevance as an alternative reaction pathway to the catalytic reactions. The nature of the results did not provide an independent verification of each of the elementary steps involved in the reaction pathway but rather integral conversion results in which the observed products are the interaction of the various pathways available. A great deal of mechanistic considerations, however, are available in the methane combustion literature and consequently we, as well as other authors (Tsang and Hampson 1986) analyzed these results in an attempt to rationalize the experimental observations.

Three approaches were taken in the analysis of the existing, fairly voluminous, literature of methane combustion. First, a simple sequential mechanism based on the product distribution as the one depicted in Fig. 2-19 was analyzed. To simulate the experimental results, mass balances for each of the species considered were written for an isothermal plug-flow reactor. The kinetics constants used were those reported in the literature by Tsang and Hampson (Tsang and Hampson 1986). In this simulation, to

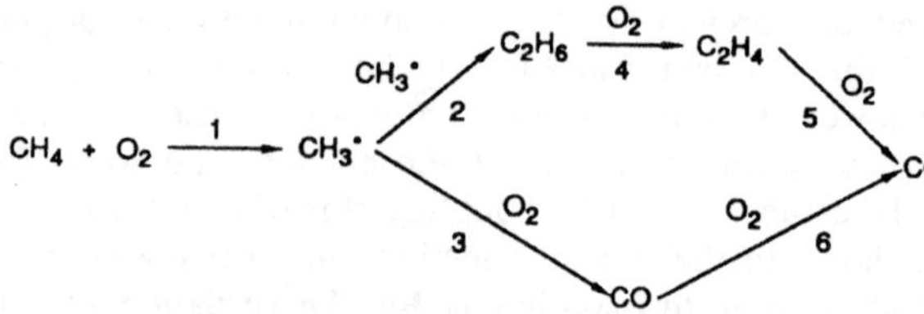

FIGURE 2-19. Assumed reaction pathway for the first simulation of the gas-phase oxidative coupling of methane.

account for the correct stoichiometry, reactions 1 and 3 of Fig. 2-19 were taken to be net reactions as follows.

Reaction 1: $CH_4 + O_2 \rightarrow 4CH_3\cdot + 2H_2O$

Reaction 3: $2CH_3\cdot + \frac{5}{2}O_2 \rightarrow 2CO + 3H_2O$

The other reactions were taken to be elementary reactions. The steady-state concentration of methyl radicals required in the kinetics of reactions 1 and 3 was obtained using the steady-state approximation and is given by

$$[P_{CH_3}\cdot] = \frac{-k_3[P_{O_2}] \pm \sqrt{4k_1k_2P_{CH_4}P_{O_2}}}{2k_2}$$

The theoretical results obtained using the reaction parameters reported by Tsang and Hampson never approached those obtained experimentally. Even when varying the reaction parameters over a broad range, a unique set of parameters valid for all the conditions used experimentally could not be determined. Because it was felt that the lack of agreement between simulation and experiments was due to the simplified chemistry considered, a second simulation involving 27 products, reactants, and radicals was developed. 140 elementary reactions included in the second simulation, along with the rate parameters reported by Tsang and Hampson, are listed in the appendix of Chapter 9 (Table 9-A1). The rate constants have an Arrhenius form and are written as the product of a preexponential factor times an exponential temperature-dependent term.

Using the parameters listed in Table 9-A1, it was found that the methane conversion rate could not be simulated with the values reported for the methyl radicals initiation reaction. Increasing the preexponential factor by a factor 2, it was possible to simulate the experimental results at 800°C. Unfortunately, the simulation failed to predict the experimental results at a lower temperature. Although it is possible to vary two or more of the key parameters from those listed in Table 9-A1 and thus to obtain a fit to the experimental values, the simulation is no longer an a priori exercise but rather one of fitting parameters, thus a true mechanism conclusion cannot be achieved from these results. It is our conclusion that the lack of agreement between the simulation using the parameters from the methane-combustion literature and our experimental results for methane coupling is due to the fact that the kinetic constants reported in Table 9-A1 were obtained during combustion experiments under conditions (high temperatures and oxygen-rich feeds) that are quite different than those used in the oxidative coupling experiments. Consequently,

Table 2-5 Possible Gas-Phase Radical Reactions

Primary mode of initiation	$CH_4 + [O_2] \rightarrow CH_3\cdot + HO_2\cdot$	(2-1)
Chain propagation	$CH_4 + HO_2\cdot \rightarrow CH_3\cdot + H_2O_2$	(2-2)
	$CH_4 + \cdot OH \rightarrow CH_3\cdot + H_2O$	(2-3)
Secondary chain initiation	$H_2O_2 + Z \rightarrow 2\cdot OH + Z$	(2-4)
	where Z is a third body	
Ethane formation by methyl radical termination	$CH_3\cdot + CH_3\cdot \rightarrow C_2H_6$	(2-5)
Ethylene formation:		
Thermally	$C_2H_6 \rightarrow C_2H_4 + H_2$	(2-6)
Radical branching and chain transfer	$C_2H_6 + [O_2] \rightarrow C_2H_5\cdot + HO_2\cdot$	(2-7)
	$C_2H_6 + Y\cdot \rightarrow C_2H_5\cdot + HY$	(2-8)
	$C_2H_5\cdot + [O_2] \rightarrow C_2H_4 + HO_2\cdot$	(2-9)
	where Y· is a $CH_3\cdot$, $HO_2\cdot$, or ·OH radicals	
Oxidation product formation	$CH_3\cdot + [O_2] \rightarrow CH_3O_2\cdot$	(2-10)
	$CH_3O_2\cdot \rightarrow CH_2O + \cdot OH$	(2-11)
	$CH_2O + [O_2] \rightarrow CHO\cdot + HO_2\cdot$	(2-12)
	$CH_2O + X\cdot \rightarrow CHO\cdot + XH$	(2-13)
	$CHO\cdot + [O_2] \rightarrow CO + HO_2\cdot$	(2-14)
	where X· could be any radical	
Secondary combustion	$C_2H_6 + [O_2] \rightarrow CO, CO_2$	(2-15)
	$C_2H_4 + [O_2] \rightarrow CO, CO_2$	(2-16)
	$CO + [O_2] \rightarrow CO_2$	(2-17)
Termination reactions	$HO_2\cdot + HO_2\cdot \rightarrow H_2O_2 + O_2$	(2-18)
	$H_2O_2 \rightarrow H_2O + \frac{1}{2}O_2$	(2-19)

extrapolation of the combustion kinetics to the oxidative coupling kinetics is not guaranteed and caution must be applied during attempts to carry out these simulations.

In order to provide a framework for discussion, it is convenient to select a minimum subset of chemical reactions that account for the formation of products. These are listed in Table 2-5 and are grouped in terms of initiation reactions, product formation, and termination reactions. Reactions (2-1)–(2-4) seem likely candidates for methyl radical initiation and have been proposed by Geisbrecht and Daubert for ethane oxidation (Geisbrecht and Daubert 1975). The rate of methane conversion via reaction (2-1) calculated from the reaction parameters reported in the literature (Tsang and Hampson 1986) is lower than that observed experimentally in this study. Consequently, additional chain propagation reactions, (2-2)–(2-4) are necessary to account for the experimental values. The rate of reaction, 11.3, is so large that only a small concentration of hydroperoxy radicals, about 30 molecules cm^{-3}, is sufficient to account for the additional chain propagation. Similar findings have been reported recently by Lunsford and co-workers when comparing the rates of methane conversion and methyl radical production and recombination (Tong, Rosynek, and Lunsford 1990).

The recombination of methyl radicals in the gas phase to yield ethane [reaction (2-5)] has been demonstrated via various techniques during catalytic studies (Ito et al. 1987; Sofranko, Leonard, and Jones 1987; Kimble and Kolts 1986; Lin et al. 1986; Labinger and Ott 1987), therefore this pathway also must be present for the gas-phase reactions. Ethane can be either dehydrogenated directly via reaction (2-6) or via oxidative dehydrogenation and secondary chain initiation [reactions (2-7)–(2-9)] to produce ethylene as experimentally demonstrated in the previous section.

Several pathways are possible for the formation of carbon oxides. Because hydroperoxy radicals have been observed in catalytic studies (Driscoll et al. 1985) and are intermediates in methane combustion (Imai and Tagawa 1986) we feel that this is an important intermediate leading to carbon oxides and to formaldehyde via reactions (2-10)–(2-14). Formaldehyde was found in trace quantities dissolved in water in our studies. Nelson and Cant (Nelson and Cant 1990) have speculated that these species were also significant in controlling the selectivity in catalytic studies. Another pathway to carbon oxide formation is the combustion of ethane and ethylene mainly to CO, as depicted by reactions (2-15) and (2-16), which was shown to take place in the gas phase and on the catalyst. Deep oxidation to CO_2 occurred mainly on the catalyst surface. Termination reactions (2-17) and (2-18) also occur as well as methane and ethane combustion, which are also assumed to take place.

Catalytic Studies

Although the gas-phase results presented here have been reproduced by other groups, the same cannot be said about the catalytic results because, due to the diversity of operating conditions and different catalyst preparation techniques, it is difficult to compare catalytic experiments. Furthermore, very little kinetic information exists about the nature of the elementary steps involved in the oxidative coupling. The various pathways have been tested separately only recently, in order to determine their relative importance in the reaction network (Feng, Niiranen, and Gutman 1991). Some of the elementary steps occurring in the gas phase also occur on the surface. It is thus not too daring to image a network, such as the one depicted in Fig. 2-19, that comprises the surface as an additional pathway for reaction previously proposed by our group based on our results (Lane, Miro, and Wolf 1989). In this simplified schematic, the surface provides the right oxygen species for the formation of methyl radicals as well as the oxidation of the various species generated by their recombination.

Although in Fig. 2-19 the surface of the catalyst can be seen in a generic manner, the rate of each of the processes involved varies with each catalyst. Although some authors (e.g., Otsuka) subscribe to a generic mechanism, it is the experience of our group that the catalytic effect can be influenced significantly by the selection of the support and the promoter. In Fig. 2-19 methane is interacting with the surface without being adsorbed. Although some authors have inferred from transient measurements that methane is adsorbed on the catalyst, direct measurements recently performed on Li/La_2O_3, La_2O_3, and Li/MgO, have failed to detect methane adsorption at temperatures similar to reaction temperatures (Moral, Ducarme, Mirodatos, and Wolf 1991). Recent results also confirm that the methyl radicals recombine in the gas phase instead of on the cotalytic surface. The oxidation of the products, although known to occur both in the gas phase and on the catalyst and although seen as the rate-determining step of the reaction, is less documented. The most significant issue is that, even though the experiments with single-component feeds prove that ethane, ethylene, and CO are oxidized on the catalyst surface, significant amounts of C_2 hydrocarbons are still produced during the cofeed operation.

Simulation of Experimental Results

After our unsuccessful attempts to simulate the experimental results presented previously using the results from methane combustion, an engineering approach was taken in order to obtain a simulation that would allow us to gain some insights into the reaction behavior and to extrapolate the results to situations not yet explored experimentally. Because empirical kinetic results for the gas phase and catalytic results had already been obtained in our group, these were incorporated in an existing computer routine that was first used to corroborate the experimental observations and then to simulate other situations.

The computational methodology has been described in detail elsewhere (Santamaria, Miro, and Wolf, 1991); thus only the main features of the model will be cited here as well as the main results. Detailed equations and reference parameters are presented in Appendix 2-A.

Kinetics

The kinetics consisted of a simple two-lumped pathway in which methane is sequentially converted to the C_2 hydrocarbons (ethane and ethylene) and to CO_x (CO and CO_2). The gas-phase kinetics were those obtained by Lane and Wolf (Lane and Wolf 1988), summarized in equations (2-A1)–(2-A3). The separate terms in equations (2-A2) and (2-A3) represent the formation of CO and CO_2 and of ethane and ethylene, respectively. For the catalytic kinetics we used the results obtained by Miro and co-workers (Miro, Santamaria, and Wolf 1990b) for the 1.6% $Na/NiTiO_3$ catalyst. Although this was not the most active catalyst it was the only one for which detailed kinetic measurements were available. The kinetics expressions are listed in equations (2-A4)–(2-A6).

Reactor Model

Many features characteristic of industrial reactors were incorporated in the model to test the effect of variables that were not studied in our laboratory reactors where the experimental information was gathered. The following assumptions were made:

1. The catalyst is in the form of spherical pellets.
2. External (interphase), as well as internal (intraphase), diffusion could occur. The transport and material properties of both gas and solid phases are assumed to be constant and independent of conversion and temperature.
3. Interparticle heat conduction and heat transfer are assumed to be less important than heat convection.
4. The reactor is assumed to be plug flow and either isothermal or adiabetic.

With these assumptions, the transient mass balance equations that incorporated the gas phase and catalytic kinetics were written. The dimensionless mass and energy balances for a packed-bed reactor affected by interphase heat and mass transport between the bulk gas phase and catalyst particles, with the proper initial and boundary conditions, are shown in equations (2-A7) and (2-A8). The corresponding equations for the mass and heat balances for transient diffusion and reaction in a single spherical particle, with initial and boundary conditions, are described in equations (2-A9) and (2-A10).

Simultaneous solution of the external field equations and particle equations were achieved by using an orthogonal collocation method coupled with a Runge–Kutta procedure. The solution involved 16 collocation points for the reactor and 3 for the catalyst particle. This enabled the concentration and temperature profiles for each point in the bed to be obtained, from which the methane conversion, C_2 selectivity, and yield were calculated.

To test the model, the parameters involved in the governing equations were calculated from existing experimental values or well-known correlations. It should be emphasized that the kinetic values are those experimental values reported by our group and are not adjustable parameters. Appendix 2-A lists also the values of the input parameters used in the program and operating conditions used experimentally. Figure 2-20 displays theoretical results (solid lines) and experimental points for the simulation of the oxidative coupling reaction on sodium-promoted nickel titanate. It is seen that a fairly good agreement between theoretical and experimental results was obtained. This is not surprising because the model parameters were obtained experimentally. The conversion is slightly underpredicted and the selectivity is slightly overpredicted due to the fact that the power-law kinetics used is an approximation of the true kinetics, which has been found to be described by an Eley–Rideal expression. It should also be noted that the selectivity–conversion behavior reflects mainly the predominance of catalytic reactions instead of the gas-phase reactions because, by experimental design, the latter reactions were minimized.

Although detailed results of the various simulations carried out using this model have been presented elsewhere, only a summary of the main results obtained is listed in Table 2-6. It should be emphasized that, although the simulations covered situations not tested experimentally, the results are nonetheless specific to the reaction conditions and the catalysts used and *should not be considered as generic for all methane coupling results*. The trends shown in Table 2-6 are those expected from general reactor engineering considerations and present no surprises. The merit of the simulation is to provide a quantitative assessment of the effect of the various variables without performing the corresponding experiments.

A particularly interesting result is the effect of simulating the oxygen concentration profile in the reactor. In a particular case, a hypothetical reactor was simulated, in which the oxygen concentration was assumed to be uniformly distributed along the reactor. This could be achieved in practice with a reactor with point feeds distributed along its axis. The

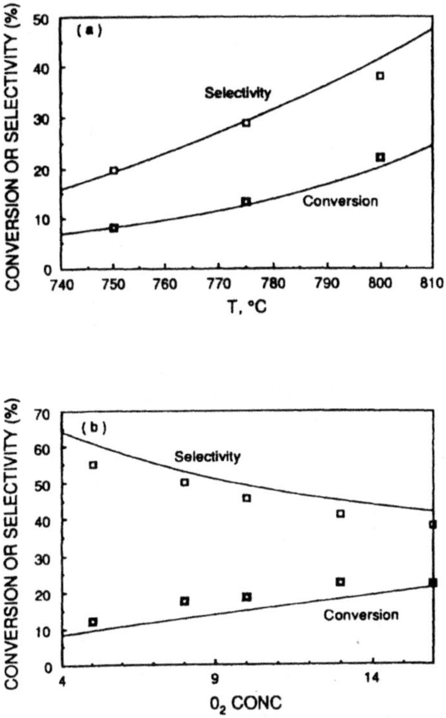

FIGURE 2-20. Comparison between theoretical predictions (solid lines) and experimental results (boxes): (a) ethane conversion and higher hydrocarbons versus temperature; (b) methane conversion and C_{2+} selectivity versus oxygen concentration.

simulation permitted us to ascertain this effect by superimposing in the program such conditions and studying their effects. The value imposed was equal to the average oxygen concentration existing (3%) in the reactor with a variable concentration equal to 8% at the reactor inlet. This enabled comparison of two reactors with the same oxygen content. The methane concentration and selectivity profiles are shown in Figure 2-21a and 2-21b, respectively, for both variable and constant oxygen concentrations. It can be seen that the two curves follow a similar trend, being first higher for the reactor with variable feed but then intersecting at the reactor exit and, in fact, being higher for the case with constant distributed feed. The C_2 selectivity, however, is higher for the reactor with constant O_2 concentration than for that with variable O_2 concentration. As a result

Table 2-6 Summary of the Oxidative Coupling of Methane Simulations

Parameter	Standard Value	Percentage of Parameter Change	CH_4 Conversion[a]	Percentage Change in CH_4 Conversion	C_2 Selectivity[b]	Percentage Change in C_2 Selectivity[b]
Reactor length[c]	0.2 m	−50	20	−40	43	−2.5
Oxygen concentration	16 vol%	−50	20	−40	43	−2.5
Particle radius	2 mm	+50	20	−10	43	+3
Effective diffusivity	5×10^{-6} $m^2\ s^{-1}$	+5000	20	+10	43	−5.0
Reactor length (exit)[d]	0.2 m	+500	20	+130	43	+24
Porosity	0.35	+100	20	−45	43	−5.0

[a,b]Conversion and selectivity obtained at reference parameters values.
[c]Conversion calculated along the reactor length.
[d]Conversion calculated at the reactor exit.

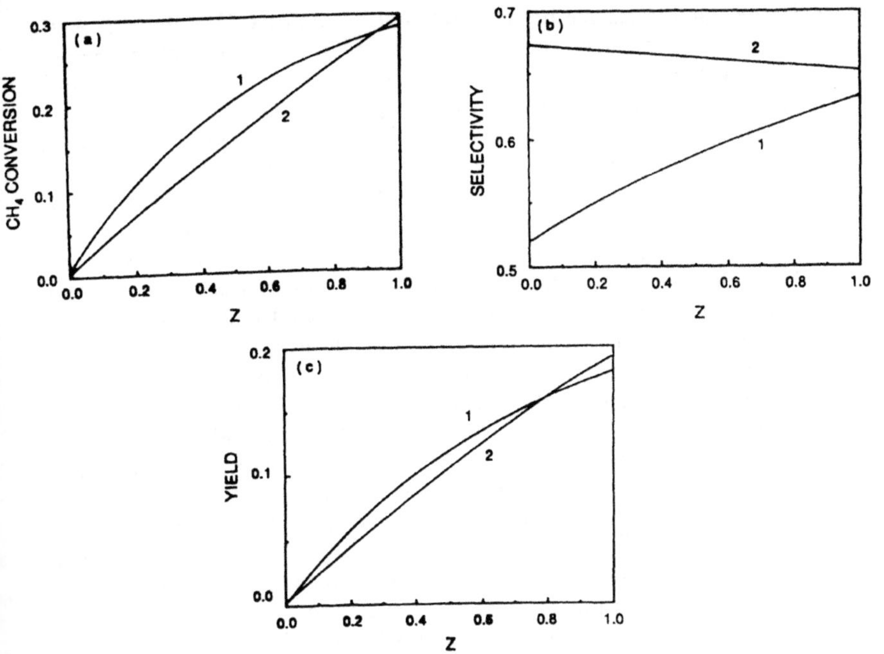

FIGURE 2-21. Distributed oxygen feed. Comparison of simulation results obtained with (1) 8% of inlet oxygen concentration and (2) 3% of constant oxygen concentration along the bed: (a) methane conversion profiles; (b) selectivity profiles; (c) yield profiles.

the yield obtained in the reactor with constant O_2 feed is higher than that obtained in a reactor with a single feed (Fig. 2-21c). Although the validity of these results is constrained to the experimental situation and catalyst used, the increase in yield makes it worthwhile to explore its experimental realization. The purpose of this section is, in fact, to demonstrate that simulation of experimental results based on simplified rate expressions can provide useful insights into situations not previously explored experimentally, and can help to design further experiments to test the validity of the theoretical predictions.

SUMMARY

Results and conclusions presented in this chapter indicate that a significant formation of C_2 hydrocarbons can take place during oxidative coupling reaction conducted without a catalyst. Under a specific set of operating conditions, methane conversion as high as 32% with 29% C_2 selectivities can be achieved. Trends observed during gas-phase experiments show that when methane conversion increases hydrocarbon selectivity decreases. In order to minimize homogeneous conversion of methane, catalytic experiments should be conducted at low reactant partial pressures, low oxygen partial pressures, and short residence times. It was also found that the presence of an inert material may significantly alter noncatalytic conversion of methane. It is reported that the gas-phase and catalytic oxidation of ethane and ethylene, the main products of the methane dimerization, starts at lower temperatures than those usually implemented during oxidative coupling, which results in the decrease of the catalytic selectivity. It is concluded that the new reactor designs may significantly improve yields of desired products; however, the search for a more selective catalyst should continue. A section of this chapter devoted to catalytic studies shows that one of the ways to avoid undesired reactions is the use of transient and isotopic techniques. Although these techniques will be discussed in detail in the following chapters of this book, it is important to realize that by avoiding undesired gas-phase reactions we are able to work under conditions of pure catalytic process. Mechanistic analysis of the gas-phase oxidative coupling suggested that the apparent activation energy for the conversion of methane and the production of ethane and ethylene is around 53 kcal mol^{-1}. It is also concluded that this reaction could possibly be the rate-limiting step and the primary mode of radical initiation. An attempt to reproduce gas-phase results by applying simulation techniques and available literature data was not fully successful, mainly because the majority of kinetic measurements available concern combustion processes conducted under oxygen-rich conditions. On the other hand, simulation of our catalytic results taking into account gas-phase kinetics allowed for an elegant analysis of different parameters and their influence on catalytic methane dimerization. It was also found that the oxygen distributed feed along the reactor axis may improve the yield of C_2 hydrocarbons.

APPENDIX 2-A

Gas-Phase Kinetics

$$-\frac{dCH_4}{dt} = (7.607 \times 10^5)e^{-27.600/T}C_{CH_4}^{1.04}C_{O_2}^{2.05}T^{3.09} \tag{2-A1}$$

$$\frac{dC_{CO_x}}{dt} = (1.29 \times 10^9)e^{-36100/T}C_{CH_4}^{0.53}C_{O_2}^{3.70}T^{4.23} + 0.622e^{-14900/T}C_{CH_4}^{-0.95}C_{O_2}^{1.33}T^{0.38} \tag{2-A2}$$

$$\frac{dC_{C_2}}{dt} = (5.367 \times 10^4)e^{-26100/T}C_{CH_4}^{1.04}C_{O_2}^{1.78}T^{2.82} + (1.464 \times 10^4)e^{-26200/T}C_{CH_4}^{1.16}C_{O_2}^{1.62}T^{2.78} \quad (2\text{-A3})$$

Catalytic Kinetics

$$-\frac{dC_{CH_4}}{dt} = (2.176 \times 10^{18})e^{-42700/T}C_{CH_4}^{1.05}C_{O_2}^{0.4} + (8.04 \times 10^9)e^{-20050/T}C_{CH_4}^{0.55}C_{O_2}^{1.04} \quad (2\text{-A4})$$

$$\frac{dC_{CO_x}}{dt} = (8.04 \times 10^9)e^{-20050/T}C_{CH_4}^{0.55}C_{O_2}^{1.04} \quad (2\text{-A5})$$

$$\frac{dC_{C_2}}{dt} = (1.088 \times 10^{18})e^{-42700/T}C_{CH_4}^{1.05}C_{O_2}^{0.4} \quad (2\text{-A6})$$

Mass and Energy Balances

$$\frac{\partial \psi_{io}}{\partial \tau} = -\frac{\partial \psi_{io}}{\partial X} - \alpha_1(\psi_{io} - \psi_{is}) - \alpha_2 r_{io}^*(T, C) \quad (2\text{-A7})$$

$$\frac{\partial \theta_o}{\partial \tau} = -\frac{\partial \theta}{\partial X} - \alpha_3(\theta_o - \theta_s) + \alpha_4 \Sigma(-\Delta H)_i r_{io}^*(T, C) \quad (2\text{-A8})$$

Boundary Conditions

$$X = 0 \qquad \psi_{io} = \theta_{io} = 1$$

and

$$\psi_{io} = \frac{C_{io}}{C_{io}|_{z=0}} \text{ (reactants)}; \qquad \psi_{io} = \frac{C_{io}}{C_{CH_4}|_{z=0}} \text{ (products)}; \qquad \theta_o = \frac{T_o}{T_o|_{z=0}}$$

$$X = \frac{Z}{L} \qquad \tau = t\frac{u}{L}$$

$$\theta = \frac{T}{T_o|_{z=0}} \qquad \gamma = \frac{E}{RT_o|_{z=0}}$$

$$\alpha_1 = \frac{k_g a_v L}{u\varepsilon_2} \qquad \alpha_2 = \frac{L}{uC_{io}|_{z=0}}$$

$$\alpha_3 = \frac{h a_v L}{u\varepsilon_2 \rho_g Cp_g} \qquad \alpha_4 = \frac{L}{u\rho_g Cp_g T_o|_{z=0}}$$

Heat and Mass Balances

$$\frac{\partial \bar{\psi}_i}{\partial \tau} = \beta_1\left(\nabla^2 \bar{\psi}_i - \beta_2 r_i(\bar{\psi}, \bar{\theta})\right) \tag{2-A9}$$

$$\frac{\partial \bar{\theta}}{\partial \tau} = \beta_3\left(\beta_4 \nabla^2 \bar{\theta} + \Sigma(\Delta H)_i r_i(\psi, \theta)\right) \tag{2-A10}$$

Boundary Conditions

$$\xi = 0 \qquad \frac{\partial \bar{\psi}_i}{\partial \xi} = \frac{\partial \bar{\theta}}{\partial \xi} = 0$$

$$\xi = 1 \qquad -\frac{\partial \bar{\psi}_i}{\partial \xi} = \mathrm{Sh}^*\left(\bar{\psi}_{is} - \psi_{io}\right)$$

$$-\frac{\partial \theta}{\delta \xi} = \mathrm{Nu}^*(\theta_s - \theta_o)$$

$$\tau = 0 \qquad \bar{\psi}_i = 0 \qquad \bar{\theta} = 1 \quad \text{for all } \xi$$

The following additional dimensionless variables and groups have been used:

$$\xi = \frac{r}{R} \qquad \bar{\psi}_i = \frac{C_i}{C_i|_{z=0}} \qquad \bar{\theta} = \frac{T}{T|_{z=0}}$$

$$\mathrm{Sh}^* = \frac{K_g R}{D_e} \qquad \mathrm{Nu}^* = \frac{LR}{K_e}$$

$$\beta_1 = \frac{D_e L}{u R^2 \varepsilon_1} \qquad \beta_2 = \frac{R^2}{D_e C_i|_{z=0}}$$

$$\beta_3 = \frac{L}{\rho_s C_{p_s} u T|_{z=0}} \qquad \beta_4 = \frac{K_e T|_{z=0}}{R^2}$$

Notation

a_v	External pellet area per unit volume of bed
C_i	Concentration of gaseous reactant i
C_p	Specific heat
D_e	Effective diffusivity
h	Heat transfer coefficient
(ΔH)	Heat of reaction
k_g	Mass transfer coefficient
K_e	Effective thermal conductivity of the catalyst
L	Reactor length
Nu*	Modified Nusselt number, Rh/K_e
r	Radial coordinate in pellet
r^*	Reaction rate
R	External radius of pellet
Sh*	Modified Sherwood number, Rk_g/D_e
t	Time
T	Absolute temperature
u	Gas velocity
z	Reactor length

Greek letters:

ε_1	Pellet void fraction
ε_2	Bed void fraction
ρ	Density

Subscripts:

g	Gas
i	Initial condition
o	Bulk gas phase
s	Surface

Reference Parameter Values

The following is a list of experimental operating conditions and values used as input parameters for the simulation program discussed in the text.

$R = 2 \times 10^{-3}$ m

$L = 0.2$ m
Oxygen concentration = 16 vol%
Methane concentration = 32 vol%
Effective diffusivity = 5×10^{-6} m^2 s^{-1}
Modified Sherwood Number, Sh* = 500
Modified Nusselt number, Nu* = 1.0
Reactor temperature = 1073 K
Reactor void fraction = 0.35
Total gas flow rate (at STP) = 10^{-4} m^3 s^{-1}
Effective thermal conductivity = 4×10^{-4} W mK^{-1}

ACKNOWLEDGMENT

The financial support provided by Amoco Research Center, Naperville, Illinois is gratefully acknowledged.

References

Aika, K.-I., T. Moriyama, N. Takasaki, and E. Iwamatsu. 1986. Oxidative dimerization over $BaCO_3$, $SrCO_3$ and these catalysts promoted with alkali. *J. Chem. Soc., Chem. Commun.* 1210.

Asami, K., S.S. Hashimoto, T. Shikada, K. Fujimoto, and H.-O. Tominaga. 1986. Selective oxidative coupling of methane over supported lead oxide catalyst. *Chem. Lett.* 1233.

Asami, K., F. Omata, K. Fujimoto, and H.-O. Tominaga. 1987. Oxidative coupling of methane in the homogeneous gas phase under pressure. *J. Chem. Soc., Chem. Commun.* 1287.

Bytyn, W., and M. Bearns. 1986. Supported PbO catalysts for the oxidative coupling of methane—The effect of surface acidity of the support on C_{2+} selectivity. *J. Appl. Catal.* 28:199.

Campbell, K.D., E. Morales, and J.H. Lunsford. 1987. Gas phase coupling of methyl radicals during the catalytic partial oxidation of methane. *J. Am. Chem. Soc.* 109:7900–901.

Driscoll, D.J., W. Martir, J.-X. Wang, and J.H. Lunsford. 1985. Formation of gas-phase methyl radicals over MgO. *J. Am. Chem. Soc.* 107:58–63.

Emesh Ali, I.T., and Y. Amenomiya. 1986. Oxidative coupling of methane over the oxides of group IIIA, IVA, and VA metals. *J. Phys. Chem.* 90:4785.

Feng, Y., J. Niiranen, and D. Gutman. 1991. Kinetic studies of the catalytic oxidation of methane, I: Methyl radical production on 1% Sr/Al_2O_3. *J. Phys. Chem.* 95:6558–63.

Geisbrecht, R.A., and T.E. Daubert. 1975. Chemical and physical processes of hydrocarbon combustion: Chemical processes. *Ind. Eng. Chem. Process Des. Dev.* 14:159.

Hatano, M., P.G. Hinson, K.S. Vines, and J.H. Lunsford. 1990. Comments on "Blank reactor corrections in studies of the oxidative dehydrogenation of methane" by D.J.C. Yates and N.E. Zlotin. *J. Catal.* 124:557.

Hinsen, W., W. Bytyn, and M. Baerns. 1984. Oxidative dehydrogenation and coupling of methane. In *Proceedings of the 8th International Congress on Catalysis, Berlin, 1984,* Vol. III, pp. 581–592.

Hutchings, G.J., M.S. Scurrell, and J.R. Woodhouse. 1988. The role of gas phase reaction in the selective oxidation of methane. *J. Chem. Soc., Chem. Commun.* 253.

Imai, H., and T. Tagawa. 1986. Oxidative coupling of methane over $LaAlO_3$. *J. Chem. Soc., Chem. Commun.* 52.

Ito, T., and J.H. Lunsford. 1985. Synthesis of ethylene and ethane by partial oxidation of methane over lithium-doped magnesium oxide. *Nature (London)* 314:721–722.

Ito, T., J.-X. Wang, C.-H. Lin, and J.H. Lunsford. 1987. Oxidative dimerization of methane over a lithium-promoted magnesium oxide catalyst. *J. Am. Chem. Soc.* 107:5062.

Iwamatsu, E., T. Moriyama, N. Takasaki, and K.-I. Aika. 1987. Importance of the specific surface area of the catalyst in oxidative dimerization of methane over promoted magnesium oxide. *J. Chem. Soc., Chem. Commun.* 19.

Kalenik, Z., E.E. Miro, J. Santamaria, and E.E. Wolf. 1989. Transient and steady state studies of the gas phase and catalytic oxidative coupling of methane. Preprints of 3B Symposium on Methane Activation, 1989 Pacifichem Meeting, Honolulu, HI, p. 109.

Kalenik, Z., and E.E. Wolf. 1990a. Comments on the effect of gas phase reactions on oxidative coupling of methane. *J. Catal.* 124:566.

Kalenik, Z., and E.E. Wolf. 1991. Methane Oxidative Coupling over Lithium Promoted Lanthanum-Titanate Oxide. *Natural Gas Conversion*. Amsterdam: Elsevier Science Publishers, pp. 97–105.

Kimble, J.B., and J.H. Kolts. 1986. Oxidative coupling of methane to higher hydrocarbons. *Energy Prog.* 6:226.

Labinger, J.A., and K.C. Ott. 1987. Mechanistic studies on the oxidative coupling of methane. *J. Phys. Chem.* 91:2682.

Lane, G.S., Z. Kalenik, and E.E. Wolf. 1989. Methane oxidative coupling over titanate catalysts. *J. Appl. Catal.* 53:183.

Lane, G.S., E.E. Miro, and E.E. Wolf. 1989. Methane oxidative coupling. II. A study of lithium-titania–catalyzed reactions of methane. *J. Catal.* 119:161.

Lane, G.S., and E.E. Wolf. 1988. Methane utilization by oxidative coupling. I. A study of reactions in the gas phase during the cofeeding of methane and oxygen. *J. Catal.* 113:144.

Lin, C.-H., K.D. Campbell, J.-X. Wang, and J.H. Lunsford. 1986. Oxidative dimerization of methane over lanthanum oxide. *J. Phys. Chem.* 90:534–7.

Lin, C.-H., T. Ito, J.-X. Wang, and J.H. Lunsford. 1987. Oxidative dimerization of methane over magnesium and calcium oxide catalysts promoted with group IA ions: The role of $[M^+O^-]$ centers. *J. Am. Chem. Soc.* 109:4808.

Martin, G.A., A. Bates, V. Ducarme, and C. Mirodatos. 1989. Oxidative conversion of methane and C_2 hydrocarbons on oxides: Homogeneous versus heterogeneous process. *Appl. Catal.* 47:287.

Matsuura, I., Y. Utsumi, M. Nakai, and T. Doi. *Chem. Lett.* 1986. Oxidative coupling of methane over lithium-promoted zinc oxide catalyst. 1981.

Miro, E.E. Z. Kalenik, J. Santamaria, and E.E. Wolf. 1990a. Transient studies on methane oxidative coupling over alkali-metal promoted titanate catalysts. *Catal. Today* 6:511.

Miro, E.E., J. Santamaria, and E.E. Wolf. 1990a. Oxidative coupling of methane on alkali metal–promoted nickel titanate. I. Catalyst characterization and transient studies. *J. Catal.* 124:451.

Miro, E.E., J. Santamaria, and E.E. Wolf. 1990b. Oxidative coupling of methane on alkali metal–promoted nickel titanate. II. Kinetic studies. *J. Catal.* 124:465.

Moriyama, T., N. Takasaki, E. Iwamatsu, and K.-I. Aika. 1986. Oxidative dimerization of methane over promoted magnesium oxide catalysts. Important factors. *Chem. Lett.* 1165.

Nelson, P.F., and N.W. Cant. 1990. Isotopic labelling studies of the mechanism of the catalytic oxidative coupling of methane. Preprints of 9th International Symposium on Natural Gas Conversion, 1990 Oslo Meeting, Oslo, Norway, p. 35.

Otsuka, K., and K. Jinno. *Inorg. Chim. Acta* 1986. Kinetic studies on partial oxidation of methane over samarium oxides. 121:237.

Otsuka, K., K. Jinno, and A. Morikawa, 1986. Active and selective catalysts for the synthesis of C_2H_4 and C_2H_6 via oxidative coupling of methane. *J. Catal.* 100:353.

Otsuka, K., and T. Komatsu. 1986. Conversion of methane to aromatic hydrocarbons by combination of catalysts. *Chem. Lett.*, 1955.

Otsuka, K., and T. Komatsu. 1987a. High catalytic activity of Sm_2O_3 for oxidative coupling of methane into ethane and ethylene. *Chem. Lett.* 483–484.

Otsuka, K., and T. Komatsu. 1987b. Active catalysts in oxidative coupling of methane. *J. Chem. Soc., Chem. Commun.* 388.

Otsuka, K., Q. Liu, M. Hatano, and A. Morikawa. 1986. Synthesis of ethylene by partial oxidation of methane over the oxides of transition elements with LiCl. *Chem. Lett.* 903.

Otsuka, K., Q. Liu, and A. Morikawa. 1986. Active and selective catalysts in oxidative coupling of methane. Nickel oxides with salts of alkali metals. *Inorg. Chim. Acta* 118:L23.

Roos, J.A., S.J. Korf, R.H.J. Veehof, J.G. Van Ommen, and J.R.H. Ross. 1989. Reaction path of the oxidative coupling of methane over a lithium-doped magnesium oxide catalyst: Factors affecting the rate of total oxidation of ethane and ethylene. *Appl. Catal.* 52:147.

Santamaria, J., Miro, E.E., and E.E. Wolf. 1991. Reactor simulation studies of methane oxidative coupling on a $Na/NItiO_3$ catalyst. *Ind. Eng. Chem. Res.* 30:1157–65.

Sofranko, J.A., J.J. Leonard, and C.A. Jones. 1987. The oxidative conversion of methane to higher hydrocarbons. *J. Catal.* 103:302.

Tong, Y., M.P. Rosynek, and J.H. Lunsford. 1990. The role of sodium carbonate and oxides supported on lanthanide oxides in the oxidative dimerization of methane. *J. Catal.* 126:291.

Tsang, W., and R.F. Hampson. 1986. Chemical kinetic data base for combustion chemistry. Part I. Methane and related compounds. *J. Phys. Chem. Ref. Data* 15:1087.

van der Wiele, K., J.W.M.H. Geerts, and J.M.N. van Kasteren. 1990. The role of heterogeneous reactions during the oxidative coupling of methane over Li/MgO catalysts. *Catal. Today* 6:497.

Yamagata, N., S.S. Tanaka, S. Sasaki, and S. Okazaki. 1987. Oxidative coupling of methane over BaO mixed with CaO and MgO. *Chem. Lett.* 81.

Yates, D.J., and N.E. Zlotin. 1990. Reply to Hatano, Hinson, Vines, and Lunsford's comments on "Blank reactor correactions in Studies of the Oxidative Dehydrogenation of Methane." *J. Catal.* 124:562.

Yates, D.J., and N.E. Zlotin. 1988. Blank reactor coreactions in studies of the oxidative dehydrogenation of methane. *J. Catal.* 111:317.

3

Partial Oxidation of Methane over Metal Oxides: Reaction Mechanism and Active Oxygen Species

Kiyoshi Otsuka and Masaharu Hatano

Department of Chemical Engineering, Tokyo Institute of Technology, Ookayama, Meguroku, Tokyo, Japan

TABLE OF CONTENTS

SUMMARY

A general picture of the mechanism of oxidative methane coupling over metal oxides has been proposed on the basis of kinetic and CH_4–CD_4 exchange studies on Sm_2O_3. The rate-determining step is the hydrogen abstraction from the methane activated on the surface. A diatomic oxygen species, probably O_2^{2-}, is responsible for the abstraction of hydrogen. This active oxygen must be in equilibrium with the gaseous oxygen. Lattice oxygen atoms do not contribute to the oxidative coupling reaction. The main products in the ethanes formed in the reaction of the mixture of CH_4, CD_4, and O_2 were CH_3CH_3, CH_3CD_3, and CD_3CD_3, indicating that the $CH_3\cdot$ radical in the gas phase or the CH_3 group on the surface is the reaction intermediate.

Calcination of the NiO added with $LiNO_3$, Li(OH), or Li_2CO_3 at > 973 K generates a Li and NiO solid solution $Li_xNi_{1-x}O$. The oxidative coupling of methane over this oxide is quite unusual. Lattice oxygen atoms of the oxide are responsible for the activation of methane. The rate of formation of C_2 products depends on the square of methane pressure. The rate-determining reaction is the coupling step of the methyl groups adsorbed on Ni^{3+}—O^{2-} pair sites. The addition of LiCl to NiO, however, did not produce a compound oxide between Li and NiO. The reaction over this LiCl-added NiO can be explained in terms of the same reaction mechanism proposed for the reaction over Sm_2O_3.

Another example of the important role of lattice oxygen was observed for the partial oxidation of methane into formaldehyde over the mixed oxide of Fe, Nb, and B. The oxygen isotope analysis of the products from CH_4 and $^{18}O_2$ has indicated that the bulk oxygen of this oxide is preferentially incorporated into the products (HCHO, CO, CO_2, and H_2O).

INTRODUCTION

The surface-catalyzed activation of C—H bonds of methane for its selective oxidative coupling into ethane and ethylene or for its partial oxidation into methanol and formaldehyde is a subject of current interest (Pitchai and Klier 1986; Otsuka 1987a; Lee and Oyama 1988; Baerns, Ross, and van der Wiele 1988; Hutchings, Scurrell, and Woodhouse 1989). Many metal oxides have been claimed to be effective catalysts for these direct conversions of methane into the functionalized chemicals. Early work by Keller and Bhasin (Keller and Bhasin 1982) focused on screening different metal oxides for the oxidative coupling of methane by lattice oxygen, using a cyclic operation in which supplies of air and methane were periodically switched in an attempt to improve selectivity for the C_2 products (C_2H_6 + C_2H_4). Of the oxides tested, oxides of Sn, Pb, Sb, Bi,

Tl, Cd, and Mn were found to be active and selective for C_2 formation. Jones and co-workers (Jones, Leonard, and Sofranko 1987a; 1987b) have described a methane-and-air cyclically fed process based on the redox of the lattice oxygen atoms of supported manganese oxide catalyst. The lattice oxygen of the lead oxide supported on magnesium oxide has been demonstrated as an active and selective oxidant for the coupling of methane (Asami et al. 1987). Thus, it seems apparent that the lattice oxygen of reducible metal oxides could activate the C—H bond of methane and convert it into partial oxidation products. However, many studies demonstrated that the methane–oxygen cyclic operation was not necessary to obtain high selectivities, and in the presence of suitable catalysts methane and oxygen cofeed also gave high C_2 yields (Pitchai and Klier 1986; Otsuka 1987a; Lee and Oyama 1988; Baerns, Ross, and van der Wiele 1988; Hutchings, Scurrell, and Woodhouse 1989). Under the cofeed conditions, that is, in the presence of gas-phase oxygen, an adsorbed oxygen species could also be important for the activation and conversion of methane into C_2 products.

The active oxygen species responsible for the catalytic conversion of methane into oxygenates (methanol and formaldehyde) is to be clarified because the reaction could proceed either in the gas phase through a chain propagation mechanism (Gesser, Hunter, and Prakash 1985) or exclusively on the surface.

Against this background, we have carried out kinetic and isotopic studies on the reaction over various metal oxides to investigate the overall reaction mechanism, especially to get information about the oxygen species participating in the reaction. In this chapter, we review our current research activities on these topics.

EXPERIMENTAL

Kinetic Studies

The kinetic studies on the partial oxidation of methane under methane–oxygen cofed conditions were carried out using a fixed-bed microreactor (3–10 mm i.d.) with a conventional gas-flow system at atmospheric pressure, using helium as a carrier gas. The kinetic equation and the apparent activation energy for the conversion of methane over different catalysts were determined under the following experimental conditions: $T = 873$–1023 K, $P(CH_4) = 6$–45 kPa, and $P(O_2) = 0.6$–14 kPa. The gas–solid reaction between methane and an oxide catalyst, the CH_4–CD_4 exchange and oxygen tracer experiments were undertaken by using a closed gas-circulation apparatus of ca. 360 cm^3 dead volume. The conversion of methane, the rates of formation of the products, the selectivities and the yields of the products were calculated on the basis of the carbon number of the methane reacted.

Catalyst Preparation

The Sm_2O_3 and NiO used were all reagent grade (purity > 99.8%) in the form of powder obtained from Asahi Chemical Co., Ltd., and Wako Pure Chemical Co., Ltd.

$LiNiO_2$ was prepared by the impregnation method using powder NiO and aqueous solution of $LiNO_3$ (Ni : Li = 1 : 1). The Li-impregnated NiO was calcined in air at 673 K for 2 h and subsequently at 1073 K for 20 h. The mixed oxide of Fe, Nb, and B (Fe : Nb : B = 1 : 2 : 2) was prepared from the Nb_2O_5 powder immersed in an aqueous solution of $Fe(NO_3)_3$ and H_3BO_3. The solution was dried and calcined at 673 and 1073 K. The mixed oxide thus prepared will be denoted as FeNbB–O hereafter. Na_2O_2 (Wako Pure Chemicals, laboratory reagent) was used without further purification.

RESULTS AND DISCUSSION

Reaction Mechanism and Active Oxygen Species Suggested from Kinetic Studies

The kinetic results on the oxidation of methane over various solid catalysts are summarized in Table 3-1. Sm_2O_3 (Otsuka, Jinno, and Morikawa 1986), Li(7 wt%)-doped MgO (Ito et al. 1985), and $BaCeO_3$ (Otsuka, Shimizu, and Komatsu 1987) are the oxides known as the selective catalysts in oxidative coupling of methane. The LiCl-added NiO and Co_3O_4 are the catalysts for the selective conversion of methane into ethylene (Otsuka, Hatano, and Komatsu 1989; Komatsu, Amaya, and Otsuka 1989). CeO_2 catalyzes deep oxidation of methane. The mixed oxide of Fe, Nb, and B (FeNbB–O) catalyzes the partial oxidation into formaldehyde (Otsuka et al. 1988). For all the catalysts in Table 3-1 (except for FeNbB–O), the

Table 3-1 Kinetic Results of Methane Oxidation for Various Catalysts

Catalyst	Main Products	Activation Energy ($kJ\ mol^{-1}$)
Sm_2O_3	C_2H_6, C_2H_4	135
Li/MgO	C_2H_6, C_2H_4	155
$BaCeO_3$	C_2H_6, C_2H_4	142
LiCl/NiO	C_2H_4	208
LiCl/Co_3O_4	C_2H_4	143
CeO_2	CO_2	142
FeNbB–O	HCHO	256

rate of conversion of methane (R_c) depends on the pressures of reactants in a very similar form,

$$R_c = k_0 \frac{k_a P(CH_4)}{k_b + k_c P(CH_4)} \frac{K_1 P(O_2)}{1 + K_1 P(O_2)} \tag{3-1}$$

where k_0, k_a, k_b, k_c, and K_1 are constants at a fixed temperature. The results that the similar rate equations were observed for the reactions over the quite different catalysts in Table 3-1 imply that the activation of methane and the oxygen species responsible for the activation are similar among the catalysts. However, in the case of FeNbB–O, R_c depended on the square root of the pressure of oxygen as follows:

$$R_c = k_0 \frac{k_a P(CH_4)}{k_b + k_c P(CH_4)} \frac{K_1 \sqrt{P(O_2)}}{1 + K_1 \sqrt{P(O_2)}} \tag{3-2}$$

This observation suggests that the oxygen species responsible for the activation of methane over this oxide is different from the one for the other catalysts in Table 3-1.

Based on these kinetic observations, we have proposed the following mechanism for the activation and conversion of methane (Otsuka et al. 1988):

$$CH_4 + S \xrightarrow{k_a} CH_4^* \cdots S \tag{3-3}$$

$$CH_4^* \cdots S \xrightarrow{k_b} CH_4 + S \tag{3-4}$$

$$CH_4^* \cdots S + CH_4 \xrightarrow{k_c} 2CH_4 + S \tag{3-5}$$

$$O_2 \overset{K_1}{\rightleftarrows} O_2^* \text{ or } 2O^*(a) \tag{3-6}$$

$$CH_4^* \cdots S + O_2^*(a)\ [\text{or } O^*(a)] \xrightarrow{k_0} CH_3 \cdot + HO_2\ (\text{or } HO) + S \tag{3-7}$$

$$2CH_3 \xrightarrow{k_d} C_2H_6 \xrightarrow{k_e} C_2H_4 \tag{3-8}$$

$$CH_3 + O^* \xrightarrow{k_f} HCHO + \tfrac{1}{2}H_2 \tag{3-9}$$

$$CH_3 + O_2 \xrightarrow{k_g} CO, CO_2 \tag{3-10}$$

The vibrationally activated CH_4^* (bending mode) generated upon the impact of the CH_4 on the surface reacts with an active diatomic oxygen or with an atomic oxygen, giving adsorbed methyl groups [or methyl radicals in the gas phase (Ito et al. 1985; Driscoll et al. 1985)] in equation (3-7). The activated CH_4^* could be either physisorbed or chemisorbed on the active sites of the catalysts. However, the lifetime of CH_4^* must be very short under such high temperatures (> 923 K). Let us assume that the concentration of CH_4^* is in steady state and that the rate-determining step is the reaction described by equation (3-7). The rate equation of methane conversion (R_c) derived from these assumptions explains well the experimental rate equations of (3-1) and (3-2) (Otsuka et al. 1988).

The kinetic isotope effect between CH_4 and CD_4 has been measured for the oxidation of methane over Sm_2O_3, $BaCeO_3$, LiCl/NiO, and CeO_2 to examine the rate-determining step described previously. The isotope effects for the conversion of methane observed were 1.3 (Sm_2O_3, 923 K), 1.5 ($BaCeO_3$, 973 K), 1.4 (LiCl/NiO, 893 K), and 1.8 (CeO_2, 973 K). These observations support the assumption that C—H bond breaking is the rate-determining step for all the catalysts examined. The results of the reactions of CH_4 and CD_4 (50 : 50) mixture with oxygen showed that the rate of H–D exchange between CH_4 and CD_4 was considerably slow compared with the conversion rate of methanes ($CH_4 + CD_4$). This observation also supports that the rate-determining step is at the C—H bond breaking. The main ethanes products formed during the reaction of the mixture of CH_4 and CD_4 were C_2H_6, $C_2H_3D_3$, and C_2D_6 (Otsuka et al. 1989). This result strongly shows that the reaction proceeds via methyl species as the reaction intermediate. The same conclusion has been reported by Cant and co-workers for the reaction over Li/MgO, $SrCO_3$, and Sm_2O_3 (Cant et al. 1988). However, whether the methyl intermediate is a methyl radical in the gas phase or a methyl group adsorbed on the surface is still an open question.

Concerning the active oxygen species, the rate equations (3-1) and (3-2) suggest that an atomic oxygen on FeNbB–O and a diatomic oxygen on the other catalysts in Table 3-1 are responsible for the activation of methane in step (3-7). On the grounds of these suggestions, we will discuss further about the active oxygen species on Sm_2O_3 and $LiNiO_2$ chosen as typical oxidative coupling catalysts and that on FeNbB–O as a catalyst for HCHO formation.

Active Oxygen on Sm_2O_3

O_2 and N_2O as Oxidant

We have reported previously the kinetic studies on the oxidative coupling of methane over Sm_2O_3 using O_2 or N_2O as an oxidant (Otsuka and Jinno 1986; Otsuka and Nakajima 1987). The rate of C_2 formation for the reaction using N_2O does not change appreciably with the partial pressures

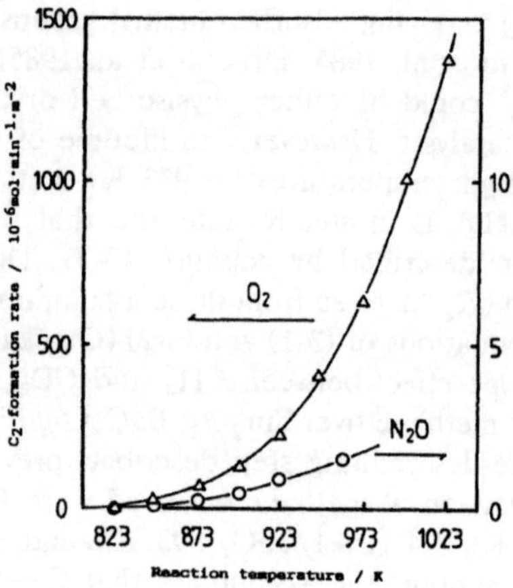

FIGURE 3-1. The rates of methane coupling using O_2 or N_2O as an oxidant: (△) CH_4 pressure = 18 kPa, O_2 pressure = 1.0 kPa, and W/F = 1.75×10^{-2} g h l^{-1}; (○) CH_4 pressure = 20 kPa, N_2O pressure = 4.5 kPa, and W/F = 0.417 g h l^{-1}.

of methane and N_2O. However, the rate of C_2 formation depends on the pressures of reactants according to equation (3-1) when O_2 is used as an oxidant. The rates of C_2 formation using O_2 and N_2O are plotted in Fig. 3-1 as functions of reaction temperatures. The rate of oxidative coupling of methane using O_2 is two to three orders of magnitude greater than that observed for the reaction using N_2O. In these studies N_2O has been utilized as a source of surface O^- oxidizing species (Naccache 1971; Liu, Iwamoto, and Lunsford 1982; Zhen et al. 1985; Hutchings, Scurrell, and Woodhouse 1987). The results in Fig. 3-1 suggest that the contribution of O^- (or atomic oxygen) to the activation of methane is negligible under methane and oxygen cofed conditions.

Reactivity of Lattice Oxygen Atoms of Sm_2O_3

There is much evidence showing the contribution of lattice oxygen atoms of metal oxides to the catalytic oxidations of olefins (Keulks, Krenzke, and Notermann 1978) and methane (Keller and Bhasin 1982; Jones, Leonard, and Sofranko 1987a; Asami et al. 1987). Here, we examine the reactivity and the selectivity of the lattice oxygen atoms of Sm_2O_3 in the partial oxidation of methane in the absence of gaseous oxygen. The gas–solid reaction between methane and the lattice oxygen atoms of Sm_2O_3 has been carried out in a closed gas-circulation apparatus at 873–998 K. The kinetic curves observed at 923 K are shown as examples in Fig. 3-2. The

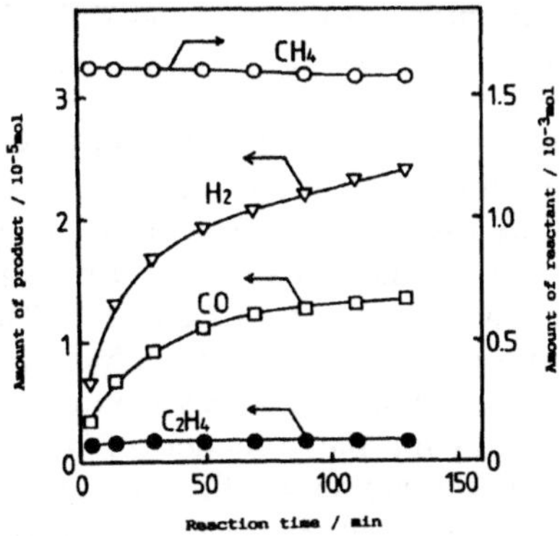

FIGURE 3-2. Oxidation of methane by lattice oxygen atoms of Sm_2O_3: Initial CH_4 pressure = 12.0 kPa, T = 923 K, and Sm_2O_3 = 5.00 g.

products were H_2, CO, C_2H_4, and H_2O (not shown). The oxygen reacted was less than 29% of the surface lattice oxygen atoms of the catalyst. It is to be noted that C_2H_4 is the only coupling product of CH_4. The formation of the C_2H_4 was completed in 30 min at all the temperatures examined. The final amounts of C_2H_4 formed did not change with temperature (0.38 ± 0.04 μmol per gram of catalyst).

In the presence of gaseous oxygen, the product distribution differed remarkably from that in Fig. 3-2. The results in the presence of gaseous oxygen are shown in Fig. 3-3. Here, the reaction was carried out at 973 K with the same gas-circulation apparatus as that used for the experiments

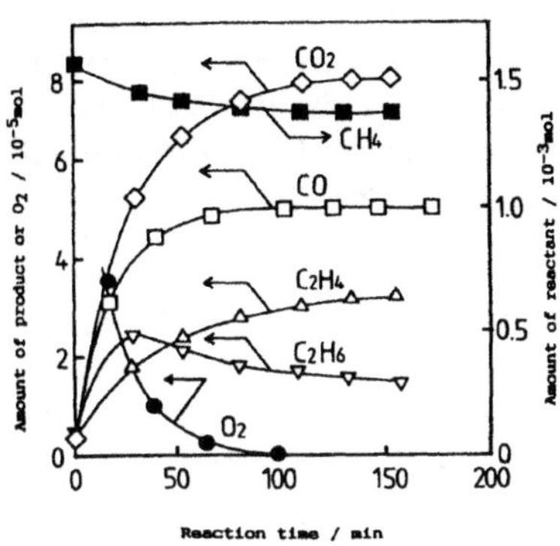

FIGURE 3-3. Oxidation of methane by adsorbed oxygens on Sm_2O_3: Initial CH_4 pressure = 12.0 kPa, initial O_2 pressure = 1.8 kPa, T = 923 K, and Sm_2O_3 = 0.010 g.

of Fig. 3-2. The kinetic curves in Fig. 3-3 show that (1) C_2H_6 is the initial coupling product, (2) only a trace of hydrogen is observed, and (3) CO_2 is the main carbon oxide, in contrast to the results in Fig. 3-2. The rates of conversion of methane were also quite different between the reaction of methane with adsorbed oxygen (in the presence of gaseous oxygen) and the reaction with lattice oxygen atoms. The initial rate per gram of catalyst for the former reaction was more than three orders of magnitude greater than the rate for the latter (see Figs. 3-2 and 3-3). Thus the reactivity of the adsorbed oxygen for the activation of methane is much greater than that of the lattice oxygen atom. These observations strongly suggest that the reaction mechanisms for the oxidation of methane caused by the lattice and adsorbed species are quite different. Furthermore, we conclude that the contribution of the lattice oxygen atoms to methane oxidation over Sm_2O_3 can be neglected under the usual methane and oxygen cofed conditions. The lattice oxygen atoms of Sm_2O_3 convert methane mainly into H_2 and CO with about a 2 : 1 ratio at all the temperatures examined. This fact implies that methane is oxidized through methanol and formaldehyde as adsorbed reaction intermediates, although these intermediates have not been observed in the gas phase during the reaction:

$$CH_4 + \text{lattice oxygen} \rightarrow CH_3OH(ad) \rightarrow HCHO(ad) + H_2$$

$$\rightarrow 2H_2 + CO \qquad (3\text{-}11)$$

The preferential formation of ethylene without a trace of ethane observed in the absence of gaseous oxygen (Fig. 3-2) strongly suggests that the ethylene is produced directly from methane. This is quite unusual because the C_2H_6 is generally believed to be the precursor of C_2H_4 in oxidative coupling of CH_4 by O_2 over metal oxides (Keller and Bhasin 1982; Otsuka, Jinno, and Morikawa 1986; Ito et al. 1985; Otsuka, Shimizu, and Komatsu 1987; Otsuka and Jinno 1986; Lin et al. 1986). The quantity of C_2H_4 formed (0.38 ± 0.04 μmol per gram of Sm_2O_3) corresponds to 0.16% of the number of surface oxygen atoms. Thus, the number of active sites converting CH_4 directly into C_2H_4 is very limited. Although the nature of the active sites is not known at the moment, we speculate that C_2H_4 may be formed through the dimerization of the $CH_2{=}$ groups generated on this active site. However, the contribution of this specific reaction can be neglected in the presence of gaseous oxygen.

Peroxide Anions as Possible Active Species

The preceding discussion suggested that the active oxygen in the oxidative coupling of methane over Sm_2O_3 is not a monatomic oxygen but diatomic

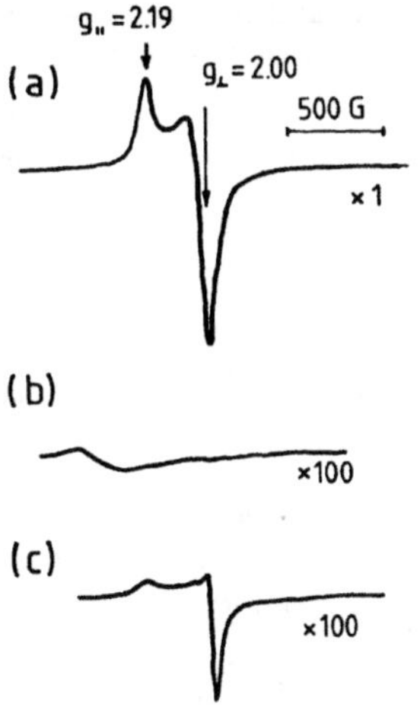

FIGURE 3-4. The ESR spectra of the Na_2O_2 samples measured after different treatments.

ones such as O_2^-, O_2^{2-}, or the O_2 chemisorbed. We have demonstrated that alkali or alkaline-earth metal peroxides can activate and convert methane into C_2 compounds at lower temperature than 673 K (Otsuka et al. 1987). Peroxide anions should be electron spin resonance (ESR) inactive. However, the Na_2O_2 sample used in this work showed the ESR spectra due to the super oxide anions O_2^- contained as impurities. Figure 3-4 shows the ESR spectra of the oxygen species observed at 77 K for the Na_2O_2 samples treated under different conditions as follows: Curve (a) is the O_2^- spectrum for the sample degassed at room temperature. The content of the O_2^- was about 10 mol% in this sample. Spectrum (b) was observed after sample (a) was degassed at 573 K for 2 h. O_2^- disappeared due to the decomposition of the impurity NaO_2. Spectrum (c) was obtained after the contact of oxygen with sample (b) at 573 K. The O_2^- was regenerated, although the amount was two orders of magnitude less than sample (a). The gas–solid reactions between CH_4 and samples (a), (b), and (c) have been examined during increasing temperatures from 298 to 673 K. The reactivities of these samples observed for the methane conversion were the same within the experimental error, suggesting that O_2^- is not reactive to CH_4. O_2^{2-} anions must be responsible for the activation of methane. The activation energy for the conversion of methane by Na_2O_2 was obtained as 127 ± 5 kJ mol^{-1}. This value is close to those observed for the catalysts in Table 3-1 (except for LiCl/NiO and FeNbB–O),

Table 3-2 Partial Oxidation of Methane by Na_2O_2 in the Presence of Gaseous Oxygen[a]

$P(O_2)$ (kPa)	$\%CH_4$ Conversion	$\%C_2H_6$ Selectivity
0	12.6	10.9
5	14.1	4.7
10	13.3	1.8
15	13.8	0.6

[a]Weight of Na_2O_2 = 0.50 g; $P(CH_4)$ = 15 kPa.

suggesting the similar activation mechanism of methane for these catalysts and Na_2O_2. Thus, it is not unreasonable to assume that O_2^{2-} is an active species for the oxidative coupling of methane at higher temperatures (> 873 K) for most of the oxides in Table 3-1.

Table 3-2 shows the results obtained when cofeeding CH_4 and O_2 over Na_2O_2 (Otsuka et al. 1990). The experiments were carried out applying a conventional pulse microcatalytic method. The results in Table 3-2 show that the conversion of methane is not affected at all by gaseous oxygen within the experimental error, but the selectivity to C_2H_6 is decreased appreciably in the presence of gaseous oxygen. These observations suggest that the initiation of methane is caused only by peroxide anions, but further oxidation of the intermediate (methyl groups on the surface or methyl radicals in the gas phase) to CO and CO_2 is enhanced strongly in the presence of gaseous oxygen.

Active Role of Lattice Oxygens in Methane Activation over $LiNiO_2$

The calcination of the NiO powder impregnated with $LiNO_3$ (Ni : Li = 1 : 1) generates a stoichiometric solid solution of Li and Ni, $Li^+Ni^{3+}(O^{2-})_2$. We have demonstrated that this oxide is also an active and selective catalyst in oxidative coupling of methane under CH_4–O_2 cofed conditions (Hatano and Otsuka 1988). However, the kinetic results obtained for this oxide are unusual compared to those for the other coupling catalysts in Table 3-1. The rate equation of methane conversion observed for $LiNiO_2$ could not be expressed by an equation similar to equation (3-1) or (3-2) (Hatano and Otsuka 1989). Figure 3-5 shows the effect of methane pressure on the rates of formations of C_2 and carbon oxides ($CO + CO_2$) over the $LiNiO_2$. The rate of C_2 formation ($C_2H_4 + C_2H_6$) depends roughly on the square of methane pressure. This is very specific because the rate of C_2 formation, in general, depends on the pressure of methane to less than first order. The rate of CO_2 formation does not change

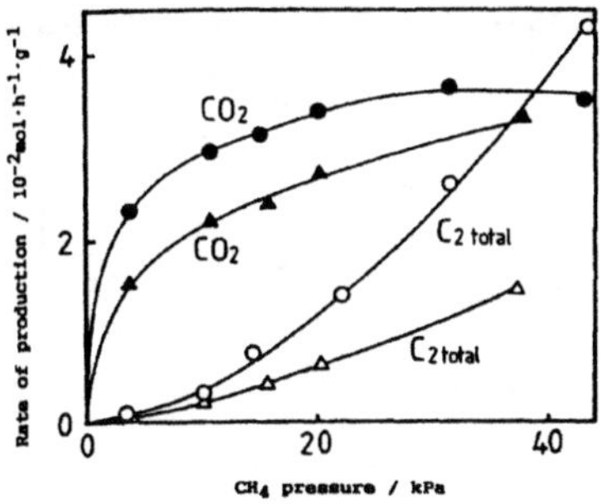

FIGURE 3-5. The rates of C_2 and CO_2 formation as functions of CH_4 pressure: O_2 pressure = 5.01 kPa, T = 973 K (●, ○) or 953 K (▲, △), $W/F = 1.67 \times 10^{-3}$ g h l^{-1}.

appreciably at pressures greater than 10 kPa. The quite different pressure effects on the rates of C_2 and CO_2 formations observed in Fig. 3-5 suggest that the two reactions proceed on different active sites.

The gas–solid reaction between methane and $LiNiO_2$ has been studied to examine the contribution of lattice oxygen atoms of the catalyst for oxidative coupling of methane. The experiments were carried out using a closed gas-circulation apparatus. The results are shown in Fig. 3-6. Here, the amount of C_2 compounds has been evaluated on the basis of the moles of methane reacted. The amount of CO_2 was calculated from the mass balance of the methane reacted and the C_2 compounds produced. As it can be seen in Fig. 3-6, C_2 compounds (mainly C_2H_6) were produced very selectively (100%) at the early stage of the reaction. The formation of CO_2 commenced after 5 min, decreasing the C_2 selectivity. The X-ray diffrac-

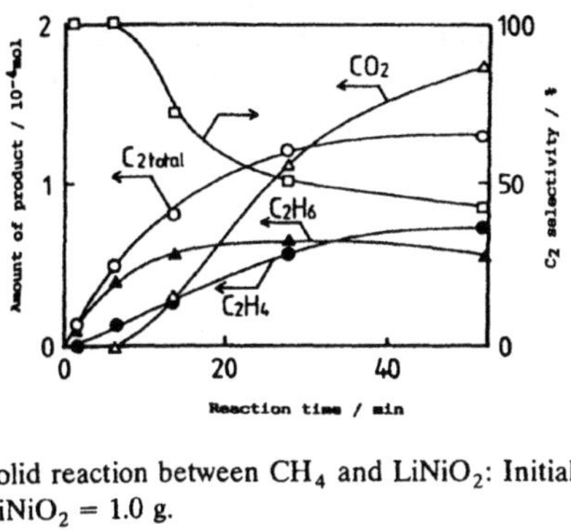

FIGURE 3-6. Gas–solid reaction between CH_4 and $LiNiO_2$: Initial CH_4 pressure = 5.3 kPa, T = 973 K, and $LiNiO_2$ = 1.0 g.

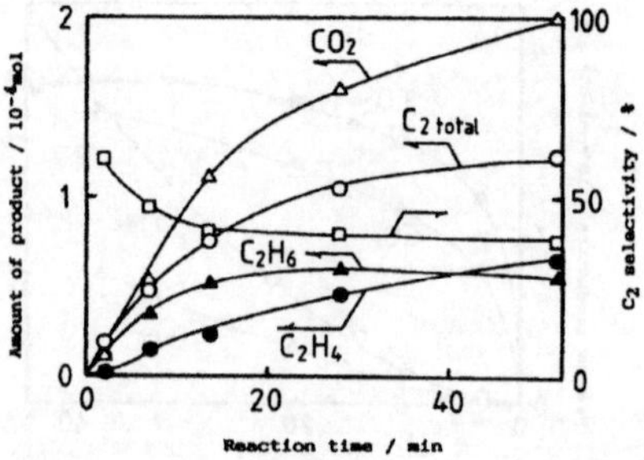

FIGURE 3-7. Gas–solid reaction between CH_4 and $LiNiO_2$ in the presence of gaseous oxygen: Initial CH_4 pressure = 5.3 kPa, initial O_2 pressure = 0.53 kPa, T = 973 K, and $LiNiO_2$ = 1.0 g.

tion (XRD) spectrum of the catalyst after this gas–solid reaction showed that a part of the $LiNiO_2$ is reduced to NiO by methane. However, after oxidation of this reduced sample with oxygen gas at 1023 K, the NiO disappeared and the diffraction peaks due to $LiNiO_2$ increased. These observations show that the oxidative coupling of methane proceeds via the redox of $LiNiO_2$ as follows (Hatano and Otsuka 1988):

$$2LiNiO_2 + 2CH_4 \rightarrow Li_2O + 2NiO + C_2H_6 + H_2O \quad (3\text{-}12)$$

$$Li_2O + 2NiO + \tfrac{1}{2}O_2 \rightarrow 2LiNiO_2 \quad (3\text{-}13)$$

The induction period for the formation of CO_2 in Fig. 3-6 can be explained by the assumption that the deep oxidation occurs on the NiO generated in equation (3-12). In fact, the gas–solid reaction between methane and NiO at 973 K produced only CO_2.

Figure 3-7 shows the kinetic curves in the presence of gaseous oxygen at the early stage of the reaction. The gaseous oxygen was consumed in 15 min. The experimental conditions for Fig. 3-7 were the same as those for Fig. 3-6, except for the addition of oxygen (initial pressure 0.53 kPa). In contrast to the results in Fig. 3-6, fast formation of CO_2 was observed at the initial stage of the reaction. However, very similar kinetic curves of C_2 formation were observed in both cases (Figs. 3-6 and 3-7), indicating that the formation of C_2 compounds is not affected by gaseous oxygen. This observation supports the assumption that the formation of C_2 compounds is caused by the lattice oxygen atoms of $LiNiO_2$ according to equation (3-12). Figure 3-8 shows the initial rate of C_2 formation measured at different methane pressures. The results in Fig. 3-8 indicate that the rate

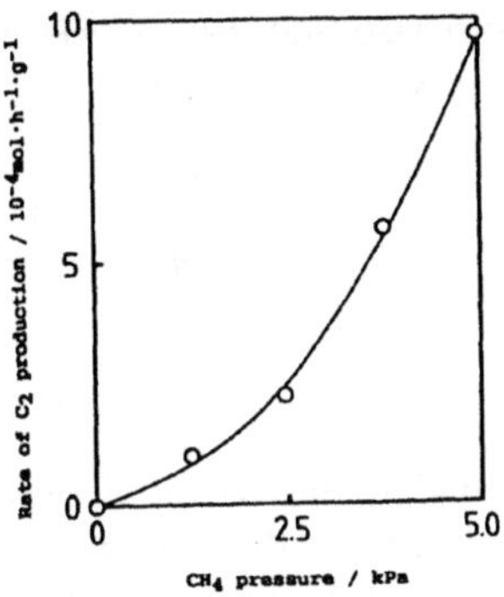

FIGURE 3-8. Initial rate of C_2 formation as a function of methane pressure (gas–solid reaction).

depends on the square of methane pressure, similar to the results shown in Fig. 3-5. Moreover, note that the rate of C_2 formation under methane–oxygen cofed conditions (Fig. 3-5) and the rate in the absence of gaseous oxygen (Fig. 3-8) are approximately the same at the same methane pressure. These observations strongly suggest that the oxidative coupling of methane over $LiNiO_2$ is initiated by the surface lattice oxygen atoms. The rapid formation of CO_2 in the presence of gaseous oxygen (Fig. 3-7) suggests that the adsorbed oxygen would oxidize methane deeply into CO_2. However, the presence of gaseous oxygen, of course, is indispensable to regenerate $LiNiO_2$ [equation (3-13)] under steady-state catalytic reactions.

The hydrogen exchange between CH_4 and CD_4 (1:1 mixture) on $LiNiO_2$ at 973 K have shown that the exchange occurs stepwise, because CHD_3 and CH_3D predominate over CH_2D_2 at the early stage of the reaction (Hatano and Otsuka 1989):

$$CH_4 \rightarrow CH_3 + H$$
$$\searrow \longrightarrow CH_3D, CD_3H \qquad (3\text{-}14)$$
$$CD_4 \rightarrow CD_3 + D$$

Moreover, CH_3CH_3, CH_3CD_3, and CD_3CD_3 are the main coupling products during the CH_4–CD_4 exchange reaction.

The observation that the gaseous oxygen has no influence on the formation of C_2 products (Figs. 3-5–3-8) strongly suggests that the coupling reaction occurs exclusively on the surface as far as $LiNiO_2$ is concerned. Let us hypothesize that the dissociative adsorption of methane occurs readily on a pair site comprising $Ni^{3+}—O^{2-}$ in $LiNiO_2$ as follows:

$$CH_4 + Ni^{3+}—O^{2-} \rightleftharpoons \underset{|}{\overset{CH_3}{}}\!\!\!Ni^{2+}—\overset{|}{O}^- \tag{3-15}$$

$$\underset{|}{CH_3} + \underset{|}{CH_3} \longrightarrow C_2H_6 \tag{3-16}$$

It is generally suggested that the vacant *d* orbital of a transition-metal cation should make electrophilic attack on σ_{CH} orbital of methane, resulting in the breaking of the C—H bond. This C—H activation on $LiNiO_2$ must be promoted by the resulting formation of OH on the partner lattice oxygen anion of Ni^{3+}. The rates of adsorption and desorption in reaction (3-15) have been roughly estimated from the kinetic curves of the CH_4–CD_4 exchange reaction (Hatano and Otsuka 1989). The results showed that the forward and reverse steps in equation (3-15) are 2 to 3 times faster than the rate of the coupling reaction [equation (3-16)]. Because this difference is the lower limit (Hatano and Otsuka 1989), it seems reasonable to assume that the coupling reaction is the rate-determining step. This assumption explains the observation that the rate of C_2 formation depends on the square of the methane pressure (Figs. 3-5 and 3-8).

Oxidative Coupling of Methane on LiCl-added NiO

As described earlier, the calcination of $LiNO_3$, Li_2CO_3, or LiOH-added NiO at 1023 K generates a solid solution of Li and NiO ($Li_xNi_{1-x}O$). However, the LiCl or LiBr-added NiO did not form any compound oxide. These results are demonstrated by the X-ray diffraction spectra of the samples prepared from various Li salts added to NiO listed in Fig. 3-9. The catalysis of the former oxide (as $LiNiO_2$) has already been described in detail. For the latter oxides, although no Li and NiO solid solution is formed, the catalytic test for the oxidative coupling of methane has shown that LiCl/NiO and LiBr/NiO are also active and selective catalysts (Otsuka, Hatano, and Komatsu 1989; Otsuka, Liu, and Morikawa 1986). Therefore, it is interesting to examine whether the lattice oxygen atoms of the Li halide-added NiO do contribute to the reaction.

Figure 3-10 shows the kinetic curves of the products for the reaction of CH_4 with LiCl/NiO at 973 K in the absence of gaseous oxygen. The reaction was carried out using a closed gas-circulation system. Before the reaction, the catalyst was treated in oxygen (13.3 kPa) at 973 K for 1 h and degassed for 10 min at the same temperature. The results in Fig. 3-10

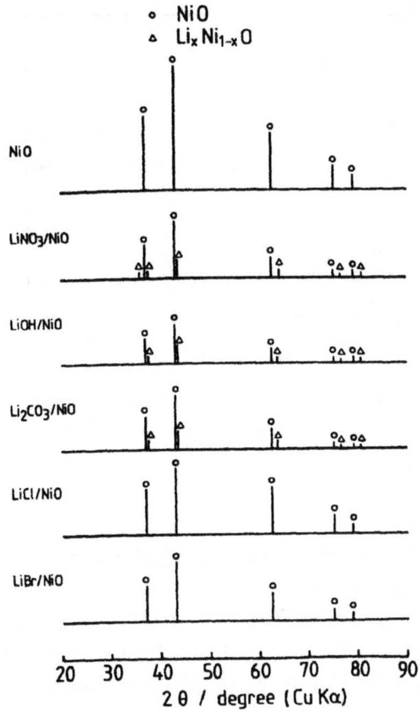

FIGURE 3-9. XRD spectra of Li-salt-added NiO: Li : NiO = 1 : 4, calcined in air at 1023 K for 2 h.

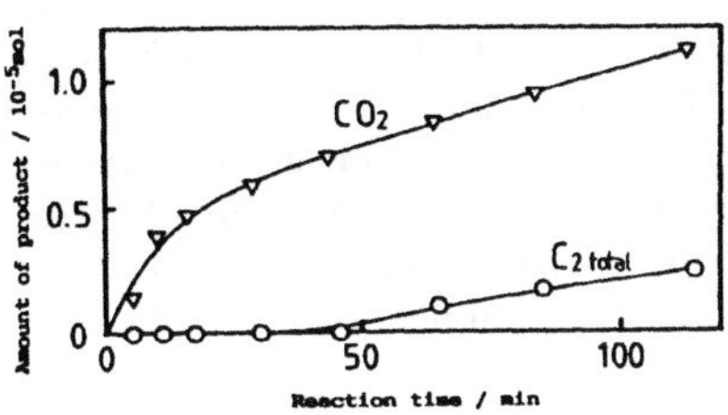

FIGURE 3-10. Oxidation of methane with LiCl/NiO in the absence of gaseous oxygen: Initial CH_4 pressure = 5.3 kPa, T = 973 K, and LiCl/NiO = 1.0 g.

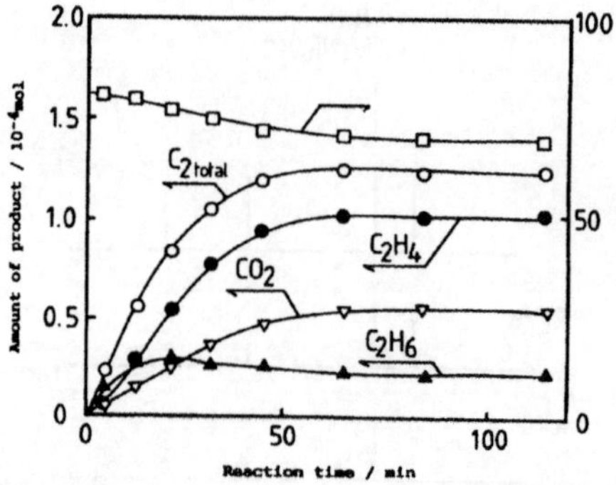

FIGURE 3-11. Oxidation of methane over LiCl/NiO in the presence of gaseous oxygen: Initial CH_4 pressure = 5.3 kPa, initial O_2 pressure = 0.53 kPa, T = 973 K, and LiCl/NiO = 1.0 g.

show that the lattice oxygen of the catalyst produces CO_2 exclusively at least at the initial stage of the reaction, in contrast to the exactly opposite results observed for $LiNiO_2$ in Fig. 3-6. Moreover, the initial rate of CO_2 formation in Fig. 3-10 is one order of magnitude less than that observed in the presence of gaseous oxygen, as is described next.

The kinetic curves obtained in the presence of gaseous oxygen were plotted in Fig. 3-11. The reaction conditions were the same as those in Fig. 3-10 except for the presence of gaseous oxygen. The initial conversion rate of methane was more than 20 times greater than that observed in the absence of gaseous oxygen. The presence of gaseous oxygen remarkably enhanced the formation rate of C_2 compounds and especially the selectivity to ethylene. These results strongly suggest that the active oxygen species responsible for the formation of C_2 compounds are not the lattice oxygen atoms but the adsorbed oxygen in the presence of gaseous oxygen. The cessation of reaction after 50 min in Fig. 3-11 can be ascribed to the consumption of all the gaseous oxygen in 50 min.

The CH_4–CD_4 exchange studies were carried out in the presence and absence of gaseous oxygen under the following conditions: $T = 973$ K, weight of LiCl/NiO = 1.00 g; (1) $P(CH_4) = P(CD_4) = 2.7$ kPa, (2) $P(CH_4) = P(CD_4) = P(O_2) = 2.7$ kPa. The hydrogen exchange did not occur at all under these reaction conditions, although the oxidative coupling occurs rapidly under the latter reaction conditions. This observation suggests that the methyl species ($CH_3\cdot$ or CH_3 groups) once formed from CH_4 go quickly to the subsequent reactions, producing C_2 hydrocarbons and carbon oxides (CO and CO_2). In other words, this means that the rate-determining step is the H atom abstraction from CH_4. Note that the CH_4–CD_4 exchange occurred faster than the coupling of methane in the case of the reaction over $LiNiO_2$. The results just described strongly suggest that the reaction mechanisms of methane activation are quite different for LiCl/NiO and $LiNiO_2$.

As described earlier, the kinetic studies on the conversion of methane over LiCl/NiO showed that the rate equation can be expressed by equation (3-1) (Otsuka, Hatano, and Komatsu 1989). We believe that the mechanism of methane activation over LiCl/NiO is the same as that proposed for Sm_2O_3.

Contribution of Lattice Oxygen Atoms in HCHO Formation

Another example of the important role of lattice oxygen atoms has been demonstrated for the partial oxidation of methane over FeNbB–O (Otsuka et al. 1988). This catalyst is composed of the mixture of $FeNbO_4$, $FeNb_{11}O_{29}$, and B_2O_3. The catalytically active component is $FeNbO_4$. The role of B_2O_3 is to reduce both decomposition and deep oxidation of the HCHO produced (Otsuka et al. 1988). As described earlier, the rate equation for the conversion of methane over this catalyst can be expressed by equation (3-2). This rate equation implies the contribution of a monatomic oxygen at the C—H bond breaking step. This oxygen could be a lattice oxygen atom of the catalyst.

In order to clarify the role of lattice oxygen atoms of FeNbB–O, the oxidation of methane with $^{18}O_2$ as carried out. The experiments were conducted using a closed gas-circulation apparatus. The approximate ratio of the quantity of oxygen atoms in the solid to that in the gas phase was

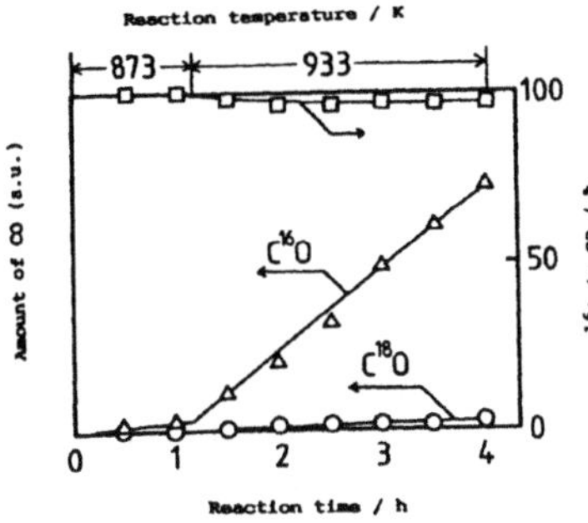

FIGURE 3-12. Oxidation of CH_4 by $^{18}O_2$ over FeNbB–O: Initial CH_4 pressure = 1.3 kPa, initial O_2 pressure = 1.3 kPa, and FeNbB–O = 0.50 g.

45. All the HCHO, CO_2, and H_2O formed were always condensed in a trap cooled at 77 K. The kinetic curves of CO (noncondensable product at 77 K) are given in Fig. 3-12. Note that only a trace of ^{18}O has been incorporated into the carbon monoxide formed. This is true also for the condensed products (HCHO, CO_2, and H_2O). The change in the $^{18}O_2:{}^{16}O_2$ ratio for the gaseous oxygen due to the gas–solid oxygen exchange can be neglected during the experiment of Fig. 3-12. These observations clearly show that the lattice oxygen atoms of FeNbB–O are responsible for the abstraction of hydrogen as well as for the formation of HCHO, CO, and CO_2.

CONCLUSION

The general picture of the reaction mechanism for oxidative coupling of methane has been demonstrated for the catalytic reaction over Sm_2O_3. The rate-determining step is at the C—H bond breaking reaction on the surface. A diatomic oxygen, probably O_2^{2-}, must be responsible for the activation of methane. However, this picture cannot be applied for the reactions over $LiNiO_2$ and FeNbB–O. The lattice oxygen atoms of these catalysts play an important role in the activation and partial oxidation of methane. The rate-determining step of the reaction on the $LiNiO_2$ is at the coupling of methyl groups adsorbed on $Ni^{3+}—O^{2-}$ pair sites. Thus, we would emphasize that the active oxygen species and the reaction mechanism depend on the types of catalysts. However, we should take into account that the reaction profile could change drastically at higher pressures of reactants (> 101 kPa) because of a serious contribution of the gas-phase reactions in the initiation and chain propagation. The mechanism for the catalytic reactions at higher pressures is the problem to be clarified in the near future.

References

Asami, K., T. Shikada, K. Fujimoto, and H. Tominaga. 1987. Oxidative coupling of methane over lead oxide catalyst: Kinetic study and reaction mechanism. *Ind. Eng. Chem. Res.* 26:2348–53.

Baerns, M., J.R.H. Ross, and K. van der Wiele. 1988. *Methane Activation.* Amsterdam: Elsevier Science Publishers. (*Catal. Today* 4:271–494.)

Cant, N.W., C.A. Lukey, P.F. Nelson, and R.J. Tyler. 1988. The rate controlling step in the oxidative coupling of methane over a lithium-promoted magnesium oxide catalyst. *J. Chem. Soc. Chem. Commun.* 766–8.

Driscoll, D.J., W. Martir, J.-X. Wang, and J.H. Lunsford. 1985. Formation of gas-phase methyl radicals over MgO. *J. Am. Chem. Soc.* 107:58–63.

Gesser, H.D., N.R. Hunter, and C.B. Prakash. 1985. The direct conversion of methane to methanol by controlled oxidation. *Chem. Rev.* 85:235–44.

Hatano, M., and K. Otsuka. 1988. Alkali metal-doped transition metal oxides active for oxidative coupling of methane. *Inorg. Chim. Acta* 146:243–7.

Hatano, M., and K. Otsuka, 1989. The oxidative coupling of methane on lithium nickelate(III). *J. Chem. Soc., Faraday Trans. 1* 85:199–206.

Hutchings, G.J., M.S. Scurrell, and J.R. Woodhouse. 1987. The role of surface O^- in the selective oxidation of methane. *J. Chem. Soc., Chem. Commun.* 1388–9.

Hutchings, G.J., M.S. Scurrell, and J.R. Woodhouse. 1989. Oxidative coupling of methane using oxide catalyst. *Chem. Soc. Rev.* 18:251–83.

Ito, T., J.-X. Wang, C.-H. Lin, and J.H. Lunsford. 1985. Oxidative dimerization of methane over a lithium-promoted magnesium oxide catalyst. *J. Am. Chem. Soc.* 107:5062–8.

Jones, C.A., J.J. Leonard, and J.A. Sofranko. 1987a. The oxidative conversion of methane to higher hydrocarbons over alkali-promoted Mn/SiO_2. *J. Catal.* 103:311–19.

Jones, C.A., J.J. Leonard, and J.A. Sofranko. 1987b. Fuels for the future: Remote gas conversion. *Energy and Fuels* 1:12–16.

Keller, G.E., and M.M. Bhasin. 1982. Synthesis of ethylene via oxidative coupling of methane. *J. Catal.* 73:9–19.

Keulks, G.W., L.D. Krenzke, and T.M. Notermann. 1978. Selective oxidation of propylene. *Adv. Catal.* 27:183–225.

Komatsu, T., T. Amaya, and K. Otsuka. 1989. LiCl doped cobalt oxide is an active catalyst for the formation of ethylene in the oxidative coupling of methane. *Catal. Lett.* 3:317–22.

Lee, J.S., and S.T. Oyama. 1988. Oxidative coupling of methane to higher hydrocarbons. *Catal. Rev.—Sci. Eng.* 30:249–80.

Lin, C.-H., K.D. Campbell, J.-X. Wang, and J.H. Lunsford. 1986. Oxidative dimerization of methane over lanthanum oxide. *J. Phys. Chem.* 90:534–7.

Liu, R.-S., M. Iwamoto, and J.H. Lunsford. 1982. Partial oxidation of methane by nitrous oxide over molybdenum oxide. *J. Chem. Soc., Chem. Commun.* 78–9.

Naccache, C. 1971. ESR study of species formed by reaction of O^- adsorbed on magnesium oxide with O_2, CO and ethylene. *Chem. Phys. Lett.* 11:323–5.

Nelson, P.F., C.A. Lukey, and N.W. Cant. 1989. Measurements of kinetic isotope effects and hydrogen/deuterium distributions over methane oxidative coupling catalysts. *J. Catal.* 120:216–30.

Otsuka, K. 1987. Direct conversion of methane to higher hydrocarbons. *Sekiyu Gakkaishi* 30:385–96.

Otsuka, K., M. Hatano, and T. Komatsu. 1989. Synthesis of C_2H_4 by partial oxidation of CH_4 over LiCl/NiO. *Catal. Today* 4:409–19.

Otsuka, K., M. Inaida, Y. Wada, T. Komatsu, and A. Morikawa. 1989. Isotopic studies on oxidative methane coupling over samarium oxide. *Chem. Lett.* 1531–4.

Otsuka, K., and K. Jinno. 1986. Kinetic studies on partial oxidation of methane over samarium oxides. *Inorg. Chim. Acta* 121:237–41.

Otsuka, K., K. Jinno, and A. Morikawa. 1986. Active and selective catalysts for the synthesis of C_2H_4 and C_2H_6 via oxidative coupling of methane. *J. Catal.* 100:353–9.

Otsuka, K., T. Komatsu, K. Jinno, Y. Uragami, and A. Morikawa. 1988. Activation of methane and synthesis of formaldehyde by partial oxidation. In *Proceedings of the 9th International Congress on Catalysis*, Vol. 2, ed. M.J. Phillips and M. Ternan, pp. 915–22. Ottawa: The Chemical Institute of Canada.

Otsuka, K., Q. Liu, and A. Morikawa. 1986. Active and selective catalysts in oxidative coupling of methane. Nickel oxides with salts of alkali metals. *Inorg. Chim. Acta* 118:L23–4.

Otsuka, K., Y. Murakami, Y. Wada, A.A. Said, and A. Morikawa. 1990. Oxidative coupling of methane, ethane, and propane with sodium peroxide at low temperatures. *J. Catal.* 121:122–30.

Otsuka, K., and T. Nakajima. 1987. Oxidative coupling of methane over samarium oxides using N_2O as the oxidant. *J. Chem. Soc., Faraday Trans. 1* 83:1315–21.

Otsuka, K., A.A. Said, K. Jinno, and T. Komatsu. 1987. Peroxide anions as possible active species in oxidative coupling of methane. *Chem. Lett.* 77–80.

Otsuka, K., Y. Shimizu, and T. Komatsu. 1987. Ba doped cerium oxides active for oxidative coupling of methane. *Chem. Lett.* 1835–8.

Pitchai, R., and K. Klier. 1986. Partial oxidation of methane. *Catal. Rev.—Sci. Eng.* 28:13–88.

Zhen, K.J., M.M. Khan, C.H. Mak, K.B. Lewis, and G.A. Somorjai. 1985. Partial oxidation of methane by nitrous oxide over molybdenum oxide. *J. Catal.* 94:501–7.

4

The Oxidative Coupling of Methane: Reaction Pathways and Their Process Implications

A. Ekstrom

CSIRO Division of Coal and Energy Technology, Lucas Heights Research Laboratories, Menai, Australia

TABLE OF CONTENTS

INTRODUCTION

Background

The development of a simple and commercially advantageous process for the direct conversion of natural gas to more-easily transportable liquid products would be a great prize. In this connection and in common with many others, we became intrigued with the results of Keller and Bhasin (Keller and Bhasin 1982) and a little later with those of Ito and co-workers (Ito et al. 1985), which suggested that it was possible to catalyze the reaction of methane with oxygen to form ethane and, more important, ethylene. This reaction appeared to offer the potential for eliminating the expensive reforming step used in conventional natural gas conversion processes as well as a new route for the production of ethylene (Preuss and Baerns 1987; Jones, Leonard, and Sofranko 1987). Our efforts, described in this chapter, formed part of a substantial CSIRO project that included the process development studies summarized in Chapter 13 of this volume (Edwards, Do, and Tyler, 1990).

One approach to the exploitation of this reaction was developed by the Atlantic Richfield Co. (Jones, Leonard, and Sofranko 1987). In an impressive development program, a variable oxidation state catalyst was used. CH_4 conversions of the order of 25% and C_{2+} selectivities of approximately 75% were achieved in a relatively complex process involving circulating a catalyst between two fluidized-bed reactors, one to oxidize the catalyst with oxygen (air), the other to carry out the coupling reaction.

Several economic and process analyses have now been published that show that the cost of liquid fuels produced by the oxidative coupling reaction, as presently conceived, would be greater than conventional technologies (Kuo 1987; Field, Niruta, and McCarthy 1987; Gray and Tomlinson 1988; Fox, Chen, and Degen 1990; Jackson, Thompson, and Whitehead 1989). These studies demonstrated the sensitivity of the process economics to the per-pass CH_4 conversion and C_{2+} selectivity, and the ability to achieve both high conversions and selectivities at elevated pressures.

Reaction Mechanisms

Our calculations using an in-house program (Turnbull and Wadsley 1988) showed that for an initial $CH_4 : O_2$ ratio of 10, CO and H_2 are the major

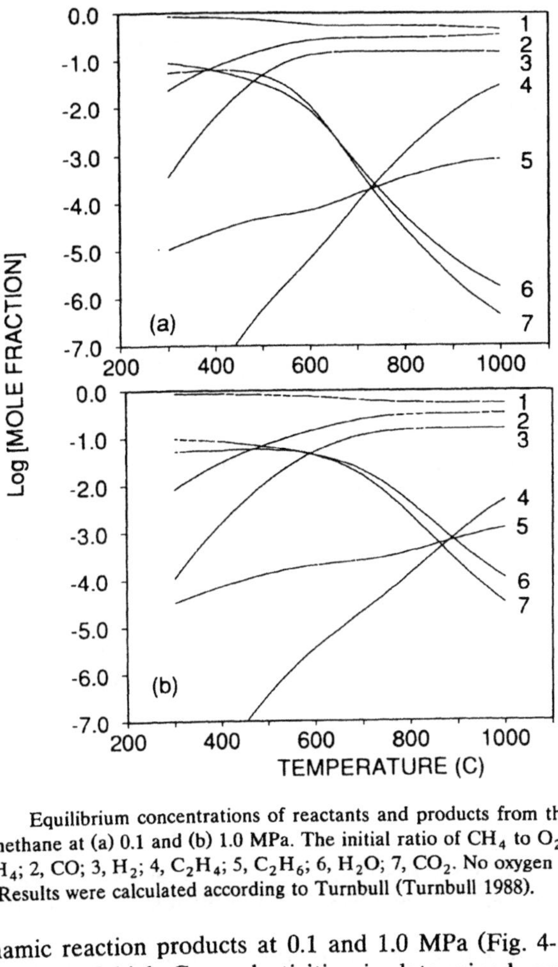

FIGURE 4-1. Equilibrium concentrations of reactants and products from the oxidative coupling of methane at (a) 0.1 and (b) 1.0 MPa. The initial ratio of CH_4 to O_2 was 10 : 1. Curves: 1, CH_4; 2, CO; 3, H_2; 4, C_2H_4; 5, C_2H_6; 6, H_2O; 7, CO_2. No oxygen remained at equilibrium. Results were calculated according to Turnbull (Turnbull 1988).

thermodynamic reaction products at 0.1 and 1.0 MPa (Fig. 4-1). Clearly, the achievement of high C_{2+} selectivities is determined *exclusively* by kinetic factors. In view of the preceding, the product distribution obtained from the catalyzed reaction is remarkable. For example, a Li/MgO catalyst, (800°C, 12.8% O_2 in CH_4, 100% O_2 consumption) gives the product distribution shown in Fig. 4-2. This corresponds to the empirical equation (4-1), in which the water yield has been calculated to give an oxygen balance.

$$CH_4 + 0.65O_2 \rightarrow 0.11H_2 + 0.026CO + 0.21CO_2 + 0.22C_2H_4 + 0.12C_2H_6 + 0.033C_3 + 0.86H_2O \quad (4\text{-}1)$$

An overall C_{2+} selectivity of 76% was obtained at 22% CH_4 conversion.

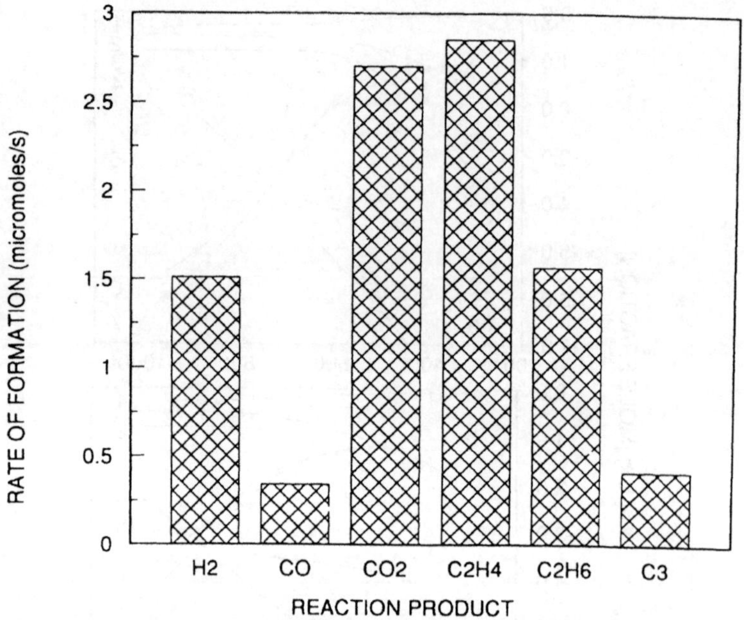

FIGURE 4-2. Typical product distribution obtained with a Li/MgO catalyst (800°C, 12.8% O_2 in CH_4), using a linear velocity of 15 cm s^{-1}. The CH_4 conversion was 24%, the O_2 consumption 98%.

Apart from the importance of the C_{2+} products, the predominance of CO_2 as the major C_1 product is noteworthy, as is the high rate of H_2 formation. As expected at these temperatures, C_2H_4 is the dominant C_2 product.

The early work of Ito and co-workers (Ito et al. 1985) using a Li_2CO_3-promoted MgO catalyst led to the conclusion that the reaction of methane on the catalyst surface resulted in the formation of methyl radicals. Convincing evidence has accumulated to show that these dimerized in the gas phase to form ethane (Driscoll, Campbell, and Lunsford 1987). The evolution of gas-phase radicals from a solid catalyst surface can be considered another example of the "spillover" effect well-established for the formation of oxygen and hydrogen atoms over a variety of catalysts (Benson, Kohn, and Boudart 1966; Ekstrom, Batley, and Johnson 1974). It was suggested that the methyl radicals were formed by the reaction sequence (4-2)–(4-4), in which Li^+O^- represents a Li^+-stabilized anion vacancy and [ ] an oxygen vacancy in the MgO lattice.

$$2Li^+O^{2-} + [\ ] + 0.5O_2 \rightarrow 2Li^+O^- + O^{2-} \qquad (4\text{-}2)$$

$$CH_4 + 2Li^+O^- \rightarrow Li^+OH^- + CH_3\cdot \qquad (4\text{-}3)$$

$$2Li^+OH^- \rightarrow Li^+O^{2-} + Li^+[\ ] + H_2O \qquad (4\text{-}4)$$

It has been amply demonstrated that the reaction of small alkanes with anion vacancies in various oxides occurs at low temperatures (Liu, Iwamoto, and Lunsford 1982; Aika and Lunsford 1977; Aika and Lunsford 1978; Ward, Lin and Lunsford 1977). Ito and co-workers (Ito et al. 1985) suggested that the oxidation of the C_2 products was not a major source of the CO_x products. Rather, it was proposed that these were formed via methoxy ions on the catalyst's surface and by the gas-phase reaction between $CH_3\cdot$ and O_2 to give the methylperoxy radical, which can undergo a complex series of reactions to form CO and CO_2.

Since these studies were first reported, a huge range of catalysts for this reaction have been discovered, and many proposals have been made regarding reaction mechanisms. Although some of this work will be cited later, this is not a literature survey, and the reader is referred to recent reviews for an overview of the topic (Lee and Oyama 1988; Hutchings, Scurrell, and Woodhouse 1989).

MAJOR ISSUES

It is now clear that a wide range of basic oxides can be effective catalysts, including many containing no additives or promoters such as Li_2CO_3. For these, cation-stabilized anion vacancies cannot be the active sites. As shown later, many catalysts result in remarkably similar conversion–selectivity results, suggesting a common active site and reaction mechanism. A frustrating aspect of the catalyst development work has been the inability to maintain the high C_{2+} selectivity as the per-pass CH_4 conversion is increased. Instead, it is found that the C_{2+} selectivity decreases almost linearly with increasing methane conversion, usually achieved by increasing the oxygen concentration in the feed. This is a fundamental impediment for the commercial development of this reaction.

Our prime interest has been in the cofeed version of the reaction, in which the oxygen is premixed with methane and fed to the catalyst, rather than added sequentially as in the Atlantic Richfield Co. process. The specific issues we sought to address in our work were the following:

- What factors control the catalyst selectivity?
- What is the origin of the CO_x products, and how important are the C_{2+} products as sources of CO_x?
- Can staged oxygen injection be used to increase the CH_4 conversion while maintaining C_{2+} selectivity?
- Can we clarify the mechanism of the methyl radical formation?
- What is the effect of pressure on the reaction?

Some of the results of this work are described and discussed in this chapter. Most were taken using a Sm_2O_3 catalyst, although other rare-earth oxides and Li/MgO were used for comparative purposes.

EXPERIMENTAL CONSIDERATIONS

It has long been a principle of catalytic research that information on reaction mechanisms and kinetics should be sought from experiments conducted under differential conditions, free from reactant and product gradients along the catalyst bed. This can be exceptionally difficult to achieve for the oxidative coupling reaction.

The importance of these gradients should be appreciated. For example, Fig. 4-3 illustrates a typical experiment, in which a 20-mg catalyst bed, perhaps 1 mm deep, results in a complete consumption of oxygen. Clearly, there are huge gradients in the O_2, C_2, and CO_x concentrations along the

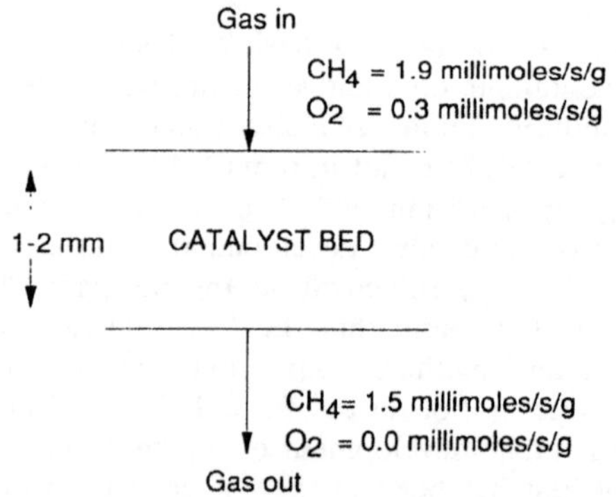

FIGURE 4-3. Illustration of the reactant and product gradients expected in a typical experiment. The data are for a run using a Sm_2O_3 catalyst (20 mg) at 800°C.

axis of the catalyst bed. It is our experience that the best C_{2+} selectivities are often obtained under conditions of complete oxygen consumption, or maximum oxygen gradients. Conversely, operation of the reaction under differential conditions with respect to oxygen often results in low C_{2+} selectivity. Consequently, changes in product distribution or conversion resulting from a change in experimental parameters may simply reflect changes in the reactant and product gradients.

It is also important to consider the very large exothermicity of the reaction, particularly when the CO_x selectivity is high. Temperatures in the catalyst bed can be much higher than in the gas phase in the immediate vicinity of the catalyst bed, making comparisons of reactivities and so on very difficult. Although it is essential to have a thermocouple in the catalyst bed, it should be remembered that this measures the temperature of the junction, not necessarily of the surface of the catalyst particles.

It should also be noted that a major product of the reaction is water [equation (4-1)] and that, at the high temperatures employed, steam-reforming of the hydrocarbons present in the system may be a further source of CO and H_2. The effects of the catalysts on this reaction are not known.

Another noteworthy aspect of the reaction is that the difference between the rates of the catalyzed and uncatalyzed, or blank reactions can be unexpectedly small. This effect, which has been noted before (Lane and Wolf 1988; Yates and Zlotin 1988; Ekstrom, Regtop, and Bhargava 1990), is illustrated in Fig. 4-4 for a Li/MgO catalyst. At the conditions employed, the major effect of the catalyst is to increase the C_{2+} formation by a factor of about 6, the CO_2 formation by a factor of nearly 26, but to *decrease* the CO formation by a factor of 2. A major function of this catalyst clearly is to enhance the rate of CO_2 formation, and in this sense Li/MgO can be regarded as a good oxidation catalyst. At higher linear gas velocities, where the importance of the blank (uncatalyzed) reaction can be reduced, much larger differences can be obtained if the catalyst activity is high enough. It is thus important to remember that at high temperatures and low space velocities the contribution of the blank reaction can seriously distort the effects of the catalyst.

OPTIMIZATION OF RARE-EARTH OXIDE CATALYSTS

Catalyst Optimization

The results of our attempts to optimize the rare-earth oxide catalysts illustrate the restrictions imposed by this reaction (Ekstrom et al. 1990). Increasing the temperature increased the conversion although the C_{2+}

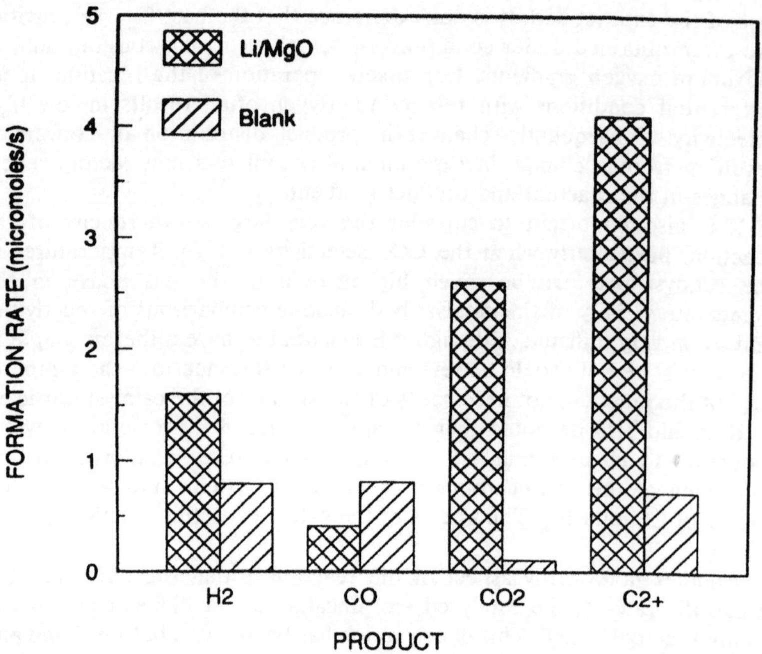

FIGURE 4-4. Comparison of the product distribution from the oxidative coupling in the absence and presence of Li/MgO catalyst. The linear velocity was 15 cm s^{-1}, 15% O_2 in CH_4 at 0.1 MPa and 850°C.

selectivity also increased substantially (Fig. 4-5a), it reached a maximum at about 800°C. Increasing the space velocity at this reaction temperature resulted in a small improvement in the C_{2+} selectivity, but the percentage of conversion remained virtually constant (Fig. 4-5b), implying that the *conversion rate* increased almost linearly with space velocity. This is an observation common to many catalysts that has intrigued us for some time. Possible explanations for this effect include mass transfer limitations and the reduction of the poisoning effect of CO_2 as the space velocity is increased.

For each rare-earth oxide, the temperature required for optimum performance was slightly different, making comparison of catalyst activity from measurements at a single temperature difficult to interpret (Otsuka, Jinno, and Morikawa 1985; Otsuka and Nakajima 1986; Otsuka and Komatsu 1987; Otsuka, Jinno, and Morikawa 1986). We believe that to a first approximation these effects are due to the presence of various

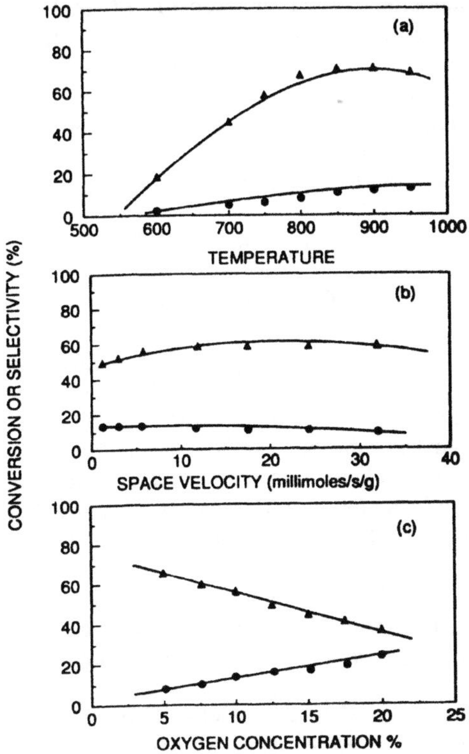

FIGURE 4-5. Results of optimization studies for the rare-earth elements: (a) effect of temperature on the C_{2+} selectivity and CH_4 conversion using a Ho_2O_3 catalyst and 10% O_2; (b) effect of space velocity at 950°C on the C_{2+} selectivity and CH_4 conversion using a Yb_2O_3 catalyst and 10% O_2; (c) the effect of O_2 concentration at 950°C on the C_{2+} selectivity and CH_4 conversion using a Yb_2O_3 catalyst. ▲, C_{2+} selectivity; ●, CH_4 conversion.

concentrations of carbonates on the catalyst surface. The decomposition temperatures of the rare-earth carbonates range from 530°C for samarium to 750°C for lanthanum. As the presence of carbonates is known to poison the reaction (Tanabe 1985; Roos 1989; Amorebieta and Colussi 1989), variations of catalyst activity at any constant temperature may simply be a reflection on the state of the oxide–carbonate equilibrium.

Not unexpectedly for an oxidation reaction of this type, increasing the oxygen concentration was the major variable that could be used to increase the conversion, but it simultaneously decreased the C_{2+} selectivity

(Fig. 4-5c). Similar effects were observed for all catalysts tested, and they resulted in the decline in the C_{2+} selectivity as the CH_4 conversion increased. Such results have been observed on numerous previous occasions (Lane and Wolf 1989).

These data suggest that at otherwise constant conditions, the O_2 concentration dependency of the C_{2+} selectivity is given by equation (4-5), and the CH_4 conversion by equation (4-6) where c is close to 0.

$$S_{C_{2+}} = a - b[O_2] \tag{4-5}$$

$$C_{CH_4} = c + d[O_2] \tag{4-6}$$

For some catalysts, for example, Sr-doped rare-earth catalysts, the constant a in equation (4-5) was close to 100%, that is, at "zero" oxygen concentrations the catalyst is almost 100% selective for C_{2+} formation. This indicates that the inherent catalyst selectivity can be very high and that the decline in selectivity with increasing oxygen concentration is due to secondary reactions. Otherwise, it must be argued that the catalyst selectivity is a function of the oxygen concentration, a less likely possibility. We think this is an important result, as the search for better catalysts is often based on the belief that the poor C_{2+} selectivity at high oxygen concentrations is the result of an inherently low catalyst selectivity rather than an indication of secondary reactions. It is likely that this belief is not warranted.

It is obvious from equations (4-5) and (4-6) that the C_{2+} yield (conversion $\times$ selectivity) will show a parabolic O_2 dependency. This effect is demonstrated in Fig. 4-6, which summarizes the C_{2+} concentration in the exit gas stream versus CH_4 conversion, using data obtained in our laboratory for 828 experiments in a fixed-bed reactor. These results are based on 25 different catalysts and were obtained under a very wide range of experimental conditions. The results show the attainment of a maximum C_{2+} concentration in the exit gas stream of about 4–5 vol% (8–10% based on carbon) at a CH_4 conversion of 15–20%.

Comments

The preceding results hint at some of the problems to be encountered in attempts to achieve simultaneously high CH_4 conversions and C_{2+} selectivities. The experimental variables (temperature, linear velocity, and oxygen concentration) that could be used appear unable to achieve the desired objectives. Furthermore, the data in Fig. 4-6 suggest a relatively

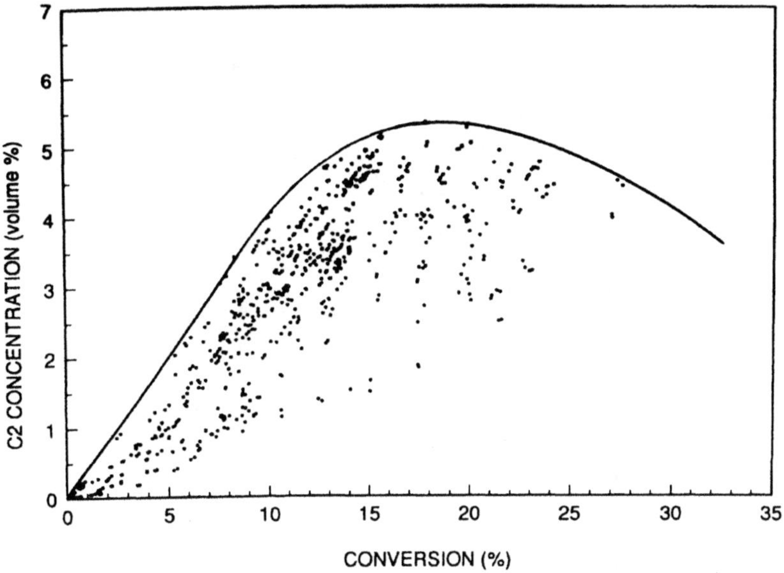

FIGURE 4-6. Effect of inlet oxygen concentration on the C_{2+} yield. Data for 828 experiments, involving 25 catalysts. Solid line is visual envelope of maximum C_2 concentration at various CH_4 conversions.

low limit in the maximum C_{2+} concentration in the exit gas. Despite extensive work, we have not so far been able to exceed this concentration.

STAGED OXYGEN INJECTION

The question we wished to explore in this part of our research was whether or not it is possible to increase the CH_4 conversion and C_{2+} selectivity by using staged oxygen injection. This technique is used commercially in the ammonia oxidation reaction and would, at the very least, help control the exotherm of the reaction if carried out in a fixed-bed reactor.

In our experiments an apparatus using three or five stages was used (Ekstrom, Kastl, and Campbell 1990). The interstage gas mixing was done in a small volume at temperatures 300–400°C lower than those used for the reaction. Mixing was accomplished using a small tangential jet. A

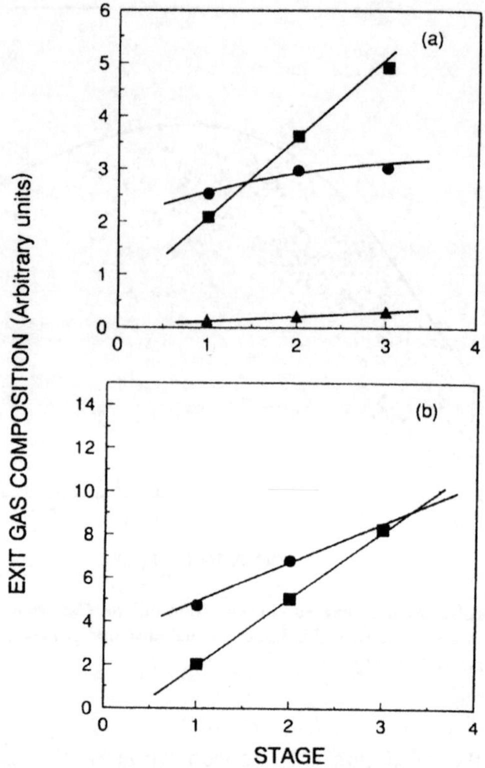

FIGURE 4-7. Effect of staged oxygen injection on the product concentrations at the exit of each stage: (a) ■, C_2H_4; ●, C_2H_6; ▲, C_{3+}. (b) ■, CO_x; ●, C_{2+}.

sample of the product gases emerging from each stage could be diverted to an on-line gas chromatograph for complete analysis.

Effect of Staged Oxygen Injection on Product Distribution

Figure 4-7 summarizes the product distribution at each stage of a three-stage oxygen injection run. It is interesting that the C_2H_4 concentration increases with the stage number, whereas the C_2H_6 concentration remains almost constant. However, the most prominent feature of these results is the rapid rise in the CO_x formation with increasing staging. This effect is

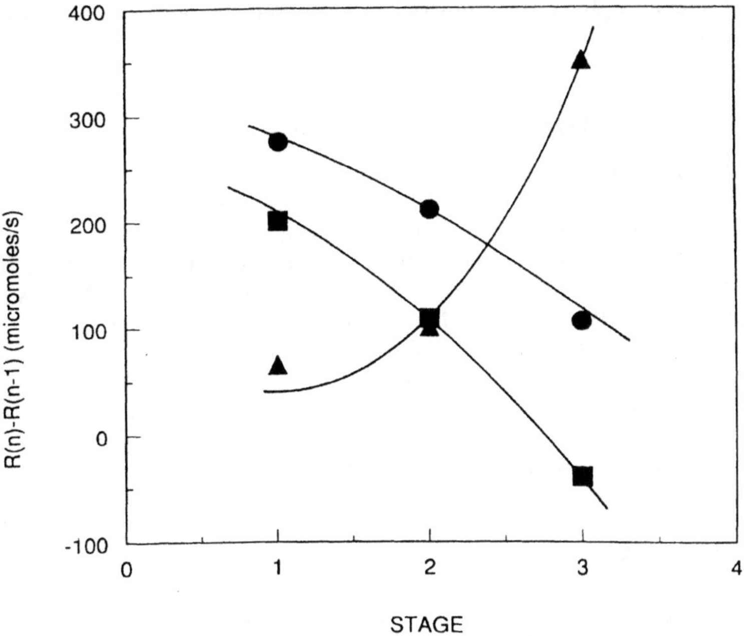

FIGURE 4-8. Effect of staged oxygen injection on the difference in the conversion–consumption rate at stage $R(n)$ and $R(n-1)$: ▲, CO_x; ●,—CH_4; ■, C_{2+}.

even more prominent when the differential rates of reaction at each stage are compared (Fig. 4-8). These latter data show a rapidly increasing rate of CO_x formation and declining rates of CH_4 conversion and C_{2+} formation. The decline in the CH_4 conversion rate is difficult to understand, unless poisoning of the catalyst by the increasing CO_2 concentration is occurring. This aspect will be discussed in more detail later.

Comments

Although the results reported here are of limited scope, they suggest that the combustion of the C_2 products is a major factor in this reaction. This is obvious from the rapid increase in the differential rate of CO_x formation and the corresponding decline in the C_{2+} formation rates (Fig. 4-7) with increasing stage number. Indeed, between stages 2 and 3, a net reduction in the ratio of C_2 formation is observed. However, it is obvious that a more sophisticated approach was needed to explore the subtleties of this effect.

MECHANISM OF CO_x FORMATION

It is apparent that an understanding of the origin and mechanism of formation of the CO_x products is of fundamental importance in this reaction. It could be argued intuitively that the C_{2+} products, particularly C_2H_4, would be oxidized faster than a stable molecule such as CH_4, and perhaps even the CH_3 radical. Nevertheless, some authors (Lin, Wang, and Lunsford 1988; Otsuka, Jinno, and Morikawa 1986) have argued for various reasons that under certain conditions the ethane and ethylene formed in the coupling reaction are not major sources of CO_x.

For the work described in this section (Ekstrom and Lapszewicz 1990), an isotope switching apparatus with a gas chromatograph–mass spectrometer system was used for the analysis of the reaction products. For a typical experiment, $^{12}CH_4$ (3700 μmol g^{-1} s^{-1}), O_2 (600 μmol g^{-1} s^{-1}) and He (70–300 μmol g^{-1} s^{-1}) were passed over 20 mg of a Sm_2O_3 catalyst at 700°C and the reaction products determined by gas chromatographic analysis. Helium was then replaced by an identical flow of $^{12}C_2H_6$ or $^{12}C_2H_4$. After the reaction reached steady state, $^{12}C_2H_6$ was replaced by a 30–120-s pulse of $^{13}C_2H_6$ or $^{13}C_2H_4$ and the reaction products were analyzed midway through the pulse, using the preceding technique (Table 4.1). The ^{12}C–^{13}C isotope effects and the reaction in the absence of the catalyst were found to be negligible. Prior to the experiments, the catalyst

Table 4-1 Conversion and Selectivities for $^{12}CH_4$, $^{13}C_2H_6$, and $^{13}C_2H_4$[a]

Reactant	$^{12}CH_4$	$^{13}C_2H_6$	$^{13}C_2H_4$
Feed rate (μmol s^{-1} g^{-1})	3619	317	278
Total conversion (%)	17	41	39
Conversion to CO_x (%)	6	15	39
	Selectivity (%)		
To C_2H_6	36	—	—
To C_2H_4	27	63	—
To CO	17	19	53
To CO_2	20	18	47
To CO_x	37	37	100

[a]Conditions: Sm_2O_3 catalyst; 700°C.

activity was stabilized by exposure to the mixture of CH_4 and O_2 at 700°C for about 20 h.

Addition of $^{13}C_2H_4$

Although the addition of $^{13}C_2H_4$ resulted in a decline in the $^{12}CO_x$ formation, the total CO_x formation rate increased significantly (Fig. 4-9). For example, at a $^{13}C_2H_4$ feed rate of 200 μmol g^{-1} s^{-1}, 50% of CO_x products were derived from the additive. Considering that the $^{12}CH_4$–$^{13}C_2H_4$ ratio in the inlet gas is approximately 12:1, these results qualitatively confirm the view of Kuo (Kuo 1987), who estimated that the C_2 products would react with oxygen 10 times faster than methane.

Analysis of the composition of the CO_x products showed that the increase in their yield was mostly due to the increase in carbon monoxide formation, because the concentration of carbon dioxide rose only marginally. The oxygen consumption increases by only 3% at the highest flow of additive (Table 4.1).

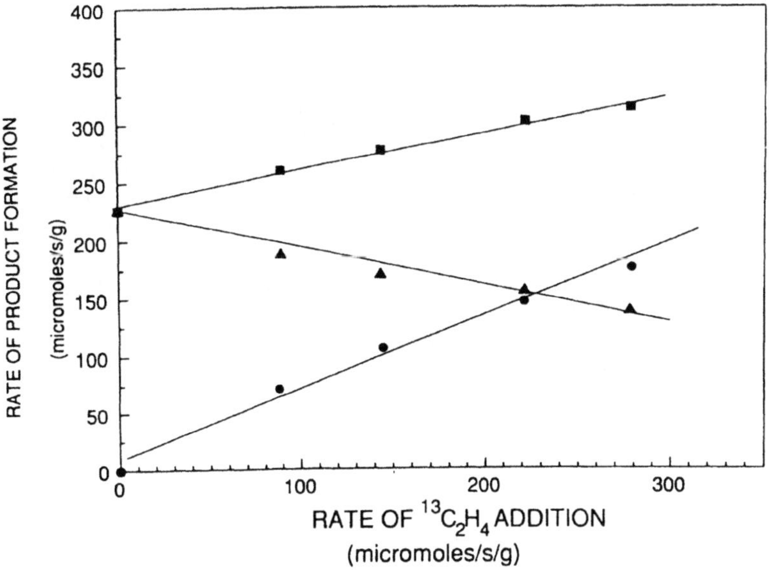

FIGURE 4-9. Effect of the addition of $^{13}C_2H_4$ on the rates of formation of (■) ($^{12}CO_x$ + $^{13}CO_x$) and (●) $^{13}CO_x$ from $^{13}C_2H_4$ and of (▲) $^{12}CO_x$ from $^{12}CH_4$.

Addition of $^{13}C_2H_4$ decreased both the $^{12}C_2H_6$ and $^{12}C_2H_4$ yields. This decline reached about 20% when 300 μmol g^{-1} s^{-1} of $^{13}C_2H_4$ was present. The net methane conversion declined by 27% at this additive concentration. Addition of ethylene to the feed is known to increase the yield of C_3 hydrocarbons (Labinger and Ott 1987; Sinev, Korchak, and Krylov 1987; Labinger 1988; Mimms et al. 1989) and to result in the formation of small amounts of acetylene. This probably occurs via the vinyl radical which can react either by elimination of hydrogen leading to acetylene, or coupling with the methyl radical giving propene [equations (4-7)–(4-9)]. These results suggest that ethylene competes with methane for the radical-forming site.

$$2\,^{12}C_2H_4 + 0.5O_2 \rightarrow 2\,^{12}C_2H_3\cdot \tag{4-7}$$

$$^{12}C_2H_3\cdot \rightarrow \,^{12}C_2H_2 + H\cdot \tag{4-8}$$

$$^{12}C_2H_3\cdot + \,^{12}CH_3\cdot \rightarrow \,^{12}C_3H_6 \tag{4-9}$$

Addition of $^{13}C_2H_6$

Addition of $^{13}C_2H_6$ resulted in a decrease in the $^{12}C_2H_6$ and $^{12}C_2H_4$ formation (Fig. 4-10), $^{13}C_2H_4$ being produced simultaneously. At a $^{13}C_2H_6$ feed rate of 300 μmol g^{-1} s^{-1}, 72% of the total C_2H_4 formed was derived from the additive. The rate of $^{13}C_2H_4$ formation in the presence of 300 μmol g^{-1} s^{-1} of $^{13}C_2H_6$ was approximately a factor of 3 higher with the catalyst present than found for the blank reaction discussed previously. The oxygen consumption decreased slightly with increasing $^{13}C_2H_6$ concentration in the feed. Most significantly, the net $^{12}CH_4$ conversion (calculated as the sum of ^{12}C products) decreased by 34% at the highest rate of ethane addition.

A possible explanation for this effect is the occurrence of reaction (4-10) in the gas phase.

$$CH_3\cdot + C_2H_6 \rightarrow CH_4 + C_2H_5\cdot \tag{4-10}$$

To explore its importance in this system, C_2D_6 was added at a rate of 300 μmol g^{-1} s^{-1} to a feed flow of 3600 μmol g^{-1} s^{-1} of $^{12}CH_4$ and 600 μmol g^{-1} s^{-1} of O_2. Neglecting any isotope effects, this addition of C_2D_6 would have been expected to reduce the net CH_4 conversion by 170 μmol g^{-1} s^{-1}. However, only 45 μmol g^{-1} s^{-1} of CH_3D was found in the reactor effluent, suggesting that at most only 25% of the reduction in CH_4 conversion could be attributed to reaction (4-10). This is probably an

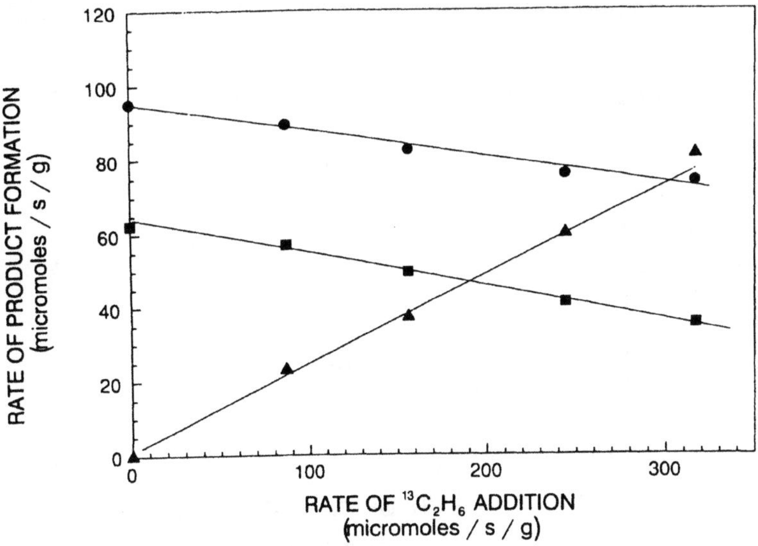

FIGURE 4-10. Effect of the addition of $^{13}C_2H_6$ on the rate of formation of $^{13}C_2$ products from $^{12}CH_4$ and $^{13}C_2H_4$ from $^{13}C_2H_6$: ■, $^{12}C_2H_4$; ●, $^{12}C_2H_6$; ▲, $^{13}C_3H_4$.

overestimate, because some isotope exchange via equations (4-11) and (4-12) would also be expected to occur (Utiyama, Hattori, and Tanabe 1978; Quanzhi and Amenomiya 1986).

$$CH_4 + C_2D_6 \rightarrow CH_3D + C_2D_5H \tag{4-11}$$

$$CH_4 + D_2 \rightarrow CH_3D + HD \tag{4-12}$$

The addition of $^{13}C_2H_6$ also resulted in a decrease in the $^{12}CO_x$ formation and a corresponding increase in the $^{13}CO_x$ formation (Fig. 4-11). It is notable that the combined ($^{13}C + ^{12}C$) CO_x formation rate remains constant and independent of the rate of addition of $^{13}C_2H_6$. For all results shown in Fig. 4-11, the CO/CO_2 ratio was constant and close to unity. At an addition rate of 320 μmol g^{-1} s^{-1}, the rates of ^{12}C and ^{13}C CO_x formation were equal, showing that at this additive concentration the rates of CO_x formation from the additive and the CH_4 were the same.

An unusual isotope effect was observed when $^{12}C_2D_6$ was used (Ekstrom and Lapszewicz 1989a). As shown in Table 4-2, large $^{12}C_2D_6$–$^{12}C_2H_6$ effects were found in (i) the C_2H_6 and C_2H_4 formation from CH_4, (ii) the

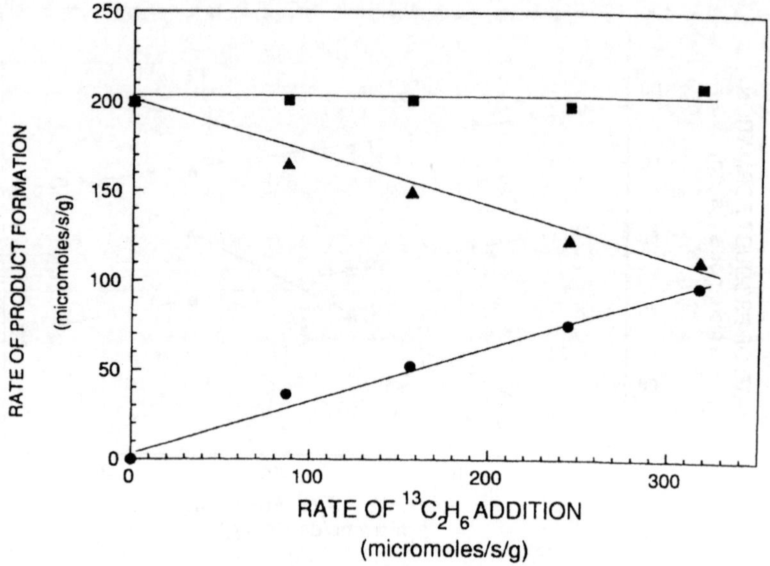

FIGURE 4-11. Effect of the addition of $^{13}C_2H_6$ on the rates of formation of (■) ($^{12}CO_x$ + $^{13}CO_x$) and (●) $^{13}CO_x$ from $^{13}C_2H_6$ and of (▲) $^{12}CO_x$ from $^{12}CH_4$.

Table 4-2 Comparison of the Effects of the Addition of He, $^{13}C_2H_6$, and $^{12}C_2D_6$ on the Oxidative Coupling Reaction[a]

	RATE OF FORMATION OR CONSUMPTION (in micromoles per second per gram of catalyst)				
Product	A1 ($^{12}CH_4$)	B1 ($^{12}CH_4$)	B2 ($^{13}C_2H_6$)	C1 ($^{13}CH_4$)	C2 ($^{12}C_2D_6$)
CO	87	64	45	38	14
CO_2	118	66	45	162	18
CO_x	205	130	96	201	35
C_2H_4	81	59	89	6	21
C_2H_6	101	93	—	24	—
$-CH_4$	571	435	—	261	—
$-C_2H_6$	—	—	138	—	40

[a]Column A1 summarizes the rates of formation of products derived from the CH_4 in the presence of He as the additive. Columns B1 and C1 contain the rates of the products formed from the CH_4 in the presence of added $^{13}C_2H_6$ and $^{12}C_2D_6$, whereas B2 and C2 contain the appropriate rates of the formation of products derived from these additives. Conditions: Sm_2O_3 catalyst, 700°C, 13.8 vol% O_2, 6.5 vol% (285 μmol s^{-1} g^{-1}) He, $^{13}C_2H_6$, or $^{12}C_2D_6$ addition.

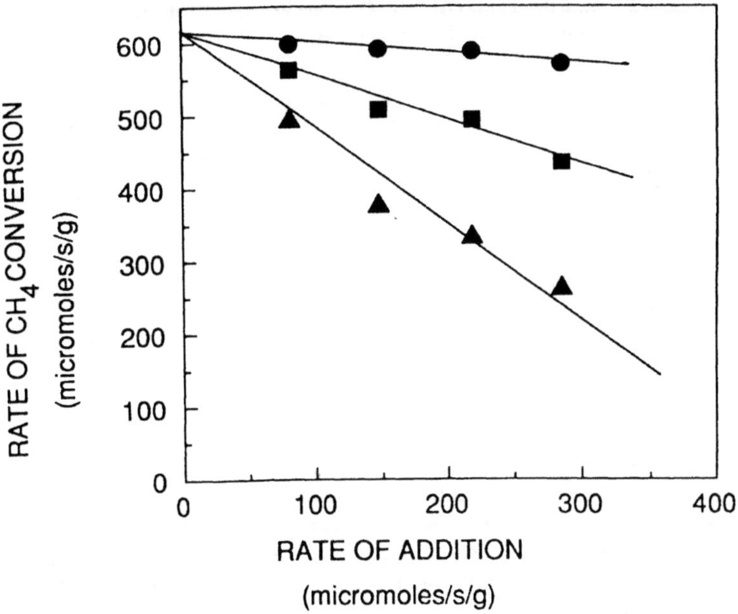

FIGURE 4-12. Effect of added He and C_2H_6 on the total CH_4 conversion: ●, $^{12}CH_4$/He; ■, $^{12}CH_4/^{13}C_2H_6$; ▲, $^{13}CH_4/^{12}C_2D_6$.

conversion of the added $^{12}C_2D_6$ and $^{13}C_2H_6$ to ethylene and CO_x, and (iii) the CH_4 and O_2 consumption. An important aspect of the results is that the total CO_x formation, that is, the sum of that derived from the CH_4 and the additive, is virtually constant for all experiments; the addition of the additives resulting only in a change in the origin of the CO_x products. This observation suggests that no significant change in the effective catalyst temperature took place on replacement of $^{13}C_2H_6$ by $^{12}C_2D_6$ and indicates that the formation of CO_x is determined by the availability of the oxidant, not the reductant (Ekstrom and Lapszewicz 1989b).

These are clearly not just simple primary isotope effects and are probably better interpreted as a reflection on a system of delicately balanced reaction rates whose minor disturbance results in disproportionately large effects in the final products. This is clearly suggested by the observation that although the overall D–H isotope effect on the slopes of the appropriate data in Fig. 4-12 is 2.3, the corresponding ratio in the C_2H_4 formation from CH_4 is almost 10, but is only 4 for the ratio of

Table 4-3 Reaction of the Oxidative Coupling Reaction Products with Steam over Sm_2O_3 Catalyst[a]

	COMPOSITION AT VARIOUS TEMPERATURES (μmol g^{-1} s^{-1})				
	INLET	EXIT			
Gas	—	700°C	800°C	900°C	1000°C
H_2	36.4	38.3	45.7	75.7	169.5
CO	15.3	12.3	10.7	13.3	34.5
CH_4	637	635	627	630	598
CO_2	15.3	18.0	20.4	21.6	23.1
C_2H_4	32.7	33.7	40.3	43.6	22.0
C_2H_6	32.3	32.0	26.2	2.9	1.6

[a] Flow of steam in the inlet gas is 150 μmol g^{-1} s^{-1}.

C_2H_4 derived from the additive. Whatever the origin of the effect, it appears that the results could complicate the analysis of isotope replacement experiments that have been interpreted (Buevskaya et al. 1987; Mimms et al. 1989; Cant et al. 1988) showing that the rate-determining step of the oxidative coupling reaction was the rupture of the C—H bond of the CH_4 molecule.

No $^{13}CH_4$ formation was detected during experiments with added $^{13}C_2H_6$, indicating that the homolytic dissociation of ethane followed by the reaction of the resulting methyl radicals with a hydrogen donor is not significant under the chosen experimental conditions. Trace amounts of acetylene and C_3 products, however, were present in the reactor effluent.

Importance of the Shift Reaction

The possible contribution of the water–gas shift reaction to the CO_x products is not usually considered, even though water is the single most important product of the reaction [equation (4-1)]. The importance of this reaction was explored by reacting a typical mix of reaction products with steam at various temperatures over a Sm_2O_3 catalyst. Not unexpectedly, the results showed (Table 4-3) that at temperatures above 700°C this reaction can make a major contribution to the total CO_x formation.

Comments

The interpretation of the preceding results is complicated by the existence of various concentration gradients of reactants and products along the

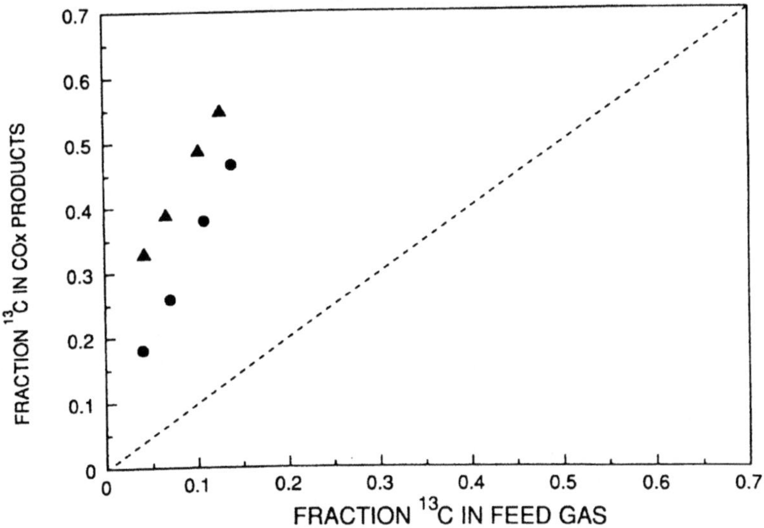

FIGURE 4-13. Effect of fraction of $^{13}C_2$ in the feed gas on the fraction of $^{13}CO_x$ in the reaction products. The broken line indicates the results expected if CH_4, C_2H_6, and C_2H_4 showed equal reactivity for conversion to CO_x; Sm_2O_3 catalyst, 700°C, gas flow rate 6000 cm^3 per minute per gram of catalyst, 10% O_2 in CH_4: ●, $^{13}C_2H_6$; ▲, $^{13}C_2H_4$ added. (Ekstrom, Campbell, and Lapszewicz 1989. Reproduced with permission of Elsevier Science Publishers Bv.)

catalyst bed axis. For example, although the system is almost differential with respect to methane, it is integral with respect to oxygen. The importance of the oxygen gradient is demonstrated by the common observation that the C_{2+} selectivity is poor when the reaction is carried out under differential conditions. It should also be noted that the concentrations of the labeled products decreases along the catalyst bed as they are consumed, but that the concentrations of the same species derived from methane increases until their rate of consumption becomes equal to the rate of formation (Ekstrom, Campbell, and Lapszewicz 1989). As a result, the labeled and nonlabeled species not only show different gradients in the reactor, but also experience different oxygen gradients. The detailed interpretation of such results is clearly not trivial.

In all experiments, the oxidation of added $^{13}C_2H_6$ or $^{13}C_2H_4$ made an increasing contribution to the total CO_x formation as the concentrations of these species in the feed increased (Fig. 4-13). The high reactivity of the C_2 products in this system highlights one of the major problems facing the

Table 4-4 Comparison of Catalyst Activity and Lattice Oxygen Exchange Rates for Cubic and Monoclinic Sm_2O_3 at 850°C

	Cubic	Monoclinic
CH_4 conversion rate (μmol s^{-1} g^{-1})	102	6
Fraction of lattice oxygen exchanged during 30-s $^{18}O_2$ pulse	0.6	0.09

commercial development of this reaction (Kuo 1987; Field, Niruta, and McCarthy 1987; Gray and Tomlinson 1988; Fox, Chen, and Degen 1990; Jackson, Thompson, and Whitehead 1989). As shown in Table 4-4, the conversions of added ethane and ethylene were 41 and 39%, respectively. At the same additive concentrations, 15% of ethane and 39% of ethylene were converted to CO_x, compared to 6% of CH_4, present at 10 times the concentration. The greater reactivity of the C_2 products is clearly evident, as is the greater selectivity to CO_x of the reaction of added C_2H_4. However, the CO_x selectivity of the C_2H_6 reaction is comparable to that of CH_4, the major product being C_2H_4. If it is assumed that the disappearance of the hydrocarbons obeys pseudo-first-order kinetics, the difference in the overall rate constants for the conversion of CH_4 and C_2H_4 to CO_x is a factor of 6.5.

There are two other major features of these results. First, the absence of any significant contribution from reaction (4-10) shows that the decrease in the net $^{12}CH_4$ conversion observed on addition of $^{13}C_2H_6$ and $^{13}C_2H_4$ must be the result of competition between CH_4 and the additives for the active sites of the catalyst. This is an entirely expected result; it would be indeed remarkable if these simple catalysts with a low surface area and almost no internal structure showed any discrimination between CH_4 and C_2H_6.

Second, the data in Fig. 4-11 in particular show that as the $^{13}C_2H_6$ in the feed was increased, the total CO_x formation was constant, although it was increasingly derived from the labeled material. As excess oxygen was present and the total oxygen consumption did not vary significantly, this seems to indicate that the formation of the CO_x products is controlled by the availability of the oxidant, not the reductant. It also shows that the oxidant is *not* molecular oxygen in the gas phase. In retrospect, this is quite logical. If it were not so, the carbon oxides would be the sole reaction products.

Although the isotope tracer studies described previously were obtained only with a Sm_2O_3 catalyst, there is now increasing evidence that the results obtained are a general phenomenon. Morales and Lunsford (Morales and Lunsford 1989) found that a Li/MgO catalyst effectively promoted the formation of C_2H_5 radicals from ethane. Similarly, Roos (Roos 1989) showed that, for the same catalyst, the CO_x products were predominantly derived from the C_2 products.

Our results are interpreted in terms of a reaction mechanism [equation (4-13)] in which CH_4 and the C_2 products compete for the catalyst active sites, which may be $[O^-]$ or similar species formed by reaction of anion vacancies on the catalyst surface with O_2.

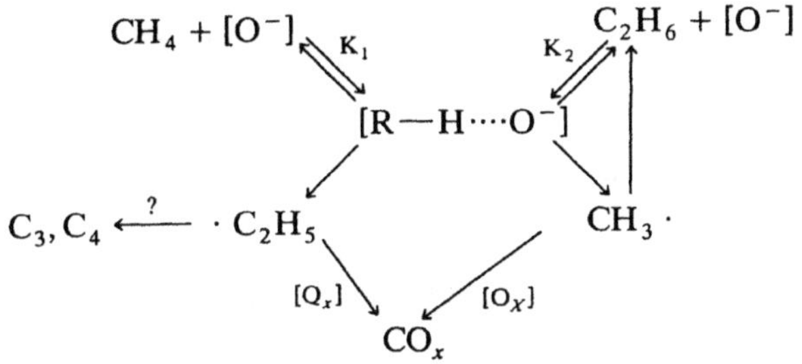

(4-13)

In this scheme, methane is converted to $CH_3\cdot$ by reaction with $[O^-]$. The $CH_3\cdot$ radical can either dimerize to C_2H_6 or be oxidized. The C_2H_6 formed can in turn also react with $[O^-]$ to form $C_2H_5\cdot$ radicals. The conversion of $C_2H_5\cdot$ to CO_x includes the intermediate formation of C_2H_4, which, from the results in Table 4-1, is oxidized faster than C_2H_6. A feature of this mechanism is that the formation of the CO_x products occurs from the C_{2+} products via the same active sites involved in the formation of the methyl radicals. As the C_{2+} concentration is increased, a steady-state C_{2+} concentration will be reached at which the rate of formation will be equal to its rate of conversion to CO_x and other products. It is important to appreciate that this mechanism implies that the origin of the CO_x may be derived largely from a C_1 species such as $CH_3\cdot$ only at low conversions. As the conversion is increased, the CO_x is increasingly derived from the C_2 products, particularly ethylene.

These considerations suggest that the similarity in the conversion–selectivity curves observed for so many catalysts is simply a reflection on the attainment of a balance between the rate of C_{2+} formation from CH_4 and the rate of their conversion to CO_x. The maximum C_{2+} concentration that can be achieved is comparable for so many catalysts because it is ultimately determined by the relative differences in catalyst activity toward CH_4, C_2H_6, and C_2H_4. It cannot be expected that the simple, low-surface-area catalysts used to date for this reaction would show large differences for the activity ratios of these very similar molecules.

The nature of the oxidizing species is not known at present. It appears from the present work that the formation of the CO_x products is determined by the availability of the oxidant, not the reductant. This suggests that the oxidizing species [B] that leads to the formation of the CO_x is formed from oxygen by some slow reaction and is present at very low concentrations, either on the catalyst or in the gas phase [equation (4-14)]. Once this species is formed, it reacts rapidly to form the carbon oxide products.

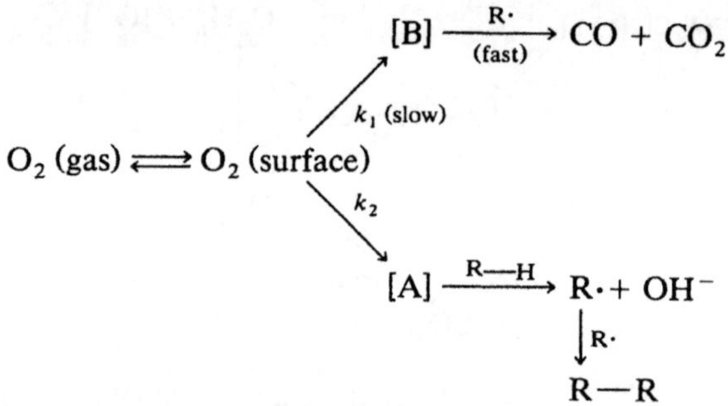

(4-14)

The other form of oxidant [A] is the precursor to the radicals and is most probably $[O^-]$ or some similar species. It is interesting that the C_{2+} selectivity is in many instances almost inversely proportional to the oxygen concentration. Such an effect would be observed if the oxidizing species is either molecular or atomic oxygen on the catalyst surface.

THE ROLE OF THE LATTICE OXYGEN

It has been known for some years that the lattice oxygen atoms in most metal oxides undergo a rapid exchange with gas-phase oxygen above 200–300°C (Winter 1968, 1969). This exchange is believed to involve anion vacancies in the oxide lattice, and the work described in this section was undertaken to explore the importance of this exchange in the oxidative coupling reaction (Ekstrom and Lapszewicz 1989a, 1989b). Sm_2O_3 was chosen as catalyst because it does not form a stable carbonate at the temperatures used for the oxidative coupling reaction. An isotope switching apparatus similar to that described previously was used for this work.

Addition of $^{18}O_2$

When $^{18}O_2$ replaced $^{16}O_2$ in helium over a Sm_2O_3 catalyst in the absence of CH_4, a slow loss of $^{16}O_2$, and a slow rise of the $^{18}O_2$ was observed (Fig.

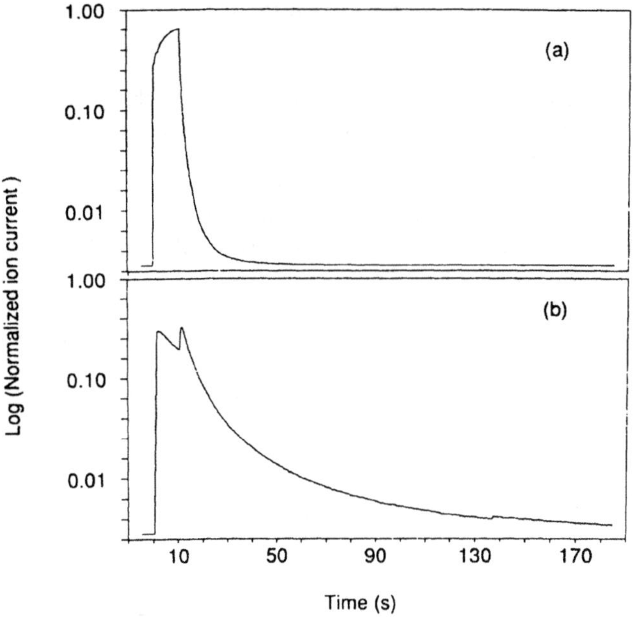

FIGURE 4-14. Transients observed when $^{18}O_2$ replaced $^{16}O_2$ flowing over a Sm_2O_3 catalyst. The switch to $^{18}O_2$ was made at "zero" time; 700°C, flow rate 78 ml s^{-1} g^{-1} of 10% O_2 in helium: (a) $m/z = 36$, $^{18}O_2$; (b) $m/z = 34$, $O\,^{18}O$. (Ekstrom and Lapszewicz 1989b. Reproduced with permission of the American Chemical Society.)

4-14). Together with the long time required to return the $^{18}O_2$ and $^{18}O\,^{16}O$ signals to their background values, these results clearly indicate the uptake of large amounts of $^{18}O_2$ during the pulse by an oxygen pool present in the catalyst. A lower limit estimate of 1×10^{21} atoms per gram is made for the number of oxygen atoms exchanged during the $^{18}O_2$ pulse. This value is comparable to the total number of oxygen atoms (5×10^{21} atoms per gram) present in Sm_2O_3 and is consistent with data by Winter (Winter 1968, 1969), who found that 1×10^{20} oxygen atoms per gram are available for exchange in Sm_2O_3 at 400°C. It is also noticeable that as the $^{16}O_2$ is replaced, the $^{16}O\,^{18}O$ formation slows as expected. Immediately after the return to $^{16}O_2$, a sharp spike in the $^{16}O\,^{18}O$ formation rate occurs, reflecting the exchange that has taken place in the mixed boundary between the two isotopic forms. As decay of the $^{18}O\,^{16}O$ formation rate continues long after $^{18}O_2$ has been replaced, the exchange must occur, at least partially, on the catalyst surface. The obvious differences in the decay rates of $^{18}O_2$ and $^{18}O\,^{16}O$ following the end of the $^{18}O_2$ pulse reflect the relative abundances of ^{18}O and ^{16}O on the catalyst surface. Very similar results have recently been reported by Peil and co-workers (Peil, Goodwin, and Marcelin 1989) using a Li/MgO catalyst.

When these experiments were repeated under synthesis conditions, that is, in the presence of methane, the ^{18}O label appeared almost exclusively in the CO_x products. It was also possible to demonstrate that the presence of Li_2CO_3 on the catalyst surface decreased the rate of the gas–lattice-oxygen exchange by about an order of magnitude. As an effect of similar magnitude was observed on the catalyst activity, the results suggest a link between catalyst activity and the rate of gas–lattice-oxygen exchange. This link is also suggested by the results in Table 4-4, which compares the conversion rates and oxygen exchange rates of cubic and monoclinic Sm_2O_3 (Bhargava and Lapszewicz 1990).

These results prompted the question, what role do the exchangeable oxygen atoms in the catalyst lattice play in the formation of carbon dioxide

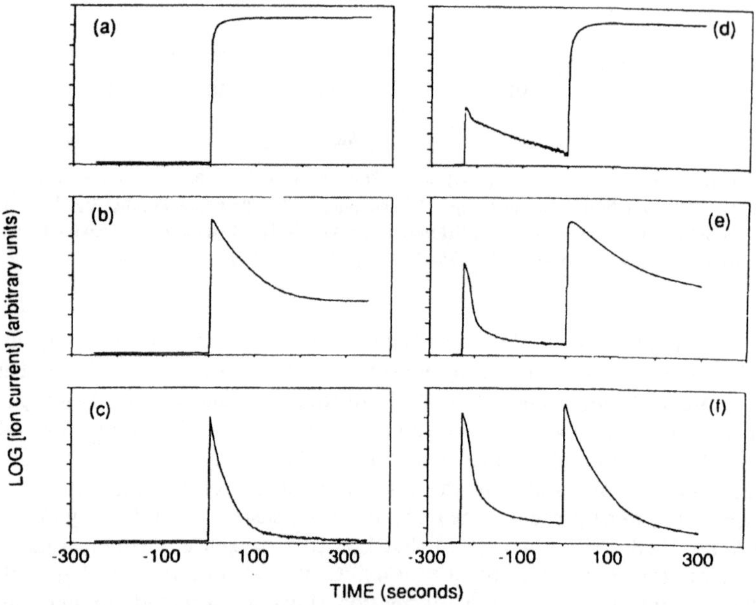

FIGURE 4-15. Transients observed when CH_4 and CH_4–O_2 mixture were admitted (a)–(c) to the Sm_2O_3 catalysts and (d)–(f) to the Pr_6O_{11} catalysts. At a time of −230 s, CH_4 was admitted to the catalysts after the remaining O_2 has been purged with He. At time zero, an O_2 flow was added to the CH_4, to give the usual 10% mix: (a) and (d) $m/z = 44$, CO_2; (b) and (e) $m/z = 46$, $CO^{18}O$; (c) and (f) $m/z = 48$, $C^{18}O_2$. (Ekstrom and Lapszewicz 1989b. Reproduced with permission of the American Chemical Society.)

and, therefore, in the determination of the C_2 selectivity? An attempt to answer this was made by first substantially (76%) exchanging the lattice oxygen atoms in Sm_2O_3. CH_4 was then admitted to the catalysts and the formation of the isotopic forms of CO_2 was monitored. For the Sm_2O_3 catalyst (Fig. 4-15a–4-15c), no reaction was observed, indicating that the lattice oxygen atoms played no role in the formation of the reaction products in the absence of molecular oxygen. Addition of $^{16}O_2$ to the CH_4 immediately led to the formation of ^{18}O labeled CO_2.

A very different result was obtained for the variable-oxidation-state catalyst Pr_6O_{11}. For this catalyst 107% (7×10^{21} atoms per gram) of the oxygen atoms were exchanged, and admission of the CH_4 led to a sharp pulse of ^{18}O-labeled CO_2 (Fig. 4-15d–4-15f). Traces of C_2H_6 also were found whose relative contribution increased with the time elapsed, the CO_2–C_2H_6 ratio decreasing from approximately 35 immediately after the admission of CH_4 to 0.3 after 85 s. It appears that, for this catalyst, the lattice oxygen can oxidize CH_4 in the absence of molecular oxygen.

Admission of $^{16}O_2$ to the methane flowing over the Pr_6O_{11} catalyst led to the formation of the expected ^{18}O-labeled CO_2. The ratio of the ^{18}O in CO_2 formed when methane only was present to that formed in the presence of oxygen was 0.2. This does not necessarily reflect the true extent of the involvement of lattice oxygen atoms in the reaction, because the formation of H_2O and CO, whose quantitative analysis by mass spectrometry is not simple, would also consume lattice oxygen atoms.

Comments

The present results confirm that a rapid exchange between the gas-phase and lattice oxygen atoms takes place on the working catalyst. According to Winter (Winter 1968, 1969), the mechanism of this reaction [shown schematically in equations (4.15a)–(4.15c), in which $[^-]$ represents an anion vacancy, $[O^-]$ an oxygen atom trapped in the vacancy, and {—O—} a lattice oxygen] is initiated by the rapid formation of the $[O^-]$ centers.

$$^{18}O_{2\,(gas)} + (\text{surface}) \Leftrightarrow {}^{18}O_{2\,(surface)} \quad (4\text{-}15a)$$

$$^{18}O_{2\,(surface)} + 2[^-] \Leftrightarrow 2[^{18}O^-] \quad (4\text{-}15b)$$

$$[^{18}O^-] + \{—^{16}O—\}_{lattice} \Leftrightarrow \{—^{18}O—\}_{lattice} + [^{16}O^-] \quad (4\text{-}15c)$$

The formation of these species requires no activation energy, the rate-determining step being the desorption of molecular oxygen from the surface, that is, the reverse of reaction (4-15a).

The importance of the exchange reaction as a means of forming [O^-] in the partial oxidation reaction may be judged from the following. First, the [O^-] species on oxide surfaces are known to be highly reactive toward C—H bond cleavage, even at temperatures much below those required for the partial oxidation reaction (Liu, Iwamoto, and Lunsford 1982; Aika and Lunsford 1977; Aika and Lunsford 1978; Ward, Lin, and Lunsford 1977). Second, there appears to be a correlation between the rates of the exchange reaction and the overall rates of the partial oxidation. For example, at 750°C, extrapolation of Winter's data gives a rate for the exchange reaction of 2×10^{19} molecules per second per gram, (assuming a surface area of 10 m^2 g^{-1}), whereas the net rate of CH_4 conversion at this temperature is in the range 10^{19}–10^{20} molecules per second per gram. The correlation observed in this work between the exchange rate and catalyst activity following addition of Li_2CO_3 to the Sm_2O_3, and the comparison of cubic and monoclinic Sm_2O_3 also suggest a link between the rate of exchange and the rate of reaction.

On the basis of the foregoing, the reaction mechanism shown in equation (4-16) was proposed (Ekstrom and Lapszewicz 1989a, 1989b), in which the reactive species is [O^-].

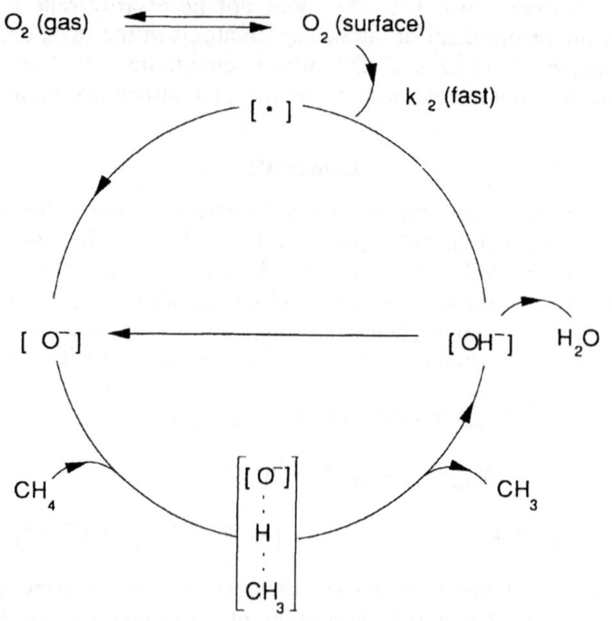

(4-16)

The reaction sequence is very similar to those suggested by Lin and co-workers (Lin et al. 1986; Liu, Wang, and Lunsford 1988) and others

(Hutchings, Scurrell, and Woodhouse 1987; Otsuka, Jinno, and Morikawa 1986), but includes Winter's mechanism for the exchange of lattice oxygen atoms to regenerate the $[O^-]$ species. The inclusion of this reaction readily accounts for the observation that C_2 formation ceases immediately after termination of the oxygen flow. The $[OH^-]$ formed as a result of the C—H bond scission may disproportionate rapidly to $[O^-]$ and the anion vacancy $[^-]$, as suggested by Lin and co-workers (Lin et al. 1986; Lin, Wang, and Lunsford 1988) for Li/MgO and Na/CaO.

EFFECT OF PRESSURE

Blank Studies

Studies undertaken to date on the effect of pressure on the reaction all suggest that increasing pressure is deleterious for the C_{2+} selectivity (Hutchings, Scurrell, and Woodhouse 1988; Asami et al. 1988; Lane and Wolf 1988; Yates and Zlotin 1988; Martin et al. 1988; Labinger 1988). It is surprising, in view of the importance of the effects of pressure for the commercialization of the reaction, that no studies on the effects of pressure under conditions of practical interest (high temperatures, high oxygen concentrations, and high per-pass CH_4 conversions) appear to have been reported for the cofeed reaction.

In one of the most comprehensive studies, Lane and Wolf (Lane and Wolf 1988) found that the CH_4 conversion for the blank reaction was given by equation (4-17), whereas Roos (Roos 1989) found that the rate of the Li/MgO-catalyzed rate was given by equation (4-18).

$$\mathrm{R}_{(\mathrm{CH}_4)} = k_{\mathrm{i}}(\mathrm{P}_{\mathrm{O}_2})^{2.05}(\mathrm{P}_{\mathrm{CH}_4})^{1.04} \tag{4-17}$$

$$R_{(\mathrm{CH}_4)} = k_{ii}(P_{\mathrm{O}_2})P_{\mathrm{CH}_4}/(1 + bP_{\mathrm{CH}_4}) \tag{4-18}$$

A similar rate law has been reported recently for a Sm_2O_3 catalyst (Amorebieta and Colussi 1988). The different oxygen and methane dependencies of these two rate laws suggest that the contribution of the uncatalyzed reaction will become progressively more important compared to the catalyzed rate as the partial pressures of oxygen and methane are increased.

This effect appears to have been observed in our study of the effects of pressure on this reaction (Ekstrom, Regtop, and Bhargava 1990). We carried out this work using conditions required for high CH_4 conversions (high O_2 concentration and high temperatures). At the higher pressure, the reaction is characterized by a substantial blank or empty reactor, reaction (Fig. 4-16). Similar results were obtained using SiC as an inert

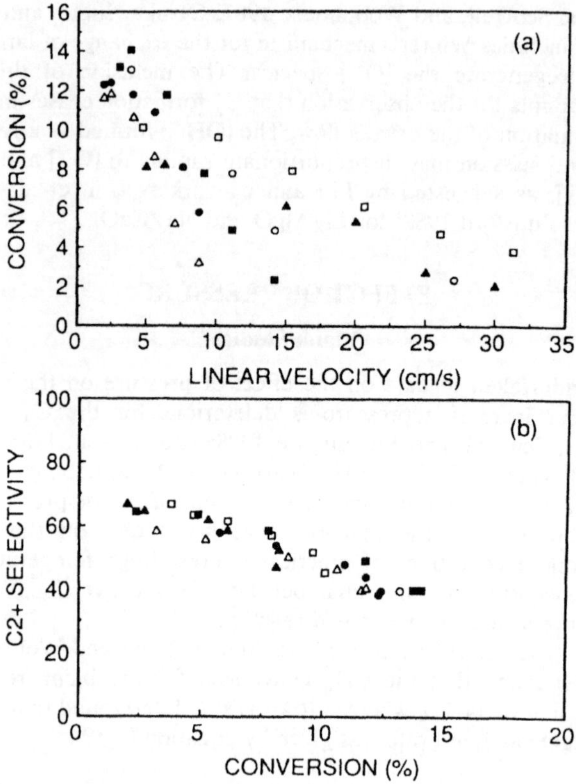

FIGURE 4-16. (a) Effect of linear velocity on the blank reaction: ○, 850°C, small tube; △, 800°C, large tube; ●, 850°C, large tube; □ 900°C; large tube with SiC; ▲, 850°C, large tube with SiC. (b) Data as before but plotted as conversion versus C_{2+} selectivity. Note that the conversion decreased as a consequence of increasing linear velocity. 0.5 MPa pressure, 10% O_2 in CH_4.

solid. An approximately threefold increase in conversion rate was observed when the pressure was increased from 0.1 to 0.5 MPa. The data also show the expected increase in conversion with increasing temperature. At the higher pressure, the conversions in the large- and small-diameter reactors were similar when compared at a fixed linear gas velocity (Fig. 4-16a). At low linear velocity (high conversion) the reaction was controlled by the oxygen concentration.

When plotted as the more familiar conversion versus C_{2+} selectivity curves, the data show surprisingly high C_{2+} selectivities of 60–70% at low

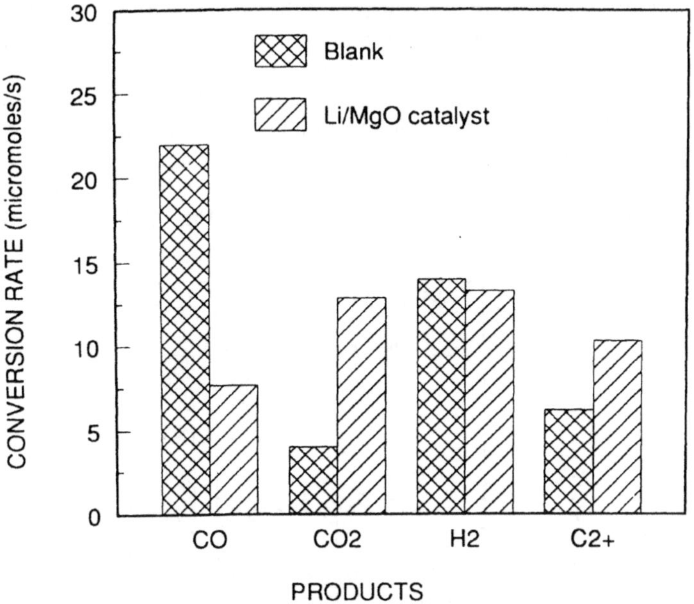

FIGURE 4-17. Comparison of the product distribution from oxidative coupling in the absence and presence of Li/MgO catalyst. The linear velocity was 15 cm s^{-1}, 15% O_2 in CH_4, 0.4 MPa at 850°C. (Ekstrom, Regtop, and Bhargava 1990. Reproduced with permission of Elsevier Science Publishers Bv.)

(< 5%) conversions (Fig. 4-16b). These results are comparable to many reported at atmospheric pressure using various catalysts (see Lane and Wolf 1988 for a summary of recent conversion–selectivity results). It is interesting to note that when plotted in this way all data fall on a common curve, suggesting that the primary factor influencing the C_{2+} selectivity is the conversion (or, perhaps more correctly, the C_{2+} concentration), irrespective of how this is obtained. Our data were in excellent agreement with those of Lane and Wolf (Lane and Wolf 1988) although the latter were obtained at lower total pressures.

Catalyzed Studies

The effect of adding a Li/MgO catalyst is illustrated in Fig. 4-17. The product distributions obtained with and without the catalyst show that under these conditions the catalyst does little more than convert CO to

CO_2. This result is in agreement with previous studies. (Hutchings, Scurrell, and Woodhouse 1988). Similar effects also have been observed for experiments at atmospheric pressure and low space velocities, where the blank reaction is important (Lane and Wolf 1988). It is clear from these data that the true catalytic reaction can be determined only at linear velocities high enough to eliminate the blank reaction.

Such data were obtained using Sm_2O_3 and Sr/Sm_2O_3 catalysts and linear velocities of 130 cm s^{-1}. As shown in Fig. 4-18 for the Sm_2O_3

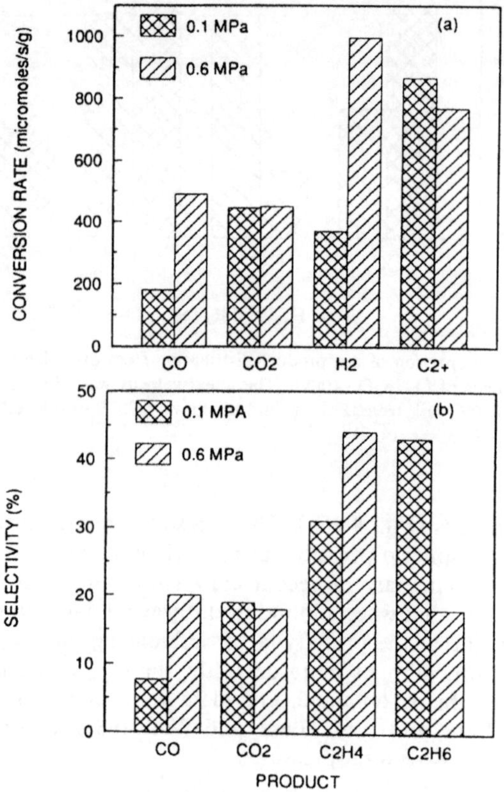

FIGURE 4-18. Comparison of the effects of temperature on the conversion and selectivity from (a) Sm_2O_3 and (b) Sr/Sm_2O_3 catalysts at 0.1 and 0.6 MPa. The flow rate was 15 ml s^{-1}, and 10% O_2 in CH_4 was used for all runs: ●, 0.1 MPa; ■, 0.6 MPa. Nominal reaction temperatures (in degrees Celsius) for selected runs are indicated by the bracketed numbers. (Ekstrom, Regtop, and Bhargava 1990. Reproduced with permission of Elsevier Science Publishers Bv.)

catalyst, increasing the pressure appeared to increase significantly the conversion and the selectivity achievable at the lowest reaction temperature of 570°C. For a given conversion, the C_{2+} selectivity was consistently lower at the higher pressure. Operation of the system under condition of constant residence time (constant linear velocity) did not result in any significant improvement, consistent with the virtual elimination of the blank reaction at these very high linear velocities.

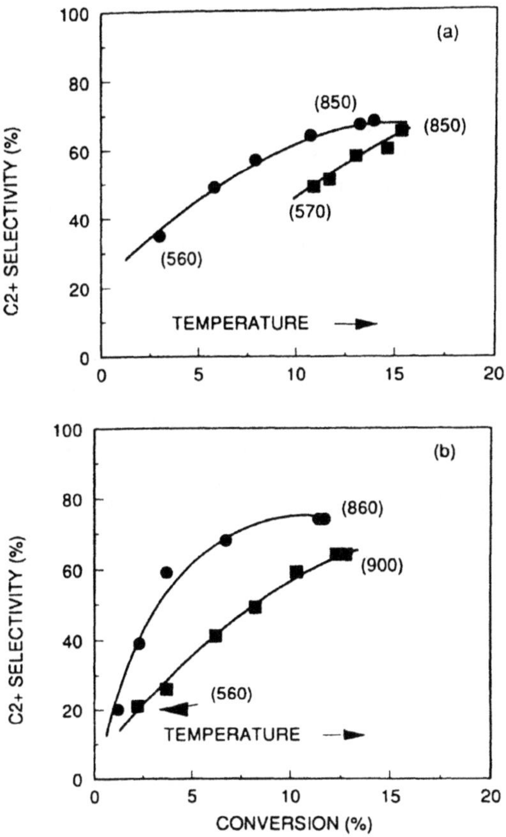

FIGURE 4-19. Comparison of the product distribution obtained with the Sr/Sm_2O_3 catalyst at 0.1 (860°C) and 0.6 MPa (900°C) pressure. See Fig. 4-18b for experimental details: (a) Rates of formation of reaction products; (b) carbon product selectivities. (Ekstrom, Regtop, and Bhargava 1990. Reproduced with permission of Elsevier Science Publishers Bv.)

Addition of Sr to the Sm_2O_3 catalysts significantly improved the C_{2+} selectivity at 0.1 MPa in agreement with previous observations and the generally beneficial effects of the alkali and alkaline earth elements reported for this reaction (Deboy and Hicks 1988a–988c). However, the effect of pressure on the conversion was much less pronounced than for Sm_2O_3, and increasing the reaction pressure from 0.1 to 0.6 MPa substantially reduced the C_{2+} selectivity at any given conversion (Fig. 4-19). The effect can be partly compensated for by increasing the reaction temperature, but even at 900°C the selectivity from Sr/Sm_2O_3 was reduced from 78% at 0.1 MPa to 63% at 0.6 MPa. The reduction in the C_{2+} selectivity obtained with the Sr/Sm_2O_3 catalyst at the high pressure (860–900°C) was again almost entirely due to the increased contribution of CO to the product spectrum.

It should be noted that at these high temperatures the effect of the catalyst and pressure on the steam reforming reaction is not known. It is interesting that the reduction in C_{2+} selectivity is largely due to the increase in the formation of CO. This, together with H_2, is of course the expected product of the steam reforming reaction.

Comments

It appears that in this reaction two quite different routes (homogeneous and surface-catalyzed) are available for the formation of one of the primary products, that is, methyl radicals. The function of the catalyst is therefore not so much to enhance the rate of the homogeneous reaction [equation (4-19)], but to introduce an altogether new reaction path [equation (4-20)], which fortuitously also results in the formation of methyl radicals.

$$CH_4 + O_2 \rightarrow CH_3\cdot + HO_2\cdot \qquad \text{(homogeneous)} \qquad (4\text{-}19)$$

$$CH_4 + O^-_{(surface)} \rightarrow CH_3\cdot + OH^- \qquad \text{(surface)} \qquad (4\text{-}20)$$

It should be noted (i) that the coproducts for the primary catalyzed and blank reactions are different, being $HO_2\cdot$ for the blank and OH^- for the catalyzed reaction (Ito et al. 1985), and (ii) that the blank reaction is not the uncatalyzed rate for reaction (4-19). The ultimate product distribution from the two reactions is similar because once the methyl radicals have been formed, their reaction rate is determined by the same factors, such as, for example, the oxygen and methyl radical concentrations. However, in the catalyzed reaction, the local methyl radical concentration may be much higher, which in turn may favor the formation of C_2 products. This could arise because the C_2 formation is second-order with respect to the methyl radical concentration, whereas the formation of CO_x should show a first-order dependence.

The true situation for the catalyzed reaction is more complicated than suggested by the preceding discussion, because the study was carried out with a reactor operating in the integral mode. Under such conditions, the partial poisoning of the catalyst by carbon dioxide will also contribute to lowering the catalyst activity. From other work, carbon dioxide is known to poison these strongly basic catalysts effectively (Roos 1989; Amorebieta and Culossi 1989). This latter conclusion is also supported by the observation that the Sr/Sm_2O_3 catalysts, which at 0.6 MPa are ineffective at low linear velocities, show a higher activity as the linear velocity is increased. At the high linear velocities, the partial pressures of CO_2 will be significantly lower.

At sufficiently high linear velocities, the blank reaction can be essentially eliminated and the effect of pressure on the activity and selectivity of active catalysts can be determined. From the present results, it appears that the effect is complex. For Sm_2O_3, the net CH_4 conversion increased, but the selectivity to C_{2+} products decreased with increasing pressure. This catalyst does not form a stable carbonate at the high reaction temperatures (Ekstrom and Lapszewicz 1989a, 1989b). By contrast, the strontium-doped Sm_2O_3 catalyst, for which the strontium will remain in the carbonate form at the reaction temperatures (Korf et al. 1988), showed only a small increase in activity, but a substantial reduction in selectivity.

The origin of these effects is not clear. However, it is clear from numerous studies that as the oxygen partial pressure is increased the C_{2+} selectivity decreases. This effect could be due to a combination of an increasing amount of oxidant on the catalyst surface and an increase in the rates of the gas-phase oxidation reactions. Some results were presented previously that suggest that gas-phase molecular oxygen is not the primary oxidant. It seems plausible, therefore, that the CO_x selectivity is related to the amount of O_2 present on the catalyst surface, as either molecular oxygen or, perhaps, oxygen atoms. Consequently, as the amount of oxygen on the surface increases with the increasing O_2 partial pressure in the gas phase, the CO_x formation will also become relatively more important.

SUMMARY

The study of this reaction has seen a remarkable development during the last five years. Much of this effort has been related to the search for better catalyst, motivated by the hope of finding catalysts of sufficient selectivity and activity to enhance the commercial prospects of the reaction. Progress on the mechanism of the reaction and the role of the catalyst has also been rapid. Although there is still room for surprises, in our view the major mechanistic features of this reaction are as follows:

1. The primary function of the catalyst is to form methyl radicals, which dimerize in the gas phase to form ethane. This can dehydrogenate, either on the surface or in the gas phase, to form ethylene.
2. A homogeneous (noncatalyzed) reaction path is available, which results in a product distribution very similar to that obtained with many catalysts. It is likely that this reaction path also involves methyl radicals, in this case formed by a homogeneous or perhaps wall-initiated reaction.
3. Depending on the experimental conditions, the catalyzed route can be only marginally faster than the blank reaction.
4. The CO_x formed from the C_2 products becomes relatively more important as the C_{2+} concentration in the system increases, and at C_2 concentrations of above approximately 10 carbon% the C_2 products become the major source of the CO_x products.
5. CH_4 and the C_2 products compete with each other for the active sites to form the corresponding radicals. As the C_2 product concentration increases, the CH_4 activation is effectively inhibited.
6. Under these conditions, the apparent limit in the C_{2+} concentration attained in the reaction is the result of the establishment of a kinetic steady state between the rates of C_2 formation from methane and its conversion to carbon oxides. The C_{2+} concentrations achieved under these conditions are thus not a function of the inherent properties of the catalyst, but of the relative rate constants controlling the reactions.

REMAINING PROBLEMS

Despite the many advances that have been made in the study of this reaction during the last five years or so, there are also aspects that remain obscure. Some of the more obvious are the following:

1. The details of the reactions leading to the formation of the CO_x products at low methane conversions are obscure at present. Surface as well as homogeneous gas-phase reactions may be involved, their relative contributions depending on the experimental conditions.
2. The question of what proportion of the total CO_x is derived from a C_1 species is also not resolved. The tracer experiments described previously do not distinguish between the $^{12}CO_x$ formed from the

$^{12}C_{2+}$ products, in turn derived from the $^{12}CH_4$, and the equivalent products derived directly from a $^{12}C_1$ species such as CH_4 or $CH_3\cdot$. This is a problem of some importance, as its elucidation could finally determine the maximum inherent selectivity of the catalysts.

3. It has been well established that the addition of the alkaline-earth elements to most catalysts improves their selectivity, although it often also decreases their activity. The mechanism by which this occurs is not understood. Possibilities include a modification of the catalyst surface, a modification of the ability of the catalyst to promote steam-reforming of hydrocarbons (if indeed this is a significant source of carbon oxide products), or the selective poisoning of catalyst sites responsible for the formation of CO_x. It should be noted that at the reaction temperatures, the alkaline-earth oxides and hydroxides have a significant vapor pressure, and it is even possible that the effect is due to the inhibition of gas-phase reactions.

ACKNOWLEDGEMENTS

The outstanding efforts of Ian Campbell, Robert Regtop, Sureyya Saricilar, George Kastl, and particularly Jacek Lapszewicz in the experimental program described in this chapter are acknowledged.

References

Aika, K., and J.H. Lunsford. 1977. *J. Phys. Chem.* 81:1393.

Aika, K., and J.H. Lunsford. 1978. *J. Phys. Chem.* 82:1794.

Amorebieta, V.T., and J. Colussi. 1989. *J. Phys. Chem.* 93:5155.

Asami, K., K. Omata, K. Fujimoto, and H. Tominaga. 1988. *Energy and Fuels* 2:574.

Benson, J.E., H.W. Kohn, and M. Boudart. 1966. *J. Catal.* 5:307.

Bhargava, S., and J.A. Lapszewicz. 1990. Unpublished.

Buevskaya, O.V., A.I. Suleimanov, S.M. Aliev, and V.D. Sokolovskii. 1987. *React. Kinet. Catal. Lett.* 33:223.

Cant, N.W., C.A. Lukey, P.F. Nelson, and R. Tyler. 1988. *J. Chem. Soc., Chem. Commun.* 766.

Deboy, J.M., and R.F. Hicks. 1988a. *Ind. Eng. Chem. Res.* 27:1577.

Deboy, J.M., and R.F. Hicks. 1988b. *J. Catal.* 113:515.

Deboy, J.M., and R.F. Hicks. 1988c. *J. Chem. Soc., Chem. Commun.* 982.

Driscoll, D.J., K.D. Campbell, and J.H. Lunsford. 1987. *Adv. Catal.* 35:139.

Edwards, J.H., K.T. Do, and R.Y. Tyler. 1991. The OXCO process: The direct conversion of natural gas to olefins and liquid fuels using fluidized-bed technology. In *Direct Methane Conversion by Oxidative Processes: Fundamental and*

Engineering Aspects, ed. E.E. Wolf, Chapter 13. New York: Van Nostrand Reinhold.
Ekstrom, A., G.E. Batley, and D.A. Johnson. 1974. *J. Catal.* 34:106.
Ekstrom, A., I. Campbell, and J.A. Lapszewicz. 1989. *Appl. Catal.* 56:29.
Ekstrom, A., I. Campbell, and S. Saricilar. 1990. Unpublished.
Ekstrom, A., I. Campbell, S. Saricilar, and R. Regtop. 1990. Unpublished.
Ekstrom, A., G. Kastl, and I. Campbell. 1990. Unpublished.
Ekstrom, A., and J.A. Lapszewicz. 1989a. *J. Am. Chem. Soc.* 111:8515.
Ekstrom, A., and J.A. Lapszewicz. 1989b. *J. Phys. Chem.* 93:5230.
Ekstrom, A., and J.A. Lapszewicz. 1990. Unpublished.
Ekstrom, A., R. Regtop, and S. Bhargava. 1990. *Appl. Catal.* 62:253.
Field, S., S.C. Niruta, and J.G. McCarthy. 1987. SRI International Project Report. An assessment of the catalytic conversion of natural gas to liquids. 6:9129-01-SQ.
Fox, J.M., T.-P. Chen, and B.D. Degen. 1990. *Chem. Eng. Progress* 86(4):42.
Gray, D., and G. Tomlinson. 1988. Direct methane conversion—An assessment. The MITRE Corporation Report No. SAND-88-7110.
Hutchings, G.H., M.S. Scurrell, and J.R. Woodhouse. 1987. *J. Chem. Soc., Chem. Commun.* 1388.
Hutchings, G.H., M.S. Scurrell, and J.R. Woodhouse. 1988. *J. Chem. Soc., Chem. Commun.* 253.
Hutchings, G.J., M.S. Scurrell, and J.R. Woodhouse. 1989. *Chem. Soc. Rev.* 18:251.
Ito, T., J.-X. Wang, C.H. Lin, and J.H. Lunsford. 1985. *J. Am. Chem. Soc.* 107:5062.
Jackson, P.J., S.C. Thompson, and I.G. Whitehead. 1989. Proceedings of Seventh Australasian Chemical Engineering Conference (CHEMECA), August 23–25; p. 719.
Jones, C.A., J.J. Leonard, and J.A. Sofranko. 1987. *Energy and Fuels* 1:12.
Keller, G.E., and M. Bhasin. 1982. *J. Catal.* 73:9.
Korf, S.J., J.A. Roos, N.A. de Bruin, J.G. van Omen, and J.R.H. Ross. 1988. *Catal. Today* 2:535.
Kuo, J.C.W. 1987. Evaluation of direct methane conversion to higher hydrocarbons and oxygenates. DOE Report DOE/PC/90009-3.
Labinger, J.A. 1988. *Catal. Lett.* 1:371.
Labinger, J.A., and K.C. Ott. 1987. *J. Phys. Chem.* 91:2682.
Lane, G.S., and E.E. Wolf. 1988. *J. Catal.* 113:144.
Lee, J.S., and S.T. Oyama. 1988. *Catal. Rev.—Sci. Eng.* 30(2):249.
Lin, C.-H., K.D. Campbell, J.-X. Wang, and J.H. Lunsford. 1986. *J. Phys. Chem.* 90:534.
Lin, C.-H., J.-X. Wang, and J.H. Lunsford. 1988. *J. Catal.* 111:302.
Liu, R.-S., M. Iwamoto, and J.H. Lunsford. 1982. *J. Chem. Soc., Chem. Commun.* 78.
Martin, G.A., A. Bates, V. Ducarme, and C. Mirodatos. 1989. *Appl. Catal.* 47:287.
Mimms, C.A., R.B. Hall, K.D. Rose, and G.R. Myers. 1989. *Catal. Lett.* 2:361.
Morales, E., and J.H. Lunsford. 1989. *J. Catal.* 118:2155.

Otsuka, K., K. Jinno, and A. Morikawa. 1985. *Chem. Lett.* 499.
Otsuka, K., K. Jinno, and A. Morikawa. 1986. *J. Catal.* 100:353.
Otsuka, K., and T. Komatsu. 1987. *Chem. Lett.* 483.
Otsuka, K., and T. Nakajima. 1986. *Inorg. Chim. Acta.* 120:L27.
Peil, K.P., J.G. Goodwin, Jr., and G. Marcelin. 1989. *Phys. Chem.* 93:5977.
Preuss, U., and M. Baerns. 1987. *Chem. Eng. Tech.* 10:297.
Quanzhi, L., and Y. Amenomiya, Y. 1986. *Appl. Catal.* 213:173.
Roos, J.A. 1989. Catalytic activation of methane. Ph.D. Dissertation to the University of Twente, Enschede, The Netherlands; Chapter 7.
Sinev, M.J., V.N. Korchak, and O.V. Krylov. 1987. *Kinetika i Kataliz* 28:1376.
Tanabe, K. 1985. Catalysis by solid bases and related topics. In *Catalysis by Acids and Bases*, ed. B. Imelik et al. Amsterdam: Elsevier Science Publishers.
Turnbull, A.G., and M.W. Wadsley. 1988. Thermodynamic calculations were carried out with the program CHEMIX of the CSIRO Thermochemistry System, Version V.
Utiyama, M., H. Hattori, and K. Tanabe. 1978. *J. Catal.* 53:237.
Ward, M.B., M.J. Lin, and J.H. Lunsford. 1977. *J. Catal.* 50:306.
Winter, E.R.S. 1968. *J. Chem. Soc.* (*A*) 2889.
Winter, E.R.S. 1969. *J. Chem. Soc.* (*A*) 1832.
Yates, D.J.C., and N.E. Zlotin. 1988. *J. Catal.* 111:317.

5

The Role of Lattice Oxygen in the Oxidative Coupling of Methane

Kevin P. Peil,[1] George Marcelin, and James G. Goodwin, Jr.

Department of Chemical and Petroleum Engineering, University of Pittsburgh, Pittsburgh, PA

TABLE OF CONTENTS

[1] Present address: Dow Chemical Company, Midland, MI 48640.

INTRODUCTION

The nature of the oxide site responsible for the oxidative coupling of methane has been the subject of much speculation with respect to the reaction mechanism. Under methane coupling conditions, various oxygen species may be present on a catalyst surface and thus the potential oxidants for methane and the intermediates formed from methane.

According to Bielanski and Haber (Bielanski and Haber 1979), oxide catalysts can be classified into two main groups according to the kind of oxygen species on the surface. One group predominantly forms adsorbed oxygen species, such as O^- and O_2^-, on the catalyst surface that are highly active and lead primarily to the total combustion products. The other group of oxide catalysts generally will not adsorb oxygen when the catalyst is in an oxidized form and lattice oxygen is the principal oxygen species available for product formation. In the latter group, the adsorption of molecular oxygen is only possible when the catalyst has been partially reduced, and then only in order to replenish the oxygen removed in the products. These catalysts can be highly selective for the production of partially oxidized products.

The participation of both adsorbed and lattice oxygen during the oxidation of methane has been confirmed for antimony-doped tin catalysts at 400°C (Parker and Patterson 1983; Brown and Patterson 1983), a catalyst that has been extensively studied for allylic oxidation reactions. Investigations demonstrated that the addition of small amount of antimony to the tin catalysts considerably reduced the amount of adsorbed oxygen on the surface. It was also determined that, for catalysts containing $> 8\%$ antimony, the product oxygen originated solely from the lattice.

The role of lattice oxygen in the oxidative coupling of methane also has been suggested by Otsuka and co-workers (Otsuka et al. 1987). They determined the number of oxygen atoms needed for oxidatively coupling methane over a Na_2O_2 catalyst to be nine times greater than the number of surface oxygen atoms.

Recently, isotopic transient techniques have been shown to be useful for unraveling the reaction path of the oxidative coupling reaction and for demonstrating the role of lattice oxygen in the production of C_2 products from methane. Additionally, this technique allows for the quantification of various parameters that are important to the oxidative coupling of methane.

This chapter presents a summary of our recent work in determining and quantifying the role of surface and lattice oxygen in reactions of methane.

STEADY-STATE ISOTOPIC TRANSIENT KINETIC ANALYSIS (SSITKA) OF CATALYTIC REACTIONS

Background

For a single, first-order elementary reaction step with intermediates present at a coverage of θ_i and exhibiting a lifetime of τ_i on the surface, the reaction rate in the form of the traditional turnover frequency (TOF) can be expressed as

$$\text{TOF} = \frac{1}{\tau_i}\theta_i = k_i\theta_i \tag{5-1}$$

where k_i (s^{-1}) is the intrinsic site turnover frequency (TOF_i) (Biloen 1983; Soong, Krishna, and Biloen 1986; Zhang and Biloen 1986; Biloen et al. 1983). Procedures for arriving at estimates of θ_i and k_i involve (i) absolute-rate-theory–based interpretations of experimentally determined preexponentials (Maatman 1980), (ii) selective poisoning experiments, or (iii) model-based interpretations of transient kinetic experiments (Furusawa, Suzuki, and Smith 1976; Boudart 1968).

The evaluation and quantification of such reaction parameters are best conducted at steady-state working conditions (Happel 1986; Tamaru 1978). The evaluation of such parameters can best be accomplished by means of steady-state isotopic transient kinetic analysis (SSITKA). This technique entails abrupt switches in the isotopic composition of one of the reactants accompanied by the continuous monitoring (e.g., by mass spectrometry) of the relaxation and evolution of labeled reactants and products (see Fig. 5-1). Values for θ_i and k_i can then be obtained from these transient curves.

Assuming that the surface behaves like a continuous stirred tank reactor (CSTR), the response of a single unidirectional reaction step to a step change in the isotopic concentration of one of the reactants is given by (Yang, Soong, and Biloen 1985)

$$\frac{dN_A}{dt} = -\frac{N_A}{\tau} = -kN_A \tag{5-2}$$

where N_A is the amount of isotopically labeled surface intermediates that form the labeled species A. The decay in N_A is observed as a decay in the

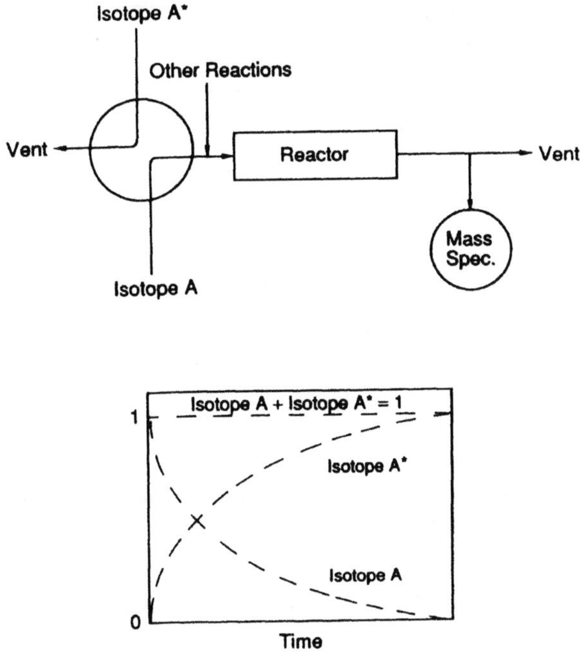

FIGURE 5-1. Switching the isotopic composition of the reactants does not disturb the steady state of the reaction.

rate of flow (R_A) of the labeled species A from the reactor. If the assumption is made that there is no isotope effect, τ is independent of N_A because the total surface concentrations are not changed and equation (5-2) can be integrated to yield

$$R_A = Re^{-t/\tau} \tag{5-3}$$

Because $R = R_A + R_{A^*} =$ constant, where R_{A^*} is the rate of flow of the unlabeled isotopic species, equation (5-3) can be normalized and written as

$$F_A(t) = \frac{R_A}{R_A + R_{A^*}} = e^{-t/\tau} \tag{5-4}$$

Integration of equation (5-4) from $t = 0$ to ∞ yields τ directly. This is represented by the shaded area in Fig. 5-2. For a series of parallel steps

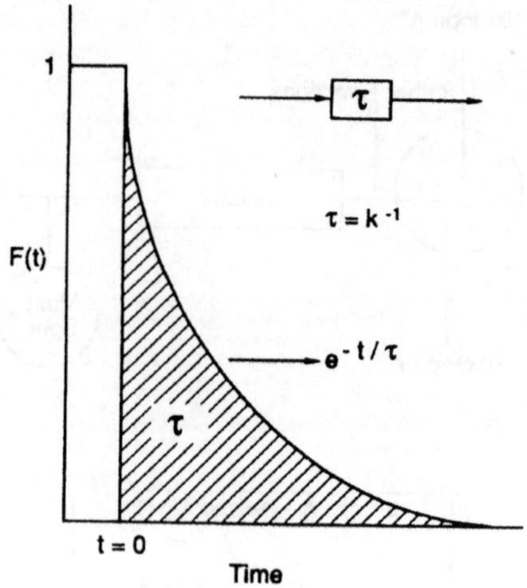

FIGURE 5-2. Response of a single, homogeneous surface pool to a step change in the isotopic concentration of one of the reactants.

(i.e., a heterogeneous surface), equation (5-4) becomes

$$F(t) = \sum_{i=1}^{N} a_i e^{-t/\tau_i} \tag{5-5}$$

where a_i is the fractional contribution of each type of site to the overall rate. Integrating equation (5-5) gives the average surface lifetime, τ_{av}. It is important to remember that the steady state of the reaction is not disturbed during the course of the preceding analysis.

Accounting for the Gas Phase and Readsorption

Included in the measured τ for all types of surface reactions are the τ for the holdup in the gas phase and the additional holdup on the surface due to readsorption:

$$\tau_{\text{measured}} = \tau_{\text{reaction}} + \tau_{\text{gas}} + \sum_{i=1}^{N} \tau_{\text{ads},i} \tag{5-6}$$

where N is the number of readsorptions, each lasting for a time $\tau_{\text{ads},i}$, and τ_{gas} is the residence time of all species in the gas phase of the reaction system.

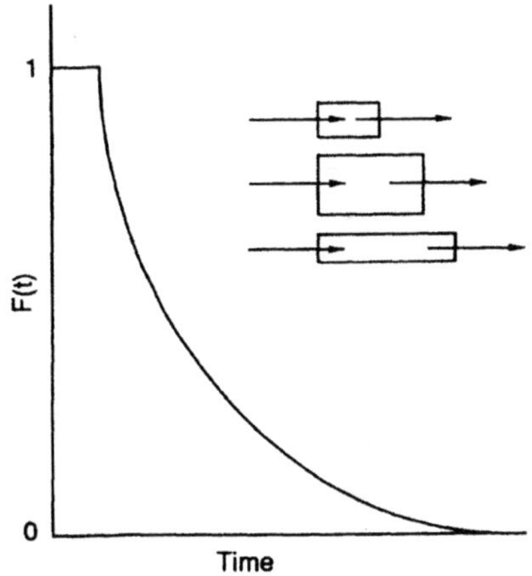

FIGURE 5-3. A heterogeneous surface exhibits a large number of parallel reaction pathways equal the number of different sites.

$\tau_{measured}$ can be corrected for τ_{gas} by including an inert gas tracer in one of the isotopes to be switched. Because the inert gas does not interact with the catalyst surface, the area under its transient is the holdup in the gas phase, τ_{gas}. Values for τ_{ads} cannot be obtained directly, but readsorption can be minimized by (i) decreasing the bed length, (ii) decreasing the holdup time in the reactor by increasing the flow rate, or (iii) adding a gas that will compete for readsorption sites. Therefore, it is possible to approximate values for the true τ of surface reaction.

Identifiability and Distinguishability of Reaction Pathways

A distinction between two possible types of mechanism can be made by examining the initial part of each transient. A parallel mechanism exists if there are multiple reaction pathways or if the catalyst has a heterogeneous surface. Figure 5-3 shows schematically the transient for a set of single step reactions. A heterogeneous surface exhibits a large number of parallel pathways equal to the number of different types of site. Therefore, the measured surface lifetime is an average, τ_{av}, of the surface lifetimes of surface intermediates along all surface reaction pathways. For parallel reaction pathways, the number of pathways would be few. For a parallel mechanism, any change in the isotopic composition of the reactants immediately appears in the products. Such a response is typical for all first-order processes (Stephanopoulos 1984).

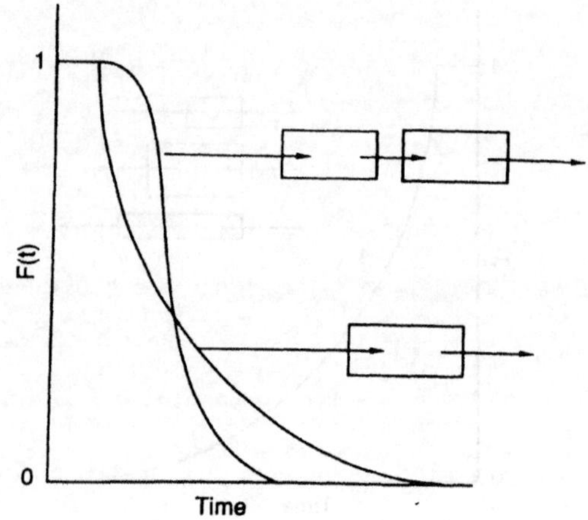

FIGURE 5-4. A comparison between the response of a single-step reaction and a multi-step reaction to an isotope switch.

Figure 5-4 shows a comparison between the response of a single-step and a multistep reaction occurring on a homogeneous surface to a step change in the isotopic composition of one of the reactants. After subtracting the effects on the transient due to gas-phase holdup and readsorption, the normalized response for the single-stage reaction, as discussed earlier, is given by $F_c(t) = e^{-t/\tau}$. The normalized response for the multistep reaction can be written as

$$F_c(t) = \sum_{i=1}^{S} \frac{\tau_i^{S-1}}{\prod_{j=1;\, j\neq i}^{S} (\tau_i - \tau_j)} e^{-t/\tau_i} \tag{5-7}$$

where S is the number of elementary reaction steps along the reaction pathway and $F_c(t)$ is the corrected normalized isotopic transient. Integrating this equation yields

$$\int_0^{\infty} F(t)\, dt = \sum_{i=1}^{S} \frac{\tau_i^{S}}{\prod_{j=1;\, j\neq i}^{S} (\tau_i - \tau_j)} = \sum_{i=1}^{S} \tau_i \tag{5-8}$$

which is equal to the total lifetime on the surface for the series of intermediates leading to the formation of the labeled product whose transient response is being followed.

For a series or multistep mechanism, a step change in the isotopic concentration of one of the reactants does not immediately produce a response in any of the products. The first pool of intermediates delays the entrance of the labeled species into the second pool. This delayed response gives the transient a sigmoidal shape. As seen in equation (5-9), the value for τ determined from the area under the normalized transient curve is the time constant for the overall series of surface reactions:

$$\tau = \sum_{i=1}^{S} \tau_i \tag{5-9}$$

Calculation of Coverages of Intermediates

In the previous sections we discussed how to obtain values for τ $(1/k)$ for the surface-generated intermediates for each of the product gases. The second basic quantity that can be determined is the abundance of the surface intermediates, N_i. All coverages are calculated directly from τ_S, where

$$\tau_S = \tau_{\text{reaction}} + \sum_{i=1}^{N} \tau_{\text{ads},\,i} = \tau_{\text{measured}} - \tau_{\text{gas}} \tag{5-10}$$

The surface concentration of reaction intermediates can be calculated by applying the formula (Yang, Soong, and Biloen 1984)

$$N_i = \tau_{Si} \cdot (\text{effluent rate of } i \text{ from the reactor})$$

Values for θ can be obtained by normalizing N_i by a monolayer coverage.

Determination of Lattice Oxygen Diffusivities

Figure 5-5 depicts an example of oxygen transients obtained when the isotopic composition of oxygen is switched from $^{16}O_2$ to $^{18}O_2$ during the oxygen exchange reaction over an oxide catalyst with varying levels of bulk oxygen exchange. The failure of the $^{16}O\,^{18}O$ transient to relax back to zero in Fig. 5-5b and 5-5c is due to the bulk oxide supplying the catalyst surface with ^{16}O atoms for exchange with the gas phase. In all cases, either the exchange reaction or the rate of lattice oxygen diffusion controls the rate of lattice oxygen exchange. The latter is eventually the rate-limiting step and is characterized by a ca. 30% decrease in the rate of $^{16}O\,^{18}O$ formation with a doubling in time.

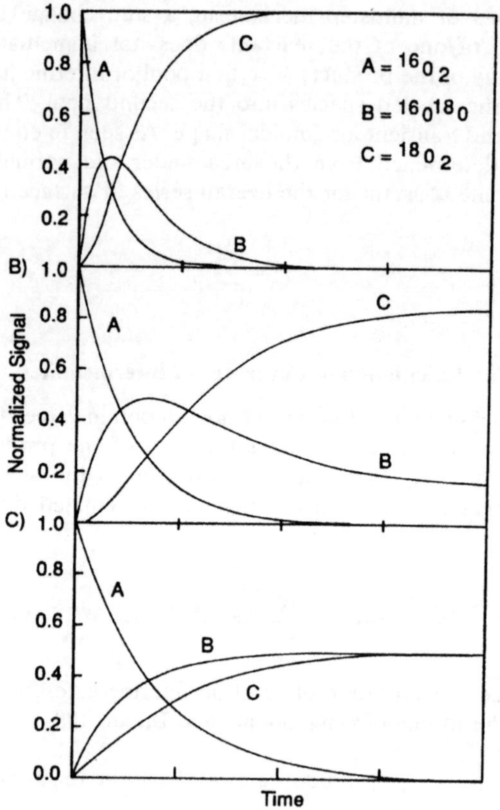

FIGURE 5-5. Theoretical oxygen exchange transients over an oxide with (a) no bulk oxygen contribution, (b) bulk oxygen involved in exchange, and (c) extremely mobile lattice oxygen resulting in an equal probability of a ^{16}O or an ^{18}O on the surface.

When lattice oxygen diffusion is rate-controlling, the diffusivity of that oxygen can be determined from the characteristics of the transient curve. If the catalyst is modeled as a solid bounded by a plane at $x = 0$ and extending to infinity in the direction of positive x, we can write (Carslaw and Jaeger 1959)

$$N_a = -D\frac{\partial C_a}{\partial x} \tag{5-11}$$

and

$$C_a = C_{a_0}\,\mathrm{erf}\left(\frac{x}{2\sqrt{Dt}}\right) \tag{5-12}$$

where N_a is the flux (in moles per square centimeter per second) of atomic oxygen from the bulk to the surface, D is diffusivity (in square centimeters per second), C_a is the concentration of atomic oxygen in the bulk (in moles per cubic centimeter), C_{a_0} is the initial concentration of atomic oxygen in the bulk (in moles per cubic centimeter), x is vertical position (in centimeters) into the bulk from the surface, and t is time (in seconds). Substituting (5-12) into (5-11) and evaluating the flux at the surface ($x = 0$) at $t > t_b$,

$$N_a(x = 0) = D\frac{C_{a_0}}{\sqrt{\pi D t}} \tag{5-13}$$

or

$$D = \pi t\left(\frac{N_a}{C_{a_0}}\right)^2 \tag{5-14}$$

N_a can be obtained by multiplying the $^{16}O\,^{18}O$ offset fraction by the O_2 flow rate and dividing by the catalyst weight and surface area. C_{a_0} is determined by dividing the density of the catalyst by its molecular weight and multiplying by the number of oxygen atoms per molecule of the bulk material and t_b is the time at which the bulk becomes the primary source of ^{16}O in the effluent stream.

OXYGEN EXCHANGE CAPABILITIES OF METAL OXIDES

Nonreducible Metal Oxides

MgO and Li / MgO

The mobility and exchangeability of lattice oxygen has been investigated for MgO and 7 wt% Li/MgO using steady-state isotopic transient kinetics (Peil, Goodwin, and Marcelin 1989; Peil, Goodwin, and Marcelin 1991a). Figures 5-6 and 5-7 show the transients obtained following a gas-phase switch of $^{16}O_2$ to $^{18}O_2$. The failure of the $^{16}O\,^{18}O$ signals to relax quickly back to zero was an indication of an oxygen source, that is, the bulk catalyst, slowly supplying the surface with ^{16}O atoms. When the same experiment was repeated in the absence of any catalyst, the $^{16}O_2$ and He (gas-phase marker) transients relaxed simultaneously.

A comparison of the relaxation profiles of the isotopes for the two catalysts clearly shows that Li/MgO is capable of a faster exchange of oxygen than MgO. The offset of the $^{16}O\,^{18}O$ signal after 300 s corresponded to about $F(t) = 0.15$ for both catalysts. Thus, 7.5% of the oxygen

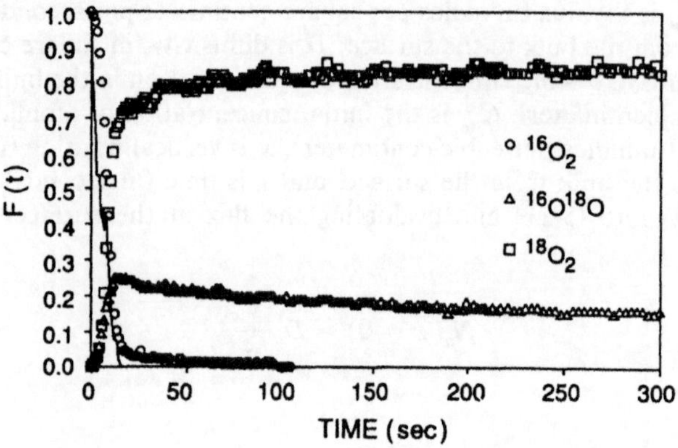

FIGURE 5-6. Oxygen exchange transients obtained over MgO at 600°C. Solid line indicates gas-phase holdup.

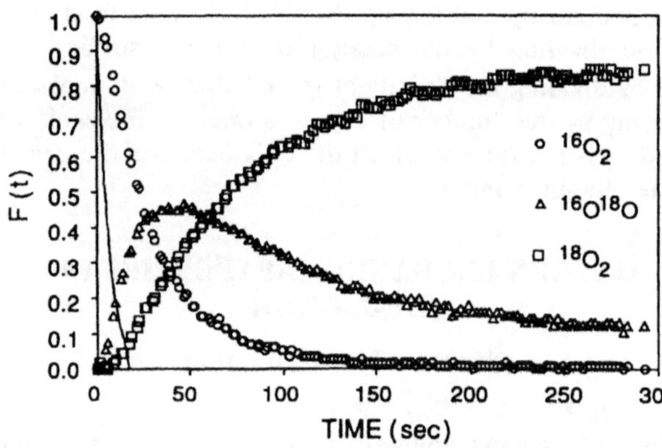

FIGURE 5-7. Oxygen exchange transients obtained over Li/MgO at 600°C. Solid line indicates gas-phase holdup.

flow ($\frac{1}{2} \times 0.15 \times 100\%$) leaving the reactor consisted of ^{16}O from the lattice of the catalysts. This offset can be viewed as a measure of the contribution by the bulk to the exchange of oxygen. The extent of this relative contribution was not affected by the addition of Li^+ to the MgO catalyst even though the rate of exchange of oxygen increased significantly.

The total contributions to the exchange were quantified by integrating the $^{16}O_2$ and $^{16}O\,^{18}O$ transient curves in Figs. 5-6 and 5-7 after correcting

Table 5-1 Oxygen Availability and Lattice Oxygen Diffusivity as Determined from Oxygen Exchange Experiments for MgO and Li/MgO

Catalyst	Temp. (°C)	Total "Surface" ^{16}O Atoms desorbing per Gram of Catalyst ($\times 10^{-20}$)	Equivalent Number of "Surface" Layers	Diffusivity of Lattice Oxygen ($cm^2\ s^{-1}$) ($\times 10^{17}$)
MgO	575	0.5	< 1[a]	> 0.64
	600	2.5	1	> 2.7
	625	2.8	1–2	> 5.1
Li/MgO	400	0.3	< 1[b]	2.1
	500	5.2	8	9.2
	550	5.9	9	13.4
	600	6.5	10	23.0
	635	7.9	12	30.9

[a] Based on a surface area of 30 $m^2\ g^{-1}$ and 12.5 $Å^2$ per oxygen atom.
[b] Based on a surface area of 8 $m^2\ g^{-1}$ and 12.5 $Å^2$ per oxygen atom.

for the ^{16}O "bulk phase" contribution (Peil, Goodwin, and Marcelin 1991a). The corrected transients yielded the total amount of ^{16}O atoms leaving the catalyst. Because the transients had been corrected for the bulk contribution, it was possible to calculate an equivalent number of atomic layers readily available for oxygen exchange (Table 5-1). These atomic layers can be considered to lie within the surface–subsurface of the catalyst and were distinct from the bulk in that their oxygen atoms were close enough to the surface to be readily available for exchange and, using SSITKA, were not distinguishable from oxygen existing solely on the external surface.

A maximum of ca. 12 of these readily accessible atomic layers was reached at the higher temperatures for Li/MgO. Of course, this number does not imply a distinct boundary and only represents the point at which the subsurface oxygen becomes indistinguishable from the bulk oxygen.

The corrected transients, that is, with no bulk phase contribution, are shown in Fig. 5-8 for the Li/MgO catalyst at four temperatures. The oxygen availability and diffusivities obtained from these measurements are summarized in Table 5-1. The diffusivities determined for MgO represent minimum values because lattice diffusion was not yet solely limiting the exchange at the end of the experiment. Oxygen exchange at 400°C for Li/MgO involved exchange solely between the gas phase and the oxide surface because the ^{16}O transient could be expressed by a single exponential and the amount of the measured "surface" oxygen was less than that predicted for one monolayer. This is presumably because at this tempera-

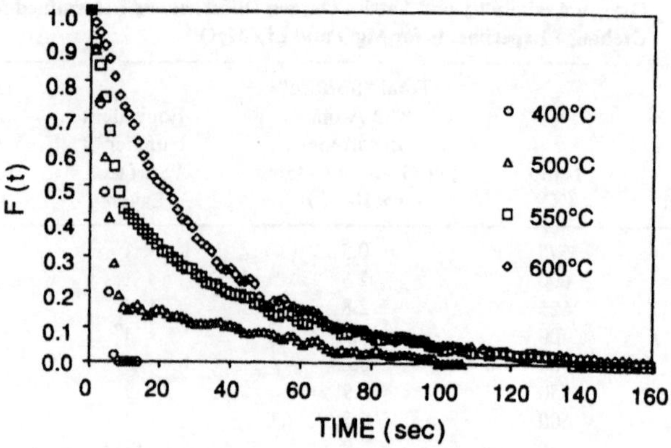

FIGURE 5-8. Total "surface" ^{16}O transients for oxygen exchange over Li/MgO at 600°C, corrected for the bulk-phase contribution.

ture only the exchange between the gas phase and the oxide surface was fast enough to be measured, as opposed to exchange with the bulk oxygen.

The transients at 500 and 550°C, however, could not be represented by a single exponential. These transients have a sudden change in slope at ca. 10 s and at least two exponentials are needed for a successful fit. As with the transient at 400°C, the initial rapid exchange was interpreted as exchange between the gas phase and the oxide surface, and the slower exchange interpreted as exchange between the gas phase and the first few subsurface atomic layers.

The transient for exchange at 600°C can again be represented with a single exponential. At this temperature the number of subsurface atomic layers participating in the exchange was large enough and the overall rate of exchange fast enough to prevent a distinction between the physical surface and the first few subsurface layers. The rate of exchange between the gas phase and the surface was equal to the rate of exchange between the surface and the first few subatomic layers.

It is important to remember that the transients in Fig. 5-8 have already been corrected for the bulk-phase contribution and the slow exchange is not a measure of the mobility of the oxygen in the oxide bulk. Rather this slower exchange is representative of subsurface atomic oxygen that is readily available for exchange due to its close proximity to the surface.

The bulk oxygen diffusivities were calculated from the exchange data based on the rate of exchange of ^{16}O with gas-phase ^{18}O as measured by the offset at large times (Peil, Goodwin, and Marcelin 1991a). Comparison of these results from Table 5-1 indicates that the addition of Li to MgO

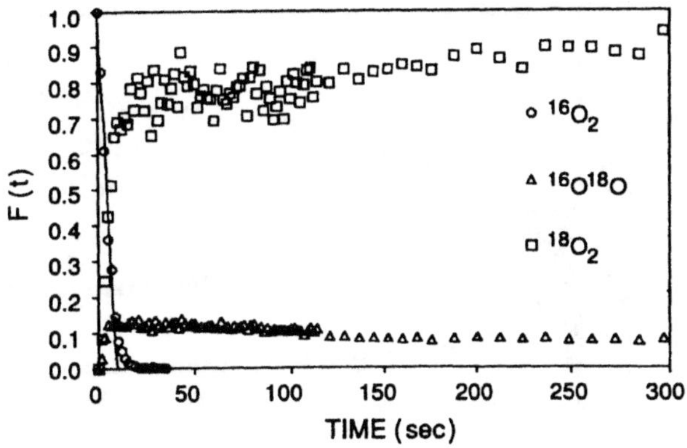

FIGURE 5-9. Oxygen exchange transients obtained over monoclinic Sm_2O_3 at 600°C. Solid line indicates gas-phase holdup.

resulted in approximately an order of magnitude increase in bulk oxygen diffusivity. Assuming an Arrhenius expression for the diffusivity as a function of temperature, values for the activation energy of bulk diffusion were obtained corresponding to 62.7 kcal mol^{-1} for MgO and 14.1 kcal mol^{-1} for Li/MgO. In comparison, literature values of 62.4 (Oishi and Kingery 1960) and 51.0 (Oishi and Ando 1984) kcal mol^{-1} have been reported for MgO. The increase in lattice oxygen mobility upon lithium addition to MgO can perhaps be attributed to the formation of lattice defects (or "holes") upon the addition of Li^+. Wang and Lunsford (Wang and Lunsford 1986) studied the addition of Li^+ to MgO and concluded that the excess of cations in the MgO matrix results in the formation of oxygen vacancies. These vacancies can apparently facilitate the rapid transport of oxygen through the lattice.

Sm_2O_3

The oxygen transients obtained during oxygen exchange at 600°C over monoclinic and cubic Sm_2O_3 are shown in Figs. 5-9 and 5-10, respectively (Peil 1990). It is evident from a comparison of the relaxation profiles of the isotopes for the two catalysts that cubic Sm_2O_3 is capable of a faster exchange of oxygen. The final offset of the $^{16}O^{18}O$ signal for cubic Sm_2O_3 was also approximately twice that for the monoclinic Sm_2O_3 and more time was needed before the transient for cubic Sm_2O_3 indicated control of oxygen exchange by lattice oxygen diffusion.

The equivalent number of atomic layers readily available for exchange along with the lattice oxygen diffusivity were calculated and are shown in

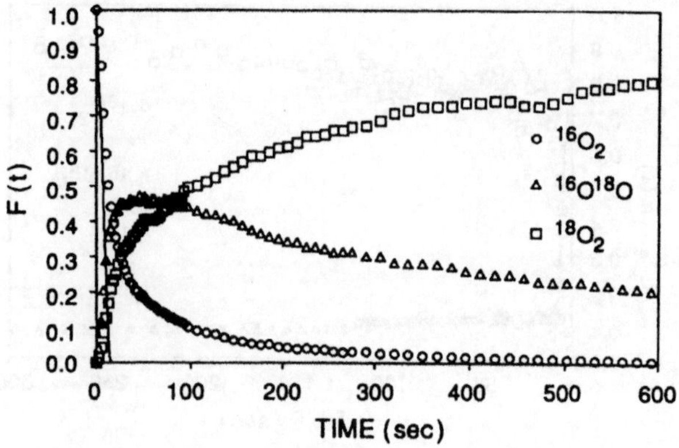

FIGURE 5-10. Oxygen exchange transients obtained over cubic Sm_2O_3 at 600°C. Solid line indicates gas-phase holdup.

Table 5-2. Surprisingly, there was a drastic difference in the lattice oxygen mobility and the equivalent number of oxide layers readily available for exchange for the two Sm_2O_3 catalysts. However, this difference can be understood by a close examination and comparison of the two crystalline structures.

The cubic Sm_2O_3 structure is stable up to 850–900°C, after which conversion to the monoclinic structure occurs (Kofstad 1972). In the cubic structure, the Sm atom is surrounded by eight equidistant oxygen atoms at the corners of a cube (Stone, Weber, and Eyring 1968; Templeton and Dauben 1954). However, one-forth of the structural oxygen atoms are actually missing, resulting in the metal atom being surrounded by six oxygen atoms and two lattice oxygen vacancies. The missing oxygen atoms

Table 5-2 **Oxygen Availability and Lattice Oxygen Diffusivity as Determined from Oxygen Exchange Experiments for Sm_2O_3**

Sm_2O_3 Catalyst	Temp. (°C)	Total "Surface" ^{16}O Atoms Desorbing per Gram of Catalyst ($\times 10^{-20}$)	Equivalent Number of "Surface" Layers	Diffusivity of Lattice Oxygen ($cm^2\ s^{-1}$) ($\times 10^{17}$)
Monoclinic	600	0.2	2[a]	886
Cubic	600	13	> 100[b]	13,600

[a] Based on a surface area of 1 $m^2\ g^{-1}$ and 19.0 Å^2 per oxygen atom.
[b] Based on a surface area of 1 $m^2\ g^{-1}$ and 17.2 Å^2 per oxygen atom.

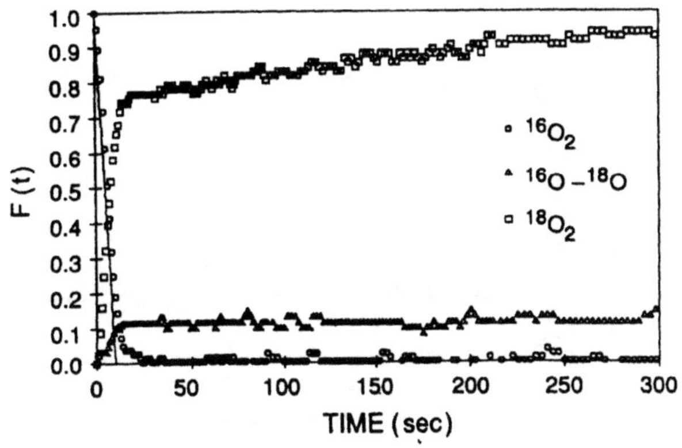

FIGURE 5-11. Oxygen exchange transients obtained over SnO_2 at 658°C. Solid line indicates gas-phase holdup.

are located either on the face diagonals or body diagonals and are ordered in such a manner that the oxygen-atom vacancies lie in a straight line through the body diagonal of the cubic crystal.

The monoclinic Sm_2O_3 structure is identical to the cubic structure except the monoclinic structure has an additional oxygen atom along a threefold axis (Cromer 1957). This seventh oxygen atom results in a distorted octahedron coordination about the Sm atoms. Not all monoclinic unit cells contain an "extra" oxygen atom, so the structure consists of both sixfold and sevenfold coordination about the metal atoms.

Because the cubic lattice is more symmetric about the metal atom than the monoclinic structure, this results in a higher degree of lattice oxygen mobility. The oxygen vacancies in the cubic structure are ordered in such a manner that they essentially provide a "pipeline" for oxygen atoms to diffuse into and out of the bulk.

Reducible Metal Oxides

SnO_2 and Sn / MgO

The mobility and exchangeability of lattice oxygen for several reducible metal oxides has also been studied. Figures 5-11 and 5-12 show the oxygen exchange transients obtained at 650°C for SnO_2 and Sn/MgO (Agarwal 1990). Comparison of these transients with those obtained for Li/MgO and Sm_2O_3 at 600°C shows a relatively small fast exchange of oxygen for the reducible oxides.

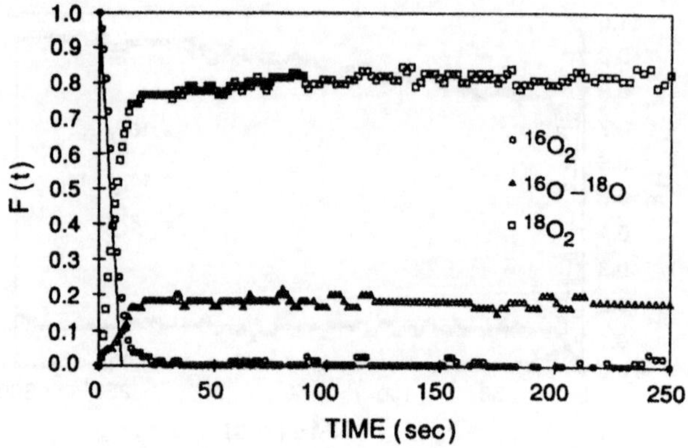

FIGURE 5-12. Oxygen exchange transients obtained over Sn/MgO at 658°C. Solid line indicates gas-phase holdup.

As shown in Table 5-3, less than a monolayer of readily available oxygen exists in these catalysts. In this case the calculated diffusivity represents only a minimum possible diffusivity of the lattice oxygen because for these catalysts the lattice oxygen diffusion was not rate-controlling by the end of the experiment. Tin oxide exhibited a lattice oxygen mobility somewhat greater than that of Sn/MgO and the latter exhibited a minimum diffusivity comparable to that of MgO. Although there was a large difference in the total amount of "surface" ^{16}O atoms desorbing per gram of catalyst, most of this can be explained by the differences in surface areas of the catalysts.

Table 5-3 Oxygen Availability and Lattice Oxygen Diffusivity as Determined from Oxygen Exchange Experiments for SnO_2, Sn / MgO, and NiO

Catalyst	Temp. (°C)	Total "Surface" ^{16}O Atoms Desorbing per Gram of Catalyst ($\times 10^{-20}$)	Equivalent Number of "Surface" Layers	Diffusivity of Lattice Oxygen ($cm^2\ s^{-1}$) ($\times 10^{17}$)
SnO_2	650	0.2	$< 1^a$	$\gg 9.9$
Sn/MgO	650	1.3	$< 1^b$	> 1.7
NiO	600	11	$> 100^c$	9770

[a]Based on a surface area of 9 $m^2\ g^{-1}$ and 12.9 $Å^2$ per oxygen atom.
[b]Based on a surface area of 36 $m^2\ g^{-1}$ and 12.5 $Å^2$ per oxygen atom.
[c]Based on a surface area of 1 $m^2\ g^{-1}$ and 12.7 $Å^2$ per oxygen atom.

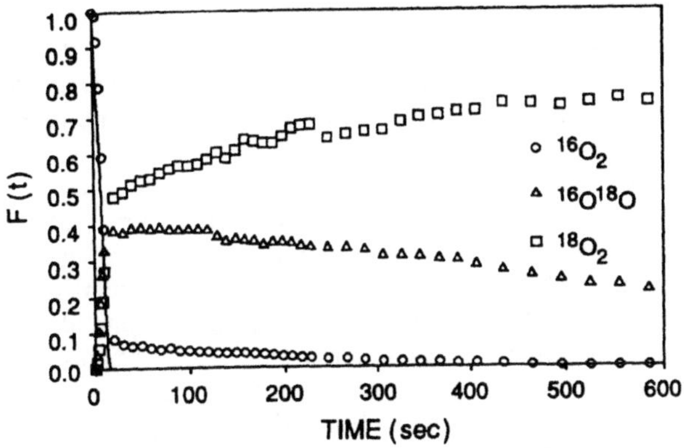

FIGURE 5-13. Oxygen exchange transients obtained over NiO at 600°C. Solid line indicates gas-phase holdup.

NiO

The mobility and exchangeability of lattice oxygen for NiO was also studied (Peil 1990). The oxygen exchange transients obtained are shown in Fig. 5-13. The total amount of "surface" ^{16}O atoms desorbing per gram of catalyst, the equivalent number of "surface" layers participating in exchange, and the diffusivity of the lattice oxygen were calculated and are given in Table 5-3.

It is interesting to note the drastic differences between MgO and NiO in their lattice oxygen diffusivities, given in Tables 5-1 and 5-3, as well as oxygen exchangeability, depicted in Figs. 5-6 and 5-13. The lattice oxygen diffusivity for NiO was at least three orders of magnitude higher than for MgO. Both MgO and NiO have identical crystal structures and the ionic radii of Ni^{2+} and Mg^{2+} are nearly the same at 0.71 and 0.65 Å, respectively (Dean 1985). Therefore, the difference in lattice oxygen mobility between these two materials cannot be simply attributed to the packing arrangement of the lattice cations and anions. This difference in oxygen diffusivity undoubtedly reflects the greater ease with which lattice vacancies can form in NiO compared to MgO (O'Keeffe and Moore 1961).

INVOLVEMENT OF LATTICE OXYGEN IN THE OXIDATIVE COUPLING OF METHANE

SSITKA Measurements

The use of isotopic switches during steady-state reaction provides information related only to the reaction pathway involving that particular isotope.

Thus, the surface residence times and surface concentrations obtained during steady-state reaction from an isotopic oxygen switch correspond to the active intermediates along the oxygen reaction pathway. No information can be obtained from such an isotopic switch on the carbon reaction pathway.

Isotopic switches under steady-state reaction conditions have provided a quantitative description of the reactive oxygen intermediates and oxygen reaction pathways over reducible and nonreducible oxides. The types of results available from these experiments include the relative contribution of the lattice oxygen during reaction, the average lifetime along the oxygen reaction pathways for the oxygen-containing intermediates for the different products, and the surface concentration of active oxygen intermediates. Information has also been obtained regarding the exchange of the oxygen in the product molecules with the catalyst surface and lattice oxygen.

Nonreducible Metal Oxides

Oxygen transients obtained during steady-state reaction over Li/MgO at 600°C are shown in Figs. 5-14–5-16 (Peil, Goodwin, and Marcelin 1989; Peil, Goodwin, and Marcelin 1991a). Molecular oxygen transients obtained during reaction for Li/MgO (Fig. 5-14) were somewhat different from the molecular oxygen transients obtained during oxygen exchange. Calculations based on oxygen exchange experiments indicate 7.9×10^{20}

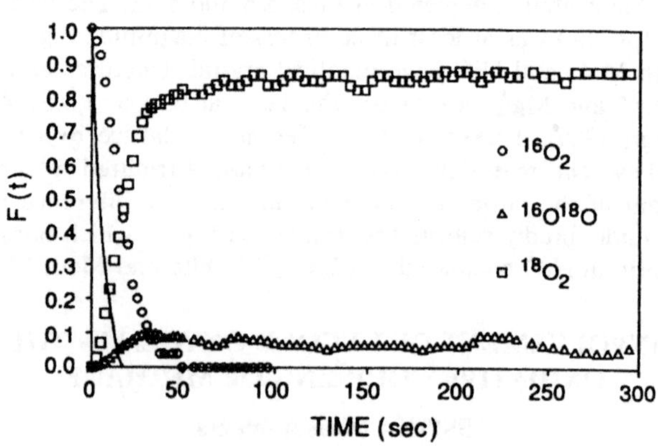

FIGURE 5-14. Molecular oxygen transients during steady-state reaction over Li/MgO at 600°C. Solid line indicates gas-phase holdup.

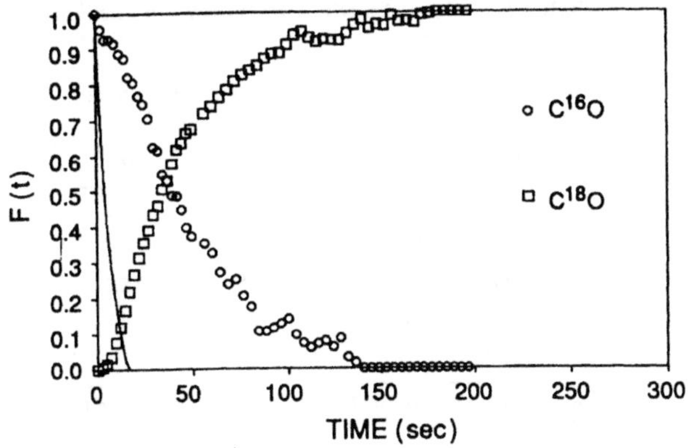

FIGURE 5-15. CO transients produced from an oxygen isotopic switch during steady-state reaction over Li/MgO at 600°C. Solid line indicates gas-phase holdup.

surface–subsurface ^{16}O atoms per gram readily available for exchange. Under reaction, the total amount of ^{16}O leaving the surface/subsurface as $^{16}O_2$, $^{16}O^{18}O$, $C^{16}O$, $C^{16}O_2$, $C^{16}O^{18}O$, and $H_2{}^{16}O$ (Table 5-4) corresponded to 7.3×10^{20} atoms. Thus, the amount of surface–subsurface ^{16}O atoms per gram leaving the catalyst as products during reaction was approximately equal to the amount of "surface" ^{16}O atoms per gram available for exchange, as determined from the exchange experiments.

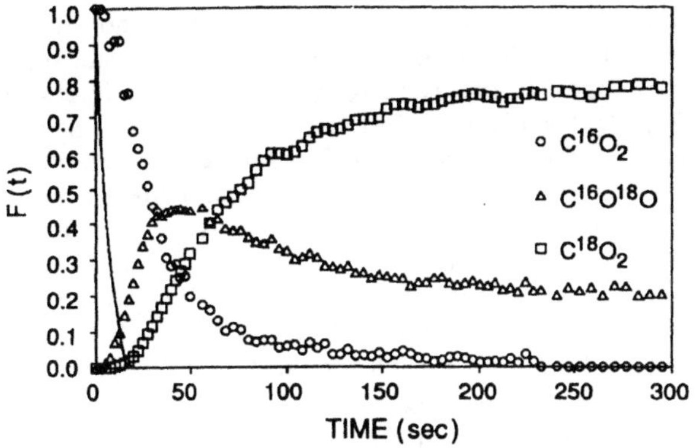

FIGURE 5-16. CO_2 transients produced from an oxygen isotopic switch during steady-state reaction over Li/MgO at 600°C. Solid line indicates gas-phase holdup.

Table 5-4 Oxygen Switch in the Presence of Methane

Species	^{16}O "SURFACE" ATOMS DESORBING PER GRAM[a] ($\times 10^{-20}$) (a)	(b)
$^{16}O_2 + {}^{16}O\,{}^{18}O$	0.6	1.4
$C\,{}^{16}O$	0.3	0.8
$H_2{}^{16}O$	4.5	5.6
$C\,{}^{16}O_2 + C\,{}^{16}O\,{}^{18}O$	1.9	4.0
Total	37.3	11.8

[a](a) 100 mg Li/MgO at 600°C; (b) 32.6 mg Li/MgO at 645°C.

However, the amount of "surface" ^{16}O desorbing as molecular oxygen during reaction was an order of magnitude less than the amount of such molecular ^{16}O desorbing during simple exchange. This difference was due to the formation of other oxygen-containing species. Continued contribution to the overall rate from the bulk oxygen was noted for $^{16}O\,{}^{18}O$ (Figure 5-14) and $C\,{}^{16}O\,{}^{18}O$ (Fig. 5-16). Figure 5-15 would appear to show no contribution of lattice oxygen for the formation of CO; however, this is simply due to the way the steady-state position of the background was normalized for plotting after the data collection. In fact, the offset at the end of the experiment was comparable to that for CO_2.

The mixing of $C\,{}^{16}O_2$, $C\,{}^{16}O\,{}^{18}O$, and $C\,{}^{18}O_2$ (Fig. 5-16) indicates that CO_2 interacted with the surface long enough to provide a statistical mixing of the oxygens. However, it is interesting to note the nonstatistical mixing of $^{16}O_2$, $^{16}O\,{}^{18}O$, and $^{18}O_2$ (Fig. 5-14) compared to the statistical mixing of the oxygen isotopes during exchange (Fig. 5-7). CO_2 interacted strongly with the Li/MgO surface by forming the carbonate species, and this probably facilitated the statistical mixing. Oxygen, on the other hand, did not show a great tendency for readsorption under these conditions.

Oxygen transients obtained during steady-state reaction over cubic Sm_2O_3 at 600°C are shown in Figs. 5-17–5-19 (Peil 1990). Continued contribution to the overall rate from the Sm_2O_3 bulk oxygen can be noted in Fig. 5-17 for the $^{16}O\,{}^{18}O$ transient and in Fig. 5-19 for the $C\,{}^{16}O\,{}^{18}O$ transient. The value of $F(t) = 0.5$ for the $C\,{}^{16}O\,{}^{18}O$ signal offset at t_{max} indicates that there was an equal probability of CO_2 desorbing with a ^{16}O or an ^{18}O atom. As was the case for Li/MgO, the formation of CO also exhibited a contribution from the lattice, although this is not evident in Fig. 5-18 for reasons discussed earlier.

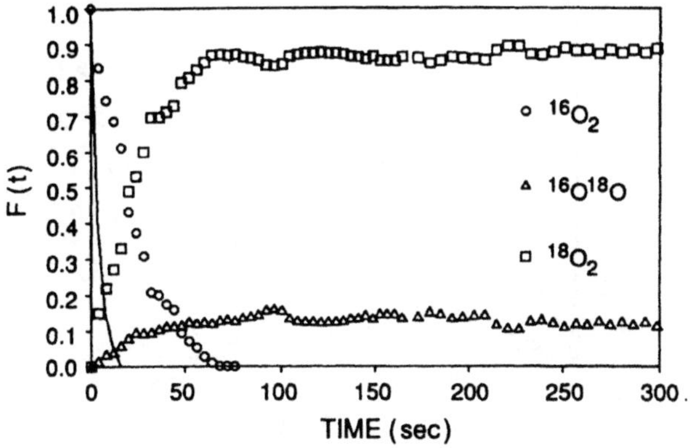

FIGURE 5-17. Molecular oxygen transients during steady-state reaction over Sm_2O_3 at 600°C. Solid line indicates gas-phase holdup.

The effect of cubic Sm_2O_3 catalyst loading on the measured surface residence times along the oxygen reaction pathway has been studied (Table 5-5) (Peil 1990). The fourfold increase in surface oxygen residence times for CO, CO_2, and H_2O with a corresponding fourfold increase in the amount of catalyst is clear indication of reexchange of the oxygen of the product molecules with the catalyst oxygen reservoir.

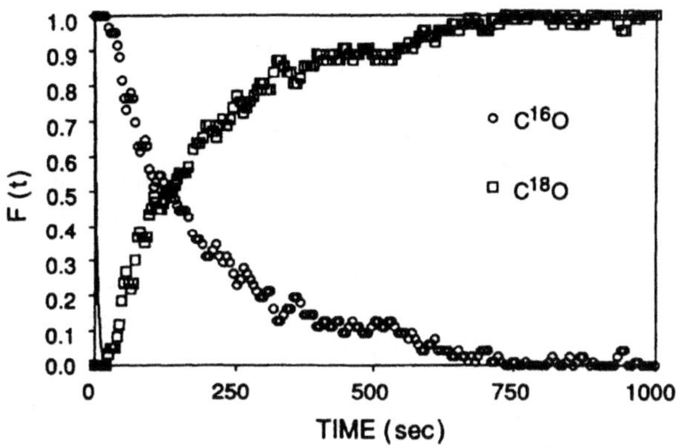

FIGURE 5-18. CO transients produced from an oxygen isotopic switch during steady-state reaction over Sm_2O_3, at 600°C. Solid line indicates gas-phase holdup.

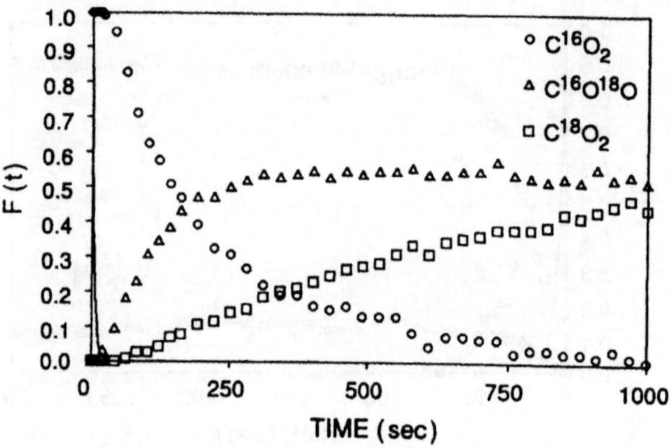

FIGURE 5-19. CO_2 transients produced from an oxygen isotopic switch during steady-state reaction over Sm_2O_3 at 600°C. Solid line indicates gas-phase holdup.

Table 5-5 Residence Times Along the Oxygen Reaction Pathway for Sm_2O_3 as a Function of Catalyst Loading; Reaction Temperature = 600°C

Species	τ_0 (IN SECONDS) 25 mg	100 mg
$^{16}O_2$	21	19
$H_2{}^{16}O$	81	294
$C^{16}O$	51	191
$C^{16}O_2$	44	226

Reducible Metal Oxides

It was previously noted that Sn/MgO did not exhibit significant oxygen exchange in the absence of methane. However, as shown in Fig. 5-20, in the presence of methane the Sn/MgO catalyst did exhibit significant oxygen exchange (Agarwal 1990). This type of behavior should be contrasted with that of the nonreducible Li/MgO catalyst in which the addition of Li to MgO served to enhance significantly the oxygen exchange capability both in the absence and presence of methane.

These differences in the bulk oxygen behavior may be explained by considering the role of the minor component and the ease of reducibility of these catalysts in the presence of methane. The high oxygen exchange exhibited by Li/MgO can, at least in part, be attributed to the formation

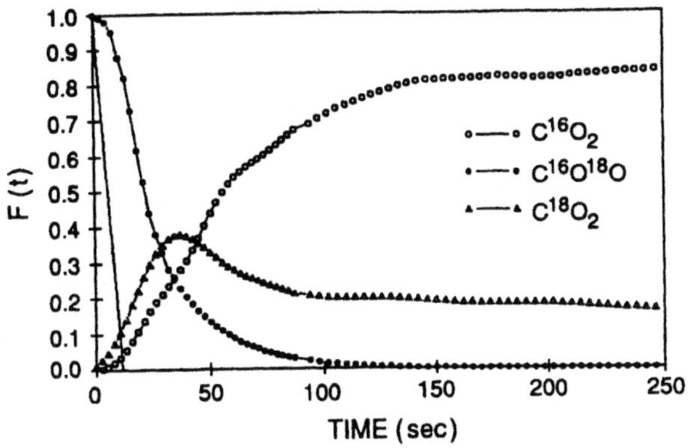

FIGURE 5-20. CO_2 transients produced from an oxygen isotopic switch during steady-state reaction over Sn/MgO at 658°C. Solid line indicates gas-phase holdup.

of Li^+O^- centers and oxygen vacancies. Similar vacancies are not formed by the addition of SnO_2 to MgO and thus the lattice oxygen mobility remains low. However, under reaction conditions methane is capable of reducing tin oxide, perhaps forming anion vacancies that in turn lead to an increase in the lattice oxygen mobility in the lattice.

Comparison of Reducible and Nonreducible Catalysts

It follows that the catalytic systems for methane coupling may be classified into at least two general types, based on the reducibility and oxygen exchange behavior, and exemplified by Li/MgO and Sn/MgO. Clear differences were found between them. Whereas Li/MgO exhibited rapid exchange of oxygen, both in the presence and absence of methane, extending up to 12 monolayers into the subsurface, Sn/MgO exhibited noticeable exchange only in the presence of methane. In both cases, however, the lattice oxygen participated in reaction.

Table 5-6 compares the catalytic behavior of the various catalysts with their lattice oxygen diffusivity. It should be noted that the reducible oxides yielded no coupled products under the reaction conditions employed in these studies whereas the nonreducible oxides were reasonably selective for C_2 formation. However, such a generalization is probably misleading because the reducible oxides also typically exhibited higher total methane conversion.

Table 5-6 Comparison of Catalyst Activity and C_2H_6 Selectivity with Lattice Oxygen Diffusivity in the Absence of Methane

	CH_4		C_2H_6			
Catalyst	Conversion (%)	Absolute Rate × 10^7 ($mol\ m^{-2}\ s^{-1}$)	Selectivity (%)	Absolute Rate × 10^7 ($mol\ m^{-2}\ s^{-1}$)	$CO_2:CO$	Lattice Oxygen Diffusivity × 10^{17} ($cm^2\ s^{-1}$)
MgO[a]	1.4	0.02	0	0	0.3	> 2.7
Li/MgO[a]	6.4	0.3	27	0.08	1.4	23.0
Mono. Sm_2O_3[a]	3.7	1.3	8	0.10	3.2	886
Cubic Sm_2O_3[a]	9.1	3.1	26	0.81	1.4	13,600
SnO_2[b]	12.2	7.5	0	0	38.8	≫ 9.9
Sn/MgO[b]	4.7	2.3	0	0	2.3	> 1.7
NiO[a]	7.2	2.5	0	0	1.8	9,770

[a]100 mg catalyst; temperature = 600°C; $Ar:CH_4:O_2 = 45:5:0.5\ cm^3\ min^{-1}$.
[b]30 mg catalyst; temperature = 658°C; $Ar:CH_4:O_2 = 43:9:2\ cm^3\ min^{-1}$.

The following oxygen scheme can be proposed that is consistent with the reported results:

$$
\begin{array}{ccccccc}
O_{gas} & \xrightarrow{k_a} & O_{surface} & \xrightarrow{k_r} & \text{Intermediates} & \xrightarrow{k_d} & \text{Products} \\
 & & \uparrow{\scriptstyle k_D} & & & & \\
 & & O_{lattice} & & & &
\end{array}
$$

where k_a, k_r, k_d, and k_D represent the rate constants for oxygen adsorption, surface reaction, desorption of products, and lattice oxygen mobility, respectively.

In the absence of methane, oxygen adsorbs and can exchange with the bulk oxygen. During reaction, methane becomes the adsorbing reactant and can absorb on the surface, where it reacts with the surface oxygen to yield intermediates that desorb as products. The relative values of k_d and k_D determine the extent of participation of lattice oxygen in the reaction. For active nonreducible oxides, such as Li/MgO, k_D is high both in the presence and absence of methane. However, for Sn/MgO, lattice diffusion is slow compared with oxygen desorption in the absence of methane so that a slower rate of exchange is observed. In a reaction mixture, however, lattice oxygen is able to participate due to its increased mobility as a result of the formation of anion vacancies during the partial reduction of the tin oxide by methane.

NON-STEADY-STATE TRANSIENTS

Further evidence for the role of lattice oxygen in the reaction is provided by experiments involving non-steady-state isotopic switches (Peil, Goodwin, and Marcelin 1991b). Figure 5-21 shows the effect of terminating the flow of oxygen on the various products during the reaction of methane and oxygen over Sm_2O_3.

The continued production of C_2H_6 and CO long after the removal of gas-phase oxygen was an indication of the contribution to reaction of the lattice oxygen of the catalyst. More important, the rate of ethane production was seen to increase when oxygen was shut off. Because this was accompanied by an immediate decrease in the rate of CO_2 production, it can be suggested that this increase in the C_2H_6 rate may have been due either to the lattice oxygen of the catalyst preferentially assisting in forming C_2H_6 or to a decrease in the homogeneous gas-phase oxidation of methyl radicals leading to CO_2 because the level of gas-phase oxygen rapidly decreased after the oxygen shutoff. However, a recent study utilizing different catalyst bed depths and SSITKA showed experimentally

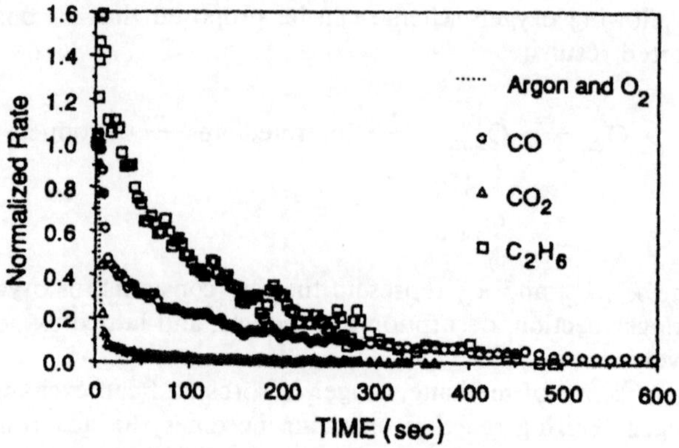

FIGURE 5-21. Transients obtained upon shutting off the O_2 flow during methane oxidation reaction over Sm_2O_3 at 600°C.

that homogeneous gas-phase oxidation does not contribute significantly at these reaction conditions. Thus, the increase in C_2H_6 formation would seem to be due to more selective partial oxidation by lattice oxygen, possibly as a result of the partial reduction of the oxide surface.

LATTICE OXYGEN MOBILITY AND CATALYST SELECTIVITY

One important result from both the steady-state and the non-steady-state studies is the direct evidence for contribution of the lattice oxygen during the oxidation of methane on these catalysts. In all cases, the lattice oxygen was involved in the oxygen reaction pathway for the formation of CO and CO_2. The ability of the lattice oxygen to diffuse readily throughout the solid oxide to a large extent determined its contribution to the reaction. A comparison of the CO_2 transients obtained during reaction reveals that the relative contribution of the lattice oxygen to the rate of CO_2 formation increased as the lattice oxygen diffusivity increased.

Catalyst activities and C_2H_6 selectivities are shown in Table 5-6 as a function of lattice oxygen mobility for the catalysts studied. Under the conditions given in Table 5-6, the catalysts with the lowest lattice oxygen diffusivity (MgO, SnO_2, and Sn/MgO) were the least selective for the production of C_2H_6. As the lattice oxygen diffusivity increased, the catalyst's ability to produce the selective product also increased. The catalyst that exhibited the highest lattice oxygen diffusivity (i.e., cubic Sm_2O_3) was found to be the most active for the formation of C_2H_6.

It is interesting to note that even though the lattice oxygen in NiO was very mobile (only cubic Sm_2O_3 exhibited a higher lattice oxygen diffusivity) no C_2H_6 formed at these conditions. However, NiO was very active for the nonselective oxidation of CH_4. This might indicate that the active surface oxygen species on NiO activates methane nonselectively or that it is too active to allow any intermediate partially oxidized products to leave the surface. At this point, however, the form of this nonselective surface oxygen is unknown.

Lattice oxygen was found to be involved in the formation of both the selective and nonselective products in the oxidative coupling of methane. As the lattice oxygen diffusivity increased, the catalyst's ability to produce the coupled product also increased. Oxygen transients obtained during reaction revealed lattice oxygen to be a part of the oxygen reaction pathway for the formation of CO_2 whereas in the absence of gas-phase oxygen the lattice oxygen is preferentially involved in the formation of C_2H_6.

CONCLUSIONS

Measurements of the oxygen exchange ability of methane coupling catalysts both in the absence of methane and during reaction have demonstrated that for active and selective catalysts there is a large reservoir of oxygen available to participate in the reaction that consists of both surface and lattice oxygen. Indeed, even at the relatively low temperatures used in these studies (600–650°C) the measured bulk oxygen diffusion is rapid enough in the active catalysts that a significant amount of subsurface oxygen can readily participate in the reaction. A comparison of relative activities and selectivities with lattice oxygen diffusivity for highly active and selective methane oxidation catalysts may suggest that the ability of a catalyst to selectively couple methane is related to an optimal availability of lattice oxygen.

Although it is difficult to discern under steady-state conditions whether the selective oxidation products are preferentially made through participation of the lattice oxygen, results from stop-oxygen flow experiments clearly show that product selectivity to coupled products increases significantly in the absence of gas-phase oxygen. So, although it is not possible to delineate different roles clearly for surface versus lattice oxygen in this reaction, it is clear that for the catalysts studied thus far lattice oxygen plays a significant and important role in the reaction pathway.

References

Agarwal, S.K. 1990. Oxidative coupling of methane: Catalytic studies over metal oxides. Ph.D. Dissertation, University of Pittsburgh.

Bielanski, A., and J. Haber. 1979. Oxygen in catalysis on transition metal oxides. *Catal. Rev.—Sci. Eng.* 19:1–41.

Biloen, P. 1983. Transient kinetic methods. *J. Mol. Catal.* 21:17–24.

Biloen, P., J.N. Helle, F.G.A. van den Berg, and W.M.H. Sachtler. 1983. On the activity of Fischer–Tropsch and methanation catalysts: A study utilizing isotopic transients. *J. Catal.* 81:450–63.

Boudart, M. 1968. *Kinetics of Chemical Processes*. Englewood Cliffs, NJ: Prentice-Hall.

Brown, I.B., and W.R. Patterson. 1983. Reactivity of tin oxide and some antimony–tin oxide catalysts for the oxidation of methane and the isotopic exchange of oxygen: An examination of the role of adsorbed and lattice oxygen in catalytic oxidation. *J. Chem. Soc., Faraday Trans. 1* 79:1431.

Carslaw, H.S., and J.C. Jaeger. 1959. *Conduction of Heat in Solids*. New York: Oxford University Press.

Cromer, D.T. 1957, The crystal structure of monoclinic Sm_2O_3. *J. Chem. Soc., Chem. Commun.* 61:753–5.

Dean, J.A., ed. 1985. *Lange's Handbook of Chemistry*, pp. 3–121. New York: McGraw-Hill.

Furusawa, T., M. Suzuki, and J.M. Smith. 1976. Rate parameters in heterogeneous catalysis by pulse techniques. *Catal. Rev.—Sci. Eng.* 13(1):43–76.

Happel, J. 1986. *Isotopic Assessment of Heterogeneous Catalysis*. New York: Academic Press.

Kofstad, P. 1972. *Nonstoichiometry, Diffusion, and Electronic Conductivity in Binary Metal Oxides*. New York: Wiley-Interscience.

Maatman, R.W. 1980. Site density and entropy criteria in identifying rate-determining steps in solid-catalyzed reactions. *Adv. Catal.* 29:97–148.

Oishi, Y., and K. Ando. 1984. Oxygen diffusion in MgO and Al_2O_3. In *Structure and Properties of MgO and Al_2O_3 Ceramics*, ed. W.B. Kingery, pp. 379–93. American Ceramic Society.

Oishi, Y., and W.D. Kingery. 1960. Oxygen diffusion in periclase crystals. *J. Chem. Phys.* 33:905–6.

O'Keeffe, M., and W.J. Moore. 1961. Diffusion of oxygen in single crystals of nickel oxide. *J. Chem. Soc., Chem. Commun.* 65:1438–9.

Otsuka, K., A.A. Said, K. Jinno, and T. Komatsu. 1987. Peroxide anions as possible active species in the oxidative coupling of methane. *Chem. Lett.* 77–8.

Parker, I.B., and W.R. Patterson. 1983. Oxidation of methane with oxygen-18: A method for the quantitative investigation of the activities of adsorbed and lattice oxygen in metal oxide catalysts. *J. Chem. Soc., Faraday Trans. 1* 79:1421–30.

Peil, K.P. 1990. Elucidation of reaction pathways and identification of active sites in the oxidative coupling of methane. Ph.D. Dissertation, University of Pittsburgh.

Peil, K.P., J.G. Goodwin, Jr., and G. Marcelin. 1989. An examination of the oxygen pathway during methane oxidation over a Li/MgO catalyst. *J. Phys. Chem.* 93:5977–9.

Peil, K.P., J.G. Goodwin, Jr., and G. Marcelin. 1991a. Surface phenomena during the oxidative coupling of methane over Li/MgO. *J. Cat.* 131:143

Peil, K.P., J.G. Goodwin, Jr., and G. Marcelin. 1991b. Steady-state vs. non-steady-state transient kinetic analysis of surface coverages during the oxidative coupling of methane. *J. Cat.* In press.

Soong, Y., K. Krishna, and P. Biloen. 1986. Catalyst aging studied with isotopic transients: Methanation over Raney nickel. *J. Catal.* 97:330–43.

Stephanopoulos, G. 1984. *Chemical Process Control: An Introduction to Theory and Practice*, p. 180. Englewood Cliffs, NJ.: Prentice-Hall.

Stone, G.D., G.R. Weber, and L. Eyring. 1968. Self-diffusion of oxygen in neodymium and samarium sesquioxide. In *Mass Transport in Oxides*, ed. J.B. Wachtman, Jr., and A.D. Franklin, pp. 179–86. National Bureau of Standards Special Publication 296. Washington, DC: US Government Printing Office.

Tamaru, K. 1978. *Dynamic Heterogeneous Catalysis*. New York: Academic Press.

Templeton, D.H., and C.H. Dauben. 1954. Lattice parameters of some rare earth compounds and a set of crystal radii. *J. Chem. Soc., Chem. Commun.* 5237–9.

Wang, J.-X., and J.H. Lunsford. 1986. Characterization of [Li^+O^-] centers in lithium doped MgO catalysts. *J. Phys. Chem.* 90:5883–7.

Yang, C.-H., Y. Soong, and P. Biloen. 1984. Abundancy and reactivity of surface intermediates in methanation, determined with transient kinetic methods. In *Proceedings of the 8th International Congress in Catalysis*, vol. 2, pp. 3–14.

Yang, C.-H., Y. Soong, and P. Biloen. 1985 A comparison of nickel- and platinum-catalyzed methanation utilizing transient-kinetic methods. *J. Catal.* 94:306–9.

Zhang, X., and P. Biloen. 1986. A transient kinetic observation of chain growth in the Fischer–Tropsch synthesis. *J. Catal.* 98:468–76.

6

The Development of Doped Li/MgO Catalyst Systems for the Low-Temperature Oxidative Coupling of Methane

S.J. Korf, J.A. Roos, and J.R.H. Ross

Faculty of Chemical Technology, University of Twente, Enschede, The Netherlands

TABLE OF CONTENTS

INTRODUCTION

The direct conversion of methane, the predominant component of natural gas, to more-useful chemicals and fuels is a topic that is attracting more and more interest, it having been recognized as one of the most technologically important and challenging problems facing industrialized nations in the latter part of the twentieth century. Although this chapter emphasizes work performed in our group, reference also is made, where appropriate, to parallel work by other groups.

GENERAL ASPECTS OF THE METHANE COUPLING REACTION OVER Li / MgO CATALYSTS

Figures 6-1 and 6-2 show a comparison of the data obtained with a methane–oxygen ratio of 10 for a reactor containing (a) quartz, (b) Li/MgO, and (c) PbO/Al_2O_3 (Roos et al. 1987). The empty reactor gave some reaction at temperatures above about 700°C and the predominant product was CO, there being a limited amount of coupling to C_2 products. The Li/MgO material gave higher conversions of both methane and oxygen over the whole temperature range and the selectivities to C_2 products were higher than with the empty reactor; both CO and CO_2 were formed as products over the whole temperature range. Ethane was the predominant product of the reaction at lower temperatures, whereas the

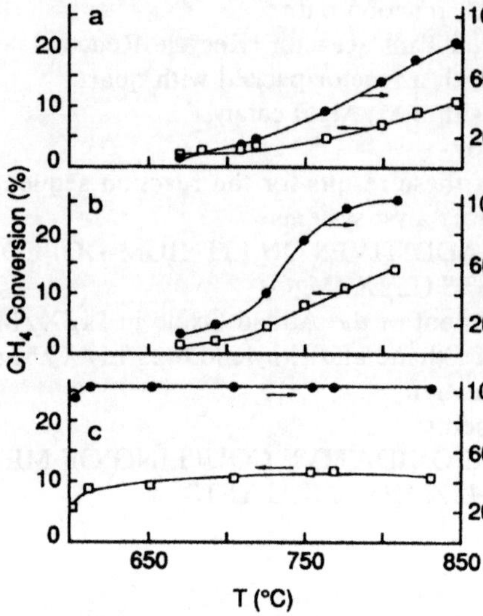

FIGURE 6-1. Conversions of methane (□) and oxygen (•) versus temperature: (A) quartz; (B) Li/MgO; (C) PbO/Al_2O_3.

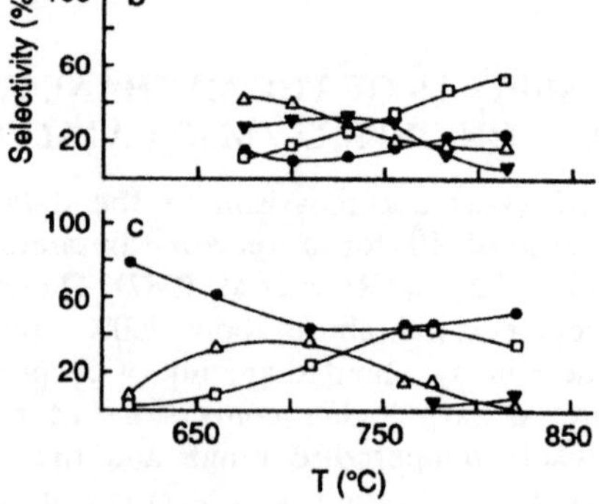

FIGURE 6-2. Selectivities corresponding to the experiments shown in Figure 6-1: (△) C_2H_6; (□) C_2H_4; (▾) CO; (•) CO_2.

proportion of ethylene found increased at higher temperatures. Increasing the amount of oxygen in the reaction mixture (to $CH_4:O_2 = 5$) led to higher conversions of methane and to lower selectivities to C_2 products.

An important observation made for both the quartz and Li/MgO experiments was that H_2 was a significant product from the coupling reaction in both cases; hydrogen was even observed under conditions in which unreacted oxygen was present. This observation was in contrast to the situation found for PbO/Al_2O_3, for which CO_2 and H_2O were the only oxygen-containing products formed at lower temperatures and for which hydrogen was only formed at higher temperatures, in the same region in which CO was first observed (Figure 6-2c). A further observation made with the lead catalysts was that significant loss of lead occurred during use. Later work has shown that similar loss of one or more catalyst component can occur for various other catalyst systems under reaction conditions; this is particularly significant for the Li/MgO system, as is discussed in the next section.

Strangely, although several authors have since also noted that hydrogen can be formed during methane coupling experiments, little attention has been paid to this observation. Our explanation of the occurrence of hydrogen was that the water–gas shift reaction (6-1) occurred whenever CO was formed as a product in the reaction;

$$CO + H_2O \Leftrightarrow CO_2 + H_2 \tag{6-1}$$

because formaldehyde was also found as a trace product in the water condensed in the exit of the reactor, we suggested that decomposition of formaldehyde formed as a parallel product of the C_2 hydrocarbons could give rise to CO and hydrogen:

$$HCHO \rightarrow H_2 + CO \tag{6-2}$$

The hydrogen could also be formed by direct dehydrogenation of ethane:

$$C_2H_6 \rightarrow C_2H_4 + H_2 \tag{6-3}$$

We now suggest that an additional route might be via the steam-reforming of the ethane (or ethylene) formed, making use of the water produced:

$$C_2H_6 + 2H_2O \rightarrow 2CO + 5H_2 \tag{6-4}$$

Steam-reforming of the methane itself cannot be ruled out, although the reactivity of methane in steam-reforming is usually lower than that of higher hydrocarbons. It is interesting to note that early steam-reforming catalysts consisted of alkali metal salts supported on refractory oxides.

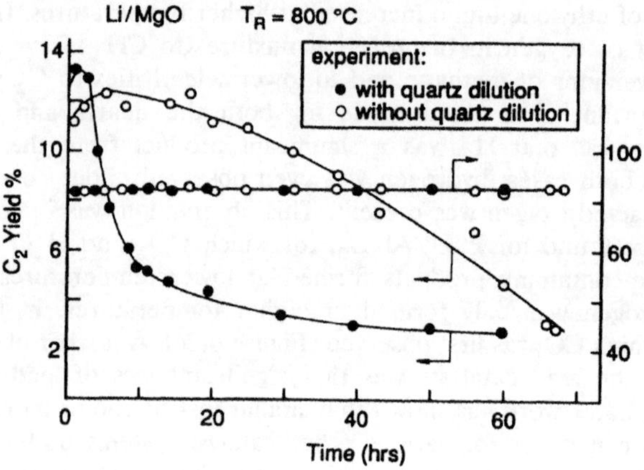

FIGURE 6-3. Aging experiments at $T_R = 800°C$ with Li/MgO, with and without quartz dilution.

The Stability of Li / MgO Catalysts

Analysis of Li/MgO catalysts prepared by impregnation techniques and tested under standard conditions showed that significant loss of lithium from the materials occurred after calcination and also during use in a continuous-flow reactor at high temperature. We therefore studied the stability of Li/MgO materials at a reaction temperature of 800°C both with and without quartz dilution (Korf et al. 1988); the results, obtained with a CH_4–O_2 ratio of 10, are shown in Fig. 6-3. With quartz dilution, the C_2 yield decreased very rapidly, as a result of a drop in conversion; the C_2 selectivity remained virtually constant throughout at a value of about 83%. We interpreted this to indicate that the nature of the active site had not altered very significantly but that the number of sites had decreased during the experiment. (The actual variation of C_2 selectivity depends on the reaction scheme and the reaction orders.) After the experiment, both the carbonate and lithium contents of the catalyst were found to have fallen to ca. 0.1 wt%. A white deposit, identified as being LiOH, was found downstream from the reactor, it was also found that Li had reacted with both the quartz diluent and the reactor itself to form a Li_2SiO_3 phase. An equivalent experiment was then carried out in which the reactant flow was stopped after 60 h and the catalyst was exposed at 780°C to a flow of CO_2 (1.67 cm^3 s^{-1} at STP) for 2 h. On recommencing the reaction, the initial behavior of the catalyst had been restored, despite the loss of Li; however, the subsequent deactivation behavior was much more rapid (Korf et al. 1987). When the aging experiment was carried out without quartz dilution (Fig. 6-3), the results were much better; the C_2 yield remained virtually constant over a period of 25 h, after which gradual deactivation started. The selectivity again remained virtually constant throughout at a value of about 83%. There was once more loss of Li from the catalyst, but this was not so significant as with the experiments with quartz diluent. It is relevant to note that a fresh catalyst containing less than 0.5 wt% Li is relatively unselective in comparison with the aged catalyst with a similar lithium content. We therefore can conclude that the form of the lithium species in the aged catalyst is much more favorable for methane coupling than its form in the fresh material, although we cannot be clear what that form is; the difference may be due to a different geometric distribution or to a different electronic interaction with the surroundings. Martin and Mirodatos (Mirodatos and Martin 1989) have confirmed these results.

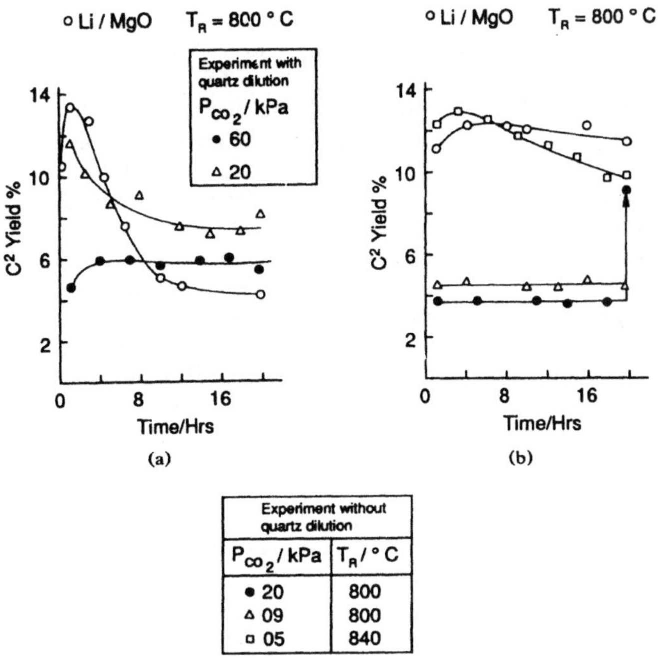

FIGURE 6-4. The influence of CO_2 on the stability of Li/MgO: (a) with quartz diluent; (b) without quartz diluent.

The Influence of CO_2 and H_2O on the Oxidative Coupling of Methane over Li / MgO

It became clear to us in the course of our work that carbonate species on the catalyst surface played an important role in determining the behavior of the Li/MgO system, an idea that is discussed more fully later. We therefore carried out a series of experiments to examine the effect of adding CO_2 and H_2O to the feed to the reactor. Figure 6-4a shows results for an experiment in which the catalyst was diluted with quartz particles and Fig. 6-4b shows equivalent results for an experiment in which there was no quartz dilution. A portion of the results of experiments without CO_2 addition (Fig. 6-3) are shown for comparison. The addition of CO_2 resulted in a reduction of the initial yield of the catalyst. In the experiment with quartz dilution (Fig. 6-4a), the use of a CO_2 concentration of 6 kPa resulted in a steady behavior with a yield of 6%, the selectivity being 85% throughout, this yield, although low at the early stages of the reaction, exceeded the yield obtained in the reaction without CO_2 addition at higher reaction times and remained constant at this value. With 2 kPa of CO_2 in the feed, the behavior was better, a steady yield of about 8% being obtained after a small initial deactivation. When the CO_2 flow was stopped after 20 h with either 2 or 6 kPa of CO_2 in the feed, the behavior with respect to C_2 production returned to that shown for the fresh catalyst in the absence of CO_2; however, the deactivation was then much more rapid. Without quartz dilution (Fig. 6-4b), addition of 2 kPa of CO_2 resulted in a major decrease in activity; the C_2 yield under these conditions was 3.8% and the C_2 selectivity was 85%. A CO_2 concentration of 0.9 kPa resulted in a C_2 yield of 4.5% and a C_2 selectivity of 83%. After 20 h, when the CO_2 flow was stopped, the yield rose, as shown by the arrow in Figure 6-4b, but the rise was to a value lower than that for a fresh catalyst in the absence of CO_2. The fresh activity could, however, be restored by a treatment at 840°C in air for 2 h. It would thus seem that the catalyst in this case was poisoned by some sort of strongly adsorbed carbonate species. It would also seem that the poisoning effect of the added CO_2 on the activity is much diminished if quartz is added to the catalyst bed. The nature of this effect is not fully understood; it is possible that the presence of quartz aids the decomposition of the surface carbonate species. That the presence of quartz affected the interaction of the Li and CO_2 was established by temperature programmed desorption (TPD) measurements, in which it was shown that CO_2 desorption occurred at lower temperatures in the presence of quartz than it does with the undiluted material. It is possible that additional catalytically active sites are created on the quartz surface that contribute to the over all activity of the catalysts as well as to their improved behavior in the presence of CO_2. We can conclude from these results that the presence of CO_2 in the gas feed has two effects: to poison reversibly the active sites for the oxidative coupling reaction, but also to stabilize them against deactivation.

The addition of water to the feed to the reactor was detrimental to the stability of the catalyst (Korf et al. 1990a). Volatile LiOH was formed and this evaporated from the catalyst, traveling in the direction of the gas stream and resulting in the formation of a white deposit (of LiOH) after the catalyst bed. The addition of CO_2 and H_2O to the gas feed influenced only the activity; the C_2 selectivities did not change significantly, indicating that the nature of the active site had not altered appreciably and that only the number of active sites had changed. This result appears to be in conflict with the findings of Kimble and Kolts (Kimble and Kolts 1986), who found that the addition of water to the gas feed increased the C_2 yield. We have also found this effect for catalysts that contained strongly bonded carbonate, which rendered the catalyst less active. Under these conditions, the improvement observed was related to more facile decomposition of the carbonate species by water (Korf et al. 1990a).

The Influence of Periodic Reversal of the Direction of Flow of the Gas Stream

Using a high gas velocity (to avoid back-mixing in the reactor, as discussed further in a later section), with a methane–oxygen ratio of 5 (0.5 bar CH_4 and 0.1 bar O_2) and an active Li/MgO catalyst, a hot spot (indicating a reaction front) was always found in the catalyst bed. This front was found to move gradually through the bed in the direction of the gas stream, indicating that the catalyst was aging, as discussed previously. The results of two aging experiments carried out using these conditions, using a bed of 1500 mg of Li/MgO catalyst at T_R = 780°C with a methane–oxygen ratio of 5, are given in Fig. 6-5. In one experiment (i) the flow was kept constant whereas in the other (ii) it was reversed periodically. The initial C_2 yield for the first experiment was 19.0% and the C_2 selectivity was 67%. The hot spot reached the end of the catalyst bed after reaction for 13.5 h, and there was then a very rapid decrease in the C_2 production; after the experiment, the catalyst was found to contain 0.18 wt% Li and 0.21 wt% CO_2. When the reaction front reached the end of the catalyst bed, a substantial amount of Li evaporated from the catalyst (as LiOH), forming the white deposit observed at the exit of the reactor. With the intention of minimizing this loss of Li from the catalyst bed and of achieving a more

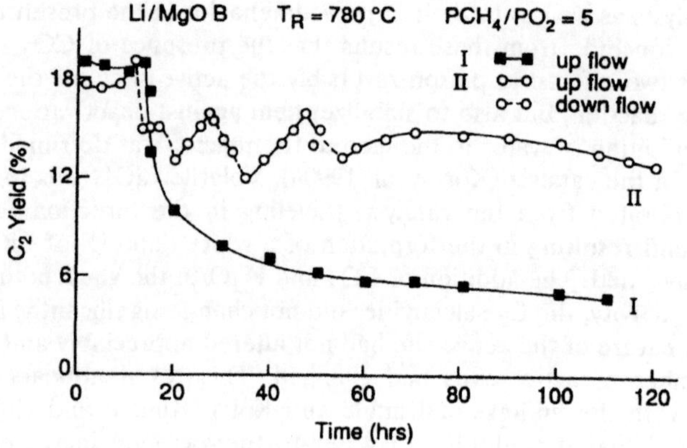

FIGURE 6-5. The yield of C_2 products as a function of time of reaction at 780°C for Li/MgO in an upflow experiment as well as for an experiment in which the reactant flow was periodically reversed.

stable behavior, the second experiment (ii) shown in Fig. 6-5 was carried out. In this, the direction of flow of the gas stream was reversed periodically. After a sequence of flow reversals, it was found that the subsequent stability of the catalyst was much improved; the yield remained much higher, even after the reversal steps were stopped, than was found in normal experiments. After run (ii) of Fig. 6-5, the catalyst contained 0.27 wt% Li, a value somewhat higher than that found in the case of the upflow experiment (i) but still very low compared with that for the fresh catalyst. This effect confirms the observation made previously that only small amounts of Li are needed to create the active sites on the surface of the Li/MgO catalyst if the Li has an optimal form; this form is encouraged by periodical reversal of the direction of flow of the gas stream.

A Reaction Model for the Li / MgO Catalyst System

From these measurements and from a study of the effects of variation of several parameters during the synthesis of a Li/MgO catalyst (Korf et al. 1988, 1990a), we proposed the model shown in Fig. 6-6 for the reactions occurring on the catalyst surface that are responsible for creating the active sites. A good Li/MgO catalyst consists of Li_2CO_3 and MgO. The active sites are created on the surface by the gradual loss of CO_2 from surface Li_2CO_3 species in the presence of oxygen; the presence of lithium

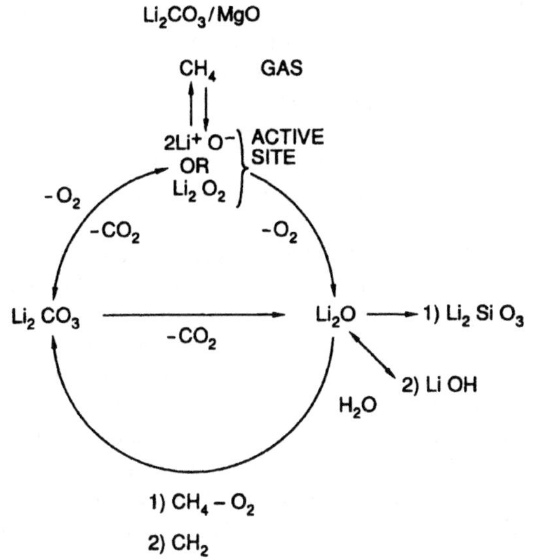

FIGURE 6-6. Scheme showing the possible interrelation between the surface species on a Li/MgO catalyst.

carbonate on the surface of the catalyst is therefore crucial for its activity. The active sites may be either Li^+O^-, as proposed by Ito and co-workers (Ito et al. 1985), or lithium peroxide entities. Reaction of methane with the active site gives rise to the formation of a methyl radical and a reduced surface site. According to the mechanism proposed by Ito and co-workers (Ito et al. 1985), this reduced site is a LiOH species; in the catalytic cycle of their model, two of these species react to give Li_2O and H_2O, followed by reoxidation of Li_2O with gas-phase oxygen to form two Li^+O^- sites. Li_2O may thus be formed as an intermediate of the oxidative coupling reaction (if the model proposed by Ito and co-workers holds), but it may also be formed as an intermediate by direct decomposition of Li_2CO_3. The Li_2O is likely to react further, it being less stable than LiOH, Li_2CO_3, or Li_2SiO_3. In the presence of a high H_2O–CO_2 ratio, volatile LiOH entities can be formed, giving rise to the loss of active species. In the presence of CO_2, the carbonate is reformed instead of the volatile hydroxide, giving an improvement of the catalyst stability and hindering the formation of the volatile LiOH species but inhibiting the coupling reaction; the CO_2 may either be that resulting from the coupling process or be added to the reaction mixture (see above). It would appear that intermediate surface OH species formed by reaction of methane with surface oxygen species, as required in the model of Ito and co-workers (Ito et al. 1985), are not immediatly lost from the catalyst. As discussed previously, deactivation of the catalyst is diminished at low H_2O–CO_2 ratios. Figure 6-6 also shows the parallel loss of Li by interaction with quartz; dilution with quartz particles and the use of quartz reaction vessels should be avoided if truly representative results for a particular catalyst system are expected, especially at high temperatures.

Lunsford and co-workers have also recently recognized the importance of carbonate species in alkali catalysts (Lunsford 1990). They found that Na/MgO was a good catalyst for the oxidative coupling of methane at relatively high temperatures ($T > 720°C$). However, they were not able to detect Na^+O^- centers on this material. They therefore came to the conclusion that although centers of the type M^+O^- are useful in interpreting catalytic results at $T < 720°C$, other species derived from the decomposition of carbonate entities are responsible for the activation of CH_4 at higher temperatures. van Kasteren and co-workers (van Kasteren, Geerts, and van der Wiele 1989) have also recently published results that confirm our model.

The Effect of the Concentration of Oxygen in the Gas Feed

A number of experiments were carried out to show the effect on the product composition of the partial pressure of oxygen in the feed; the experiments used a Li/MgO catalyst in a reactor with a small post catalytic volume. Figure 6-7 shows the partial pressures of the C_2 products and total oxidation products as a function of the partial pressure of oxygen in the feed for reactor temperatures of 700, 750, and 800°C. Of particular importance is the observation that the C_2H_4–C_2H_6 ratio increased with increasing temperature and partial pressure of oxygen. Little CO was formed at low partial pressures of oxygen, but the proportion increased as the oxygen partial pressure was increased. The amounts of CO and CO_2 increased as the partial pressure of oxygen was increased, but the total C_2 product concentration remained constant; the proportion of ethylene in the products increased with increase oxygen partial pressure, indicating that the ethylene probably was produced from ethane by oxidative dehydrogenation as has also been concluded by other authors (Otsuka, Jinno, and Morikawa 1986; Hinsen, Bytyn, and Baerns 1984). We shall show in the following section that this increase in the degree of total oxidation is likely to be due to a high rate of oxidation of the C_2 compounds encouraged by the high partial pressures of oxygen.

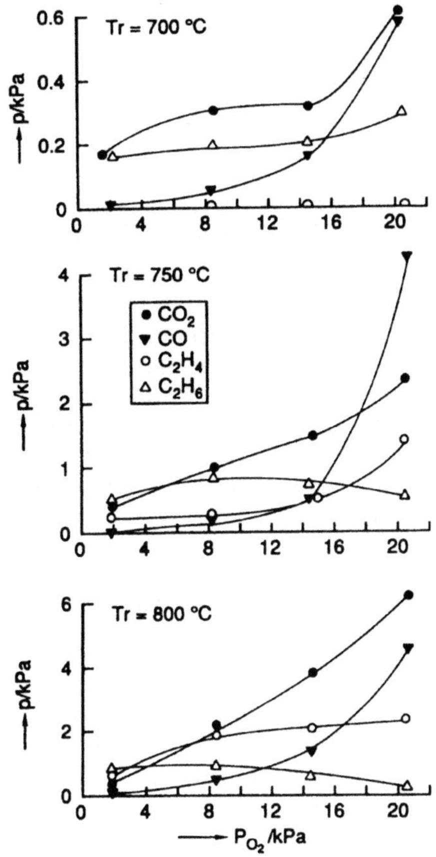

FIGURE 6-7. Product concentrations as a function of the partial pressure of oxygen in the reactor feed containing 50 kPa CH_4, balance He, and total pressure 101 kPa.

The Effect of Back-Mixing

We have demonstrated the importance of sequential oxidation reactions of the desired C_2 hydrocarbons by examining the effect on the product distribution of back-mixing in the reactor. From a reactor engineering standpoint, it is to be expected that the selectivity of a reaction system in which there is the possibility of sequential reactions will be highest in a plug-flow reactor. [In an ideal plug-flow reactor, back-mixing is absent (Westerterp, van Swaaij, and Beenackers 1984)]. In the reaction under

Table 6-1 Results Showing the Effect of Back-Mixing Using an External Recirculation Pump[a]

R	CH_4 Conversion (%)	O_2 Conversion (%)	C_2 Selectivity (%)	C_2 Yield (%)
1	25	95	67	16.4
10	17	82	54	9.0

[a]500 mg of Li/MgO catalyst (2.8 wt% Li); recycle ratio $R = 1$ indicates no recirculation, $R = 10$ indicates ideally mixed reactor. Reactor feed: $P_{CH_4} = 50$ kPa, $P_{O_2} = 10$ kPa, flow rate = 0.83 $cm^3\ s^{-1}$, $T_r = 800°C$.

discussion, back-mixing of the C_2 products (unstable relative to methane) will lead to transport of these products to those regions of the reactor, nearer to the entrance, where a higher partial pressure of oxygen exists; this gives an increased chance of further oxidation of the C_2 products and consequently a lower C_2 selectivity (Roos et al. 1988).

The results of two experiments involving different degrees of back-mixing (but with the same amount of catalyst and the same residence time in the reactor) are given in Table 6-1. In the first experiment, the reactor was used in a single-pass mode (approaching plug flow) as in the experiments described previously. In the second experiment, it was used in a recycle mode, with a recycle ratio of 10, the experiment being carried out with an external recirculation pump in such a way that the conditions approached those of an ideally mixed reactor. Because the other process conditions were exactly the same in both experiments, any differences between the two experiments can only be caused by the different degree of back-mixing and the consequently different concentration profiles throughout the reactor. Table 6-1 shows that there were distinct differences between the results of the two experiments. For the single-pass experiment, both the conversion and selectivity were superior. The lower conversion obtained with the well-mixed reactor ($R = 10$) is as would be expected for this type of reactor for a reaction with an order greater than zero. The low C_2 selectivity in the second experiment compared with the first is consistent with the suggestion that sequential oxidation of the C_2 products to CO_x occurs. The following section summarizes further evidence for this suggestion.

Addition of Ethylene and Ethane to a Gas Mixture Containing Methane and Oxygen

Figure 6-8 shows the results of an experiment in which ethylene was added in various concentrations to the methane–oxygen feed to a continuous-flow reactor; the consumption of oxygen was complete for all the conditions

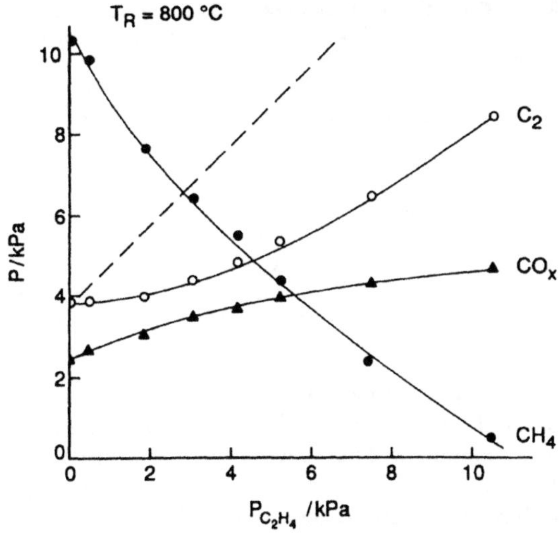

FIGURE 6-8. Product concentrations in the reactor effluent and the absolute methane conversion (ΔCH_4) as a function of the ethylene concentration in the reactor feed.

shown. The dotted line gives the C_2 concentration that would be expected if there were no reaction of the added ethylene and if the coupling reaction were undisturbed. Comparison of this line with the experimentally found C_2 curve shows that there was a reduction in C_2 product concentration. Figure 6-8 also shows the change in partial pressure of the methane through the reactor, it is clear that the addition of ethylene caused a significant decrease in the conversion of methane and we must conclude that the ethylene competes with the methane for the oxygen, the reactivity of ethylene with oxygen being much higher than that of methane.

From similar experiments involving the addition of ethane to the reactor feed, we concluded that the reactivity of ethane with oxygen is also higher than that of methane. The order of reactivity of the three molecules was found to be as follows: ethylene > ethane > methane. This order of reactivity has also been reported by Martin and co-workers (Martin et al. 1989).

DETERMINATION OF THE REACTION PATH

Continuous Stirred Tank Reactor (Recycle Reactor)

Table 6-2 shows the results of five separate experiments for a Li/MgO catalyst, carried out in the recirculation reactor, in which the effect on the

Table 6-2 Rate of Formation of CO and CO_2 in Oxidation Experiments in Which the Reactor Was Filled with a Li / MgO Catalyst; $T = 720°C$

	HYDROCARBONS IN REACTOR FEED				
	C_2H_4 (1)	CH_4–C_2H_4 (2)	C_2H_6 (3)	CH_4–C_2H_6 (4)	CH_4 (5)
P_{O_2} (kPa)	5.4	5.7	5.8	5.5	5.6
P_{CH_4} (kPa)	0.3	103.3	0.5	100.9	113.5
$P_{C_2H_6}$ (kPa)	0.0	1.4	7.6	7.5	2.1
$P_{C_2H_4}$ (kPa)	6.3	6.3	6.1	6.5	1.9
R_{CO_x} (10^{-7} mol s^{-1})	3.9	4.0	4.0	4.5	2.1

rate of formation of the total oxidation products of a variation of the partial pressures of CH_4, C_2H_6, and C_2H_4 (at a constant partial pressure of oxygen) was examined. The reaction temperature was chosen to be 720°C, as previous experiments (Roos et al. 1989b) had shown that the rates of gas-phase oxidation reactions of CH_4, C_2H_6, and C_2H_4 were negligible under these conditions.

The first four columns of Table 6-2 give the partial pressures of O_2, CH_4, C_2H_6, and C_2H_4 measured at the exit of the reactor; because the reactor is practically gradientless, these concentrations can be taken to be those throughout the catalyst bed. The last column of Table 6-2 gives the rate of formation of CO_x for each of the experiments.

The rate of formation of CO_x in each of the experiments shown in the first four columns was approximately the same, regardless of whether or not methane was present in the reaction mixture; these rates were twice as high as the rate observed in the absence of ethane or ethylene, shown in the last column. We therefore concluded that the contribution of the formation of CO_x directly from methane was negligible in the case of the Li/MgO catalyst under these conditions of relatively high conversions. Comparison of the results of column 1 with those of column 3 or of column 2 with those of column 4 shows that the reaction of ethane to give CO_x was negligible in comparison with that of ethylene: the rate of production of CO_x was not affected by increasing the partial pressure of ethane, either in the presence of methane or without methane, as long as the partial pressure of ethylene was the same. We therefore concluded that the main route to the formation of total oxidation products is the oxidation of ethylene and that the reaction in the presence of a Li/MgO catalyst in a recycle reactor at $T_R = 720°C$ can be simplified to a sequential reaction scheme:

$$CH_4 \rightarrow C_2H_6 \rightarrow C_2H_4 \rightarrow CO_x$$

Table 6-3 Rates of Production of CO and CO_2 in Oxidation Experiments in Which the Reactor Was Filled with Quartz; T = 800°C

	HYDROCARBONS IN REACTOR FEED	
	C_2H_4 (1)	CH_4–C_2H_4 (2)
P_{O_2} (kPa)	3.0	2.9
P_{CH_4} (kPa)	1.4	117.8
$P_{C_2H_6}$ (kPa)	0.1	0.3
$P_{C_2H_4}$ (kPa)	3.3	3.3
P_{CO_x} (10^{-7} mol s^{-1})	6.2	0.8

At a reaction temperature of 800°C, gas-phase oxidation reactions appear to be more important than when the reaction temperature is 720°C and thus homogeneous reactions compete with the catalytic reactions (Roos et al. 1989a). We demonstrated, as is shown later, that the addition of methane had an effect on the rates of the total oxidation reactions of C_2H_6 and C_2H_4 in the gas phase (Roos et al. 1989d) and this makes it difficult to reach such clear conclusions concerning the reaction network at that temperature as was possible for the reaction temperature of 720°C.

Experiments with a Reactor Packed with Quartz

The effect of excess methane on the rate of oxidation of ethylene at 800°C in the recycle system in the presence of quartz, but in the absence of a catalyst, is shown in Table 6-3. It should be noted that the partial pressures of the reactants were not the same as those of the equivalent experiments in Table 6-2. The experiment of column 1 shows that the rate of oxidation of ethylene to CO_x in the gas phase was relatively high if the methane concentration was low. However, when the concentration of methane was high (column 2), the concentration of oxygen and ethylene being kept the same, the rate of CO_x formation was significantly lower and also was considerably lower than the values given in Table 6-2. Analogous results to those shown in Table 6-3 were obtained with ethane in place of ethylene. We can conclude from these experiments that the methane has an inhibiting effect on the oxidation of the C_2 products, probably by a reduction in the number of chain carriers, if there is no catalyst present. It is thus dangerous to conclude from experiments without added methane that the homogeneous oxidation of C_2 molecules plays an important role in the production of CO_x at higher temperatures under methane coupling conditions.

Table 6-4 Rate of Formation of CO and CO_2 in Oxidation Experiments in Which the Reactor Was Filled with a Li / MgO Catalyst; $T = 800°C$

	HYDROCARBONS IN REACTOR FEED	
	C_2H_4 (1)	CH_4–C_2H_4 (2)
P_{O_2} (kPa)	2.2	2.2
P_{CH_4} (kPa)	0.4	119.1
$P_{C_2H_6}$ (kPa)	0.1	1.6
$P_{C_2H_4}$ (kPa)	3.7	3.7
R_{CO_x} (10^{-1} mol s^{-1})	6.4	4.7

Experiments with a Li / MgO Catalyst

Table 6-4 shows experiments similar to those given in Table 6-3 (T_R = 800°C) when the reactor contained the Li/MgO catalyst. Comparison of the rates of CO_x formation in the two experiments of Table 6-4 shows that methane again had an inhibiting effect but that the effect was much less significant than for the data of Table 6-3. Similar results were obtained when ethane was used in place of ethylene.

These results can be explained if one postulates that the rate of the surface reaction of the C_2 molecules with oxygen at 800°C is affected less by the addition of methane than is the gas-phase reaction at the same temperature; as shown previously, the addition of methane had no appreciable effect at a temperature of 720°C and the same is likely to be the case at higher temperatures. Because the surface reaction has a greater chance of proceeding in the bed when the reactor is filled with catalyst than when it is filled with silica, the gas-phase reaction will play a lesser part in the experiments with catalyst and so the effect of added methane will be less.

Plug-Flow Reactor

In an attempt to obtain further evidence for the reaction pathway found for Li/MgO in a recycle reactor, the effect on reactant and product concentrations of varying the residence time (W/F) was studied in a plug-flow reactor, holding the flow rate F constant at 1.67 $cm^3\ s^{-1}$ (at STP) and varying the catalyst weight W (Korf et al. 1990b). As discussed in the preceding section, we showed, using Li/MgO, that back-mixing of the reaction mixture had a negative effect on the C_2 selectivity: Because ethane and ethylene are susceptible to oxidation, the optimum selectivity was reached under plug-flow conditions. When a flow rate F of 1.67 cm^3

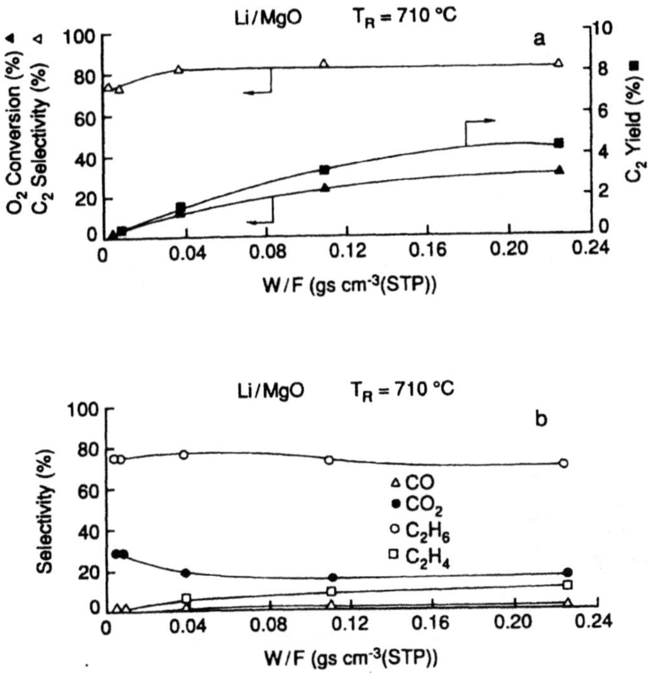

FIGURE 6-9. Product selectivity, oxygen conversion, and C_2 yield as a function of W/F at T_R = 710°C for Li/MgO.

s^{-1} (at STP) is used, the reactor thus can be considered to be operating under plug-flow conditions (Korf et al. 1989a). Furthermore, reactions occurring in the so-called postcatalytic volume (deep oxidation of the C_2 products) may be considered to be negligible with this rate of gas flow (Korf et al. 1989a).

Figure 6-9 shows that only C_2H_6 and CO_2 were formed at low residence times, the selectivity to C_2H_6 being high; these results show that CO and C_2H_4 were not primary products with this catalyst under these conditions (Korf et al. 1990b). There was a small increase in C_2 selectivity with increasing residence time at low values, whereas the C_2 yield increased over the whole range of W/F as a result of increasing conversions. The CO_2 selectivity decreased with increasing residence times and a small amount of CO was formed. Figure 6-10 shows equivalent results for the effect of residence time at the higher temperature of 780°C for the Li/MgO sample. The selectivity to C_2H_6 was again very high at low

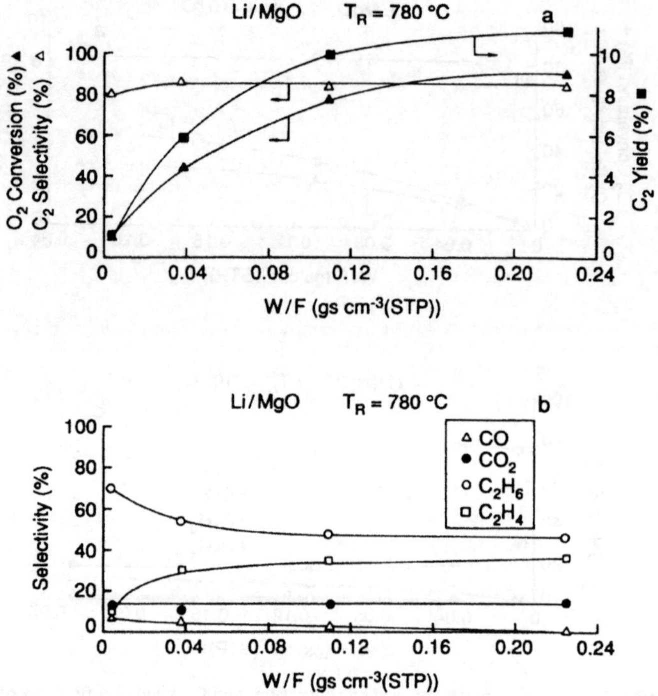

FIGURE 6-10. Product selectivity, oxygen conversion, and C_2 yield as a function of W/F at $T_R = 780°C$ for Li/MgO.

residence times but it decreased markedly and the ethylene selectivity increased with increasing W/F; the CO selectivity decreased with increasing residence time, whereas the CO_2 selectivity first decreased slightly and then increased. The C_2 selectivity was virtually constant for all values of the residence time.

Significance of these Results for the Reaction Sequence and Comparison with Other Catalyst Systems

It is generally accepted for the methane coupling reaction that C_2H_6 is produced by combination of two $CH_3\cdot$ species, formed in the rate-determining step; the combination probably occurs either at the surface or in the gas phase near the surface (Ito et al. 1985):

$$2CH_3\cdot \rightarrow C_2H_6 \tag{6-5}$$

According to reaction (6-5), the rate of formation of the C_2 compounds should be proportional to the square of the $CH_3\cdot$ concentration; if one assumes that CO and CO_2 are formed from single $CH_3\cdot$ species, one can conclude that the rate of formation of these species will have a lower order with respect to the $CH_3\cdot$ concentration than does the coupling reaction (Ito et al. 1985). An increase in the amount of catalyst, W (keeping F constant), and an increase in temperature will both give rise to the production of more methyl species; thus this in turn will favor the formation of C_2 products. This increased formation of C_2 species will be appreciable only if further oxidation of the C_2 products to CO and CO_2 is relatively insignificant. The results given in Figs. 6-9 and 6-10 indeed show that increases in W/F (with F constant) or temperature resulted in an increase in C_2 selectivity and C_2 yield for Li/MgO.

It is widely agreed (see also above) that ethane is a primary product of methane activation and that ethylene is formed by (oxidative) dehydrogenation of ethane (Ito et al. 1985; Otsuka, Jinno, and Morikawa 1986; Hinsen, Bytyn, and Baerns 1984). That conversion of ethane into ethylene occurs at the higher reaction temperatures with a Li/MgO catalyst can also be concluded from the results in Fig. 6-10.

In the previous section, we suggested (Roos et al. 1989d) that the oxides of carbon are formed predominantly from the C_2 products over Li/MgO catalysts. The sequential oxidation becomes more important with increasing temperature, increasing oxygen partial pressure, and increasing partial pressures of ethane and ethylene. Using a continuous stirred tank reactor (CSTR) (Roos et al. 1989d), we have shown that the reactions taking place during the oxidative coupling of methane over Li/MgO at 720°C follow a sequential reaction scheme.

Under the conditions of the CSTR [high $W/F = 3.76$ g s cm^{-3} (STP)], a high concentration of $CH_3\cdot$ species and thus high concentrations of ethane and ethylene are formed; hence, the formation of CO_x from the C_2 products will also be very significant. Thus, under the conditions of the CSTR, the rate of oxidation of the methyl species and the oxidation of ethane will be negligible in relation to the rate of oxidation of ethylene.

In the comparative studies that we have carried out to date, the highest C_2 selectivities were obtained with Li/MgO. All the rare-earth oxides that we have studied (Korf et al. 1989a) (i.e., Sm_2O_3, La_2O_3, Dy_2O_3, Gd_2O_3, and Tb_4O_7) gave lower selectivies than found with the Li/MgO sample. However, the oxides of Sm, Dy, La, and Gd were active at lower reaction temperatures than was Li/MgO (Korf et al. 1989a). We have also found that Sm_2O_3, La_2O_3, Dy_2O_3, and Ca/Sm_2O_3 are more stable than Li/MgO (Korf et al. 1989a). In an attempt to find catalysts with which the active component is not lost during reaction, we also studied the use of the more-stable carbonate $BaCO_3$, supported on $Ca(OH)_2$ (Ba/CaO). We chose $BaCO_3$ for study because $BaCO_3$ cannot be volatized under the reaction conditions used for methane coupling. This catalyst was also found to be more stable but less selective than was Li/MgO (Korf et al. 1990b). From measurements in a plug-flow reactor of the effect of residence time on the product distribution of these catalysts, we concluded that the direct oxidation of methane (or of intermediate methyl species) to give CO and CO_2 plays a less significant role over Li/MgO than over Ba/CaO, Sm_2O_3 (cubic and monoclinic), Na/Sm_2O_3, and La_2O_3 (Korf et al. 1990b; Korf, van Ommen, and Ross 1990). The Li/MgO material is thus unique in this respect.

THE EFFECT OF ADDITIVES ON LITHIUM-DOPED MAGNESIUM OXIDE CATALYSTS (Li / X / MgO)

As we have shown previously a significant disadvantage of the Li/MgO system is that the highest C_2 selectivities and yields are reached at relatively high temperatures (780°C). Under these conditions, the active sites created are not stable but are destroyed as a result of the loss of lithium. Homogeneous gas-phase reactions also begin to play a significant role at these high temperatures, with the consequence that control of the selectivity of the catalyst by modification of its formulation is difficult to achieve. An ideal catalyst would have high activities at relatively low temperatures while maintaining relatively high selectivies. With the aim of improving the activities and stabilities of materials based on the Li/MgO

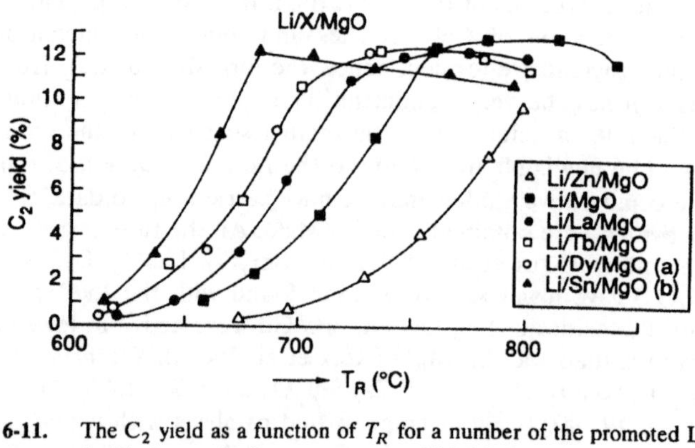

FIGURE 6-11. The C_2 yield as a function of T_R for a number of the promoted Li/MgO catalysts compared with the results for an unpromoted Li/MgO sample.

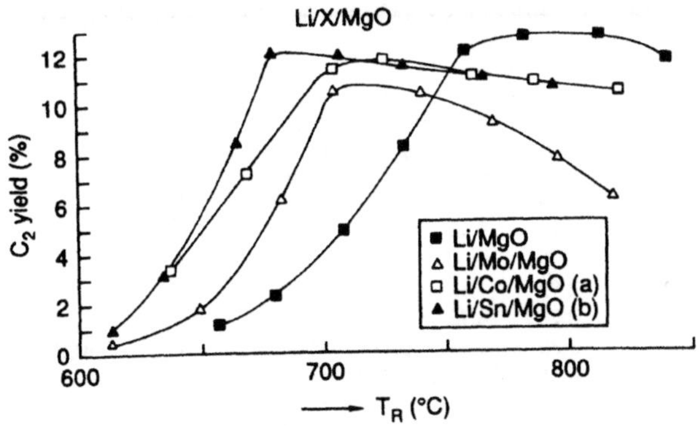

FIGURE 6-12. The C_2 yield as a function of T_R for a number of the promoted Li/MgO catalysts compared with the results for an unpromoted Li/MgO sample.

system while maintaining the same C_2 selectivities, we therefore examined the effect of adding a number of different oxides to a standard sample of Li/MgO in the hope that they would increase the activity and allow operation at lower temperatures. A number of the additives tested (e.g., SnO_2, TiO_2, Dy_2O_3, La_2O_3, and Tb_4O_7) caused little or no difference in the selectivity to C_2 products achieved with the resultant catalysts. However, the temperatures required to give the optimum yields of C_2 products were lowered considerably in a number of cases. The results for various three-component Li/MgO systems are given in Figs. 6-11 and 6-12 and in Tables 6-5 and 6-6 (Korf et al. 1989b).

The behavior of the selectivity of a number of the samples with respect to the reaction temperature were similar to the behavior of the unmodified Li/MgO. At very low temperatures, the only reaction products were CO and CO_2; however, C_2 production started at low O_2 conversions (ca. 3–4%) and with increasing temperature the C_2 selectivity soon reached the steady levels given in Table 6-6. Samples promoted by Zn, Ca, La, Dy, and Tb showed similar selectivities to those given by the unpromoted material whereas the selectivities of the materials promoted by V, Ni, Ti, Mo, and Bi were slightly lower; in the case of the sample containing Bi, considerable loss of a volatile Bi species took place. For the Sn-promoted material, the selectivity was slightly lower at the temperature required to give 50% oxygen conversion but reached the level observed with the unpromoted material when the conversion was higher (Table 6-6).

Addition of CoO gave a material that still had 100% selectivity to CO_2 at 50% oxygen conversion. However, when the temperature was increased to give an oxygen conversion of 90%, the C_2 selectivity approached that of

Table 6-5 Analysis, Surface Area, and Structure for the Different Catalyst Samples Used in This Work

Catalyst	Li (wt%)	Li (mmol g^{-1})	X (mmol g^{-1})	CO_2 (mmol g^{-1})	Area (m^2 g^{-1})	XRD[a]
Li/MgO	2.34	3.37	abs	1.53	2.4	*
Li/Zn/MgO	4.28	6.17	0.229	2.68	—	—
Li/Ca/MgO	2.78	4.00	0.314	1.56	—	*
Li/V/MgO	2.99	4.31	0.324	1.61	0.5	*
Li/Ni/MgO	3.28	4.72	0.261	2.25	0.5	*
Li/Na/MgO	3.85	5.55	0.211	2.64	1.5	$La(OH)_3$, $LiLaO_2$
Li/Tb/MgO	4.36	6.28	0.233	2.69	—	—
Li/Dy/MgO						
(a)	4.24	5.97	0.579	2.04	—	Dy_2O_3, $LiDyO_2$
(b)	5.74	8.27	1.414	2.55	—	*
Li/Sn/MgO						
(a)	4.77	6.87	0.029	2.30	—	*
(b)	3.99	5.75	0.206	2.80	5.0	*
(c)	3.91	5.63	0.460	1.94	—	*
Li/Ti/MgO	3.37	4.85	0.221	—	1.9	*
Li/Co/MgO						
(a)	3.98	5.73	0.238	2.36	1.5	*
(b)	4.77	6.86	0.973	2.13	—	—
(c)	4.74	6.83	2.343	1.68	—	—
Li/Mn/MgO	3.91	5.63	0.781	2.18	3.7	MnO_2, Li_2MnO_3
Li/Mo/MgO	4.31	6.21	0.322	1.69	—	—
Li/Bi/MgO	3.88	5.59	0.208	2.60	—	—
Li/Pb/MgO	4.73	6.82	0.226	2.59	—	—

[a] *: Li_2CO_3 and MgO; abs, absent; —, not measured.

a number of the other promoted catalysts. On cooling once more, there was a small increase in C_2 selectivity. We think that it is likely that the change in behavior with heating is related in some way to the mobility or volatility of the Li species. At the very low reaction temperature of 610°C, the mobility is likely to be very low and bare CoO sites will be accessible, these giving total oxidation; when the temperature is increased, the lithium species become mobile and possibly also volatile (Korf et al. 1990b), this having a beneficial effect on the selectivity. After cooling, the lithium is probably better spread on the surface of the catalyst. These effects are only seen at very low conversions with the other catalysts because these materials do not become significantly active until temperatures in the range in which the lithium becomes mobile.

It is unlikely that the presence of new surface compounds involving Li is responsible for the improved properties of the majority of the materials, whose behavior is given in Figs. 6-11 and 6-12 and Table 6-5 and 6-6, because only in a few cases the formation of lithium compounds in the promoted materials was shown by X-ray diffraction (see Table 6-5).

The results of Table 6-5 also show that an increase of surface area is unlikely to be directly responsible for the improvement of the catalytic behavior; for example, although the V- and Ni-promoted materials were active at lower temperatures, their areas (0.5 m^2 g^{-1}) were somewhat lower than that of the unpromoted material (2.4 m^2 g^{-1}).

Other research groups have also reported the effect of promoters of the Li/MgO system. Larkins and Nordin (Larkins and Nordin 1988) found that addition of Mn, Fe, or Co at concentrations of 2 mol% (relative to Li) resulted in an increased conversion and an increased C_2 selectivity at T_R = 800°C. Bi and co-workers (Bi et al. 1988) have shown that it was possible to improve the activity and selectivity of the Li/MgO catalyst by the addition of cerium oxide. This result has been confirmed by Bartsch and co-workers (Bartsch and Hofmann 1990), who have studied the system in more detail.

Table 6-6 Catalytic Results for Li / X / MgO Catalysts, Showing the Temperatures for Oxygen Conversions of 50% and 90%, the Corresponding Difference in Temperature (ΔT) between these Results and those for Li / MgO, and the Corresponding C_2 Selectivities

	50% O_2 CONVERSION			90% O_2 CONVERSION		
Catalyst	T (°C)	ΔT (°C)	S_2 (%)	T (°C)	ΔT (°C)	S_2 (%)
Li/Zn/MgO	773	48	85	806	48	86
Li/MgO	725	—	85	758	—	85
Li/Ca/MgO	725	0	83	758	0	84
Li/V/MgO	712	−13	78	745	−13	79
Li/Ni/MgO	700	−25	78	740	−18	81
Li/La/MgO	693	−32	84	724	−34	85
Li/Dy/MgO (a)[a]	678	−47	83	711	−47	84
Li/Tb/MgO	675	−50	82	701	−57	82
Li/Sn/MgO (b)[a]	655	−70	78	680	−78	84
Li/Ti/MgO	655	−70	70	690	−68	78
Li/Co/MgO (a)[a]	610	−115	0	690	−68	78
Li/Mo/MgO	680	−45	68	725	−33	71
Li/Mn/MgO	—	—	—	640	−118	39
Li/Pb/MgO	640	−85	23	672	−80	50
Li/Bi/MgO	712	−13	67	750	−8	65

[a]See Table 6-5.

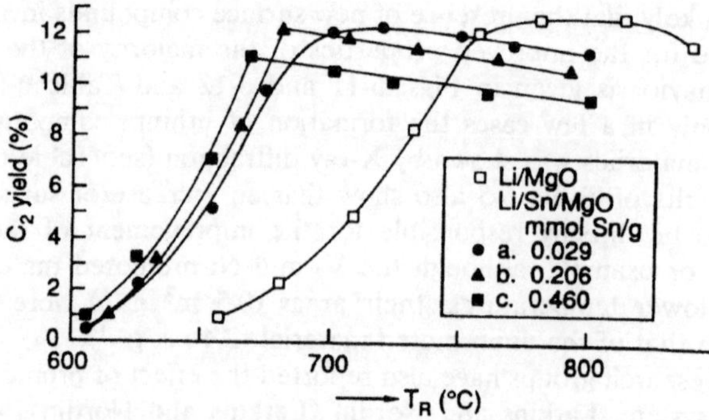

FIGURE 6-13. The C_2 yield as a function of T_R for Li/Sn/MgO materials with different tin contents; the results for the unpromoted material are shown for comparison.

The Effect of Content of the Added Oxide in Li / X / MgO Catalysts

The addition of only a small amount of Sn was sufficient to give a large increase in activity. As shown in Figure 6-13, the addition of more Sn resulted in a further small increase in activity at relatively low reaction temperature T_R; however, at higher values of T_R there was a decrease in C_2 production, caused by a decrease in C_2 selectivity. With increasing Sn content the maximum C_2 yield that could be reached decreased and total oxidation became more important. Li/Dy/MgO and Li/Co/MgO catalysts gave results analogous to those for Li/Sn/MgO.

The Oxidation of Ethane and Ethylene over Li / X / MgO Catalysts

The same sequences of activity were found for the methane coupling reaction, for the oxidative dehydrogenation of ethane, and for the oxidation of ethylene using the catalysts Li/Sn/MgO (b), Li/Dy/MgO (a), and Li/MgO. As discussed previously, we have proposed a serial reaction scheme to explain the results for Li/MgO. A similar scheme seems also to apply to the promoted Li/MgO systems. The observation that the addition of promoters does *not* give an increase in the maximum C_2 yield reached would seem to indicate that the promoters affect all of the constituent reactions of this scheme in the same way and to approximately the same extent.

The Function of the Promoter in the Li / X / MgO Catalyst System

The oxides that we have added as promoters for the Li/MgO catalyst all adsorb oxygen more strongly than do (almost) nonconducting materials such as Li/MgO and Li/CaO. The interaction of oxygen with the catalysts will be enhanced by the addition of the oxides, resulting in an improved oxygen adsorption. On account of this enhanced oxygen adsorption, the interaction of methane with the catalyst would also be improved. Kinetic measurements (see below) indeed showed an improved oxygen and methane adsorption of the Li/Sn/MgO catalyst when compared with the Li/MgO catalyst.

The mobile Li_2CO_3 probably has two functions:

1. It inhibits total oxidation by preventing strong oxygen adsorption on the added promoter oxide.
2. It creates active sites on the surface in the presence of oxygen by the gradual loss of CO_2 from surface lithium carbonate species ($T > T_d$) (Korf et al. 1987).

An alternative role of the promoter can be to stabilise the Li in an active form.

The Stability of Li / X / MgO Catalysts

Effect of Calcination

Table 6-5 gives the Li contents of the unused catalysts after calcination at 850°C. The promoted materials contained in all cases more Li than did the unpromoted Li/MgO material after calcination. Because the same uncalcined Li/MgO precursor was used in the preparation of all the samples, we concluded that the presence of the added oxide prevented the loss of lithium. This result clearly shows that there is a relatively strong interaction between the promoter and the Li species in the catalyst and that this interaction exists even under conditions in which no catalytic reaction is carried out.

Aging Experiments

The promoted materials behaved much better in aging experiments than did the unpromoted materials. This is shown in Figs. 6-14 and 6-15. The addition of the promoter brought about a large decrease in the loss of Li. The temperature chosen for these comparative aging experiments was that at which the optimum behavior was achieved for the unpromoted Li/MgO

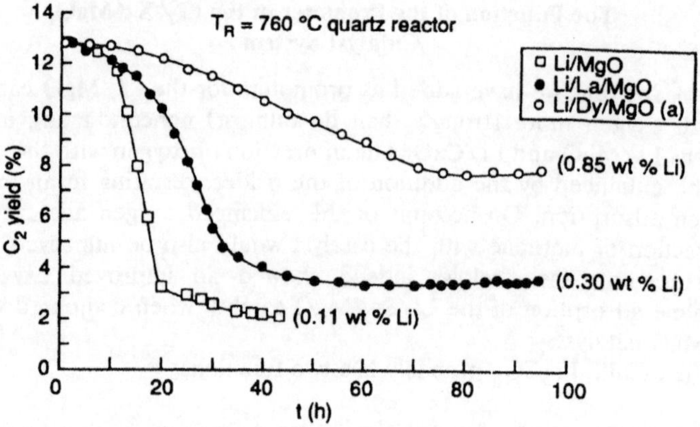

FIGURE 6-14. The C_2 yield as a function of time on stream at T_R = 760°C in a quartz reactor for Li/MgO, Li/La/MgO, and Li/Dy/MgO (a).

sample; it is clearly well above the temperature at which the optimum behavior was found with the promoted Li/X/MgO samples. It could be concluded that the improved retention of the lithium was responsible for the more-stable behavior exhibited by these catalysts. As was discussed previously for the Li/MgO material, the C_2 selectivity for those materials did not change whereas the yield decreased. A tentative explanation for the increased stability of the Li species in the doped Li/MgO catalysts is

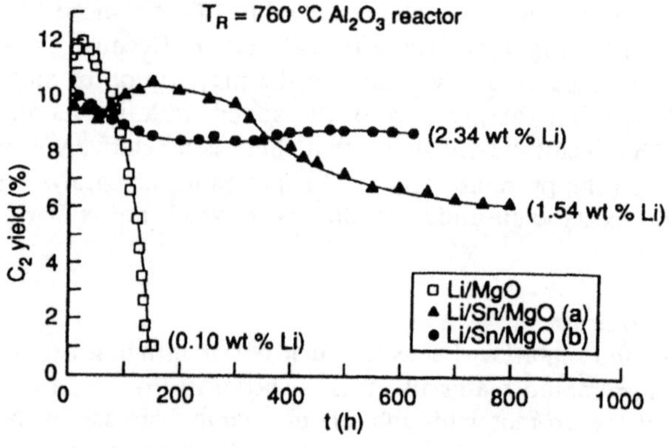

FIGURE 6-15. The C_2 yield as a function of time on stream at T_R = 760°C in an alumina reactor for Li/MgO, Li/Sn/MgO (a), and Li/Sn/MgO (b).

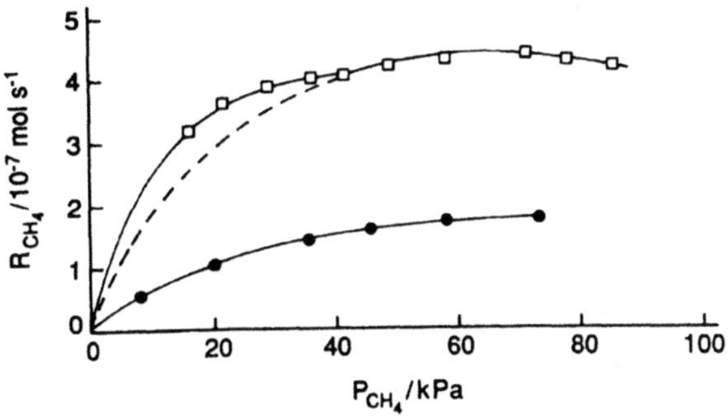

FIGURE 6-16. The rate of reaction of CH_4 as a function of the partial pressure of CH_4 for Li/Sn/MgO (□) and Li/MgO (•); $T = 720°C$, $P_{O_2} = 2.2$ kPA, and $P_{CO_2} = 7.8$ kPa.

possibly to be found in the increased interaction of oxygen with the catalyst. This may lead to a faster reoxidation of the catalyst and thus a shorter residence time of the volatile LiOH on the surface of the catalyst.

KINETICS OF THE OXIDATIVE COUPLING OF METHANE OVER A Li / Sn / MgO AND A Li / MgO CATALYST

We have studied the rate of reaction of methane with oxygen in the presence of a Li/Sn/MgO catalyst at $T_R = 720°C$ as a function of the partial pressures of CH_4, O_2, and CO_2, using a well-mixed reaction system that is practically gradientless with respect to gas-phase concentrations (Korf et al. 1990). The results are shown in Figs. 6-16–6-18. For comparison, the calculated reaction rates for Li/MgO (Roos et al. 1989c) under the same reaction conditions are also shown in Fig. 6-16. From these measurements and from the measurements on Li/MgO, it was concluded that the rate-determining step on both the promoted and unpromoted Li/MgO involves reaction of a molecule of CH_4 adsorbed on the catalyst surface with an adsorbed diatomic oxygen species. For both parts of the curve shown in Fig. 6-16 ($P_{CH_4} < 41$ kPa for part 1; $P_{CH_4} > 41$ kPa for part 2) the kinetics are consistent with a Langmuir–Hinshelwood type of mechanism involving competitive adsorption of CH_4, O_2, and CO_2 on a single type of site. Table 6-7 gives the adsorption constants and the k_r values for this model. This table also shows, for comparison, the results for Li/MgO (Korf et al. 1990c, Roos et al. 1989c).

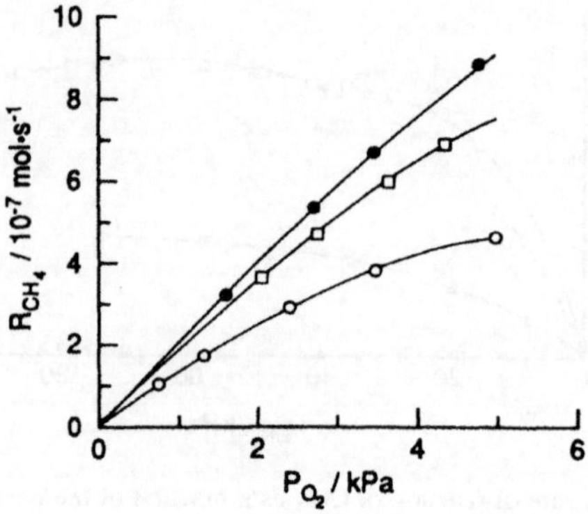

FIGURE 6-17. The rate of reaction of CH_4 as a function of the partial pressure of O_2 for three partial pressures of methane: P_{CH_4} = 16 (○), 39 (□), and 52 (●) kPa; P_{CO_2} = 7.8 kPa, T = 720°C.

For Li/MgO, the reaction rate had a first-order dependency on the partial pressure of oxygen ($P_{O_2} < 11$ kPa), so independent values of b_{O_2} and k_r could not be calculated for this catalyst (Roos et al. 1989c). However as Table 6-7 shows, the values of b_{O_2} for Li/Sn/MgO could be calculated from measurements in which P_{O_2} was varied up to 5 kPa (Fig. 6-17) for both parts of the methane curve (Fig. 6-16). That this was

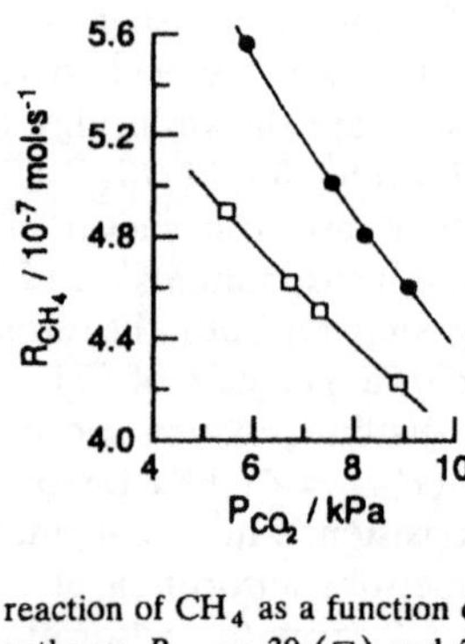

FIGURE 6-18. The rate of reaction of CH_4 as a function of the partial pressure of CO_2 for two partial pressures of methane: P_{CH_4} = 39 (□) and 52 (●) kPa; P_{O_2} = 2.2 kPa, T = 720°C.

Table 6-7 Kinetic Parameters Calculated for a Langmuir–Hinshelwood Mechanism Involving Competitive Adsorption of CH_4, O_2, and CO_2; $T_R = 720°C$

Part	b_{CH_4} (kPa^{-1})	b_{O_2} (kPa^{-1})	b_{CO_2} (kPa^{-1})	k_r (10^{-6} mol s^{-1})	$k_r \cdot b_{O_2}$ [a] (10^{-7} mol $(s^{-1}\ kPa)^{-1}$)
Li/Sn/MgO					
1	0.041	0.095	0.072	13.6	12.9
2	0.023	0.041	0.077	32.7	13.4
Li/MgO					
	0.016	—	0.070	—	4.8

[a] The values of $k_r \cdot b_{O_2}$ and k_r are calculated from the series of experiments in which P_{CH_4} was varied.

possible indicates that the interaction of oxygen with the Li/MgO catalyst is enhanced by the addition of SnO_2. The adsorption constants given in Table 6-7 also show that the methane is more strongly adsorbed on the Li/Sn/MgO catalyst than on the Li/MgO catalyst; this is so for both parts of the curve showing the dependence on methane partial pressure but particularly in the first part of the curve ($P_{CH_4} < 41$ kPa). It can also be concluded from the results of Table 6-7 that the poisoning effect of CO_2 is less for the Li/Sn/MgO catalyst than for the Li/MgO catalyst: The CO_2 thus competes less effectively with the more strongly bonded oxygen and methane present on the Li/Sn/MgO catalyst. However, the effect is far less pronounced for the second part of the curve showing the dependence on P_{CH_4}, as the b_{CH_4} and b_{O_2} values are closer to those for Li/MgO.

Applying a Langmuir–Hinshelwood model to describe adequately the kinetic measurements, we assume that

$$CH_4\,(a) + O_2\,(a) \rightarrow CH_3\cdot + HO_2\cdot$$

is the rate-determining reaction step, a step that is equivalent to that required in the gas-phase activation of methane. The suggestion recently made by Lunsford (Lunsford 1990) that O_3^- species may be involved would appear to be consistent with these observations; it implies thus that adsorption of oxygen occurs on O^- sites in a reversible manner prior to the rate-determining step. The suggestion that molecular oxygen is involved in the rate-determining step is also consistent with the exchange studies of Cant and co-workers (Cant et al. 1988) and Mims and co-workers (Mims et al. 1989).

In order to try to establish the reaction scheme over Li/Sn/MgO, we have also studied the rate of reaction of ethylene with oxygen in the presence of this material as a function of the partial pressures of C_2H_4, O_2, and CO_2 at $T_R = 720°C$, these measurements being analogous to those carried out with methane at 720°C. From these kinetic measurements, we obtained further evidence for a serial reaction scheme over a Li/Sn/MgO catalyst:

$$CH_4 \rightarrow C_2H_6 \rightarrow C_2H_4 \rightarrow CO_x$$

CONCLUSIONS

It was possible to improve the activity and stability of Li/MgO catalysts by the addition of small amounts of various oxides. For a number of these oxides this had little or no influence on the C_2 selectivity of the resultant catalyst. A comparison of the various systems showed that Li/Sn/MgO is a promising catalyst system for the oxidative coupling of methane. More research is needed to elucidate the role of Sn.

References

Bartsch, S., and H. Hofmann. 1990. *Catal. Today* 6:527.

Bi, Y.-L., K.-J. Zhen, Y.-T. Jiang, C.-W. Teng, and X.-G. Yang. 1988. *Appl. Catal.* 39:185.

Cant, N.W., C.A. Lukey, P.F. Nelson, and R.J. Tyler. 1988. *J. Chem. Soc., Chem. Commun.* 766.

Hinsen, W., W. Bytyn, and M. Baerns. 1984. *Proc. 8th Int. Congr. Catal. Berlin* 3:581.

Ito, T., J.-X. Wang, C.-H. Lin, and J.H. Lunsford. 1985. *J. Am. Chem. Soc.* 107:5062.

Kimble, J.B., and J.H. Kolts. 1986. *Energy Prog.* 6:226.

Korf, S.J., J.A. Roos, N.A. de Bruijn, J.G. van Ommen, and J.R.H. Ross. 1987. *J. Chem. Soc., Chem. Commun*, 1433.

Korf, S.J., J.A. Roos, N.A. de Bruijn, J.G. van Ommen, and J.R.H. Ross. 1988. *Catal. Today* 2:535.

Korf, S.J., J.A. Roos, N.A. de Bruijn, J.G. van Ommen, and J.R.H. Ross. 1990a. *Appl. Catal.* 58:131.

Korf, S.J., J.A. Roos, J.W.H.C. Derksen, J.A. Vreeman, J.G. van Ommen, and J.R.H. Ross. 1990b. *Appl. Catal.* 59:291.

Korf, S.J., J.A. Roos, J.M. Diphoorn, R.H.J. Veehof, J.G. van Ommen, and J.R.H. Ross. 1989a. *Catal. Today* 4:279.

Korf, S.J., J.A. Roos, L.J. Veltman, J.G. van Ommen, and J.R.H. Ross. 1989b. *Appl. Catal.* 56:119.

Korf, S.J., J.A. Ross, J.A. Vreeman, J.W.H.C. Derksen, J.G. van Ommen, and J.R.H. Ross. 1990c. *Catal. Today* 6:417.

Korf, S.J., J.G. van Ommen, and J.R.H. Ross. 1990. American Chemical Society Meeting, Boston.

Larkins, F.P., and M.R. Nordin. 1988. In *Methane Conversion*, ed. D.M. Biddy, C.D. Chang, R.F. Howe, and S. Yurchak, Amsterdam: Elsevier Science Publishers. p. 409.

Lunsford, J.H. 1990. *Catal. Today* 6:235.

Martin, G.A., A. Bates, V. Ducarme, and C. Mirodatos. 1989. *Appl. Catal.* 47:287.

Mims, C.A., R.B. Hall, K.D. Rose, and G.R. Myers. 1989. *Catal. Lett.* 2:361.

Mirodatos, C. and G.A. Martin. 1989. *Catal. Today* 4:301.

Otsuka, K., K. Jinno, and A. Morikawa. 1986. *J. Catal.* 100:353.

Roos, J.A., A.G. Bakker, H. Bosch, J.G. van Ommen, and J.R.H. Ross. 1987. *Catal. Today* 1:133.

Roos, J.A., S.J. Korf, A.G. Bakker, N.A. de Bruijn, J.G. van Ommen, and J.R.H. Ross. 1988. In *Methane Conversion*, ed. D.M. Bibby, C.D. Chang, R.F. Howe, and S. Yurchak, pp. 427–. Amsterdam: Elsvier Science Publishers.

Roos, J.A., S.J. Korf, J.J.P. Biermann, J.G. van Ommen, and J.R.H. Ross. 1989a. Contribution to the Second European Workshop Meeting on New Developments in Selective Oxidation, Rimini, Italy, September 1989.

Roos, J.A., S.J. Korf, R.H.J. Veehof, J.G. van Ommen, and J.R.H. Ross. 1989b. *Catal. Today* 4:441.

Roos, J.A., S.J. Korf, R.H.J. Veehof, J.G. van Ommen, and J.R.H. Ross. 1989c. *Appl. Catal.* 52:131.

Roos, J.A., S.J. Korf, R.H.J. Veehof, J.G. van Ommen, and J.R.H. Ross. 1989d. *Appl. Catal.* 52:147.

van Kasteren, J.M.N., J.W.M.H. Geerts, and K. van der Wiele. 1989. Contribution to the Second European Workshop Meeting on New Developments in Selective Oxidation, Rimini, Italy, September 1989.

Westerterp, K.R., W.P.M. van Swaaij, and A.A.C.M. Beenackers. 1984. *Chemical Reactor Design and Operation*. Chichester: John Wiley and Sons.

7

Studies of the Mechanism of the Oxidative Coupling of Methane Using Oxide Catalysts

Graham J. Hutchings

Leverhulme Centre for Innovative Catalysis, Department of Chemistry, University of Liverpool, Liverpool United Kingdom

Michael S. Scurrell

Catalysis Programme, Division of Energy Technology, CSIRO, Pretoria, South Africa

TABLE OF CONTENTS

INTRODUCTION

Following the oil crisis of the 1970s a renewed effort was given to research on various aspects of synfuels production. Initially, this effort was focused on improving the existing pathways based on the hydrogenation of carbon monoxide (Hutchings 1986; Biloen and Sachtler 1981) but subsequently led to the development of a new commercial process based on methanol conversion (Hutchings 1986, 1987; Chang 1988; Hutchings and Hunter 1990). Softer oil prices have of late resulted in a reassessment of the commercial viability of these routes to synfuels and consequently the research attention given to synfuels production decreased significantly. In its place, the research impetus has now been directed at the utilization of natural gas as a source of chemicals and fuels. The major reason for this is that the current reserves of natural gas are considered to be a large and underutilized energy resource, but the majority of these reserves are in remote locations and require chemical conversion to liquid products in order to facilitate transportation. Environmental issues have also influenced this research, because methane emissions, or flaring, contribute significantly to the production of pollutants responsible for global warming. In this case, the availability of commercially viable chemical conversion routes for methane could alleviate the problem.

The conversion of natural gas via oxidation to obtain more-valuable products, (e.g., methanol and ethene) is not new, and considerable research was completed in the 1920s and 1930s on this topic (Gesser, Hunter, and Prakash 1985). By far the majority of the research carried out into the oxidative couplings of methane using oxide catalysts has been directed at the formation of higher hydrocarbons, mainly ethane and ethene, and only limited effort has been given to the synthesis of methanol (Foster 1985; Pitchai and Klier 1986).

The aim of this chapter is to highlight the salient features of the most recent literature concerning the oxidative coupling of methane, particularly that published since 1986. In particular, this chapter seeks to discuss the research approach we have adopted to elucidate the mechanism of methane oxidation over catalysts and to consider experimental approaches that could enable the design of an improved catalyst. In addition, a survey on published conversion–selectivity data, categorized according to catalyst type is presented.

EVALUATION OF OXIDE CATALYSTS USING VARIOUS OXIDANTS

Molecular Oxygen as Oxidant

The early studies at Union Carbide (Keller and Bhasin 1982; Bhasin 1988) were carried out on the basis that nonselective methane oxidation, either gas-phase or catalyzed, could dominate the reactions occurring when oxygen and methane were cofed over a catalyst. To minimize any possible adverse effects (particularly total oxidation), they operated the reaction cyclically. Methane and air were fed separately to the reactor, interspaced by a short purge of an inert gas. In this way the methane reacted stoichiometrically with oxygen retained on the oxide surface, and once all the readily reacted oxygen was depleted the reaction ceased. A range of oxides were identified that became active above 600°C and produced ethene and ethane as major products; manganese oxide was found to give the best results.

Later studies demonstrated (Hinsen and Baerns 1983; Hinsen, Bytyn, and Baerns 1984) that the cyclic operation regime was not necessary in order to obtain high selectivities. Hence, in the presence of suitable catalysts, cofed methane and oxygen gave high selectivities to ethene and ethane. Subsequently, Ito and Lunsford (Ito and Lunsford 1985) made the important discovery that lithium-promoted magnesium oxide gave a significantly improved catalytic performance. All these early studies demonstrated that the oxidative coupling of methane to form ethene, ethane, and higher hydrocarbons was a high-temperature reaction. Temperatures in excess of 600°C were required in order to observe selective methane activation, whereas CO_2 and CO were observed as the only carbon-containing products at lower temperatures. It is important to note that these early studies failed to identify that significant levels of hydrogen can also be formed at these elevated reaction temperatures. This is surprising, because hydrogen was known to be a major product of the earlier studies involving noncatalytic gas-phase activation of methane and could therefore have been expected (Gesser, Hunter, and Prakash 1985). However, subsequent studies noted that significant levels of hydrogen can be present

(Hutchings, Scurrell, and Woodhouse 1988c). In general, the products observed in the high-temperature catalytic activation of methane are CO_2, H_2, C_2H_6, C_2H_4, CO, H_2O, and, to a lesser extent, higher hydrocarbons (few studies report these products because, typically, C_3 hydrocarbons are formed at an order of magnitude lower concentration than the C_2 hydrocarbons). Only trace levels of oxygenated products (e.g., CH_3OH and CH_2O) are reported in most studies, and this represents a major difference from the earlier noncatalytic studies when, at high pressures and lower temperatures, these products predominated (Gesser, Hunter, and Prakash 1985).

These earlier studies using cyclic and cofeed experimental regimes give rise to an interesting question: It has to be considered whether the methane activation process is the same for the two processes. In the absence of gas-phase oxidant, oxides such as PbO and Mn_2O_3 are readily reduced and lose lattice oxygen, O^{2-}, probably from low-coordination defect sites, which acts as a stoichiometric oxidizing agent. In recent experiments (Hutchings, Scurrell, and Woodhouse 1989b), in the absence of gas-phase oxygen both MgO and Li/MgO have been found to be inactive for methane oxidation, in agreement with the findings of Keller and Bhasin (Keller and Bhasin 1982) that Li_2O and MgO were poor oxidizing agents for the cylic experimental mode. This indicates that lattice O^{2-} may not be involved in methane activation on MgO and Li/MgO catalysts. Hence, superficially, it could be concluded that different mechanisms for the methane activation step could be occurring for these different experimental regimes. However, in the presence of a gas-phase oxidant, an adsorbed species could become important for manganese oxide catalysts, and further experimental work is required to address this problem.

Since the early studies of Lunsford and of Keller and Bhasin, a vast number of catalyst formulations have been investigated in an extensive screening and evaluation exercise involving a large number of research groups. As a consequence, most simple oxides have been tested as catalysts for the partial oxidation of methane. Tables 7-1–7-5 summarize representative data from these studies for different classifications of metal oxides. Although hundreds of oxide formulations have been tested, the striking feature of these studies is that the product distribution remains invariant (i.e., C_2H_6, C_2H_4, CO, CO_2, H_2, H_2O, and lesser amounts of C_{3+} hydrocarbons); oxygenated products (e.g., CH_3OH, and CH_2O) have only been reported for very few catalysts (Otsuka et al. 1988; Mac Giolla Coda et al. 1989; Kasztelan and Moffat 1987). It is possible that at the elevated temperature ($> 600°C$) required to observe hydrocarbon formation, the oxygenated products are unstable and decompose, possibly to

Table 7-1 Methane Oxidation with Molecular Oxygen Using Group IIA Metal Oxide Catalysts

Catalyst	CH_4–O_2 Mole Ratio	GHSV (h^{-1})	Temp. (°C)	CONVERSION CH_4 (%)	O_2 (%)	C_2 HYDROCARBON Selectivity (%)	Yield (%)	Ethene / Ethane	Reference	Ref. No.[a]
7% Li/MgO	1.6	2990	720	42.8	—	45.4	19.4	1.75	Ito and Lunsford 1985; Ito et al. 1985	—
	2.0	2990	720	37.8	—	50.3	19.0	1.61	Ito and Lunsford 1985; Ito et al. 1985	1
	3.1	2990	720	29.1	—	58.1	16.9	1.52	Ito and Lunsford 1985; Ito et al. 1985	2
BeO	9.3	31.1[b]	740–750	9	99	22	2.0	—	Carreiro et al. 1988	—
MgO	9.3	0.2[b]	740–750	12	93	47	5.6	—	Carreiro et al. 1988	—
CaO	9.3	0.1[b]	740–750	11	94	55	6.1	—	Carreiro et al. 1988	—
SnO	9.3	23.1[b]	740–750	8	88	72	5.8	—	Carreiro et al. 1988	4
BaO	9.3	14.5[b]	740–750	14	98	70	9.8	—	Carreiro et al. 1988	5
3% Li/CaO	9.3	0.4[b]	740–750	14	97	72	10.1	—	Carreiro et al. 1988	6
3% Na/CaO	9.3	0.3[b]	740–750	16	95	76	12.2	—	Carreiro et al. 1988	7
3% K/CaO	9.3	0.4[b]	740–750	14	83	72	10.1	—	Carreiro et al. 1988	8
MgO	200	12[b]	800	—	—	—	0.1	1.79	Aika and Nishiyama 1988a	—
20% Bi/MgO	100	12[b]	800	—	—	—	0.3	0.33	Aika and Nishiyama 1988a	—
15% K/MgO	100	12[b]	800	—	—	—	2.3	1.64	Aika and Nishiyama 1988a	—
MgO	3.5	667	550	3.0	—	0	0	—	Hutchings, Scurrell, and Woodhouse 1988a, 1989b	—
									Hutchings, Scurrell, and Woodhouse 1988b, 1987a	—
	3.5	667	720	5.2	100	58.1	3.0	0.55	Hutchings, Scurrell, and Woodhouse 1988a, 1989b	9
									Hutchings, Scurrell, and Woodhouse 1988b, 1987a	—
5% Li/MgO	3.3	648	550	2.0	—	0	0	0.29	Hutchings, Scurrell, and Woodhouse 1988a, 1989b	—

									Hutchings, Scurrell, and Woodhouse 1988b, 1987a	—
	3.3	648	720	2.7	37.0	58.1	3.0	0.68	Hutchings, Scurrell, and Woodhouse 1988b, 1987a	—
3% Li/BeO	3.0	0.61^b	640	19.8	69.6	51.0	10.1	—	Doi, Utsumi, and Matsuura 1988	11
	3.0	0.61^b	680	27.5	96.6	52.0	14.3	—	Doi, Utsumi, and Matsuura 1988	12
	3.0	0.61^b	760	28.4	93.3	58.5	16.6	—	Doi, Utsumi, and Matsuura 1988	13
MgO	20	0.75^b	750	11.3	—	39.1	4.4	0.27	Burch, Squire, and Tsang 1988	—
5–10% Li/MgO	20	0.75^b	750	7.5	—	76.6	5.7	0.37	Burch, Squire, and Tsang 1988	14
Li/MgO	1.7	0.1^b	700	2	—	58	1.16	0.38	Martin et al. 1989	—
	1.7	0.1^b	800	> 15	—	76	> 11.4	0.80	Martin et al. 1989	—
MgO	3.71	2.4^b	650	2.0	7.9	18.0	0.36	0.50	Peng, Richards, and Stair 1990	—
0.02% Li/MgO	3.71	2.4^b	650	4.1	—	7.6	0.31	0.22	Peng, Richards, and Stair 1990	—
0.05% Li/MgO	3.71	2.4^b	650	6.2	—	9.6	0.60	0.39	Peng, Richards, and Stair 1990	—
0.10% Li/MgO	3.71	2.4^b	650	7.6	37.0	17.5	1.33	0.44	Peng, Richards, and Stair 1990	—
0.15% Li/MgO	3.71	2.4^b	650	9.3	44.4	25.7	2.39	0.53	Peng, Richards, and Stair 1990	—
0.20% Li/MgO	3.71	2.4^b	650	8.2	38.3	35.7	2.93	0.56	Peng, Richards, and Stair 1990	—
0.25% Li/MgO	3.71	2.4^b	650	9.6	36.8	30.8	2.96	0.53	Peng, Richards, and Stair 1990	—
1.00% Li/MgO	3.71	2.4^b	650	9.1	44.8	36.4	3.49	0.68	Peng, Richards, and Stair 1990	—
5.00% Li/MgO	3.71	2.4^b	650	8.0	30.4	34.5	2.76	0.43	Peng, Richards, and Stair 1990	—
7.00% Li/MgO	3.71	2.4^b	650	6.0	27.4	30.6	1.84	0.21	Peng, Richards, and Stair 1990	—

Table 7-1 (*Continued*)

Catalyst	CH_4–O_2 Mole Ratio	GHSV (h^{-1})	Temp. (°C)	CONVERSION CH_4 (%)	O_2 (%)	C_2 HYDROCARBON Selectivity (%)	Yield (%)	Ethene/Ethane	Reference	Ref. No.[a]
10.00% Li/MgO	3.71	2.4[b]	650	7.6	—	35.6	2.71	0.62	Peng, Richards, and Stair 1990	—
Li/MgO[c]	10.7	0.16[b]	860	15	99	72.0	10.8	1.75	Edwards, Tyler, and White 1990	16
Li/MgO[d]	7.84	0.16[b]	875	15.6	97	61.2	9.55	4.17	Edwards, Tyler, and White 1990	16
MgO	3.5	0.01[e]	710	3.64	13.4	24.3	0.89	0.23	Burch et al. 1989a, 1989b	—
	3.5	0.016[e]	710	5.26	14.2	26.8	1.41	0.28	Burch et al. 1989a, 1989b	—
	3.5	0.032[e]	710	7.12	27.8	26.0	1.85	0.28	Burch et al. 1989a, 1989b	—
	3.5	0.065[e]	710	11.2	35.1	23.2	2.60	0.38	Burch et al. 1989a, 1989b	—
	3.5	0.196[e]	710	20.0	71.5	21.8	4.36	0.79	Burch et al. 1989a, 1989b	—
	3.5	0.295[e]	710	19.2	92.0	51.3	9.85	0.76	Burch et al. 1989a, 1989b	17
	3.5	0.446[e]	710	21.9	98.2	47.4	10.38	0.89	Burch et al. 1989a, 1989b	—
	3.5	0.889[e]	710	19.3	98.0	30.9	5.96	0.51	Burch et al. 1989a, 1989b	—
	3.5	0.775[e]	710	19.5	97.5	26.6	5.19	0.65	Burch et al. 1989a, 1989b	—
La/MgO	2.0	51,360	800	27.9	73.8	43.3	12.1	7.69	Choudhary et al. 1989	—
	3.0	51,360	800	26.5	95.2	49.8	13.2	2.94	Choudhary et al. 1989	—
	4.7	51,360	800	21.7	99.1	63.8	13.8	1.70	Choudhary et al. 1989	18
	8.0	51,360	800	15.1	93.2	74.4	11.2	0.80	Choudhary et al. 1989	18
	3.0	51,360	700	28.6	99.5	50.3	14.4	2.33	Choudhary et al. 1989	—
	3.0[f]	51,360	700	32.0	99.1	59.2	18.9	1.59	Choudhary et al. 1989	18

	3.0[f]	51,360	700	33.6	99.3	62.0	20.8	1.49	Choudhary et al. 1989	18
	3.0	51,360	750	28.1	99.4	49.5	13.9	2.50	Choudhary et al. 1989	—
	3.0[f]	51,360	750	31.6	98.9	58.6	18.5	1.79	Choudhary et al. 1989	—
	3.0[f]	51,360	750	35.7	99.0	60.6	21.6	1.79	Choudhary et al. 1989	18
	4.0	51,360	800	22.6	90.1	62.4	14.1	1.79	Choudhary et al. 1989	18
	4.0[f]	51,360	800	25.3	86.6	70.3	17.8	1.21	Choudhary et al. 1989	18
	8.0	51,360	850	17.7	98.9	74.5	13.2	1.30	Choudhary et al. 1989	—
	8.0[f]	51,360	850	20.1	98.7	80.4	16.2	1.21	Choudhary et al. 1989	18
	4.0	102,700	750	22.6	85.1	64.7	14.6	1.30	Choudhary et al. 1989	18
	4.0[f]	102,700	750	29.1	98.7	70.0	20.3	1.21	Choudhary et al. 1989	18
	4.0	102,700	850	23.1	95.0	61.0	14.1	2.33	Choudhary et al. 1989	18
	4.0[f]	102,700	850	25.6	94.2	70.0	17.9	1.00	Choudhary et al. 1989	—
MgO	2.11	0.12[b]	700	13.0	47.2	15.0	1.95	0.74	Matsuura et al. 1989b	—
1 mol% Li/MgO	2.11	0.12[b]	700	38.5	94.8	49.0	18.9	1.92	Matsuura et al. 1989b	—
3 mol% Li/MgO	2.11	0.12[b]	700	38.3	91.4	55.0	21.1	1.96	Matsuura et al. 1989b	19
5 mol% Li/MgO	2.11	0.12[b]	700	30.4	68.4	54.1	15.6	1.85	Matsuura et al. 1989b	19
10 mol% Li/MgO	2.11	0.12[b]	700	16.1	43.4	56.2	9.0	1.64	Matsuura et al. 1989b	19
15 mol% Li/MgO	2.11	0.12[b]	700	5.8	12.8	54.0	3.1	1.49	Matsuura et al. 1989b	19
Li/MgO	2.0	0.12[b]	715	0.058	—	29.0	0.02	0.16	Mims et al. 1989	—
	2.0	0.18[b]	730	0.054	—	61.0	0.03	0.17	Mims et al. 1989	20
	2.0	0.29[b]	702	3.1	—	65	2.02	0.12	Mims et al. 1989	—
Li/MgO(?)[d]	?	?	850	20	—	75	15.0	?		21

[a] For use with Fig. 7-1.
[b] Denotes W/F in gram-seconds per milliliter.
[c] Fixed bed.
[d] Fluidized bed.
[e] Residence time s.
[f] O_2 distribution system used. For details see original paper; CH_4–O_2 ratio is the overall ratio used.

Table 7-2 Methane Oxidation with Molecular Oxygen Using Lanthanide Metal Oxide Catalysts

				CONVERSION		C_2 HYDROCARBON				
Catalyst	CN_4–O_2 Mole Ratio	GHSV (h^{-1})	Temp. (°C)	CH_4 (%)	O_2 (%)	Selectivity (%)	Yield (%)	Ethene / Ethane	Reference	Ref. No.[a]
La_2O_3	4.6	1.0[b]	650	10	74	8	0.8	0.59	Otsuka, Jinno, and Morikawa 1985; Otsuka and Nakajima 1986	—
CeO_2	4.6	1.0[b]	650	9	100	3	0.3	0.00	Otsuka, Jinno, and Morikawa 1985; Otsuka and Nakajima 1986	—
Pr_6O_{11}	4.6	1.0[b]	650	11	100	20	2.2	0.44	Otsuka, Jinno, and Morikawa 1985; Otsuka and Nakajima 1986	—
Nd_2O_3	4.6	1.0[b]	650	14	99	37	5.2	0.33	Otsuka, Jinno, and Morikawa 1985; Otsuka and Nakajima 1986	—
Sm_2O_3	4.6	1.0[b]	650	15	96	37	5.6	0.42	Otsuka, Jinno, and Morikawa 1985; Otsuka and Nakajima 1986	—
Dy_2O_3	4.6	1.0[b]	650	13	91	24	3.1	0.37	Otsuka, Jinno, and Morikawa 1985; Otsuka and Nakajima 1986	—
La_2O_3	10.0	0.22[b]	740	18.8	—	40	7.5	—	Roos et al. 1989	—
Sm_2O_3	10.0	0.22[b]	780	12.2	—	54	6.6	—	Roos et al. 1989	22
Gd_2O_3	10.0	0.22[b]	770	12.1	—	48	5.8	—	Roos et al. 1989	—
Tb_4O_7	10.0	0.22[b]	820	7.8	—	28	2.2	—	Roos et al. 1989	—
Dy_2O_3	10.0	0.22[b]	793	12.9	—	53	6.8	—	Roos et al. 1989	23
La_2O_3	37.0	2.4[b]	725	3.6	88	71.0	2.5	0.67	Lin et al. 1986	24
	8.8	2.4[b]	725	9.4	91	46.6	4.4	1.0	Lin et al. 1986	—
	3.4	2.4[b]	725	19.7	93	23.7	4.7	1.25	Lin et al. 1986	—
La_2O_3	3.0	900	710	22.7	100	15.8	3.6	0.46	Hutchings, Scurrell, and Woodhouse 1989b	—

5% Li/La_2O_3	3.3	700	710	10.3	33	49.9	5.1	0.46	Hutchings, Scurrell, and Woodhouse 1989b	—
Sm_2O_3	3.1	592	710	19.1	100	15.3	2.9	0.07	Hutchings, Scurrell, and Woodhouse 1989b	—
5% Li/Sm_2O_3	3.0	857	710	24.5	87	52.9	13.0	1.11	Hutchings, Scurrell, and Woodhouse 1989b	25
La_2O_3	6.0	37,500	750	19.6	100	56.6	11.1	0.83	De Boy and Hicks 1988a	26
1% Li/La_2O_3	6.0	37,500	800	21.6	100	70.2	15.2	0.83	De Boy and Hicks 1988a	27
1% Na/La_2O_3	6.0	37,500	800	20.0	100	64.6	12.9	0.91	De Boy and Hicks 1988a	28
1% K/La_2O_3	6.0	37,500	800	20.6	100	60.4	12.4	1.00	De Boy and Hicks 1988a	29
1% Sr/La_2O_3	6.0	37,500	750	20.9	100	63.6	13.3	0.83	De Boy and Hicks 1988a, 1988b	30
Sm_2O_3	6.0	37,500	750	12.9	100	53.0	9.5	0.77	De Boy and Hicks 1988a	—
Sm_2O_3	20.0	0.75[b]	750	13.3	—	52.3	7.0	0.42	Burch, Squire, and Tsang 1988	—
5–10% Sm_2O_3	20.0	0.75[b]	750	9.3	—	64.0	6.0	0.67	Burch, Squire, and Tsang 1988	31
Sm_2O_3	2.5	0.6[b]	750	29.0	100	40.0	11.6	—	Otsuka et al. 1987a	—
10% Li/Sm_2O_3	2.5	0.6[b]	750	37.0	100	57.0	21.1	—	Otsuka et al. 1987a	32
10% Na/Sm_2O_3	2.5	0.6[b]	750	37.0	100	50.0	18.5	—	Otsuka et al. 1987a	33
10% K/Sm_2O_3	2.5	0.6[b]	750	37.0	100	52.0	19.2	—	Otsuka et al. 1987a	34
10% Rb/Sm_2O_3	2.5	0.6[b]	750	37.0	100	46.0	17.0	—	Otsuka et al. 1987a	—
10% Cs/Sm_2O_3	2.5	0.6[b]	750	37.0	100	45.0	16.6	—	Otsuka et al. 1987a	—
CeO_2	2.0	0.56[b]	750	—	—	0	0	—	De Boy and Hicks 1988a	—
20% Li/CeO_2	2.0	0.56[b]	750	28.6	—	50.0	14.3	—	De Boy and Hicks 1988a	35
20% Na/CeO_2	2.0	0.56[b]	750	33.4	—	41.0	13.7	—	De Boy and Hicks 1988a	—
20% K/CeO_2	2.0	0.56[b]	750	56.6	—	23.0	13.0	—	De Boy and Hicks 1988a	—
Cs_2CO_3/Sm_2O_3	5.0	230,000	750	22.2	90.4	40	8.88	2.08	Inui, Ishihara, and Kamachi 1989	—
	5.0	230,000	750	19.8	77.4	50	9.90	1.39	Inui, Ishihara, and Kamachi 1989	36
Li_2CO_3/Sm_2O_3	5.0	230,000	750	19.7	98.9	38	7.49	1.11	Inui, Ishihara, and Kamachi 1989	—
	5.0	230,000	750	18.4	95.0	19	3.50	0.59	Inui, Ishihara, and Kamachi 1989	—
Sm_2O_3	5.0	230,000	750	18.4	99.0	50	9.20	1.49	Inui, Ishihara, and Kamachi 1989	37

[a]For use with Fig. 7-2.
[b]Denotes W/F in gram-seconds per milliliter.

Table 7-3 Methane Oxidation with Molecular Oxygen Using Transition-Metal Oxides

				CONVERSION		C₂ HYDROCARBON				
Catalyst	CN_4–O_2 Mole Ratio	GHSV (h^{-1})	Temp. (°C)	CH_4 (%)	O_2 (%)	Selectivity (%)	Yield (%)	Ethene/Ethane	Reference	Ref. No.[a]
MnO_2/α-Al_2O_3	[a]	96	800	11	—	45	5	—	Keller and Bhasin 1982	—
CdO/α-Al_2O_3	[a]	96	800	10	—	40	4	—	Keller and Bhasin 1982	—
TiO_2/α-Al_2O_3	[a]	96	800	1.7	—	< 6	0.1	—	Keller and Bhasin 1982	—
α-Al_2O_3	[a]	96	800	1.8	—	11	0.2	—	Keller and Bhasin 1982	—
1% Li/TiO_2	2	0.42[b]	800	16.7	42	75	12.5	0.83	Lane and Wolf 1988a	38
15% MnO_2/SiO_2	[a]	600	800	30.0	—	50	15.0	0.63	Jones, Leonard, and Sofranko 1987	39
	[a]	860	850	22.0	—	61	13.4	3.33	Sofranko et al. 1988	40
	5	4800	900	22.0	—	59	13.0	1.96	Sofranko et al. 1988	41
$NaCl/MnO_2$	2	0.23[b]	750	32.0	98	49	15.7	5.88	Otsuka and Komatsu 1987	—
10% MnO_2/SiO_2	3.3	1000	750	20.6	—	30.6	6.3	1.32	Minachev et al. 1987	—
MnO_2	20 —	0.75[b]	750	6	—	8.9	0.5	∞	Burch, Squire, and Tsang 1988	
15% MnO_2/SiO_2	20	0.75[b]	750	6	—	16.8	1.0	0.20	Burch, Squire, and Tsang 1988	—
$NaCl$/15% MnO_2/SiO_2	20	0.75[b]	750	7	—	36.5	2.6	0.24	Burch, Squire, and Tsang 1988	—
NiO	2.0	0.6[b]	750	29.9	81	0	0	—	Otsuka et al. 1986	—
$LiCl/NiO$	2.0	0.6[b]	750	25.9	57	71.8	18.6	3.70	Otsuka et al. 1986	42
Li_2CO_3/NiO	2.0	0.6[b]	750	25.9	70	48.9	12.7	1.18	Otsuka et al. 1986	—

$LiNO_3/NiO$	2.0	0.6[b]	750	26.1	75	55.8	14.6	0.75	Otsuka et al. 1986	43
NaCl/NiO	2.0	0.6[b]	750	17.9	44	34.7	6.2	0.14	Otsuka et al. 1986	—
12% Li/ZnO	2	0.167[b]	680	16.7	42.5	56.2	9.4	—	Matsuura et al. 1986	44
	2	0.167[b]	720	30.6	81.6	60.5	18.5	—	Matsuura et al. 1986	45
	2	0.167[b]	760	34.6	89.9	59.5	20.6	—	Matsuura et al. 1986	46
0.18% Li/ZnO	2	4.8[b]	650	5.1	11.7	51.1	2.6	0.26	Zhan et al. 1988	47
	2	4.8[b]	700	13.9	25.3	59.0	8.2	0.87	Zhan et al. 1988	48
	2	4.8[b]	720	21.4	32.4	52.6	11.3	0.87	Zhan et al. 1988	49
	2	4.8[b]	750	28.2	55.5	53.3	15.0	1.18	Zhan et al. 1988	50
$LiCl/Co_3O_4$	1.98	2.4[b]	550	~ 5	—	~ 20	~ 1	0.93	Komatsu, Amaya, and Otsuka 1989	—
	1.98	2.4[b]	650	~ 27	—	~ 36	~ 17	3.23	Komatsu, Amaya, and Otsuka	—
	1.98	2.4[b]	750	~ 47	—	~ 47	~ 22	13.3	Komatsu, Amaya, and Otsuka 1989	—
Li_2O/TiO_2	2.75	0.6[b]	750	22.2	61.8	51.5	11.4	1.18	Fen-Chang et al. 1989	51
$LiCl/TiO_2$	2.72	0.6[b]	750	40.1	84.6	40.8	16.3	17.54	Fen-Chang et al. 1989	—
Na_2O/TiO_2	2.75	0.6[b]	750	17.5	48.3	53.4	9.3	2.76	Fen-Chang et al. 1989	51
$NaCl/TiO_2$	2.75	0.6[b]	750	34.8	100.0	65.5	22.7	6.06	Fen-Chang et al. 1989	51
K_2O/TiO_2	2.75	0.6[b]	750	20.8	78.1	36.2	7.53	1.11	Fen-Chang et al. 1989	—
KCl/TiO_2	2.75	0.6[b]	750	32.0	79.8	54.5	17.4	4.35	Fen-Chang et al. 1989	51
Na_2CO_3/TiO_2	2.75	0.6[b]	750	18.2	63.2	24.4	4.44	0.45	Fen-Chang et al. 1989	—
NaH_2PO_4/TiO_2	2.75	0.6[b]	750	14.3	31.5	28.6	4.09	0.74	Fen-Chang et al. 1989	—
NaCl	2.75	0.6[b]	750	14.2	18.5	9.5	1.35	14.3	Fen-Chang et al. 1989	—
$MnO/NaCl/TiO_2$	2.75	0.6[b]	750	42.7	100	61.0	26.0	6.1	Fen-Chang et al. 1989	51
	2.75	0.6[b]	750	43.0	100	56.7	24.4	5.4	Fen-Chang et al. 1989	51
	2.75	0.6[b]	750	42.9	100	55.4	23.8	52.6	Fen-Chang et al. 1989	51

[a] For use with Fig. 7-3.
[b] Denotes W/F in gram-seconds per milliliter.

Table 7-4 **Methane Oxidation with Molecular Oxygen Using Group IIIA, IVA, and VA Metal Oxide Catalyst**

Catalyst	CH_4–O_2 Mole Ratio	GHSV (h^{-1})	Temp. (°C)	CONVERSION CH_4 (%)	CONVERSION O_2 (%)	C_2 HYDROCARBON Selectivity (%)	C_2 HYDROCARBON Yield (%)	Ethene/Ethane	Reference	Ref. No.[a]
Tl_2O_3/α-Al_2O_3	[a]	96	800	6.8	—	44	3.0	—	Keller and Bhasin 1982	—
SnO_2/α-Al_2O_3	[a]	96	800	8.4	—	4.8	0.4	—	Keller and Bhasin 1982	—
PbO/α-Al_2O_3	[a]	96	800	4.4	—	43	2.0	—	Keller and Bhasin 1982	—
Bi_2O_3/α-Al_2O_3	[a]	96	800	4.4	—	9.1	0.4	—	Keller and Bhasin 1982	—
5% SnO_2/SiO_2	[a]	600	700	3.2	—	20.9	0.7	0.53	Jones, Leonard, and Sofranko 1984a	—
	[a]	600	800	10.1	—	29.3	3.0	1.16	Jones, Leonard, and Sofranko 1984a	—
5% Sb_2O_3/SiO_2	[a]	600	700	1.2	—	11.8	0.1	0.90	Jones, Leonard, and Sofranko 1984b	—
	[a]	600	800	2.7	—	21.0	0.6	0.89	Jones, Leonard, and Sofranko 1984b	—
24% PbO/α-Al_2O_3	5	29,800	751	14.2	—	38.2	5.4	—	Sofranko et al. 1988	—
6.5% PbO/Al_2O_3	10	0.11[b]	740	—	99	12.8	—	0.35	Hinsen, Bytyn, and Baerns 1984; Bytyn, and Baerns 1986	—
36% PbO/Al_2O_3	10	0.22[b]	740	—	95.1	56.2	—	0.46	Hinsen, Bytyn, and Baerns 1984; Bytyn, and Baerns 1986	—
7% Bi_2O_3/Al_2O_3	10	0.011[b]	640	6.6	87	23.9	1.6	—	Emesh and Amenomiya 1986	—
7% PbO/Al_2O_3	10	0.011[b]	640	4.2	63	7.6	0.3	—	Emesh and Amenomiya 1986	—
7% SnO_2/Al_2O_3	10	0.011[b]	640	6.4	85	14.7	0.9	—	Emesh and Amenomiya 1986	—

PbO	100	2^{b}	750	—	—	—	0.1	—	Aika and Nishiyama 1988b	—
$Bi_2O_3/P_2O_5/K_2O/MgO$	8	—	710	12.9	100	59	7.6	—	Keulks and Yu 1987	52
SiO_2	4.5	0.33^{b}	780	3.5	—	25	0.9	0.19	Agarwal, Migone, and Marcelin 1989	—
0.34% K/SiO_2	4.5	0.33^{b}	780	1.3	—	27	0.4	0.13	Agarwal, Migone, and Marcelin 1989	—
$Sb–O/SiO_2$	4.5	0.33^{b}	780	1.0	—	48	0.5	0.14	Agarwal, Migone, and Marcelin 1989	—
2.5% $Li–Sb–O/SiO_2$	4.5	0.33^{b}	780	1.1	—	63	0.7	0.07	Agarwal, Migone, and Marcelin 1989	53
5% $Li–Sb–O/SiO_2$	4.5	0.33^{b}	780	2.0	—	68	1.4	0.13	Agarwal, Migone, and Marcelin 1989	53
20% $Li–Sb–O/SiO_2$	4.5	0.33^{b}	780	1.4	—	56	0.8	0.10	Agarwal, Migone, and Marcelin 1989	53
5% $K–Sb–O/SiO_2$	4.5	0.33^{b}	780	1.1	—	57	0.6	0.12	Agarwal, Migone, and Marcelin 1989	53
5% $Cs–Sb–O/SiO_2$	4.5	0.33^{b}	780	1.9	—	61	1.2	0.15	Agarwal, Migone, and Marcelin 1989	53
SiO_2	7.0	3.0^{b}	775	0.9	~ 10	26.5	0.2	0.02	Ahmed and Moffat 1989	—
SiO_2 (+CCl_4)	7.0	3.0^{b}	775	2.1	~ 15	52.0	1.1	1.00	Ahmed and Moffat 1989	54
Ba/SiO_2	7.0	3.0^{b}	775	4.8	~ 45	13.0	0.6	0.91	Ahmed and Moffat 1989	—
Ba/SiO_2 (+CCl_4)	7.0	3.0^{b}	775	11.9	~ 85	54.5	6.5	3.00	Ahmed and Moffat 1989	54
Cs/SiO_2	7.0	3.0^{b}	775	3.9	~ 40	12.0	0.5	5.00	Ahmed and Moffat 1989	—
Cs/SiO_2 (+CCl_4)	7.0	3.0^{b}	775	13.3	~ 95	59.0	7.9	2.78	Ahmed and Moffat 1989	54
Mn/SiO_2	3.67	30^{b}	775	11.8	~ 93	15.0	1.8	0.50	Ahmed and Moffat 1989	—
Mn/SiO_2 (+CCl_4)	3.67	3.0^{b}	775	14.5	~ 100	31.5	4.6	4.17	Ahmed and Moffat 1989	—
K/SiO_2	7.0	3.0^{b}	775	0.5	~ 5	29.0	0.2	0.16	Ahmed and Moffat 1989	—
K/SiO_2 (+CCl_4)	7.0	3.0^{b}	775	3.2	~ 15	58.0	1.9	1.08	Ahmed and Moffat 1989	54
Ca/SiO_2	7.0	3.0^{b}	775	6.3	~ 65	20.0	1.3	1.00	Ahmed and Moffat 1989	—
Ca/SiO_2 (+CCl_4)	7.0	3.0^{b}	775	8.7	~ 95	33.0	2.9	6.25	Ahmed and Moffat 1989	—
Co/SiO_2	7.0	3.0^{b}	775	5.9	~ 90	12.0	0.7	0.41	Ahmed and Moffat 1989	—
Co/SiO_2 (+CCl_4)	7.0	3.0^{b}	775	20.5	~ 100	3.0	0.6	0	Ahmed and Moffat 1989	—

Table 7-4 (*Continued*)

				CONVERSION		C₂ HYDROCARBON				
Catalyst	CH_4–O_2 Mole Ratio	GHSV (h^{-1})	Temp. (°C)	CH_4 (%)	O_2 (%)	Selectivity (%)	Yield (%)	Ethene / Ethane	Reference	Ref. No.[a]
Bi/SiO_2	7.0	3.0[b]	775	1.0	~ 10	35.5	0.4	0.15	Ahmed and Moffat 1989	—
Bi/SiO_2 (+CCl_4)	7.0	3.0[b]	775	5.6	~ 70	23.0	1.3	1.19	Ahmed and Moffat 1989	—
Pb/SiO_2	7.0	3.0[b]	775	5.4	~ 35	80.0	4.3	0.42	Ahmed and Moffat 1989	54
Pb/SiO_2 (+CCl_4)	7.0	3.0[b]	775	1.4	~ 5	70.5	1.0	1.14	Ahmed and Moffat 1989	54
Tl/SiO_2	7.0	30[b]	775	14.9	~ 100	74.0	11.0	0.70	Ahmed and Moffat 1989	54
Ti/SiO_2 (+CCl_4)	7.0	3.0[b]	775	5.0	~ 55	30.5	4.6	1.27	Ahmed and Moffat 1989	—
SiO_2	1.0	3.0[b]	775	4.9	20	15.0	0.8	0.36	Ahmed and Moffat 1989	—
SiO_2 (+CCl_4)	1.0	3.0[b]	775	8.3	25	19.5	1.6	2.00	Ahmed and Moffat 1989	—
Ba/SiO_2	1.0	3.0[b]	775	9.5	15	5.0	0.5	0.25	Ahmed and Moffat 1989	—
Ba/SiO_2 (+CCl_4)	1.0	3.0[b]	775	20.0	32	36.5	7.3	5.00	Ahmed and Moffat 1989	—
Cs/SiO_2	1.0	3.0[b]	775	12.2	17	2.5	0.3	0.25	Ahmed and Moffat 1989	—
Cs/SiO_2 (+CCl_4)	1.0	3.0[b]	775	36.5	57	13.5	4.9	7.69	Ahmed and Moffat 1989	—
Mn/SiO_2	1.0	3.0[b]	775	22.2	42	13.5	3.0	0.80	Ahmed and Moffat 1989	—
Mn/SiO_2 (+CCl_4)	1.0	3.0[b]	775	45.0	75	20.0	9.0	9.09	Ahmed and Moffat 1989	—
Ca/SiO_2	1.0	3.0[b]	775	19.8	27	13.5	2.7	0.93	Ahmed and Moffat 1989	—
Ca/SiO_2 (+CCl_4)	1.0	3.0[b]	775	26.0	43	6.5	1.7	5.56	Ahmed and Moffat 1989	—
Pb/SiO_2	1.0	3.0[b]	775	13.3	25	43.5	5.8	0.81	Ahmed and Moffat 1989	—
Pb/SiO_2 (+CCl_4)	1.0	3.0[b]	775	4.5	12	31.5	1.4	1.52	Ahmed and Moffat 1989	—

[a] For use with Fig. 7-4.
[b] Denote W/F in gram-seconds per milliliter.

CO_x and H_2, via rapid gas-phase reactions (Chou and Albright 1978). However, formation of oxygenates at lower temperatures is not reported, and in studies where low-temperature activation of methane has been attempted (Hutchings, Scurrell, Woodhouse 1988b; Tashiro, Ito, and Toi 1990), no enhancement in CH_3OH or CH_2O has been noted, and H_2, H_2O, and CO_x predominate as products. Although the product distribution remains invariant, product selectivity is markedly affected by the choice of metal oxide as catalyst; however, even for the most selective oxides the C_2 yield rarely exceeds 25%, which is an observation consistent with the 100% rule previously cited by Sofranko and co-workers (Sofranko et al. 1988). The alkene–alkane ratio also varies markedly for all the oxides studied. However, this cannot be related solely to a feature of the specific catalytic selectivity of an oxide surface, because a variation in the ethene–ethane ratio can be caused by the design of the reactor, particularly the volume of the heated zone immediately downstream of the catalyst bed (van Kasteren, Geerts, and van der Wiele 1988), and for most of the published studies summarized in Tables 7-1–7-5 the precise details of the reactor are not given. However, a recent detailed study of oxyhalide catalysts (Thomas et al. 1989; Hutchings 1989) has shown that the ethene–ethane ratio increases with increasing methane or oxygen conversion, indicating that some ethene is generated from the steam cracking of ethane, a gas-phase noncatalyzed reaction that occurs downstream of the catalyst bed. Another general feature of these studies is that catalyst activity is related to basicity (Aika and Nishiyama 1989; Carreiro et al. 1988). This is most readily observed from the addition of metal additives to single oxides, such as MgO or CaO (Table 7-1), which results in catalysts that are significantly more active for C_2 hydrocarbon production.

It is important to note that a very wide range of experimental conditions has been used in the data summarized in Tables 7-1–7-5. For example, studies have utilized CH_4–O_2 mole ratios from 2 : 1 to 200 : 1, and in some studies an inert diluent has been included. Hence, it is very difficult to draw any broad conclusions from these studies concerning a relative catalyst activity and selectivity. It is apparent that the metal oxides studied exhibit a wide range of catalytic activities and to some extent the variation is due solely to the intrinsic surface area and, hence, the density of surface sites of the oxide. The importance of surface area has been noted (Iwamatsu et al. 1988) and low-surface-area materials seem to be preferable, this being one of the major reasons for the large-scale use of alkali metal compound additives, because these result in loss of surface area caused by sintering. We therefore consider that it is important to compare catalysts in terms of their specific activity (moles of product per square meter per hour) and these data are summarized for the most active catalysts in Table 7-6. These data can then be used as the basis for a

Table 7-5 Methane Oxidation with Molecular Oxygen Using Complex Oxide Catalysts

Catalyst	CH_4–O_2 Mole Ratio	GHSV (h^{-1})	Temp. (°C)	CONVERSION CH_4 (%)	CONVERSION O_2 (%)	C_2 HYDROCARBON Selectivity (%)	C_2 HYDROCARBON Yield (%)	Ethene/Ethane	Reference	Ref. No.[a]
$LaAlO_3$	5.0	2000	710	25.3	94.7	48.4	12.2	—	Imai and Tagawa 1986a; Imai, Tagawa, and Kamide 1987; Imai et al. 1988	—
$PbMoO_4$	8.3	33.6	740	1	3	19	0.2	—	Carreiro and Baerns 1987; Carreiro et al. 1988	—
$PbWO_4$	8.3	55.3	740	8	100	5	0.4	—	Carreiro and Baerns 1987; Carreiro et al. 1988	—
$PbCrO_4$	8.3	16.1	740	10	100	4	0.4	—	Carreiro and Baerns 1987; Carreiro et al. 1988	—
$Pb_3(PO_4)_2$	8.3	13.6	740	9	87	51	4.6	—	Carreiro and Baerns 1987; Carreiro et al. 1988	—
$PbSO_4$	8.3	52.4	740	8	63	63	5.0	—	Carreiro and Baerns 1987; Carreiro et al. 1988	55
$Bi_2Sn_2O_7$	4.8	0.22[b]	508	13	—	25	3.3	—	Roos et al. 1987	—
$SrCeO_3$	2.0	0.72[b]	750	44.2	—	49.3	21.8	0.85	Machida and Enyo 1987	—
$SrCe_{0.9}Yb_{0.1}O_{2.95}$	2.0	0.72[b]	750	52.6	—	60.1	31.6	1.52	Machida and Enyo 1987	56
$Ce_{0.95}Yb_{0.1}O_{1.95}$	2.0	0.72[b]	750	25.1	—	2.9	0.7	∞	Machida and Enyo 1987	—
CeO_2	2.0	0.72[b]	750	31.5	—	12.7	4.0	9.09	Machida and Enyo 1987	—
SrO	2.0	0.72[b]	750	25.3	—	62.4	15.8	2.70	Machida and Enyo 1987	56
Yb_2O_3	2.0	0.72[b]	750	28.3	—	44.5	12.6	1.02	Machida and Enyo 1987	—
$AgCrO_2$	6	1500	697	10.5	99.0	8.4	0.9	∞	Unger, Zhang, and Lambert 1988	—
$NaFeO_2$	6	1500	697	9.1	99.2	9.1	0.8	0.13	Unger, Zhang, and Lambert 1988	—
$LiNiO_2$	6	1500	697	25.9	46.6	46.7	12.1	0.67	Unger, Zhang, and Lambert 1988	—
$LiFeO_2$	6	1500	697	8.5	99.1	9.1	0.8	0.21	Unger, Zhang, and Lambert 1988	—

$LiYO_2$	6	1500	697	22.0	100	58.2	12.8	0.91	Unger, Zhang, and Lambert 1988	57
Na_2CeO_3	6	1500	697	16.7	99.8	43.2	7.2	0.23	Unger, Zhang, and Lambert 1988	—
$LiTiO_3$	6	1500	697	11.5	98.0	16.2	1.9	0.08	Unger, Zhang, and Lambert 1988	—
Y_2O_3	6	1500	697	10.0	100	37.2	3.7	0.67	Unger, Zhang, and Lambert 1988	—
$Pb_2Mn_2Si_2O_9/SiO_2$	7.7	1300	650	4.4	—	42.2	1.9	—	Thomas, Zhang, and Stachurski 1988	—
	7.7	1300	700	9.4	—	61.9	5.8	—	Thomas, Zhang, and Stachurski 1988	58
	7.7	1300	750	17.5	—	63.9	11.2	—	Thomas, Zhang, and Stachurski 1988	58
	7.7	1300	800	30.0	—	67.5	20.3	—	Thomas, Zhang, and Stachurski 1988	58
$LiBi_3O_4Cl_2$	2.0	82.8[b]	750	22.1	—	77.1	17.1	3.45	Ueda and Thomas 1988a, 1988b	59
$NaBi_3O_4Cl_2$	2.0	82.8[b]	750	22.6	—	83.1	18.3	3.45	Ueda and Thomas 1988a, 1988b	59
$KBi_3O_4Cl_2$	2.0	82.8[b]	750	23.8	—	82.3	19.6	3.84	Ueda and Thomas 1988a, 1988b	59
$AgBi_3O_4Cl_2$	2.0	82.8[b]	750	23.7	—	65.0	15.4	5.00	Ueda and Thomas 1988a, 1988b	59
$PbBiO_2Cl$	2.0	82.8[b]	750	13.7	—	72.9	10.0	1.79	Ueda and Thomas 1988a, 1988b	59
$BiOCl$	2.0	82.8[b]	750	23.0	—	71.1	16.4	4.35	Ueda and Thomas 1988a, 1988b	59
Bi_2O_3	2.0	82.8[b]	750	9.1	—	31.1	2.8	0.30	Ueda and Thomas 1988a, 1988b	59
$La_2O_3 \cdot 2TiO_2$	4	6000	650	17.3	97.0	30	5.2	0.58	Lane, Kalenik, and Wolf 1989	—
	4	6000	750	20.0	100	45	9.0	0.95	Lane, Kalenik, and Wolf 1989	—
	4	6000	850	20.0	100	42	8.4	1.79	Lane, Kalenik, and Wolf 1989	—
$MgTiO_3$	4	6000	750	2.0	9.0	53	1.1	0.20	Lane, Kalenik, and Wolf 1989	60
$NiTiO_3$	4	6000	750	3.0	11.0	48	1.4	0.23	Lane, Kalenik, and Wolf 1989	—
K_2TiO_3	4	6000	750	15.0	72.0	50	7.5	0.86	Lane, Kalenik, and Wolf 1989	60

Table 7-5 (*Continued*)

Catalyst	CH_4–O_2 Mole Ratio	GHSV (h^{-1})	Temp. (°C)	CONVERSION CH_4 (%)	CONVERSION O_2 (%)	C_2 HYDROCARBON Selectivity (%)	C_2 HYDROCARBON Yield (%)	Ethene/Ethane	Reference	Ref. No.[a]
$CaTiO_3$	4	6000	750	10.0	50.0	27	2.7	0.59	Lane, Kalenik, and Wolf 1989	—
$Y_2Sn_2O_7$	2.0	0.11[b]	727	21.3	—	11.0	2.3	0.24	Ashcroft et al. 1989	—
$La_2Sn_2O_7$	2.0	0.11[b]	727	29.7	—	15.0	4.5	0.52	Ashcroft et al. 1989	—
$Pr_2Sn_2O_7$	2.0	0.11[b]	727	4.3	—	17.3	0.7	0.19	Ashcroft et al. 1989	—
	2.0	0.04[b]	877	44.1	—	21.4	9.4	4.76	Ashcroft et al. 1989	—
$Nd_2Sn_2O_7$	2.0	0.11[b]	727	4.6	—	27.3	1.3	2.38	Ashcroft et al. 1989	—
$Sm_2Sn_2O_7$	2.0	0.11[b]	727	3.9	—	45.6	1.8	0.38	Ashcroft et al. 1989	—
	2.0	0.04[b]	877	42.4	—	21.4	9.1	4.55	Ashcroft et al. 1989	—
$Eu_2Sn_2O_7$	2.0	0.11[b]	727	2.0	—	35.8	0.7	0.20	Ashcroft et al. 1989	—
	2.0	0.04[b]	877	33.4	—	28.4	9.5	3.33	Ashcroft et al. 1989	—
$Gd_2Sn_2O_7$	2.0	0.11[b]	727	28.1	—	9.9	2.8	0.48	Ashcroft et al. 1989	—
$Tb_2Sn_2O_7$	2.0	0.11[b]	727	4.4	—	29.8	1.3	0.45	Ashcroft et al. 1989	—
	2.0	0.04[b]	877	40.1	—	18.3	7.3	4.17	Ashcroft et al. 1989	—
$Dy_2Sn_2O_7$	2.0	0.11[b]	727	30.4	—	21.9	6.7	1.09	Ashcroft et al. 1989	—
$Ho_2Sn_2O_7$	2.0	0.11[b]	727	26.5	—	10.9	2.9	0.24	Ashcroft et al. 1989	—
$Er_2Sn_2O_7$	2.0	0.11[b]	727	24.9	—	11.5	2.9	0.31	Ashcroft et al. 1989	—
$Tm_2Sn_2O_7$	2.0	0.11[b]	727	2.1	—	28.9	0.6	0.21	Ashcroft et al. 1989	—
	2.0	0.04[b]	877	21.2	—	24.8	5.3	2.08	Ashcroft et al. 1989	—
$Yb_2Sn_2O_7$	2.0	0.11[b]	727	2.0	—	43.0	0.9	0.21	Ashcroft et al. 1989	—
	2.0	0.04[b]	877	38.0	—	24.9	9.5	4.55	Ashcroft et al. 1989	—
$Lu_2Sn_2O_7$	2.0	0.11[b]	727	5.8	—	30.9	1.8	0.41	Ashcroft et al. 1989	—
Sm_2O_3	2.0	0.11[b]	727	37.3	—	31.4	11.7	1.12	Ashcroft et al. 1989	—
	2.0	0.04[b]	877	44.1	—	9.4	4.2	2.94	Ashcroft et al. 1989	—
SmOCl	6.6	2.4[b]	700	11.1	100	24.3	2.7	0.63	Williams et al. 1989	—
LaOCl	6.6	2.4[b]	700	12.9	100	13.4	1.7	0.29	Williams et al. 1989	—
BiOCl	6.6	2.4[b]	700	13.8	100	19.4	2.7	62.5	Williams et al. 1989	—
$(Bi_{0.5}La_{0.5})OCl$	6.6	2.4[b]	700	9.9	100	20.1	2.0	1.33	Williams et al. 1989	—
$(Bi_{0.5}Sm_{0.5})OCl$	6.6	2.4[b]	700	10.6	100	20.0	2.1	0.67	Williams et al. 1989	—
$LiMnO_2$	2.0	3.6[b]	760	30.8	—	43.6	13.4	—	Matsuura et al. 1989b	—
2.5 mol% Li_2O–$LiMnO_2$	2.0	3.6[b]	780	32.2	—	50.1	16.1	—	Matsuura et al. 1989b	61

$LiFeO_2$	2.0	3.6[b]	820	34.5	—	38.4	13.2	—	Matsuura et al. 1989b	—
2.5% $Li^+/LiFeO_2$	2.0	3.6[b]	800	40.1	—	54.7	21.9	—	Matsuura et al. 1989b	61
$LiNiO_2$	2.0	3.6[b]	740	40.0	—	53.8	21.5	—	Matsuura et al 1989b	61
2.5% $Li^+/LiNiO_2$	2.0	3.6[b]	760	34.5	—	68.4	23.6	—	Matsuura et al. 1989b	61
$LiInO_2$	2.0	3.6[b]	760	32.0	—	26.0	8.3	—	Matsuura et al. 1989b	—
2.5% $Li/LiInO_2$	2.0	3.6[b]	780	35.4	—	56.8	20.1	—	Matsuura et al. 1989b	61
Li_2TiO_3	2.0	3.6[b]	740	30.9	—	26.6	8.3	—	Matsuura et al. 1989b	—
2.5% Li^+/Li_2TiO_3	2.0	3.6[b]	780	29.7	—	61.8	18.4	—	Matsuura et al. 1989b	61
Li_2GeO_3	2.0	3.6[b]	800	32.5	—	46.9	15.2	—	Matsuura et al. 1989b	—
2.5% Li^+/Li_2GeO_3	2.0	3.6[b]	840	31.8	—	49.9	15.9	—	Matsuura et al. 1989b	—
$LiZrO_3$	2.0	3.6[b]	740	34.7	—	36.6	9.0	—	Matsuura et al. 1989b	—
2.5% $Li^+/LiZrO_3$	2.0	3.6[b]	800	36.5	—	55.5	20.3	—	Matsuura et al 1989b	61
Li_2SnO_3	2.0	3.6[b]	760	31.0	—	28.2	8.7	—	Matsuura et al. 1989b	—
2.5% Li^+/Li_2SnO_3	2.0	3.6[b]	780	29.2	—	63.3	18.5	—	Matsuura et al. 1989b	61
Li–B–Mn–Mg–O	—	0.8–49.[c]	850	10	—	90.0	9.0	?	Sofranko and Jubin 1989	62
Li–B–Mn–Mg–O	—	0.8–4.9[c]	850	20	—	80.0	16.0	?	Sofranko and Jubin 1989	62
Li–B–Mn—Mg–O	—	0.8–4.9[c]	850	30	—	70.0	21.0	?	Sofranko and Jubin 1989	62
Li–B–Mn–Mg–O	—	0.8–4.9[c]	850	40	—	60.0	24.0	?	Sofranko and Jubin 1989	62
Li–B–Mn–Mg–O	—	0.8–4.9(?)[d]	850	10	—	~ 98.0	~ 9.8	?	Sofranko and Jubin 1989	63
Li–B–Mn–Mg–O	—	0.8–4.9(?)[d]	850	20	—	~ 95.0	~ 19.0	?	Sofranko and Jubin 1989	63
Li–B–Mn–Mg–O	—	0.8–4.9(?)[d]	850	30	—	82	24.6	?	Sofranko and Jubin 1989	63
Li–B–Mn–Mg–O	—	0.8–4.9(?)[d]	850	40	—	72	28.8	?	Sofranko and Jubin 1989	63

[a] For use with Fig. 7-5.
[b] Denotes W/F in gram-seconds per milliliter.
[c] CH_4 and O_2 cofeeding.
[d] CH_4 and O_2 alternative feeding (redox option).

Table 7-6 Specific Activity Data for Methane Oxidation with Molecular Oxygen Using Oxide Catalysts

Catalyst	Mole Ratio CH_4–O_2	Temp. (°C)	Methane Feed Rate [mol (g cat.)$^{-1}$ h]	10^4 Specific activity (in moles of product per square meter per hour)							Reference
				H_2	CO_2	CO	C_2H_4	C_2H_6	Total C_2	Total C[a]	
5% Li/MgO	3	710	0.0616[b]	7.7	16.1	1.3	2.5	8.0	10.5	38.4	Hutchings, Scurrell, and Woodhouse 1989a
MgO	3	710	0.0616[b]	3.7	4.5	1.3	0.8	0.8	1.6	9.0	Hutchings, Scurrell, and Woodhouse 1989a
5% Li/La_2O_3	3	710	0.0616[b]	20	24	5.1	1.8	8.9	10.7	50.5	Hutchings, Scurrell, and Woodhouse 1989a
La_2O_3	3	710	0.0616[b]	11	17	1.8	4.2	4.9	9.1	37.0	Hutchings, Scurrell, and Woodhouse 1989a
5% Li/Sm_2O_3	3	710	0.0616[b]	16	97	3.9	29.0	42.0	71.0	242.9	Hutchings, Scurrell, and Woodhouse 1989a
Sm_2O_3	3	710	0.0616[b]	19	28	3.5	5.0	6.8	11.8	55.1	Hutchings, Scurrell, and Woodhouse 1989a
MgO	2.1	700	0.0148	—	—	—	0.02	0.06	0.08	1.4	Lin et al. 1987
7% Li/MgO	2.1	700	0.0185	—	—	—	3.0	3.0	6.0	16.5	Lin et al. 1987
20% Na/MgO	2.1	700	0.0185	—	—	—	0.21	0.43	0.64	4.2	Lin et al. 1987
30% K/MgO	2.1	700	0.0185	—	—	—	0	0.02	0.02	3.0	Lin et al. 1987
CaO	2.1	700	0.0185	—	—	—	0.05	0.54	0.59	7.9	Lin et al. 1987
5% Li/CaO	2.1	700	0.0185	—	—	—	2.8	3.9	6.7	16.7	Lin et al. 1987
15% Na/CaO	2.1	700	0.0185	—	—	—	4.9	5.5	10.4	30.8	Lin et al. 1987
23% K/CaO	2.1	700	0.0185	—	—	—	3.5	4.9	8.4	28.7	Lin et al. 1987
Y_2O_3	46	700	0.0122	—	—	—	0.20	0.46	0.66	1.7	Otsuka, Jinno, and Morikawa 1986
La_2O_3	46	700	0.0122	—	—	—	0.15	0.28	0.43	1.0	Otsuka, Jinno, and Morikawa 1986

Sm_2O_3	46	700	0.0122	—	—	—	1.1	2.3	3.4	7.3	Otsuka, Jinno, and Morikawa 1986
Gd_2O_3	46	700	0.0122	—	—	—	0.37	0.83	1.2	3.0	Otsuka, Jinno, and Morikawa 1986
Ho_2O_3	46	700	0.0122	—	—	—	0.24	1.16	1.4	3.3	Otsuka, Jinno, and Morikawa 1986
Yb_2O_3	46	700	0.0122	—	—	—	0.1	0.5	0.6	11.4	Otsuka, Jinno, and Morikawa 1986
PbO	46	700	0.0122	—	—	—	0.2	2.3	2.5	—	Otsuka, Jinno, and Morikawa 1986
Bi_2O_3	46	700	0.0122	—	—	—	0.3	1.6	1.9	5.1	Otsuka, Jinno, and Morikawa 1986
$LaAlO_3$	1	710	0.179	—	—	—	—	—	43.1	178	Imai and Tagawa 1986b
La_2O_3	1	710	0.179	—	—	—	—	—	8.7	37.8	Imai and Tagawa 1986b
Sm_2O_3	1	710	0.179	—	—	—	—	—	22.5	120	Imai and Tagawa 1986b
CeO_2	2	750	0.0298	—	—	—	1.5	0.2	1.7	26.8	Machida and Enyo 1987
Yb_2O_3	2	750	0.0298	—	—	—	2.2	2.2	4.4	19.7	Machida and Enyo 1987
$Ce_{0.95}Yb_{0.1}O_{1.95}$		750	0.0298	—	—	—	0	0.2	0.2	83.4	Machida and Enyo 1987
SrO	2	750	0.0298	—	—	—	5.6	2.0	7.6	24.4	Machida and Enyo 1987
$SrCeO_3$	2	750	0.0298	—	—	—	12.5	14.7	27.2	110.0	Machida and Enyo 1987
$SrCe_{0.9}Yb_{0.1}O_{2.95}$		750	0.0298	—	—	—	20.2	13.4	33.6	112.0	Machida and Enyo 1987

[a]Based on methane converted in carbon-containing products.

comparison for the relative activities of these oxides. At 700–710°C, catalysts can be ranked in order of decreasing specific activity for C_2 production as follows:

$$Li/Sm_2O_3 > Na/CaO > K/CaO > LaAlO_3 > Sm_2O_3$$
$$\sim Li/CaO > PbO > Bi_2O_3 > Ho_2O_3 > Gd_2O_3 > Li/MgO$$
$$\sim Yb_2O_3 \sim Y_2O_3 > Na/MgO \sim CaO > MgO$$

Although catalyst selectivities for all these oxides may be similar, it is clear that there exists a wide range of specific activities spanning almost two orders of magnitude. Studies by Machida and Enyo (Machida and Enyo 1987) indicate that $SrCeO_3$ may exhibit a higher specific activity than Li/Sm_2O_3 at 750°C. This study demonstrates the potential that exists to improve the specific activity further by careful selection of catalyst preparation methods, but to date no detailed study has been published concerning optimization of the specific activity.

Notwithstanding the difficulties associated with drawing conclusions as to which catalyst types exhibit the most promise regarding activity and selectivity, an attempt has been made to depict published achievements regarding C_2 yields for the classes of catalyst distinguished as in Tables 7-1–7-5. The data are represented in Figs. 7-1–7-5 and the conversion–selectivity boundaries applicable to the body of literature assessed in this review are summarized in Fig. 7-6. Broadly, the results suggest that complex oxides are associated with the highest selectivities and yields, but that catalysts based on group IIA oxides are not far behind. Yields in excess of 30% have been found only for complex oxides. Group III, IV, and V oxides have exhibited yields of less than 10%, although only relatively few data are available for this class of catalyst.

Nitrous Oxide as Oxidant

Although the greater majority of catalyst studies have used oxygen as oxidant, a number have utilized N_2O as oxidant because of the specific selectivity effects that can be achieved. Early studies by Liu and co-workers (Liu et al. 1984) and Zhen and co-workers (Zhen et al. 1985) demonstrated that Mo and V catalysts oxidized methane to methanol with N_2O as oxidant. Solymosi and co-workers (Solymosi, Tombacz, and Katson 1985) showed that with Bi_2O_3 as catalyst formaldehyde could be obtained from methane with N_2O as oxidant; addition of SnO_2 enhanced the effect, which was ascribed to enhanced decomposition of N_2O. Subsequent studies using Sm_2O_3 (Otsuka and Nakajima 1987; Hutchings, Scurrell, and

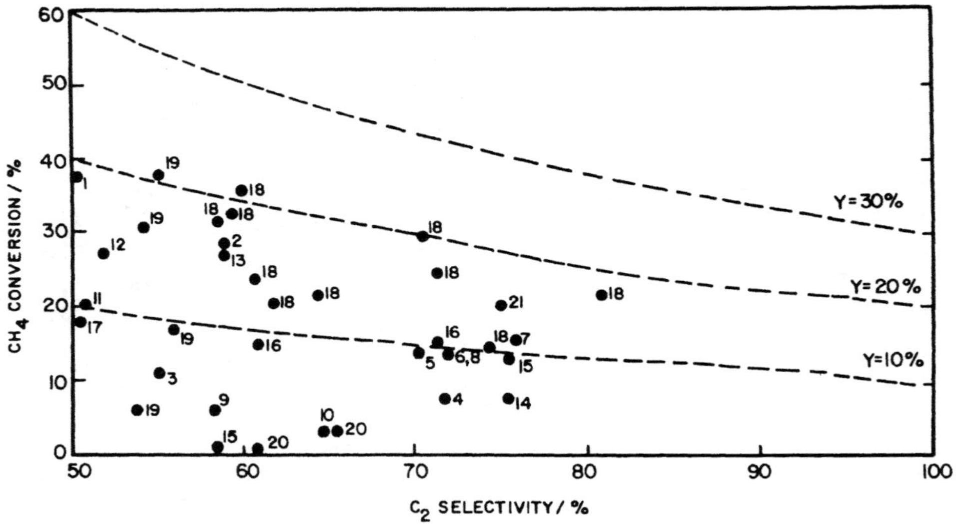

FIGURE 7-1. Methane oxidation with molecular oxygen, using group IIA metal oxides. Data reference numbers are for use with Table 7-1.

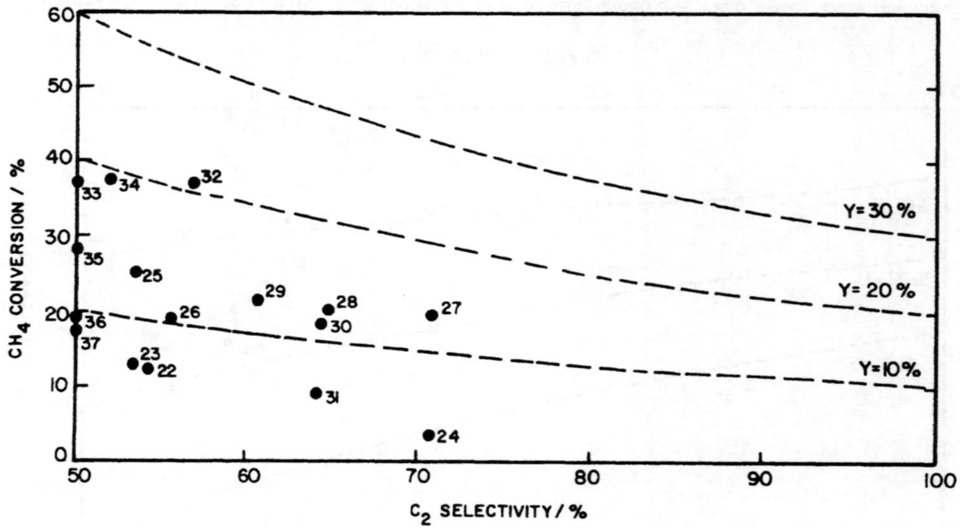

FIGURE 7-2. Methane oxidation with molecular oxygen, using lanthanide metal oxide catalysts. Data reference numbers are for use with Table 7-2.

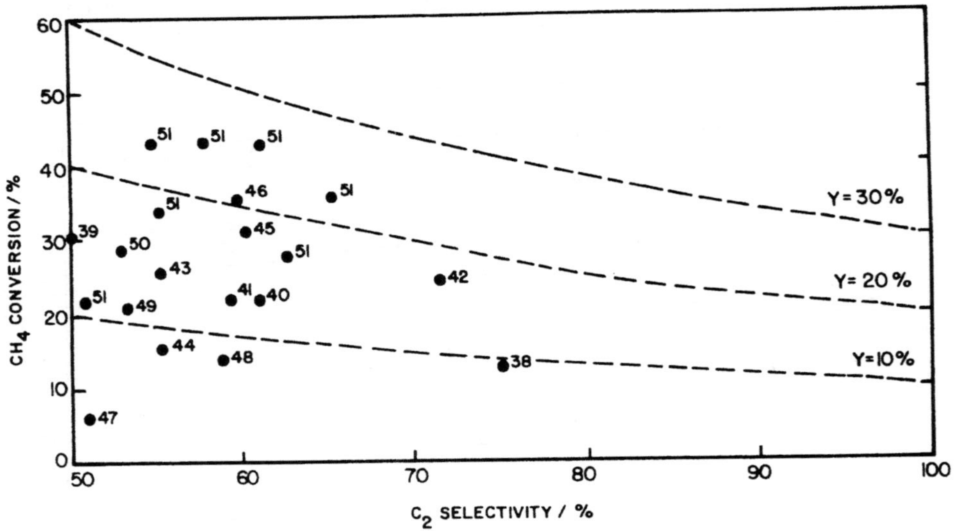

FIGURE 7-3. Methane oxidation with molecular oxygen, using transition metal oxides. Data reference numbers are for use with Table 7-3.

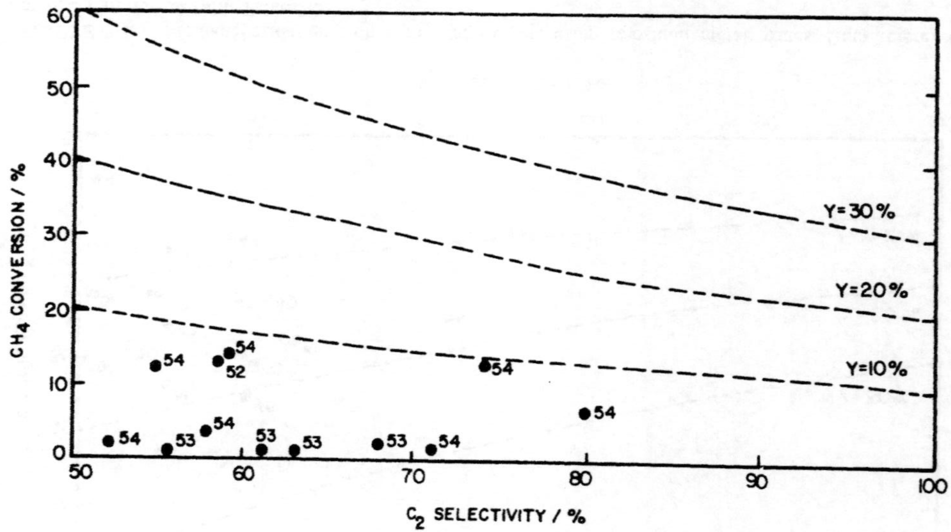

FIGURE 7-4. Methane oxidation with molecular oxygen, using group IIIA, IVA, and VA metal oxide catalysts. Data reference numbers are for use with Table 7-4.

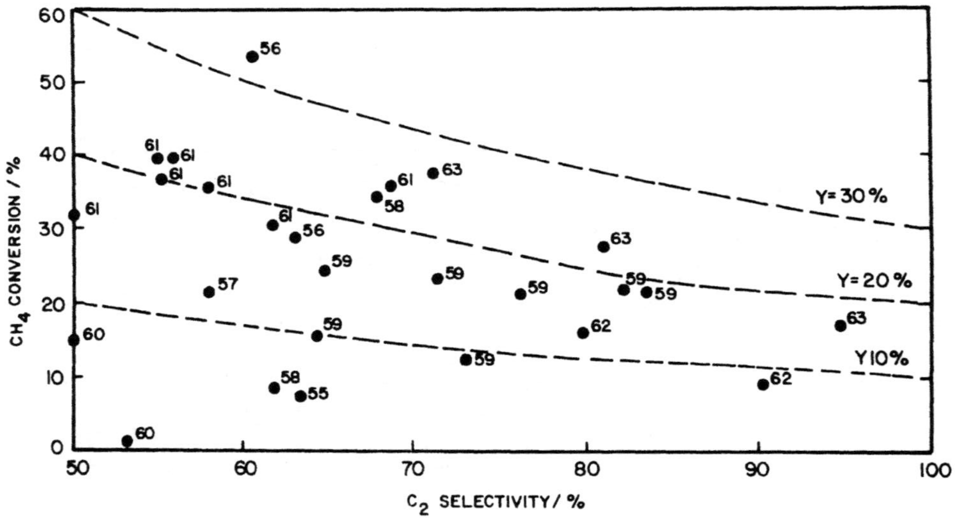

FIGURE 7-5. Methane oxidation with molecular oxygen, using complex metal oxides. Data references numbers are for use with Table 7-5.

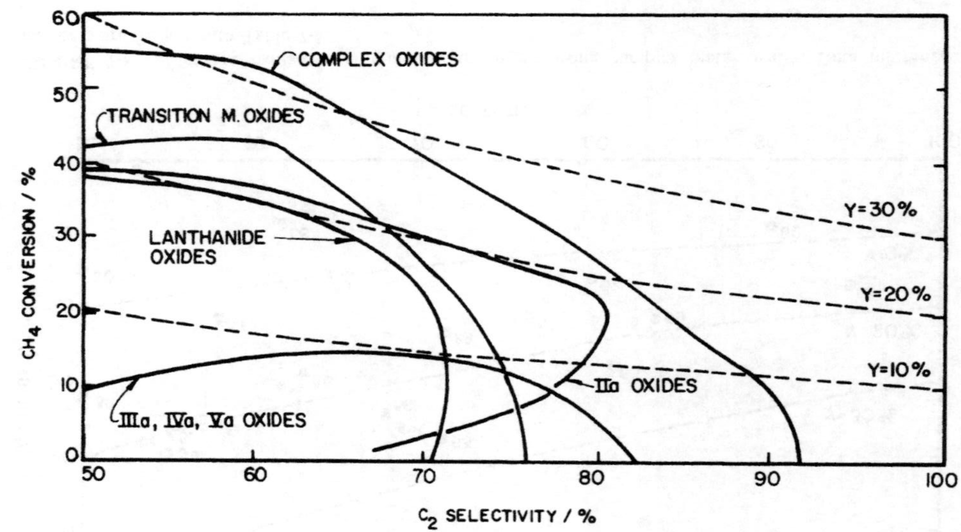

FIGURE 7-6. Boundaries of conversions versus selectivities achieved using data depicted in Figs. 7-1–7-5.

Woodhouse 1989b), MgO (Hutchings, Scurrell, and Woodhouse 1987a), and Li/MgO (Hutchings, Scurrell, and Woodhouse 1987a) have shown that N_2O, when used as oxidant, does not enhance the yield of CH_3OH or CH_2O with these catalysts and, hence, this is not a general feature of this oxidant. This has been confirmed in a recent study by Kasztelan and Moffat (Kasztelan and Moffat 1987) using silica as a catalyst when oxygen as oxidant gave significant yields of CH_2O, but with N_2O as oxidant only complete oxidation was observed.

A comparison of the relative efficacy of N_2O and O_2 as oxidants has been attempted in some studies and these are summarized in Table 7-7. At comparable reaction conditions with N_2O as oxidant, significantly lower methane conversions are observed with increased C_2 selectivity, particularly to ethane. The selectivity increases are most marked at low reaction temperatures. Because it is a general feature of methane oxidation that C_2 selectivity decreases with increasing conversion (Lane and Wolf 1988b), it is feasible that the marked increase in C_2 selectivity may just be a result of this selectivity–conversion relationship. However, a comparative study (Hutchings, Scurrell, and Woodhouse 1989b) of N_2O versus O_2 as oxidants at similar methane conversion levels (Fig. 7-7) demonstrates that N_2O is always more selective for C_2 formation than O_2 as oxidant. The decreased oxidation activity observed with N_2O is probably due to a combination of two factors. First, decomposition of N_2O may be slow on the oxide surface and, hence, limits the formation of the selective activation species for methane; indeed, experiments by the authors have shown that only partial decomposition of N_2O occurs for MgO and Li/MgO under typical reaction conditions. Second, even at complete decomposition, N_2O on a molar basis, can be considered to have only half the potential for stoichiometric reaction compared with O_2. However, experiments where the latter factor has been allowed for (Otsuka and Nakajima 1987) have demonstrated that N_2O remains less active than O_2 and, hence, the former factor may be the more important.

The main factor that has influenced the use of N_2O as an oxidant is its ability to deliver O^- as a surface species (Naccache 1971). Confirmation of this is obtained from the summary of data given in Table 7-7, which shows that with N_2O CO is oxidized to CO_2, which is produced in significantly higher quantities than with O_2 as oxidant. Based on a detailed analysis of reaction selectivities at 550 and 710°C, it has been shown (Hutchings, Scurrell, and Woodhouse 1987a) that O^- has two distinct oxidizing roles for these catalyst systems. The first is a selective oxidation role, which is most marked at the lower temperature, the second is a nonselective role, which becomes most apparent at the higher temperature. The mechanistic significance of these results will be discussed subsequently.

Table 7-7 Comparison of Nitrous Oxide and Molecular Oxygen as Oxidants for Methane

Catalyst	Oxidant	CH_4–Oxidant Mole Ratio	Temp. (°C)	GHSV (h^{-1})	CONVERSION (mol%) CH_4	Oxidant	PRODUCT SELECTIVITY[a] (mol%) CO_2	CO	C_2H_4	C_2H_6	Total C_2	Reference
MgO	O_2	3.5	550	667	3.0	—	66.2	31.8	0.4	1.6	2.0	Hutchings, Scurrell, and Woodhouse 1987a, 1989b
	N_2O	3.3	550	563	1.8	—	96.1	2.0	0.3	1.6	1.9	Hutchings, Scurrell, and Woodhouse 1987a, 1989b
	O_2	3.5	710	667	5.2	100	50.0	3.2	28.4	18.4	46.8	Hutchings, Scurrell, and Woodhouse 1987a, 1989b
	N_2O	3.3	710	545	3.0	98	88.9	6.8	2.7	1.6	4.3	Hutchings, Scurrell, and Woodhouse 1987a, 1989b
Li/MgO	O_2	3.3	550	648	2.0	—	59.8	17.7	5.0	17.5	22.5	Hutchings, Scurrell, and Woodhouse 1987a, 1989b
	N_2O	3.4	550	638	0.2	—	17.2	0	7.0	75.8	82.8	Hutchings, Scurrell, and Woodhouse 1987a, 1989b
	O_2	3.3	710	648	2.7	37	33.0	1.6	26.5	38.9	65.4	Hutchings, Scurrell, and Woodhouse 1987a, 1989b
	N_2O	3.4	710	638	0.3	2.8	51.9	0	9.3	38.9	48.3	Hutchings, Scurrell, and Woodhouse 1987a, 1989b
Sm_3O_3	O_2	3.1	710	592	19.1	100	82.4	2.3	1.0	14.3	15.3	Hutchings, Scurrell, and Woodhouse 1989b
	N_2O	3.0	710	720	14.6	99	64.4	3.7	14.6	17.3	31.8	Hutchings, Scurrell, and Woodhouse 1989b

5% Li/ Sm_2O_3	O_2	3.0	710	857	24.5	87	45.8	1.3	27.0	25.0	52.9	Hutchings, Scurrell, and Woodhouse 1989b
	N_2O	2.4	710	756	3.0	13	29.1	0	11.5	60.3	71.8	Hutchings, Scurrell, and Woodhouse 1989b
La_2O_3	O_2	4.6	550	3660	2	25	—	—	0	0	0	Otsuka and Nakajima 1986; Otsuka 1987
	N_2O	4.6	550	3660	Trace	4	—	—	0	57	57	Otsuka and Nakajima 1986; Otsuka 1987
	O_2	4.6	700	3660	10	75	—	—	2	6	8	Otsuka and Nakajima 1986; Otsuka 1987
	N_2O	4.6	700	3660	4	24	—	—	15	70	85	Otsuka and Nakajima 1986; Otsuka 1987
Sm_2O_3	O_2	4.6	550	3660	11	93	—	—	1	10	11	Otsuka Nakajima 1986; Otsuka 1987
	N_2O	4.6	550	3660	0.5	5	—	—	3	82	85	Otsuka and Nakajima 1986; Otsuka 1987
	O_2	4.6	700	3660	15	98	—	—	10	21	31	Otsuka and Nakajima 1986; Otsuka 1987
	N_2O	4.6	700	3660	5	35	—	—	10	68	78	Otsuka and Nakajima 1986; Otsuka 1987

[a]Selectivities calculated on carbon basis.

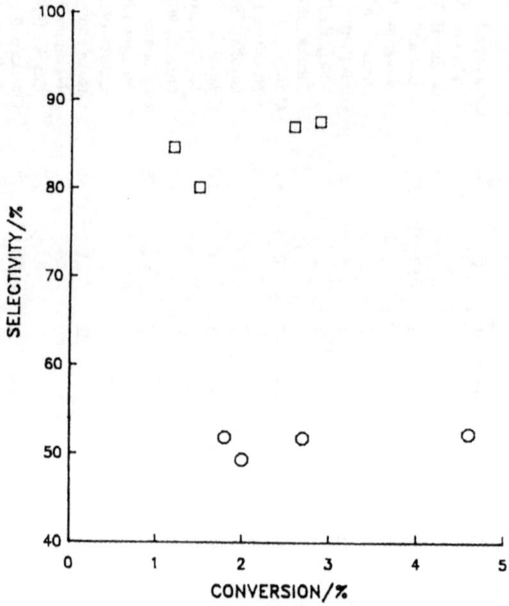

FIGURE 7-7. Comparison of C_2 selectivity over Li/MgO; 710°C; mole ratio CH_4–oxidant = 3: ○, O_2 as oxidant, □, N_2O as oxidant; data taken from Hutchings, Scurrell and Woodhouse 1989a.

Ozone as Oxidant

Foster in a review of the early methane oxidation literature noted that ozone, by virtue of its enhanced oxidizing activity, could be active for methane oxidation at temperatures lower than those observed for molecular oxygen (i.e., < 550°C). However, a comparative study of ozone and oxygen as oxidants (Hutchings, Scurrell, and Woodhouse 1986b) has shown that the use of ozone does not enhance the production of either C_2 hydrocarbons or oxygenates at lower temperatures (Fig. 7-8). Ozone was indeed observed to be more reactive than oxygen at the lower temperatures, but the only products observed were H_2, H_2O, and CO_2. At higher temperatures (i.e., > 500°C) oxygen and ozone gave similar methane conversions and product selectivities, but this was concluded to be due to the rapid decomposition of ozone to oxygen that would occur at these elevated temperatures.

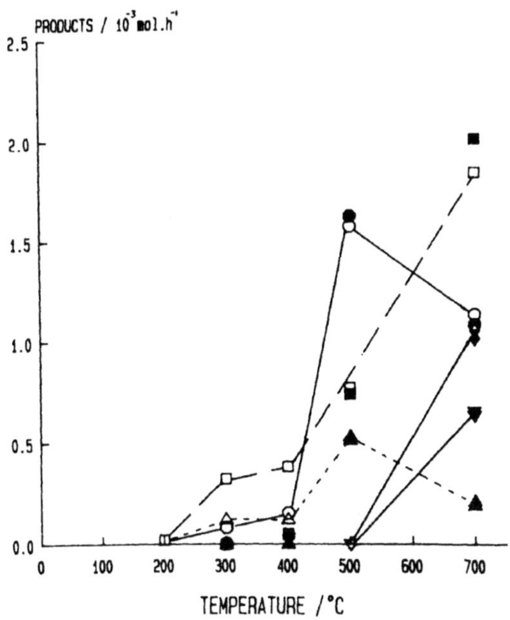

FIGURE 7-8. Comparison of ozone and oxygen as oxidants over Li/MgO; mole ratio CH_4–oxidant = 3.3, 70 × 10^{-3} mol h^{-1} inlet CH_4; open symbols, O_3/O_2, 5 mol% O_3; closed symbols, O_2: ◇, C_2H_4; ▽, C_2H_6; □, CO_2; △, CO; ○, H_2.

Reaction of CH_4–Oxidant–NO and CH_4–Oxidant–CO_2 Mixtures

The introduction of 2.5 mol% NO in N_2 to CH_4–N_2O reactants (mole ratio CH_4–N_2O = 3) over Li/MgO as catalyst has been observed (Hutchings, Scurrell, and Woodhouse 1989c, 1990) to have an immediate and significant effect on the methane coupling reaction (Fig. 7-9). A sharp decrease in C_2 production with a concomitant increase in CO_2 was observed. On removal of NO–N_2 from the reactants, the production of C_2 hydrocarbons immediately increased, whereas the production of CO_2 decreased. Significantly, the C_2 production increased to a level higher than that observed immediately preceding NO treatment of the catalyst and then declined slowly to the expected value; a similar but opposite trend was observed for CO_2 production. Similar results were observed (Hutchings, Scurrell, and Woodhouse 1990) for reaction of CH_4–O_2 mixtures in the presence of NO–N_2 over Li/MgO.

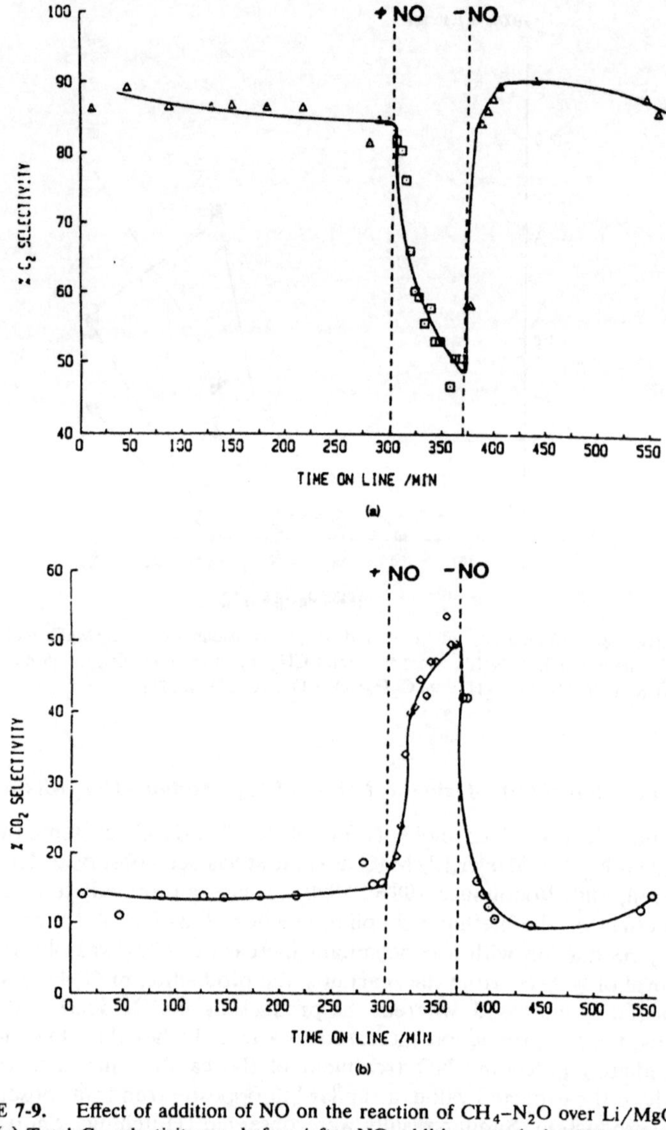

FIGURE 7-9. Effect of addition of NO on the reaction of CH_4–N_2O over Li/MgO at 650°C: (a) Total C_2 selectivity; △, before/after NO addition; □, during NO addition; (b) CO_2 selectivity; ○, before and after NO addition; ◇, during NO addition; (c) CH_4 conversion.

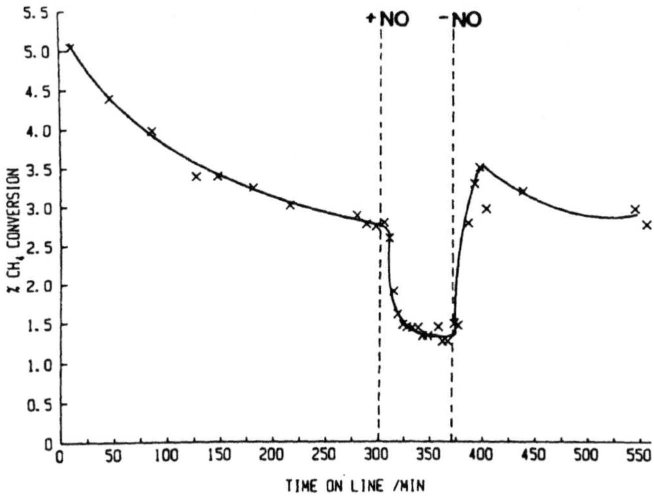

(c)

FIGURE 7-9 (*Continued*).

In the presence of NO with Li/MgO as catalyst, the production of C_2 hydrocarbons was not completely inhibited (Fig. 7-9). For the reaction of CH_4–N_2O mixtures, NO was found to completely inhibit the production of ethene (Fig. 7-10) and the ethene selectivity declined more rapidly than the ethane selectivity until only ethane was observed as product. A marked difference was apparent when O_2 was used as oxidant, because a small but significant ethene selectivity was still observed (Fig. 7-10). Subsequent experiments (Hutchings, Scurrell, and Woodhouse 1990) showed that NO inhibited ethane oxidation to H_2, CO_2, and C_2H_4 when both N_2O and O_2 were used as oxidants and when either Li/MgO or MgO were used as catalysts. It has been shown (Hutchings, Scurrell, and Woodhouse 1989a) that oxidation of ethane is largely gas-phase in nature and involves a monatomic oxygen species, and hence, inhibition of this reaction by the monoradical scavenger NO can be expected.

For MgO as catalyst different trends were observed (Hutchings, Scurrell, and Woodhouse 1990) on addition of NO. For CH_4–O_2 addition of NO did not affect the production of CO_2, and only a slight and nonimmediate decrease in C_2 production was observed (Fig. 7-11). These results indicate that significant differences exist for the mechanism of the activation of methane on MgO and Li/MgO catalysts.

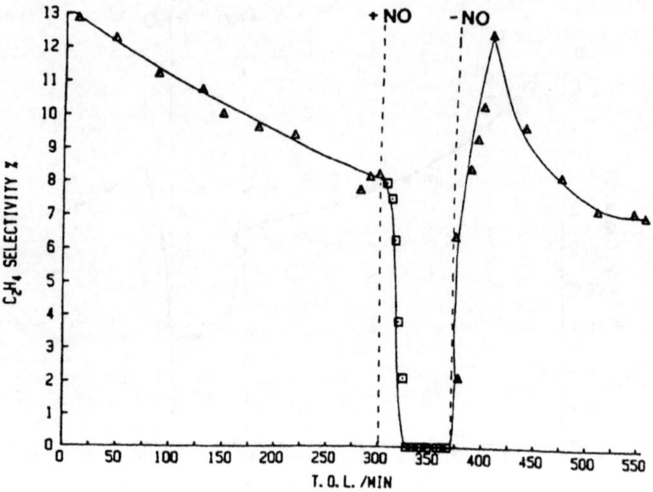

(a)

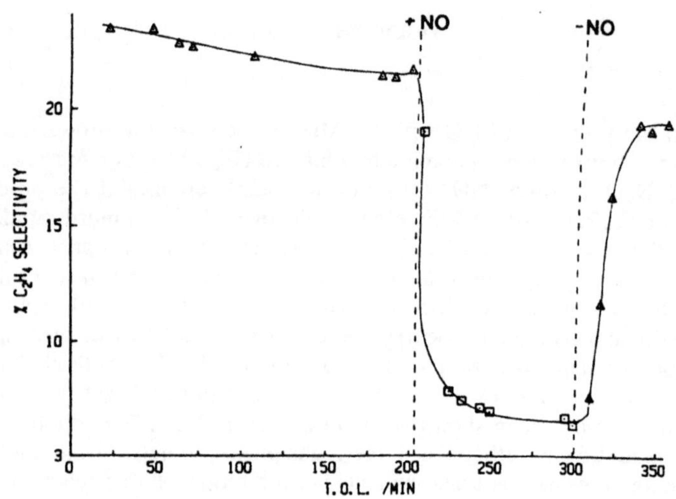

(b)

FIGURE 7-10. Effect of NO addition on ethene production: (a) CH_4–N_4O, Li/MgO at 650°C; (b) CH_4–O_2, Li/MgO at 650°C.

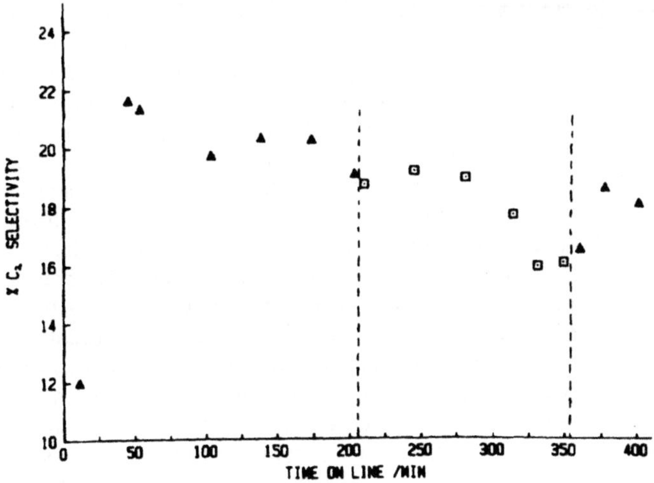

(a)

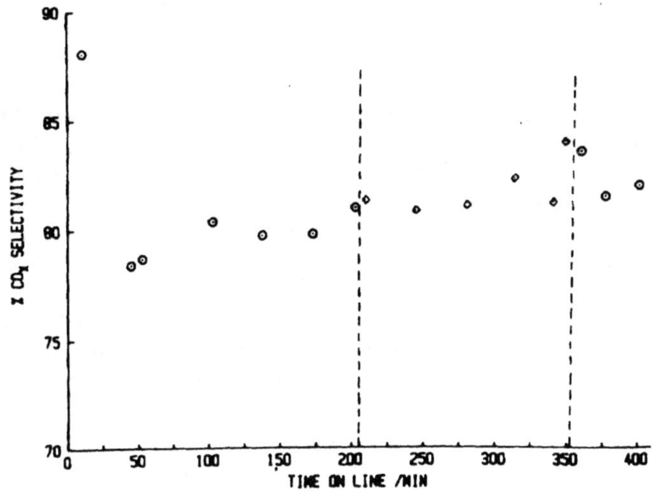

(b)

FIGURE 7-11. Effect of NO addition on the reaction of CH_4–O_2 over MgO at 650°C: (a) total C_2 selectivity; ▲, before and after NO addition; □, during NO addition; (b) CO_2 selectivity; ⊙, before and after NO addition; ◇ during NO addition.

Aika and Nishiyama (Aika and Nishiyama 1988b) have reported that addition of CO_2 to CH_4–O_2 reactant mixtures can significantly enhance the yield of C_2 hydrocarbons for a broad range of catalysts, in particular with PbO–MgO catalysts. It was proposed by these workers that CO_2 was involved in these reactions as an oxidant, providing a monatomic oxygen species via degradation to carbon monoxide. No isotopic labeling studies have been provided to substantiate this claim, and it is considered that a more probable role of CO_2 is to enhance or maintain the formation of surface carbonate species and thereby influence product selectivity. Some evidence for this possibility is provided by other studies (Korf et al. 1987), which have indicated that CO_2 can play a significant role in the reactivation of Li/MgO catalysts, presumably via carbonate formation. It was noted in a previous study (Hutchings, Scurrell, and Woodhouse 1990) that CH_4 oxidation with N_2O or O_2 as oxidant with Li/MgO as catalyst in the presence of NO leads to a significant increase in CO_2 selectivity, and on removal of NO the selectivity to C_2 hydrocarbons becomes significantly enhanced. The effects were not observed for MgO as catalyst, and it is possible that the effect of enhancement in C_2 selectivity observed for Li/MgO may be related to an effect of CO_2, rather than NO. Close inspection of the data (Fig. 7-9) indicates that the yield of CO_2 is not significantly affected on addition of NO (the CO_2 increases marginally from 0.3 to 0.39 vol%). It is considered improbable that a small increase in CO_2 could effect a marked increase in C_2 selectivity. The effect of CO_2 as a co-reactant has been investigated (Hutchings, Scurrell, and Woodhouse 1990) (Fig. 7-12). For CH_4–N_2O using Li/MgO as catalyst, the addition of CO_2 (10 vol%) caused an immediate and total inhibition of C_2 production. On removal of CO_2 a slow increase in C_2 selectivity occurred and no enhancement in C_2 selectivity above that previously attained was noted. Similar trends for the reaction of CH_4–O_2 were observed, except that the C_2 selectivity was more rapidly established. Lower levels of CO_2 (ca. 0.4 vol%) were not observed to have any significant effect on C_2 selectivity. Hence, from this study it was concluded that CO_2 was not involved in the enhancement in C_2 selectivity observed when NO is removed from CH_4–N_2O or CH_4–O_2 reactant mixtures for the Li/MgO catalyst. However, it is apparent that addition of NO or CO_2 to CH_4–oxidant mixtures causes immediate inhibition of the methane coupling reaction. It has been demonstrated (Zecchina, Lofthouse, and Stone 1975) that for MgO there exists a competitive adsorption between adsorbates, and the strength of adsorption increases in the order $O_2 < N_2O < NO < CO_2$; it can be expected that a similar trend exists for Li/MgO. Competitive adsorption between CO_2 and either N_2O or O_2 can therefore readily explain the observed effect of CO_2 addition (Fig. 7-12). For NO addition, a similar

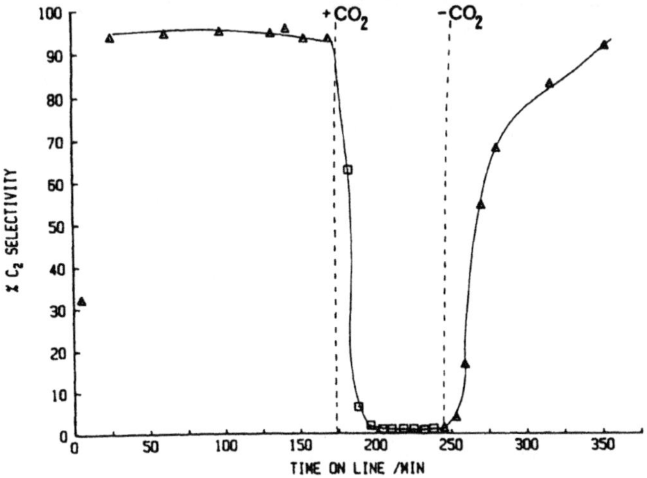

FIGURE 7-12. Effect of CO_2 addition on the reaction of CH_4–N_2O; Li/MgO, 650°C.

but less pronounced effect would be expected, hence, it was concluded (Hutchings, Scurrell, and Woodhouse 1990) that inhibition by NO not only involved competitive adsorption, but also involved NO scavenging of a selective surface oxidizing species.

STUDIES ON THE REACTION MECHANISM

Structure–Activity Relationships: Nature of the Active Site and the Role of Alkali Promoters

Group IIA Metal Oxides Promoted by Group IA Cations

From the preceding section it is clear that a considerable effort has been applied to the identification of active catalyst formulations. However, only a few studies have attempted to understand the relationship between catalyst structure and activity. Early studies concluded that MgO, in the absence of alkali promoters, was not particularly active (Ito et al. 1985) or, in some cases, showed no selective activity (Driscoll et al. 1985). A consequence of this was that unpromoted MgO was relatively little studied, which is unfortunate, because the structure of MgO has been particularly well-studied (Moodie and Warble 1971; Jones et al. 1984). Indeed, a theoretical study (Mehandru, Anderson, and Bradzil 1988) has indicated that oxygen atoms of different coordination numbers within the MgO

Table 7-8 Comparison of the Catalytic Performance of MgO Prepared by Different Methods[a] (Joyner 1989; Hargreaves, Hutchings, and Joyner 1990)

Catalyst	Flow Rate[b] (ml min^{-1})	Temp. (°C)	CONVERSION CH_4 (%)	O_2 (%)	PRODUCT SELECTIVITY[c] C_2H_6 (%)	C_2H_4 (%)	CO (%)	CO_2 (%)
MgO (ex b.c.)	50	737	26.3	100	25.9	23.0	11.1	40.0
	100	740	27.5	100	29.2	20.7	12.2	37.9
	200	743	28.8	100	30.6	19.2	16.8	33.4
	400	749	37.9	97.8	31.9	31.1	14.7	22.3
MgO (ribbon residue)	50	733	17.9	100	20.0	10.5	15.1	54.4
	100	733	16.0	85.9	20.9	8.4	21.3	49.4

[a] CH_4–O_2 = 5.75; CH_4 = 46.6 kPa.
[b] Total inlet flow rate.
[c] Based on moles of C converted.

lattice could exhibit marked differences in their efficacy of reaction with methane. The catalytic performance of MgO prepared by two methods: thermal decomposition of magnesium hydroxycarbonate [ex (b.c.)] and burning Mg in air (ex ribbon residue) has been compared (Table 7-8). MgO (ex b.c.) exhibits a higher surface area than MgO (ex ribbon residue) and consequently is more active; however, under conditions of either complete or incomplete oxygen conversion, the MgO (ex b.c.) catalyst was found to be more selective in the formation of C_2 hydrocarbons. Comparison of the structure of the two MgO samples using transmission electron microscopy (TEM) (Fig. 7-13) demonstrated that the MgO (ex ribbon residue) was comprised mainly of regular cubes with varying size distribution, as has been noted in earlier studies (Jones et al. 1984). In contrast, the MgO (ex b.c.) was much less well defined and the material consisted of agglomerates of smaller crystallites. Neither sample of MgO was observed to contain defects within the individual crystallites. These findings were in agreement with the other detailed study of similar materials (Moodie and Warble 1971), which concluded that the MgO (ex b.c.) should reach the density of high-index mean planes and consequently would have a higher concentration of low coordinate oxide ions. The catalytic performances of the two samples of MgO therefore can be ascribed to the distinctly different morphologies, and demonstrates the importance of crystal morphology with respect to the oxidative coupling reaction. The results are in contrast to those of an earlier electron microscopy study (Lunsford et al. 1989) of MgO and Li/MgO that concluded that catalyst morphology was of limited importance in contrast to the addition of catalyst additives. In addition, it is apparent that MgO (ex b.c.) can give high selectivities to C_2 hydrocarbons (> 50%) in the absence of a promoter, a finding that

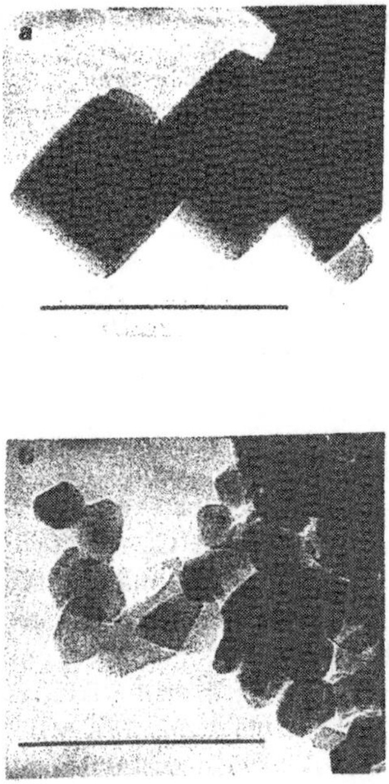

FIGURE 7-13. Transmission electron micrograph; linear scale represents 0.25 μm: (a) MgO (ex ribbon residue); (b) MgO (ex b.c.).

requires a fuller mechanistic investigation. Since these results were initially disclosed (Joyner 1989; Hargreaves, Hutchings, and Joyner 1990) others (Carreiro and Baerns 1989) have noted similar findings for MgO prepared by different routes.

By far the greater majority of studies have concentrated on the elucidation of the role of alkali promoters. Addition of Li^+ to MgO has been shown (Ito et al. 1985; Hutchings, Scurrell, and Woodhouse 1987a) to decrease the surface area, to increase specific activity, both for methane oxidation and the selective production of C_2 hydrocarbons (Table 7-6), and to increase significantly C_2 selectivity, particularly for C_2H_6. Most of these studies have been based on addition of Li^+ to commercial MgO samples for which the initial morphology is not defined. In a recent study (Hutchings, Scurrell, and Woodhouse 1991) we have reevaluated the effect of Li^+ addition to MgO (ex b.c.), which has previously been shown (Joyner 1989; Hargreaves, 1990) to be the preferred structural form. Addition of very low amounts of Li^+, particularly sub-monolayer concen-

Table 7-9 **Specific Activity Data for Li-Doped MgO Prepared by Thermal Decomposition on Basic Carbonate; 700°C, 1200 h^{-1}, CH_4–O_3 = 3 (Hutchings, Scurrell, and Woodhouse 1990)**

Catalyst	Surface Srea[a] ($m^2\ g^{-1}$)	10^4 SPECIFIC ACTIVITY (IN MOLES OF PRODUCT PER SQUARE METER PER HOUR) H_2	CO_2	CO	C_2H_4	C_2H_6
MgO	19	3.4	8.1	0.8	0.3	0.4
0.2% Li_2CO_3–MgO	6.5	6.2	10.3	0.5	2.4	5.7
5% Li_2CO_3–MgO	1.3	8.2	19.2	1.5	3.6	8.3

[a]Determined after reaction.

trations of 0.2% (Cunningham 1989), significantly affect the overall catalyst performance (Table 7-9) and C_2 selectivity (Fig. 7-14). It is apparent that addition of Li significantly enhances C_2 selectivity and although the 2% Li_2CO_3–MgO catalyst gave a slightly enhanced selectivity when compared with the other lithium-doped materials, the highest yield of C_2 hydrocarbons was observed with the lowest doping level due to surface area effects. However, specific activity increases with increasing [Li^+], which is of significance in considering the mechanism of activation of methane.

Lunsford and co-workers (Ito and Lunsford 1985; Ito et al. 1985; Driscoll et al. 1985; Driscoll and Lunsford 1983; Driscoll et al. 1986; Driscoll and Lunsford 1985; Lunsford et al. 1988) have extensively studied the Li/MgO catalyst using electron paramagnetic resonance (EPR) spec-

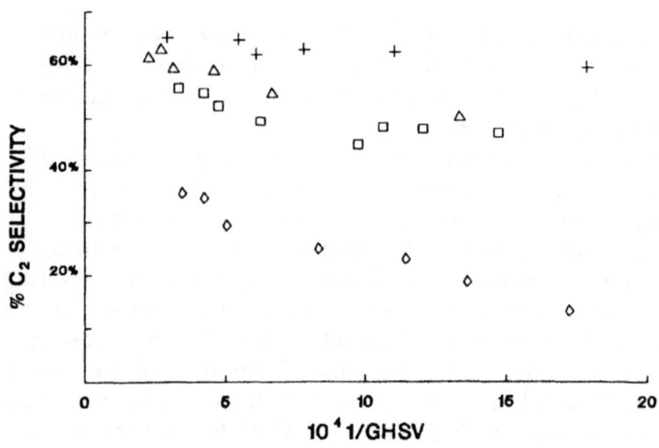

FIGURE 7-14. Selectivity of total C_2 hydrocarbon versus reciprocal space velocity at 700°C: ◇, MgO; △, 0.2% Li_2CO_3–MgO; +, 2.0% Li_2CO_3–MgO; □, 5% Li_2CO_3–MgO.

troscopy and have shown that in the presence of O_2 this material stabilizes O^- as a surface species by the formation of $[Li^+O^-]$ centers. Previously, studies by Abraham and co-workers (Abraham et al. 1986; Chen et al. 1977; Lacy et al. 1978; Boldu, Abraham, and Chen 1979; Olson et al. 1980) demonstrated that $[Li^+O^-]$ centers can be formed in both Li-doped MgO or CaO, either by thermal treatment or by high-intensity electron bombardment. It has been shown (Driscoll et al. 1986) that a direct correlation exists between the concentration of $[Li^+O^-]$ centers and the concentration of CH_3 radicals generated from the reaction of CH_4–O_2 over Li-doped MgO samples. Therefore, it was concluded that the CH_3 radical was produced from the interaction of CH_4 with the O^- of the $[Li^+O^-]$ center.

For MgO, doping with Na^+ is not found to be as effective as Li^+, whereas, for CaO, doping with Na^+ is most effective (Table 7-6). This has been explained in terms of the ionic radii of Li^+, Na^+, Mg^{2+}, and Ca^{2+}, which are 68, 99, 66, and 97 pm, respectively. Hence, although Li^+ can readily substitute for Mg^{2+} in MgO, Na^+ is too large to be readily accommodated, whereas in CaO the Na^+ readily substitutes for Ca^{2+}.

Confirmation of the role of O^- in the selective oxidation of methane have come from photoluminescence studies (Anpo et al. 1988), together with the use of N_2O as oxidant (Table 7-7), a known source of surface O^- (Naccache 1971). In addition, model studies using X-ray photoelectron spectroscopy (Au and Roberts 1987) on oxidant Mg for the analogous activation of N—H bonds in NH_3 have also demonstrated the importance of the O^- species. Hence, there appears to be a general consensus that O^- on the surface of MgO or CaO, stabilized by the addition of Li^+ or Na^+, is the species responsible for the initial activation of methane via hydrogen atom abstraction to form a methyl radical. The radical nature of the methane coupling reaction over Li/MgO is confirmed by the NO scavenging experiments previously described. However, it must be noted that in the absence of alkali dopant, MgO and CaO are also active for methane coupling and for MgO, in contrast to Li/MgO, addition of NO had no significant effect (Fig. 7-11). Isotopic exchange experiments have shown (Martin and Mirodatos 1987) that with undoped catalysts there exists only a short delay between the CO–CO_2 isotopic response and that of CH_4, whereas with alkali-doped catalysts a long delay was noted. All these observations point to the possibility that a different mechanism of methane activation may be occurring on the undoped oxides. This aspect clearly requires further detailed study, but it has been noted previously that catalyst basicity (Carreiro et al. 1988) is an important factor and, hence, activation may be via the formation of CH_3 via a heterolytic (i.e., two-electron) process on the basic sites of the surface.

The results previously discussed for the effect of NO addition during CH_4–N_2O oxidation over Li/MgO are considered to be of mechanistic importance. The increase in C_2 to levels higher than those observed immediately prior to NO addition (Fig. 7-9), and the slow attainment of C_2 and CO_2 selectivities similar to those observed, prior to NO addition, is indicative that a second surface oxygen species, responsible for nonselective oxidation, is established more slowly than the selective oxidizing species. These studies clearly demonstrate the presence of two oxidizing species or sites on the surface of Li/MgO, only one of which is involved in the selective oxidation process. This is consistent with the previous observation that identified two roles for N_2O as oxidant (Hutchings, Scurrell, and Woodhouse 1987b).

Lanthanide Metal Oxides

The rate of methyl radical formation on La_2O_3 is reported (Lin et al. 1986) to be higher than on Li/MgO under comparable conditions. Superoxide, O_2^-, was observed to be the most abundant species, but it has been noted that the broad EPR spectrum of O_2^- may have masked the presence of O^-. It is interesting to note that addition of Li^+ to La_2O_3 does not appear to enhance the specific activity (Table 7-6), which is in direct contrast to the effect of Li^+ addition to MgO or Sm_2O_3. These observations could indicate that O^- is not the active species on La_2O_3. However, the studies do indicate that methane activation is via radical formation and is therefore analogous to Li/MgO.

Sm_2O_3 has also been extensively studied; in particular, the morphology of this oxide has been found to be significant, because it exists in either monoclinic or cubic forms. It was initially concluded (Korf et al. 1989) that the cubic form was most active, but more recently it has been observed (Sokolovskii et al. 1990) that catalysts containing mixtures of the two forms are most active and this effect is ascribed to the high mobility of oxygen vacancies in the mixed system. From a kinetic (Otsuka and Nakajima 1987; Otsuka and Jinno 1986) and a model compound (Otsuka et al. 1987b) investigation, Otsuka proposed that O_2^{2-} could be the active species in Sm_2O_3. It has been noted (Lee and Oyama 1988) that O_2^{2-} can be regarded as a dimer of O^-, (i.e., O^-—O^-), and the formation of O^- from O_2^{2-} has been noted previously in the definitive reviews of Che and Tench (Che and Tench 1982, 1983). Further evidence that O^- is the active species on Sm_2O_3 is provided by both the effect of Li^+ doping (Table 7-6), which gives a significant enhancement in specific activity and the use of N_2O as oxidant (Table 7-7), which significantly enhances C_2H_6 formation. These observations are similar to those observed for Li/MgO and, hence, by analogy it can be concluded that O^- is the selective oxidizing species for Sm_2O_3 and Li/Sm_2O_3.

Primary Hydrocarbon Product Formation

Identification of Primary Hydrocarbon Products

Knowledge of the nature of the primary reaction products is essential if a detailed understanding of the reaction mechanism is to be achieved. The conventional method of determining primary products is to measure the product selectivity as a function of contact time [reciprocal gas hourly space velocity (GHSV)] at constant temperature and CH_4–oxidant ratio. Initial studies (Hinsen and Baerns 1983; Otsuka, Jinno, and Morikawa 1986) concluded that ethane was the only primary hydrocarbon product; however, a more detailed investigation by the present authors (Hutchings, Scurrell, and Woodhouse 1989a, 1990b) has provided evidence that a small, but significant, primary selectivity to ethene may be observed with some oxide catalysts. Typical plots of selectivity versus 1/GHSV are shown in Fig. 7-15 for MgO and Li/MgO for the reaction of CH_4–O_2. Similar plots have been obtained for the reaction of CH_4–N_2O (Hutchings, Scurrell, and Woodhouse (1989a) and also for Sm_2O_3, Li/Sm_2O_3, La_2O_3, and Li/La_2O_3 (Hutchings, Scurrell, and Woodhouse 1989b). From these plots it is possible to conclude that ethane was a primary product and also that addition of Li^+ to an oxide significantly affects the ethene–ethane product ratio. However, the situation is less clear concerning ethene as a primary product, because, although linear extrapolation of ethene selectivity for Fig. 7-15b would indicate a positive intercept on the ordinate, it is concluded (Hutchings 1989; Hutchings, Scurrell, and Woodhouse 1991) that the shortest residence time utilized in these early experiments was too long to draw such conclusions. This point has been recently reemphasized by others (Burch et al. 1989a), who repeated our earlier studies, as shown in Fig. 7-15a, using a commercial sample of MgO. In general, the main problem associated with such plots is that the methane conversion is too high to make definitive statements concerning primary products (e.g., the lowest conversion in Fig. 7-15a is ca. 5% and the lowest conversion in the study of Burch et al. 1989a was 3.6%).

More recently the group of Ross (Korf et al. 1990) have commented on the data used in the construction of these early plots; the main criticism made (Korf et al. 1990) is that such data may not be obtained under plug-flow conditions and consequently secondary reactions could be making a significant contribution. The data shown in Fig. 7-15 were obtained under plug-flow conditions (catalyst particles ca. 0.5 mm, reactor bed length = 30 mm, reactor bed diameter = 6.5 mm) and therefore this criticism is invalid. Moreover, further detailed studies (Hutchings, Scurrell, and Woodhouse 1991) have been carried out for the Li/MgO catalyst for the reaction of CH_4–O_2. These studies involved a small catalyst bed volume with a minimized postreactor heated zone to limit the oxidative

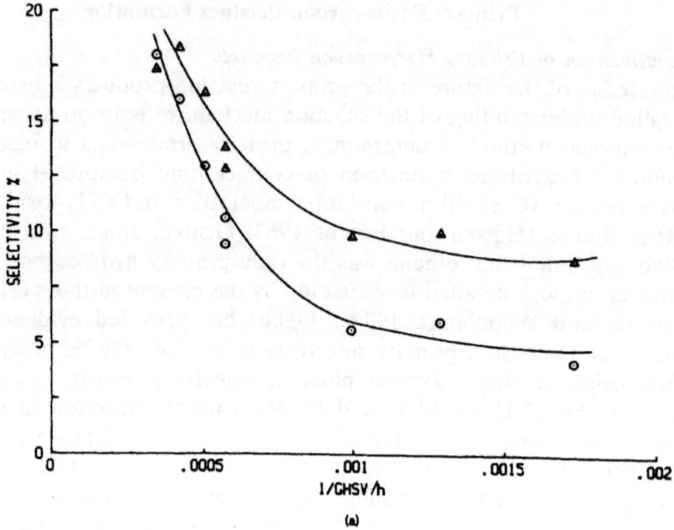

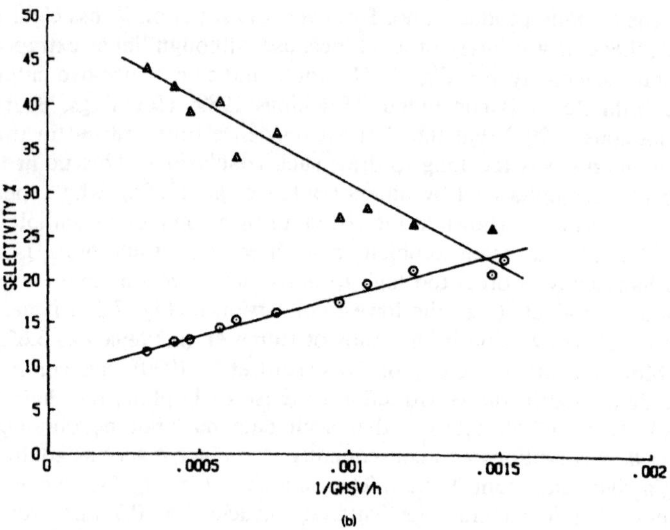

FIGURE 7-15. Selectivity versus reciprocal gas hourly space velocity: (a) MgO/O_2; (b) $Li/MgO/O_2$.

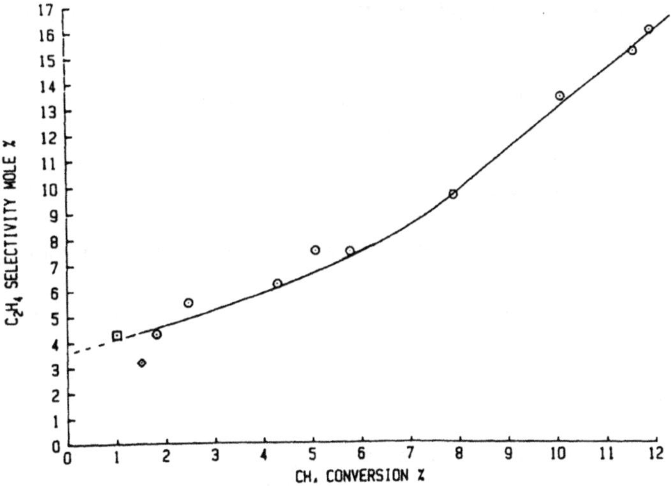

FIGURE 7-16. Ethene selectivity as a function of CH_4 conversion; Li/MgO, 710°C: mole ratios CH_4–O_3 = 3 (○), 5 (◇), and 10 (□).

dehydrogenation of ethane downstream of the catalyst (van Kasteren, Geerts, and van der Wiele 1988); in addition, it is essential that a sensitive flame ionization detector is utilized. It was observed that for CH_4–O_2 = 3, a significant ethene selectivity of ca. 4% could be obtained by extrapolation to zero methane conversion (Fig. 7-16). Furthermore, these studies indicate that ethene selectivity is not a linear relationship with methane conversion in the conversion range 0–10%, and extrapolation from data at conversions greater than 5% could give misleading results. In addition, at low methane conversion (1–1.5%) the selectivity to ethene is found to be largely invariant with changes in oxygen concentration (Hutchings, Scurrell, and Woodhouse 1989a, 1991) although the ethane selectivity was markedly affected. It was observed that this finding strongly supported the conclusion that ethene was a minor primary product, because if ethene were to be formed solely from oxidation dehydrogenation of ethane or from steam cracking of ethane, as postulated in earlier studies (Hinsen and Baerns 1983; Lee and Oyama 1988; Otsuka, Jinno, and Morikawa 1986), then variation in $[O_2]$ or $[C_2H_6]$ would have been expected to affect the selectivity to ethene, and this was not observed.

Mechanism of Primary Hydrocarbon Product Formation

Based on the preceding discussion it is apparent that ethane is the major primary hydrocarbon product and on some oxide catalysts a minor primary product of ethene can also be observed. As previously noted there is general consensus (Hutchings, Scurrell, and Woodhouse 1987a; Driscoll et al. 1985; Mehandru, Anderson, and Bradzil 1988; Lunsford et al. 1988; Anpo et al. 1988; Au and Roberts 1987) that surface O^-, derived from O_2 and stabilized by Li^+-doping as $[Li^+O^-]$ centers or from N_2O decomposition, is the selective oxidant for the formation of ethane. Interaction of O^- with methane, either in the gas phase or surface associated [because isotopic labeling studies (Ekstrom and Lapszewicz 1988a; 1988b) have indicated that a significant amount of methane is associated with the surface even at reaction temperatures], leads to the formation of $CH_3\cdot$ radicals, which on dimerization (in the gas phase) give ethane. $CH_3\cdot$ radicals are formed via hydrogen atom abstraction, and recent results (Cant et al. 1988) show that this step is rate-determining, because a kinetic isotope effect is seen. The radical nature of this reaction has been further confirmed by the NO scavenging experiments (Hutchings, Scurrell, and Woodhouse 1990).

The presence of Li^+ dopant in MgO, which aids the stabilization of O^-, has significantly different effects on ethane and ethene selectivities. Based on this observation, it is concluded that the O^- cannot be the sole oxidizing species responsible for the formation of the ethene primary product (Hutchings, Scurrell, and Woodhouse 1989a). Additional evidence concerning the importance of different oxidants is obtained from the observation that NO completely inhibits ethene formation on Li/MgO for the reaction of CH_4–N_2O (Fig. 710a), but this is not observed with CH_4–O_2 under similar conditions (Fig. 7-10b). Furthermore, due to the diversity of ethene primary selectivities observed for oxide catalysts, it has been proposed (Hutchings, Scurrell, and Woodhouse 1989a) that primary ethene formation is probably a surface-catalyzed process. Carbene may be involved as a reactive intermediate in methane oxidation (Martin and Mirodatos 1987, 1988). Their initial proposal that ethane was formed via carbene insertion into a C—H bond of methane has been discounted by reaction of CD_4–CH_4–O_2 reaction mixtures (Nelson, Lukey, and Cant 1988), because no CD_2HCD_3 or similar isotopically labeled ethane was observed. However, these isotopic experiments cannot discount the involvement of carbene in the formation of ethene. Such a species is an attractive possibility for primary ethene formation and could react via a number of pathways, for example,

$$2CH_2 \rightarrow C_2H_4 \quad \text{or} \quad CH_2 + CH_3\cdot \rightarrow C_2H_5\cdot \rightarrow C_2H_4 + H\cdot$$

It has been concluded that the carbene intermediate could not be formed from a second interaction of a $CH_3\cdot$ radical with O^-, that is, a second hydrogen atom abstraction, because this pathway would not be consistent with the changes observed when $[O^-]$ is influenced by addition of Li^+ or use of N_2O. However, a $CH_3\cdot$ radical could interact with an alternative surface or gas-phase oxidant and this possibility has not yet been investigated. However, it is well known (Che and Tench 1982, 1983) that MgO can stabilize a number of surface oxidants, including O_2^- and O_2^{2-}. Because O_2 provides the selective oxidizing species for ethene formation relative to N_2O, this has led to the conclusion that a dioxygen surface oxidant is probably an important intermediate and it has been proposed (Hutchings, Scurrell, and Woodhouse 1989a) that the peroxide species, O_2^{2-}, is the most probable. It is interesting to note that this species has been proposed by Otsuka for Sm_2O_3, a catalyst that gives very high ethene selectivities (Tables 7-2 and 7-6).

Secondary Reactions of Ethane and Ethene and Nonselective Oxidation Reactions

Secondary Conversion of Ethane to Ethene

Ethene can be formed from the primary ethane product via two reaction pathways:

$$C_2H_6 \rightleftarrows C_2H_4 + H_2 \tag{7-1}$$

$$C_2H_6 + \tfrac{1}{2}O_2 \rightarrow C_2H_4 + H_2O \tag{7-2}$$

Reaction (7-1) is essentially nonoxidative ethane cracking reaction that is mediated by steam, a reaction by-product, and is expected (Stull, Westrum, and Sinke 1969) to occur as a gas-phase reaction at the temperatures used in the coupling reaction in the absence of oxygen. In the greater majority of catalyst evaluation studies, oxygen conversions of 100% have been utilized and it is anticipated that, under these conditions, steam cracking of ethane can become significant. This has been noted as a possible reaction in the heated zone downstream of a reaction (van Kasteren, Geerts, and van der Wiele 1988), and, in addition, it has been observed (Hutchings 1989) that the ethene–ethane ratio for the product with oxyhalide catalysts is proportional to the amount of water formed as a reaction by-product. In the presence of an oxidant, reaction (7-1) has been observed to occur only to a limited extent (Hutchings, Scurrell, and Woodhouse 1989a) and, hence, reaction (7-2) is the major pathway for secondary ethene formation under conditions of nontotal oxidant conversion. Comparison of C_2H_6–O_2 and C_2H_6–N_2O have indicated that N_2O is always more selective than O_2 for ethene formation; hence, it has been proposed that a monoatomic oxygen species is involved in reaction (7-2). Studies (Hutchings 1989; Roos et al. 1989; Geerts, van Kasteren, and van der Wiele 1989) have provided evidence that reaction (7-2) is a gas-phase reaction and the presence of a catalyst only leads to enhance the parallel total oxidation pathway. These studies also indicate that reaction (7-2) is facile under coupling conditions and must be considered to be the major route for ethene production at significant methane conversions (i.e., > 2%).

Primary and Secondary Total Oxidation Processes

Studies using N_2O as oxidant (Hutchings, Scurrell, and Woodhouse 1987a) at high and low reaction temperatures have indicated that the O^- species has two distinct roles: (a) it is the selective oxidizing species for the formation of methyl radicals, the precursor to ethane, and (b) it participates in nonselective methane oxidation via a parallel pathway to selective methane oxidation. In addition, the studies involving NO as a radical scanvenger have indicated that a second oxidizing species other than O^- may also be significant for nonselective methane oxidation. The nature of this second species or site has not yet been elucidated.

A number of studies have been conducted concerning the secondary oxidation of ethane and ethene. Based on a detailed comparison of N_2O and O_2 as oxidants with Li/MgO as catalyst for ethane (Hutchings, Scurrell, and Woodhouse 1989a), it has been concluded that the nonselective oxidation of ethane involves mainly a surface-catalyzed reaction involving a dioxygen species. The same study also indicated that ethene total oxidation involved both surface-catalyzed and gas-phase reactions and that a dioxygen species was also involved.

Importance of Gas-Phase Reactions

Based on the preceding discussion, it is apparent that a number of the reactions involved in the oxidation of methane can occur as homogeneous gas-phase reactions in the absence of a catalyst. It appears that the prime function of the catalyst surface is to activate methane via interaction with a surface O^- species, which generates methyl radicals. These radicals are released into the gas phase and subsequently dimerize to form ethane. At the high reaction temperatures required for the coupling reaction (> 600°C) the reaction between dioxygen and a methyl radical to form a methyl peroxy radical is not favored, and this is the prime reason why oxygenate formation is not observed to any significant extent at high temperatures. Early studies (Boomer and Thomas 1937) demonstrated that high pressures can be used to activate methane via a radical pathway in the absence of a catalyst acting as a radical initiator. Reaction pressure is therefore a parameter that required evaluation, and yet this was not actively studied until comparatively recently. An initial study by the authors (Hutchings, Scurrell, and Woodhouse 1988d) demonstrated that increasing the reaction pressure from 85 to 585 kPa negated the requirement for a catalyst, because at this pressure significant gas-phase radicals can be thermally generated. Further, it was observed that the presence of a catalyst could be deleterious because it tended to catalyze nonselective reaction pathways. A number of subsequent definitive studies (Lane and Wolf 1988b; Yates and Zlotin 1988; Onsager et al. 1989) have confirmed these initial findings and have shown that gas-phase reactions must be recognized as playing a major role in the oxidative coupling reaction. A number of recent catalyst formulations that have been claimed to be highly active are those involving the addition of halogens, for example, LiCl/NiO (Otsuka et al. 1986). Considerable debate has ensued (Burch et al. 1989b; Burch, Squire, and Tsang 1990) as to the involvement of gas-phase halogen radicals in the activation of methane with these materials and it is clear that this cannot be ruled out with these systems.

Formation of Hydrogen

Under most reaction conditions hydrogen is a major reaction product (Hutchings, Scurrell, and Woodhouse 1988d; Roos et al. 1987), although early studies failed to observe this due to analytical difficulties. Hydrogen formation is particularly marked with MgO as catalyst (Table 7-6) and on doping with Li^+ the selectivity of H_2 relative to C_2 hydrocarbons decreases. Hydrogen can be formed via a number of pathways, including steam cracking of ethane, the water–gas shift reaction, the decomposition of formaldehyde, and steam reforming reactions (Roos et al. 1987). In addition, hydrogen is also known to be a significant product from homogeneous gas-phase methane oxidation (Gesser, Hunter, and Prakash 1985), and it has been suggested (Kimble and Kolts 1988) that hydrogen could be formed by decomposition of ethyl radicals, although the recombination of hydrogen atoms is not energetically favored. A detailed study of the origin of hydrogen in the methane coupling reaction over MgO has been completed (Hargreaves, Hutchings, and Joyner 1991). It was found that the water gas shift reaction is an important source, but thermal cracking of ethane to ethene and partial oxidation of hydrocarbons to CO and H_2 are also involved. It was observed that partial oxidation becomes more significant at high flow rates and low O_2 conversions.

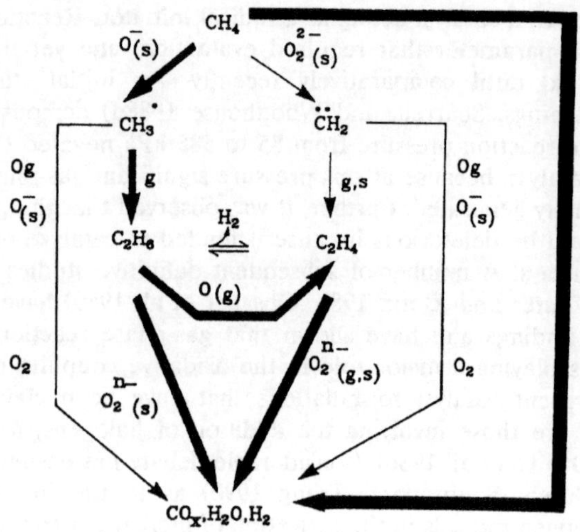

FIGURE 7-17. Proposed reaction mechanism for the oxidative coupling of methane over nonreducible oxide catalysts; heavy arrow denotes main pathway; light arrow denotes minor pathway.

Overall Reaction Mechanism

Against the background of the reviewed literature and the basis of a detailed study (Hutchings, Scurrell, and Woodhouse 1989a), we have proposed a reaction pathway (Fig. 7-17) for the oxidation of methane over Li/MgO. It is apparent that for the range of nonreductible oxide catalysts investigated (Tables 7-1–7-5) the product distributions are broadly similar; hence, it is considered that a similar mechanism may occur for all such simple oxides and alkali-doped oxides. In this mechanistic scheme the formation of ethene from methane occurs only as a minor pathway. However, it should be noted that Li/Sm_2O_3 exhibits a specific activity for ethene formation that is over an order of magnitude higher than Li/MgO (Table 7-6). It is anticipated that it should be possible to enhance the formation of ethene directly from methane, but further studies are required to confirm if this can be achieved.

CONCLUDING REMARKS

The essential aim of the research we have carried out on the oxidative coupling of methane has been aimed at the elucidation of the mechanism by comparison of the different oxidizing efficacies of O_2, N_2O, and

O_3/O_2. This information was then to be utilized in the design of improved catalysts. However, it is apparent that these and other studies have identified that this reaction experiences two major problems: (a) high temperatures are required for the activation of methane via a radical pathway, and (b) at these elevated temperatures increasing reaction pressure may be a more effective means of achieving selective methane activation in comparison with most of the catalysts identified to date. The identification of catalyst formulations capable of activating methane at low reaction temperatures represents a major challenge in catalyst design. It should be noted that alkali-doped catalysts, particularly Li/Sm_2O_3 have not been optimized. This therefore represents a challenge to both the solid-state and surface scientist to identify catalyst structures with an enhanced concentration of active sites for low-temperature methane activation. Another aspect that has been emphasized in our studies has been the indication that specific oxidizing species (e.g., O^-) can be involved in both selective and nonselective oxidation pathways. In addition, these studies have identified the existence of an oxidizing species or catalyst surface site that is involved specifically with total oxidation, but as yet this has not been identified. It can be anticipated that identification and control of this species or site could be of great significance in the design of improved catalysts. The observation that the greatest yields of selective oxidation products apparently have been obtained with complex oxides may well reflect that it is these solids that can provide the delicate balance between competing reactions involved in methane oxidation. Yet it is likely that the complex nature of these catalysts will provide a serious obstacle to those engaged in attempting to correlate surface structure and catalytic action and that for such work more progress is expected if studies are confined to "simpler" oxide systems.

References

Abraham, M.M., Y. Chen, L.A. Boatner, and R.W. Reynolds. 1986. *Phys. Res. Lett*. 37:849.

Agarwal, S.K., R.A. Migone, and G. Marcelin. 1989. *Appl. Catal.* 53:71.

Ahmed, S., and J.B. Moffat. 1989. *Catal. Lett*. 2:309.

Aika, K.I., and T. Nishiyama. 1988a. In *Proceedings of the 9th International Congress on Catalysis, Calgary*, Vol. 2, ed. M.J. Phillips, and M. Ternan, p. 907. Ottawa: Chemical Institute of Canada.

Aika, K.I., and T. Nishiyama. 1988b. *J. Chem. Soc., Chem. Commun*. 70.

Aika, K.I., and T. Nishiyama. 1989. *Catal. Today* 4:271.

Anpo, M., M. Sunamoto, T. Doi, and I. Matsuura. 1988. *Chem. Lett*. 701.

Ashcroft, A.T., A.K. Cheetham, M.L.H. Green, C.P. Grey, and P.D.F. Vernon. 1989. *J. Chem. Soc., Chem. Commun*. 1667.

Au, C.T., and M.W. Roberts. 1987. *J. Chem. Soc., Faraday Trans. 1* 83:2047.
Bhasin, M.M. 1988. *Stud. Surf. Sci. Catal.* 36:343.
Biloen, P., and W.M.H. Sachtler. 1981. *Adv. Catal.* 31:165.
Boldu, J.L., M.M. Abraham, and Y. Chen. 1979. *Phys. Rev. B* 19:4421.
Boomer, E.H., and V. Thomas. 1937. *Can. J. Res. Sect. B* 15:401.
Burch, R., E.M. Crabb, G.D. Squire, and S.C. Tsang. 1989. *Catal. Lett.* 2:249.
Burch, R., A.A. Dost, D.A. Rice, and G. Webb. 1989a. *Catal. Lett.* 3:331.
Burch, R., G.D. Squire, and S.C. Tsang. 1988. *Appl. Catal.* 43:105.
Burch, R., G.D. Squire, and S.C. Tsang. 1990. *Catal. Today.* 6:503.
Bytyn, W., and M. Baerns. 1986. *Appl. Catal.* 28:199.
Cant, N.W., C.A. Lukey, P.F. Nelson, and R.J. Tyler. 1988. *J. Chem. Soc., Chem. Commun.* 766.
Carreiro, J.A.S.P., and M. Baerns. 1987. *React. Kinet. Catal. Lett.* 35:349.
Carreiro, J.A.S.P., and M. Baerns. 1989. *J. Catal.* 117:250.
Carreiro, J.A.S.P., G. Follmer, L. Lehman, and M. Baerns. 1988. In *Proceedings of the 9th International Congress on Catalysis, Calgary,* Vol. 2, ed. M.J. Phillips and M. Ternan, p. 905. Ottawa: Chemical Institute of Canada.
Chang, C.D. 1988. *Stud. Surf. Sci. Catal.* 36:127.
Che, M. and A.J. Tench. 1982. *Adv. Catal.* 32:77.
Che, M., and A.J. Tench. 1983. *Adv. Catal.* 32:1.
Chen, Y., H.T. Tohrer, J. Narayan, and M.M. Abraham. 1977. *Phys. Rev. B* 16:5535.
Chou, T.C., and L.F. Albright. 1978. *Ind. Eng. Chem. Process Des. Dev.* 17:454.
Choudhary, V.R., S.T. Chaudhari, A.M. Rajput, and V.H. Rane. 1989. *J. Chem. Soc., Chem. Commun.* 1526.
Cunningham, J. 1989. *Faraday Discuss., Chem. Soc.* 87.
De Boy, J.M., and R.F. Hicks. 1988a. *J. Chem. Soc., Chem. Commun.* 982.
De Boy, J.M., and R.F. Hicks. 1988b. *J. Catal.* 113:517.
Doi, T., Y. Utsumi, and I. Matsuura. 1988. In *Proceedings of the 9th International Congress on Catalysis, Calgary*, Vol. 2, ed. M.J. Phillips and M. Ternan, p. 937. Ottawa: Chemical Institute of Canada.
Driscoll, D.J., and J.H. Lunsford. 1983. *J. Phys. Chem.* 87:301.
Driscoll. D.J., and J.H. Lunsford. 1985. *J. Phys. Chem.* 89:4415.
Driscoll, D.J., W. Martir, J.X. Wang, and J.H. Lunsford. 1985. *J. Amer. Chem. Soc.* 107:58.
Driscoll, D.J., W. Martir, J.X. Wang, and J.H. Lunsford. 1986. In *Adsorption and Catalysis on Oxide Surfaces*, ed. M. Che and G.C. Bond, p. 403. Amsterdam: Elsevier.
Edwards, J.H., R.J. Tyler, and S.D. White. 1990. *Energy and Fuels* 4:85.
Ekstrom, A., and J.A. Lapszewicz. 1988a. *J. Chem. Soc., Chem. Commun.* 797.
Ekstrom, A., and J.A. Lapszewicz. 1988b. *Prepr.—Am. Chem. Soc. Div. Pet. Chem.* 33:430.
Emesh, I.T.A., and Y. Amenomiya. 1986. *J. Phys. Chem.* 90:4785.
Fen-Chang, W., Z.L. Chang, C.T. Au, and K.R. Tsai. 1989. Paper 26, presented at the Symposium on Methane Activation, Conversion and Utilization, Honolulu.

Foster, N.R. 1985. *Appl. Catal.* 19:1.
Geerts, J.W.M.H., J.N.M. van Kasteren, and K. van der Wiele. 1989. *Catal. Today* 4:453.
Gesser, H.D., N.R. Hunter, and C.B. Prakash. 1985. *Chem. Rev.* 85:235.
Hargreaves, J.S.J., G.S. Hutchings, and R.W. Joyner. 1990. *Catal. Today* 6:481.
Hargreaves, J.S.J., G.J. Hutchings, and R.W. Joyner. 1991. *Stud. Surf. Sci. Catal.* 61:155.
Hinsen, W., and M. Baerns. 1983. *Chem.-Ztg.* 107:223.
Hinsen, W., W. Bytyn, and M. Baerns. 1984. In *Proceedings of the 8th International Congress on Catalysis, Berlin*, Vol. 3, p. 581. Frankfurt au Main. Verlag-Chemie.
Hutchings, G.J. 1986. *S. Afr. J. Chem.* 39:65.
Hutchings, G.J. 1987. *Chem. Brit.* 23:762.
Hutchings, G.J. 1989. *Faraday Disc. Chem. Soc.* 87:62.
Hutchings, G.J., and R. Hunter. 1990. *Catal. Today* 6:279.
Hutchings, G.J., M.S. Scurrell, and J.R. Woodhouse. 1987a. *J. Chem. Soc., Chem. Commun.* 1388.
Hutchings, G.J., M.S. Scurrell, and J.R. Woodhouse. 1987b. *J. Chem. Soc., Chem. Commun.* 1862.
Hutchings, G.J., M.S. Scurrell, and J.R. Woodhouse. 1988a. In *Proceedings of the 9th International Congress on Catalysis, Calgary*, Vol. 2, ed. M.J. Phillips and M. Ternan, p. 293. Ottawa: Chemical Institute of Canada.
Hutchings, G.J., M.S. Scurrell, and J.R. Woodhouse. 1988b. *Stud. Surf. Sci. Catal.* 36:415.
Hutchings, G.J., M.S. Scurrell, and J.R. Woodhouse. 1988c. *Appl. Catal.* 38:157.
Hutchings, G.J., M.S. Scurrell, and J.R. Woodhouse. 1988d. *J. Chem. Soc., Chem. Commun.* 253.
Hutchings, G.J., M.S. Scurrell, and J.R. Woodhouse. 1989a. *J. Chem. Soc., Faraday Trans.* 1 85:2507.
Hutchings, G.J., M.S. Scurrell, and J.R. Woodhouse. 1989b. *Catal. Today* 4:371.
Hutchings, G.J., M.S. Scurrell, and J.R. Woodhouse. 1989c. *J. Chem. Soc., Chem. Commun.* 765.
Hutchings, G.J., M.S. Scurrell, and J.R. Woodhouse, 1990. *Catal. Today* 6:399.
Hutchings, G.J., M.S. Scurrell, and J.R. Woodhouse, 1991. *Catal. Lett.* Catal. Lett. 8:115.
Imai, H., and T. Tagawa. 1986a. *J. Chem. Soc., Chem. Commun.* 52.
Imai, H., and T. Tagawa. 1986b. *J. Chem. Soc., Chem. Commun.* 525.
Imai, H., T. Tagawa, and N. Kamide. 1987. *J. Catal.* 106:394.
Imai, H., T. Tagawa, N. Kamide, and S. Wada. 1988. In *Proceedings of the 9th International Congress on Catalysis, Calgary*, Vol. 2, ed. M.J. Phillips and M. Ternan, p. 952. Ottawa: Chemical Institute of Canada.
Inui, T., Y. Ishihara, and K. Kamachi. 1989. Paper 27, presented at the Symposium on Methane Activation, Conversion and Utilization, Honolulu.
Ito, T., and J.H. Lunsford. 1985. *Nature (London)* 314:721.
Ito, T., J.-X. Wang, C.-H. Lin, and J.H. Lunsford. 1985. *J. Am. Chem. Soc.* 107:5062.

Iwamatsu, E., T. Mosiyama, N. Takasaki, and K.I. Aika. 1988. *Stud. Surf. Sci. Catal.* 36:373.

Jones, C.A., J.J. Leonard, and J.A. Sofranko. 1984a. U.S. Patent 4444 984.

Jones, C.A., J.J. Leonard, and J.A. Sofranko. 1984b. U.S. Patent 4443 664.

Jones, C.F., R.L. Segall, R.S.C. Smart, and P.S. Turner. 1984. *J. Mat. Sci. Lett.* 3:810.

Jones, C.A., J.J. Leonard, and J.A. Sofranko. 1987. *J. Catal.* 103:311.

Joyner, R.W. 1989. *Faraday Disc., Chem. Soc.* 87.

Kasztelan, S., and J.B. Moffat. 1987. *J. Chem. Soc., Chem. Commun.* 1663.

Keller, G.E., and M.M. Bhasin. 1982. *J. Catal.* 73:9.

Keulks, G.W., and M. Yu. 1987. *React. Kinet. Catal. Lett.* 35:361.

Kimble, J.B., and J.H. Kolts. 1988. *CHEMTECH* 501.

Komatsu, T., T. Amaya, and K. Otsuka. 1989. *Catal. Lett.* 3:317.

Korf, S.J., J.A. Roos, N.A. de Bruijn, J.G. van Ommen, and J.R.H. Ross. 1987. *J. Chem. Soc., Chem. Commun.* 1433.

Korf, S.J., J.A. Roos, J.M. Diphoorn, R.J.H. Veehof, J.E. van Ommen, and J.R.H. Ross. 1989. *Catal. Today* 4:279.

Korf, S.J., J.A. Roos, J.W.H.C. Derksen, J.A. Vreeman, J.G. van Ommen, and J.R.H. Ross. 1990. *Appl. Catal.* 59:291.

Lacy, J.B., M.M. Abraham, J.L. Boldu, Y. Chen, J. Narayan, and H.T. Tohver. 1978. *Phys. Rev. B* 18:4236.

Lane, G.S., Z. Kalenik, and E.E. Wolf. 1989. *Appl. Catal.* 53:183.

Lane, G.S., and E.E. Wolf. 1988a. In *Proceedings of the 9th International Congress on Catalysis, Calgary*, Vol. 2, ed. M.J. Phillips and M. Ternan, p. 944. Ottawa: Chemical Institute of Canada.

Lane, G.S., and E.E. Wolf. 1988b. *J. Catal.* 113:144.

Lee, J.S., and S.T. Oyama. 1988. *Catal. Rev.—Sci. Eng.* 30:249.

Lin, C.-H., K.D. Campbell, J.-X. Wang, and J.H. Lunsford. 1986. *J. Phys. Chem.* 90:534.

Lin, C.-H., T. Ho, J.-X. Wang, and J.H. Lunsford. 1987. *J. Am. Chem. Soc.* 109:4808.

Liu, H.F., R.S. Liu, K.Y. Liew, R.W. Johnson, and J.H. Lunsford. 1984. *J. Am. Chem. Soc.* 106:4117.

Lunsford, J.H., M.D. Cisneros, P.G. Hinsen, V. Tong, and H. Zhang. 1989. *Faraday Disc., Chem. Soc.* 87.

Lunsford, J.H. C.-H. Lin, J.-X. Wang, and K.D. Campbell. 1988. In *Microstructure and Properties of Catalysts*, ed. M.M.J. Treacy, J.M. Thomas, and J.M. White, *Materials Research Society Symposium Proceedings*, Vol. 3, p. 305. Pittsburgh: Materials Research Society.

Mac Giolla Coda, E., R. van Hoek, E. Mulhall, and B.K. Hodnett. 1989. *Catal. Today* 4:383.

Machida, K.I., and M. Enyo. 1987. *J. Chem. Soc., Chem. Commun.* 1639.

Martin, G.A., A. Bates, V. Ducarme, and C. Mirodatos. 1989. *Appl. Catal.* 47:287.

Martin, G.A., and C. Mirodatos. 1987. *J. Chem. Soc., Chem. Commun.* 1393.

Martin, G.A., and C. Mirodatos. 1988. In *Proceedings of the 9th International Congress on Catalysis, Calgary*, Vol. 2, ed. M.J. Phillips and M. Ternan, p. 905. Ottawa: Chemical Institute of Canada.

Matsuura, I., Y. Utsumi, T. Doi, and K. Ohyachi. 1989a. Paper 92, presented at the Symposium on Methane Activation, Conversion and Utilization, Honolulu.

Matsuura, I., Y. Utsumi, T. Doi, and Y. Yoshida. 1989b. *Appl. Catal.* 47:299.

Matsuura, I., Y. Utsumi, M. Nakai, and T. Doi. 1986. *Chem. Lett.* 1981.

Mehandru, S.P., A.B. Anderson, and J.F. Bradzil, 1988. *J. Am. Chem. Soc.* 110:1715.

Mims, C.A., R.B. Hall, K.D. Rose, and G.R. Myers. 1989. *Catal. Lett.* 2:361.

Minachev, Kh.M., N.Ya. Usachev, Yu.S. Khodakov, L.L. Kozlov, V.N. Udut, and O.A. Fomin. 1987. *Izv. Akad. Nauk SSSR, Ser. Khim.* 504.

Mirodatos, C., A. Holmen, R. Mariscal, and G.A. Martin. 1990. *Catal. Today* 6:601.

Moodie, A.F., and C.E. Warble. 1971. *J. Cryst. Growth* 10:26.

Naccache, C. 1971. *Chem. Phys. Lett.* 11:323.

Nelson, P.F., C.A. Lukey, and N.W. Cant. 1988. *J. Phys. Chem.* 92:6176.

Olson, J.L., V.M. Orera, Y. Chen, and M.M. Abraham. 1980. *Phys. Rev. B* 20:1258.

Onsager, O.-T., R. Lodeng, P. Soraker, A. Anundskaas, and B. Helleborg. 1989. *Catal. Today* 4:355.

Otsuka, K. 1987. *J. Jpn. Pet. Inst.* 30:385.

Otsuka, K., and K. Jinno. 1986. *Inorg. Chim. Acta* 121:237.

Otsuka, K., K. Jinno, and A. Morikawa. 1985. *Chem. Lett.* 499.

Otsuka, K., K. Jinno, and A. Morikawa. 1986. *J. Catal.* 100:353.

Otsuka, K., and T. Komatsu. 1987. *J. Chem. Soc., Chem. Commun.* 338.

Otsuka, K., T. Komatsu, K. Jinno, Y. Uragami, and A. Morikawa. 1988. *Proceedings of the 9th International Congress on Catalysis, Calgary*, Vol. 2, ed. M.J. Phillips and M. Ternan, p. 905. Ottawa: Chemical Institute of Canada.

Otsuka, K., Q. Liu, M. Hatamo, and M. Morikawa. 1986. *Inorg. Chim. Acta* 118:L24.

Otsuka, K., Q. Liu, M. Hatano, and A. Morikawa. 1987a. *Chem. Lett.* 1835.

Otsuka, K., and T. Nakajima. 1986. *Inorg. Chim. Acta* 120:L27.

Otsuka, K., and T. Nakajima. 1987. *J. Chem. Soc., Faraday Trans. 1* 83:1315.

Otsuka, K., A.A. Said, K. Jinno, and T. Komatsu. 1987b. *Chem. Lett.* 77.

Peng, X.D., D.A. Richards, and P.C. Stair. 1990. *J. Catal.* 121:99.

Pitchai, R., and K. Klier. 1986. *Catal. Rev.—Sci. Eng.* 28:13.

Roos, J.A., A.G. Bakker, H. Bosch, J.G. van Ommen, and J.R.H. Ross. 1987. *Catal. Today* 1:133.

Roos, J.A., S.J. Korf, R.J.H. Veehof, J.G. van Ommen, and J.R.H. Ross. 1989. *Catal. Today* 4:441.

Sofranko, J.A., and J.C. Jubin. 1989. Paper 165, presented at Symposium on Methane Activation, Conversion and Utilization, Honolulu.

Sofranko, J.A., J.J. Leonard, C.A. Jones, A.M. Gaffney, and H.P. Withers. 1988. *Catal. Today* 3:127.

Sokolovskii, V.D., O.V. Buyevskaya, L.M. Plyasova, G.S. Litvak, and N.Ph. Uvarov. 1990. *Catal. Today* 6:489.

Solymosi, F., I. Tombacz, and G. Katson. 1985. *J. Chem. Soc., Chem. Commun.* 1455.

Stull, R., E.F. Westrum, Jr., and G.C. Sinke. 1969. *The Chemical Thermodynamics of Organic Compounds*. New York: John Wiley.

Tashiro, T., T. Ito, and K. Toi. 1990. *J. Chem. Soc., Faraday Trans. 1* 86:1139.

Thomas, J.M., W. Ueda, J. Williams, and K.D.M. Harris. 1989. *Faraday Disc., Chem. Soc.* 87:33.

Thomas, J.M., X. Zhang, and J. Stachurski. 1988. *J. Chem. Soc., Chem. Commun.* 162.

Ueda, W., and J.M. Thomas. 1988a. In *Proceedings of the 9th International Congress on Catalysis, Calgary*, Vol. 2, ed. M.J. Phillips and M. Ternan, p. 960. Ottawa: Chemical Institute of Canada.

Ueda, W., and J.M. Thomas, 1988b. *J. Chem. Soc., Chem. Commun.* 1148.

Unger, R.K., X. Zhang, and R.M. Lambert. 1988. *Appl. Catal.* 42:L1.

van Kasteren, J.M.N., J.W.M.H. Geerts, and K. van der Wiele. 1988. *Proceedings of the 9th International Congress on Catalysis, Calgary*, Vol. 2, ed. M.J. Phillips and M. Ternan, p. 930. Ottawa: Chemical Institute of Canada.

Williams, J., R.H. Jones, J.M. Thomas, and J. Kent. 1989. *Catal. Lett.* 3:247.

Yates, D.Y.C., and N.E. Zlotin. 1988. *J. Catal.* 111:317.

Zecchina, A., M.G. Lofthouse, and F.S. Stone. 1975. *J. Chem. Soc., Faraday Trans. 1* 71:1476.

Zhan, H.-S., J.X. Wang, D.J. Driscoll, and J.H. Lunsford. 1988. *J. Catal.* 112:366.

Zhen, K.J., M.M. Khan, C.H. Mak, K.B. Levis, and G.A. Somorjai. 1985. *J. Catal.* 94:501.

8

Elementary Reactions and Kinetic Modeling of the Oxidative Coupling of Methane

K. van der Wiele, J.W.M.H. Geerts,
and J.M.N. van Kasteren

*Eindhoven University of Technology,
Laboratory of Chemical Process Technology, Eindhoven,
The Netherlands*

TABLE OF CONTENTS

INTRODUCTION

The oxidative coupling of methane is typically carried out at temperatures of 650–950°C, using a methane-rich mixture of methane and oxygen or air, and with an oxidic catalyst of low porosity. The process is very complex in the sense that reactions at the surface of the catalyst strongly interfere with reactions in the homogeneous gas phase.

The main products of the coupling reactions are ethane and ethylene. At suitable conditions for C_2 formation, only traces of higher hydrocarbons and only very small amounts of acetylene are formed. However, combustion and pyrolysis are significant side reactions, leading to substantial amounts of CO, CO_2 and H_2O, and some H_2.

It is generally accepted that the most important function of the catalyst is to accelerate the production of methyl radicals, which subsequently couple to form ethane (Baerns, van der Wiele, and Ross 1990). Ethylene is formed next, by abstraction of hydrogen from ethane. Combustion, initiated by the formation of a carbon–oxygen bond, may start from several reaction intermediates and may be induced by a variety of radical reactions as well as by interaction with the catalyst. This, in particular, complicates the understanding of the factors that influence the selectivity of the process.

The aim of this chapter is to give more insight into the nature and the significance of various elementary steps and to make a first attempt at "overall modeling" of the oxidative coupling process.

The first section of the chapter deals with purely gas-phase oxidative coupling reactions, studied in the absence of a catalyst. The understanding and quantification of these reactions is of utmost importance: At the process conditions used for the catalytic process, these reactions inevitably occur besides catalytic steps and in part their role is essential.

The next section deals with the catalytic process, using lithium-promoted magnesium oxide (Li/MgO) as the catalyst. After exploring essential factors for the selectivity of the process in general, specific experiments with products, intermediates, and mixtures are presented and discussed. Included are experiments at elevated pressures and experiments at very reduced pressures (below 100 Pa). The results of this work reveal in particular what the main reaction pathways are and what the predominant role of the catalyst is.

The last section deals with kinetic modeling. The transient analysis of steady-state isotopic switching experiments using $^{18}O_2$ and $^{13}CH_4$ reveals the extent to which oxygen and methane interact with the catalyst at reaction conditions. Differential kinetic measurements further quantify the influence of oxygen and methane partial pressure. A complete quantitative picture of the kinetics of all elementary reaction steps, with the interaction between gas-phase radical reactions and catalytic reactions is still beyond reach. The chapter concludes with some results of a strongly simplified kinetic overall model, assuming a limited number of rate-determining steps.

GAS-PHASE OXIDATIVE COUPLING

Introduction

The study of the purely gas-phase oxidative coupling is of great interest because it produces the same products as the catalytic process, albeit at a different rate (the catalyst strongly accelerates the reaction) and with a different selectivity (the catalyst is more selective toward ethane and ethylene, particularly at a higher conversion level).

In the early stages of the research on methane oxidative coupling, it was suggested that the catalyst would merely act as a "methane activator" that effectively produces methyl radicals and thus initiates a process that is further occurring in the gas phase. This unfortunately is an oversimplification of the process, although the main coupling step (methyl radicals to ethane) is most probably a homogeneous step. Obviously one should study the homogeneous reactions, as they essentially contribute to the catalytic process.

Some essential features of the homogeneous partial oxidation are explained first in this section, by presenting the results of experiments in an empty tubular flow reactor carried out at various temperatures and conversion levels.

Next the kinetic modeling is considered. A computer program was developed to handle the vast amount of data on elementary reactions available in the literature and to analyze the predominant reaction pathways. Thus a reaction model with a reduced number of equations was derived and verified with experimental results.

The last part of this section concerns experiments in a tubular reactor with gas recirculation, particularly suitable for investigating the effect of specific components (CO_2, H_2O, H_2, and C_2H_6) on the course of the reaction between methane and oxygen.

Experiments in an Empty Tubular Reactor

A standard microflow setup was used to carry out homogeneous gas-phase experiments. The reactor consisted of a quartz tube with an inner diameter of 11 mm. The tube was heated in an electric oven over a length of 10 cm. Further experimental details have been described elsewhere (Geerts 1990). The gas mixture consisted of methane, oxygen, and helium, approximately in the ratio 10 : 1 : 10. The experiments were carried out at various temperatures and flow rates.

The influence of the reaction temperature will be considered first. Figure 8-1 shows the results of experiments in which the temperature was varied at a constant flow rate (and hence reaction time). Besides the products shown, water and hydrogen also are formed. An increase in temperature results in steeply increasing methane and oxygen conversions and a shift in the product distribution from CO_2 to carbon monoxide. Furthermore, the ethylene selectivity shows a maximum at a steadily decreasing ethane selectivity. This can be explained by assuming that ethane is converted into ethylene at higher temperatures. However, at higher temperatures the combustion of C_2 products plays an increasingly important role, which finally results in a lower yield of higher hydrocarbons.

The pronounced temperature influence on conversion and the long residence times required to obtain high conversions indicate a high activation energy of the reaction. From the selectivity patterns in Fig. 8-1, one could mistakenly conclude that the temperature has a strong influence on

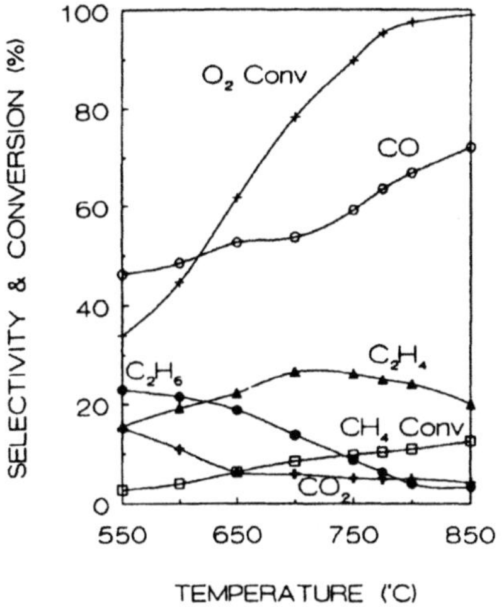

FIGURE 8-1. Temperature influence on the uncatalyzed methane oxidative coupling. Conditions: $CH_4/O_2 = 10$, $P_{tot} = 1$ bar, $P_{CH_4} = 0.91$ bar, $\tau = 67$ s.

the selectivity, with a clearly negative effect at higher temperatures. However, the most important parameter in fact is the conversion level. This is proven by experiments in which the oxygen conversion was kept constant at 72% by adapting the residence time (increasing the flow rate) with increasing temperature. As Fig. 8-2 shows, the C_2 selectivity is now slightly increasing with temperature, whereas the main effect is a shift from ethane to ethylene. The latter now clearly does not yet reach a maximum at 900°C. Figures 8-1 and 8-2 clearly demonstrate that comparison of selectivities is only valid at a similar conversion level, whenever consecutive reactions may occur.

To determine initial selectivities of the methane partial oxidation, experiments were carried out at a very low methane conversion level. This was achieved by using a narrower reaction tube (9-mm diameter), a high flow rate, and a low oxygen partial pressure ($CH_4/O_2 = 20$). Moreover, ultrapure methane was used to enable accurate determination of the small quantities of ethane and ethylene produced. The results, depicted in Fig. 8-3 show that surprisingly high C_2 selectivities are obtained at low methane conversion. Note that it is possible to extrapolate the ethylene selectivity curve to zero and the ethane selectivity curve to 100%, as shown by the dashed lines in Fig. 8-3. This clearly demonstrates that ethane is the primary product and that it is rapidly dehydrogenated into ethylene.

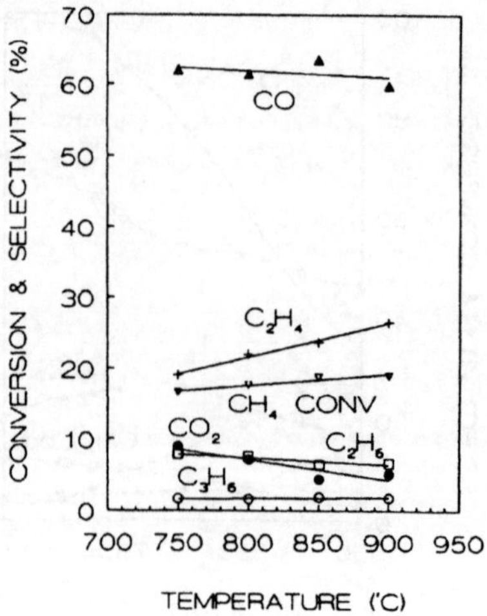

FIGURE 8-2. Temperature effect of the homogeneous gas-phase methane oxidative coupling at a constant oxygen conversion of 72%. Conditions: $CH_4/O_2 = 10$, $CH_4/He = 1.25$, $P_{tot} = 1$ bar.

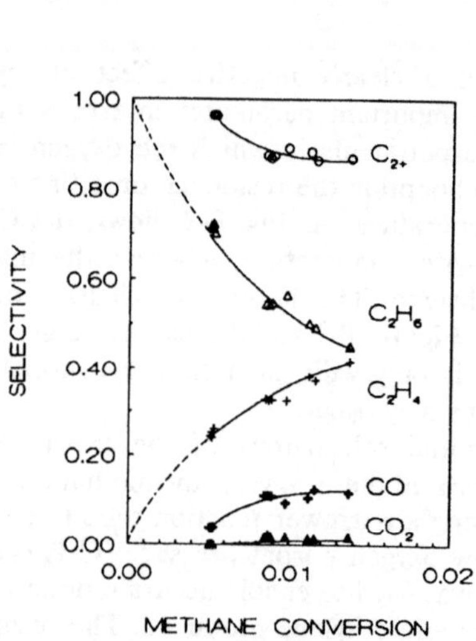

FIGURE 8-3. Selectivity–conversion profile for the homogeneous gas-phase coupling of methane: $CH_4/O_2 = 20$, $P_{tot} = 1$ bar, $T = 800°C$.

In conclusion, ethylene synthesis from methane by partial oxidation can take place in the homogeneous gas phase. No (catalyst) surface is required to activate the methane. The temperature is the key factor that determines the reaction rate, but it has only minor influence on the selectivity, which is greatly dependent on the conversion level and, hence, on the oxygen partial pressure. Low methane conversion and low oxygen partial pressure result in high C_2 selectivities. The initial C_2H_6 selectivity approaches 100%, indicating that ethylene is a secondary product that stems from ethane.

Kinetic Modeling

The homogeneous gas-phase partial oxidation process is well known to take place via free radical reactions, and a vast amount of data on elementary reactions is available in the literature. It is obvious therefore to base a kinetic model for the methane oxidation on the kinetics of a network of elementary reactions. Because hundreds of reactions may occur, a computer program is required to handle the data and to carry out the calculations. We developed our own program, including (1) reaction sensitivity analysis and (2) quantification of predominant reaction pathways. Basically simultaneous rate equations were solved by a numerical integration procedure, and thus the results of a reaction in a batch, or plug-flow, reactor were simulated.

Initially a set of over 400 elementary reactions was used. The reaction rate for each reaction is described by an extended Arrhenius expression of the form

$$k = AT^b \exp(-E_{act}/RT)$$

in which A = frequency factor, b = nonlinearity coefficient (−), T = absolute temperature (K), E_{act} = energy of activation (kJ mol^{-1}), R = gas constant (kJ mol^{-1} K^{-1}), and the units for AT^b are s^{-1} for a first-order reaction, m^3 mol^{-1} s^{-1} for a second-order reaction, and m^6 mol^{-2} s^{-1} for a third-order reaction.

An important feature of the program we developed is its option for a kinetic sensitivity analysis. It determines the influence of each elementary reaction on the product distribution. The principle of the method is that the frequency factor (k_j) of each reaction in turn is varied and that the effect of this variation on the concentration of each product (c_i) is quantified by a set of sensitivity factors (s_{ij}), calculated from

$$s_{ij} = \Delta \ln c_i / \Delta \ln k_j$$

As a result of the analysis, a set of over 400 reactions, see Fig. 8-4, could

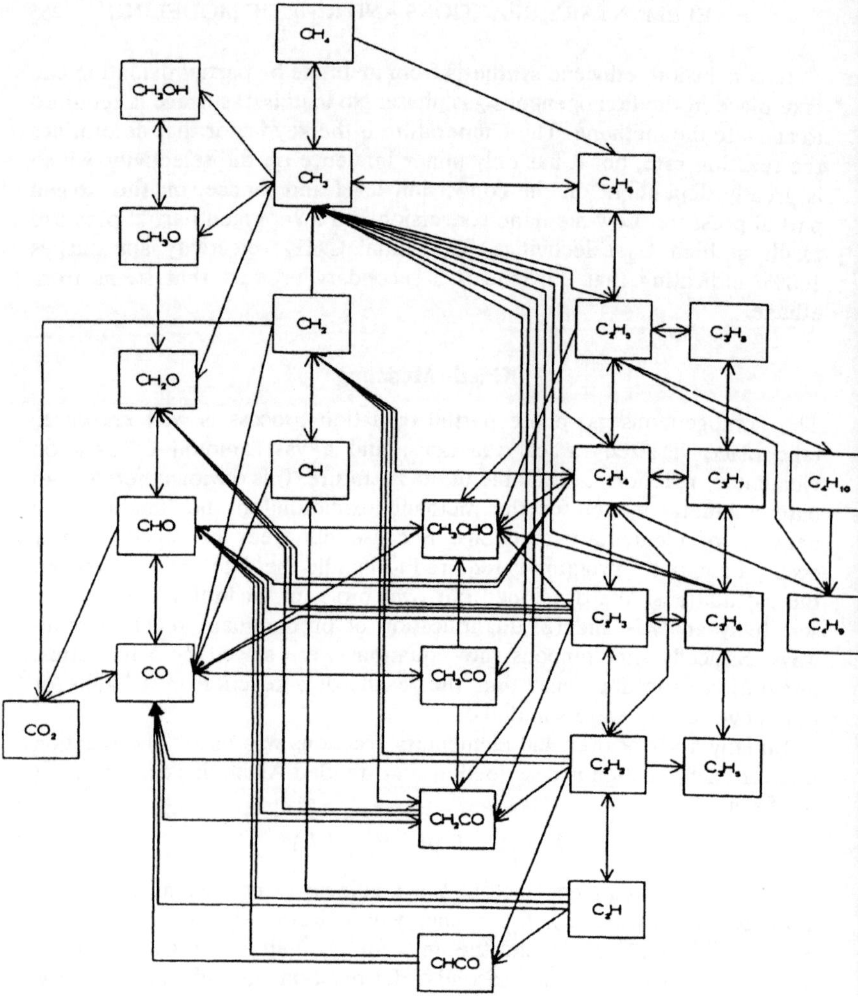

FIGURE 8-4. Reaction scheme of carbon-containing species of the homogeneous gas-phase partial oxidation of methane: downward = H abstraction; to the right = methyl radical addition; down to the left = O addition to that species.

Table 8-1 **Set of Elementary Radical Reactions as Used in the Model (See Text for Units; M = Inert)**

No.	Reaction	A	b	E_{act}
1	$CH_4 \rightarrow CH_3 + H$	3.700E + 15	0.000E + 00	4.340E + 02
2	$CH_4 + O_2 \rightarrow CH_3 + HO_2$	4.000E + 07	0.000E + 00	2.380E + 02
3	$CH_4 + H \rightarrow CH_3 + H_2$	2.200E − 02	3.000E + 00	3.660E + 01
4	$CH_4 + O \rightarrow CH_3 + OH$	1.200E + 01	2.100E + 00	3.190E + 01
5	$CH_4 + OH \rightarrow CH_3 + H_2O$	6.000E + 07	0.000E + 00	3.850E + 01
6	$CH_4 + HO_2 \rightarrow CH_3 + H_2O_2$	1.800E + 05	0.000E + 00	7.770E + 01
7	$CH_3 + H_2 \rightarrow CH_4 + H$	6.600E − 04	3.000E + 00	3.240E + 01
8	$CH_3 + H \rightarrow CH_4$	6.000E + 10	−1.00E + 00	0.000E + 00
9	$CH_3 + OH \rightarrow CH_4 + O$	1.300E − 01	2.100E + 00	1.960E + 01
10	$CH_3 + HO_2 \rightarrow CH_4 + O_2$	3.600E + 06	0.000E + 00	0.000E + 00
11	$CH_3 + H_2O_2 \rightarrow CH_4 + HO_2$	1.100E + 06	0.000E + 00	1.670E + 01
12	$CH_3 + H_2O \rightarrow CH_4 + OH$	2.900E − 01	2.100E + 00	7.030E + 01
13	$CH_3 + O_2 \rightarrow CH_3O + O$	2.000E + 12	−1.57E + 00	1.220E + 02
14	$CH_3 + OH \rightarrow CH_3O + H$	2.000E + 10	0.000E + 00	1.150E + 02
15	$CH_3 + HO_2 \rightarrow CH_3O + OH$	7.943E + 06	0.000E + 00	0.000E + 00
16	$CH_3 + CH_3 \rightarrow C_2H_6$	1.500E + 07	0.000E + 00	0.000E + 00
17	$CH_3 + CH_4 \rightarrow C_2H_6 + H$	8.000E + 07	0.000E + 00	1.672E + 02
18	$CH_3 + CH_3 \rightarrow C_2H_5 + H$	8.000E + 07	0.000E + 00	1.110E + 02
19	$CH_3 + CH_3 \rightarrow C_2H_4 + H_2$	1.000E + 10	0.000E + 00	1.340E + 02
20	$CH_3 + CH_2 \rightarrow C_2H_4 + H$	4.200E + 07	0.000E + 00	0.000E + 00
21	$CH_3O + OH \rightarrow CH_3 + HO_2$	1.000E + 07	0.000E + 00	0.000E + 00
22	$CH_3O + O \rightarrow CH_3 + O_2$	3.020E + 08	0.000E + 00	3.100E + 00
23	$CH_3O \rightarrow CH_2O + H$	1.580E + 14	0.000E + 00	1.150E + 02
24	$CH_3O + O_2 \rightarrow CH_2O + HO_2$	6.600E + 04	0.000E + 00	1.090E + 01
25	$CH_2O + O_2 \rightarrow CHO + HO_2$	2.000E + 07	0.000E + 00	1.630E + 02
26	$CH_2O + H \rightarrow CHO + H_2$	2.500E + 07	0.000E + 00	1.670E + 01
27	$CH_2O + O \rightarrow CHO + OH$	3.500E + 07	0.000E + 00	1.470E + 01
28	$CH_2O + OH \rightarrow CHO + H_2O$	3.000E + 07	0.000E + 00	5.000E + 00
29	$CH_2O + HO_2 \rightarrow CHO + H_2O_2$	1.000E + 06	0.000E + 00	3.350E + 01
30	$CH_2O + CH_3 \rightarrow CHO + CH_4$	5.500E − 03	2.810E + 00	2.450E + 01
31	$CH_2 + H \rightarrow CH + H_2$	4.000E + 07	0.000E + 00	0.000E + 00
32	$CH_2 + O_2 \rightarrow CO + H + OH$	1.300E + 07	0.000E + 00	6.300E + 00
33	$CH + O_2 \rightarrow CO + OH$	2.000E + 07	0.000E + 00	0.000E + 00
34	$CH + O \rightarrow CO + H$	4.000E + 07	0.000E + 00	0.000E + 00
35	$CHO + H_2 \rightarrow CH_2O + H$	1.800E + 00	2.000E + 00	7.460E + 01
36	$CHO + HO_2 \rightarrow CH_2O + O_2$	3.000E + 08	0.000E + 00	1.260E + 01
37	$CHO + H_2O_2 \rightarrow CH_2O + HO_2$	1.000E + 05	0.000E + 00	2.900E + 01
38	$CHO + CH_4 \rightarrow CH_2O + CH_3$	7.300E − 03	2.850E + 00	9.420E + 01
39	$CHO + M \rightarrow CO + H + M$	5.000E + 15	−2.14E + 00	8.550E + 01
40	$CHO + O_2 \rightarrow CO + HO_2$	3.600E + 06	0.000E + 00	0.000E + 00
41	$CHO + CH_3 \rightarrow CO + CH_4$	1.200E + 08	0.000E + 00	0.000E + 00
42	$CO + H + M \rightarrow CHO + M$	6.300E + 08	−1.82E + 00	1.540E + 01
43	$CO + O_2 \rightarrow CO_2 + O$	2.500E + 06	0.000E + 00	2.000E + 02
44	$CO + O + M \rightarrow CO_2 + M$	6.200E + 02	0.000E + 00	1.260E + 01
45	$CO + OH \rightarrow CO_2 + H$	3.036E + 05	0.000E + 00	3.388E + 00
46	$CO + HO_2 \rightarrow CO_2 + OH$	1.500E + 08	0.000E + 00	9.890E + 01
47	$CO_2 + H \rightarrow CO + OH$	1.500E + 08	0.000E + 00	1.110E + 02
48	$C_2H_6 \rightarrow CH_3 + CH_3$	3.200E + 22	−1.79E + 00	3.810E + 02
49	$C_2H_6 \rightarrow C_2H_5 + H$	5.000E + 16	0.000E + 00	3.720E + 02
50	$C_2H_6 + O_2 \rightarrow C_2H_5 + HO_2$	4.000E + 07	0.000E + 00	2.130E + 02
51	$C_2H_6 + H \rightarrow C_2H_5 + H_2$	5.500E − 04	3.500E + 00	2.180E + 01
52	$C_2H_6 + OH \rightarrow C_2H_5 + H_2O$	8.800E + 03	1.040E + 00	7.600E + 00

Table 8-1 (*Continued*)

No.	Reaction	A	b	E_{act}
53	$C_2H_6 + HO_2 \rightarrow C_2H_5 + H_2O_2$	2.900E + 05	0.000E + 00	6.250E + 01
54	$C_2H_6 + CH_3 \rightarrow C_2H_5 + CH_4$	5.500E − 07	4.000E + 00	3.470E + 01
55	$C_2H_6 + CH_2 \rightarrow C_2H_5 + CH_3$	2.200E + 07	0.000E + 00	3.630E + 01
56	$C_2H_5 + H \rightarrow CH_3 + CH_3$	3.000E + 07	0.000E + 00	0.000E + 00
57	$C_2H_5 + HO_2 \rightarrow CH_3 + CH_2O + OH$	2.400E + 07	0.000E + 00	0.000E + 00
58	$C_2H_5 + H_2 \rightarrow C_2H_6 + H$	3.000E − 06	3.600E + 00	3.540E + 01
59	$C_2H_5 + H \rightarrow C_2H_6$	3.600E + 07	0.000E + 00	0.000E + 00
60	$C_2H_5 + OH \rightarrow C_2H_6 + O$	3.000E + 05	0.000E + 00	2.500E + 01
61	$C_2H_5 + HO_2 \rightarrow C_2H_6 + O_2$	3.000E + 05	0.000E + 00	0.000E + 00
62	$C_2H_5 + H_2O_2 \rightarrow C_2H_6 + HO_2$	8.700E + 03	0.000E + 00	4.100E + 00
63	$C_2H_5 + CH_4 \rightarrow C_2H_6 + CH_3$	8.600E − 08	4.140E + 00	5.260E + 01
64	$C_2H_5 + CH_2O \rightarrow C_2H_6 + CHO$	5.500E − 03	2.810E + 00	2.450E + 01
65	$C_2H_5 + CHO \rightarrow C_2H_6 + CO$	1.200E + 08	0.000E + 00	0.000E + 00
66	$C_2H_5 \rightarrow C_2H_4 + H$	2.000E + 13	0.000E + 00	1.660E + 02
67	$C_2H_5 + O_2 \rightarrow C_2H_4 + HO_2$	8.400E + 05	0.000E + 00	1.620E + 01
68	$C_2H_4 + H \rightarrow C_2H_5$	8.400E + 02	1.490E + 00	4.100E + 00
69	$C_2H_4 + HO_2 \rightarrow C_2H_5 + O_2$	1.320E + 05	0.000E + 00	5.730E + 01
70	$C_2H_4 + O_2 \rightarrow C_2H_3 + HO_2$	3.000E + 06	0.000E + 00	1.230E + 02
71	$C_2H_4 + H \rightarrow C_2H_3 + H_2$	1.500E + 08	0.000E + 00	4.270E + 01
72	$C_2H_4 + O \rightarrow C_2H_3 + OH$	1.300E + 05	6.300E − 01	5.700E + 00
73	$C_2H_4 + OH \rightarrow C_2H_3 + H_2O$	7.000E + 07	0.000E + 00	1.260E + 01
74	$C_2H_4 + CH_3 \rightarrow C_2H_3 + CH_4$	1.500E + 06	0.000E + 00	4.970E + 01
75	$C_2H_4 + O \rightarrow CH_3 + CHO$	1.320E + 02	1.550E + 00	1.800E + 00
76	$C_2H_4 + OH \rightarrow CH_3 + CH_2O$	1.000E + 07	0.000E + 00	0.000E + 00
77	$C_2H_4 + O \rightarrow CH_2 + CH_2O$	2.510E + 07	0.000E + 00	2.090E + 01
78	$C_2H_4 + HO_2 \rightarrow CH_3CHO + OH$	6.000E + 03	0.000E + 00	3.330E + 01
79	$C_2H_3 + H_2 \rightarrow C_2H_4 + H$	3.000E − 02	2.630E + 00	3.570E + 01
80	$C_2H_3 + H_2O \rightarrow C_2H_4 + OH$	4.800E − 04	2.900E + 00	6.220E + 01
81	$C_2H_3 + CH_4 \rightarrow C_2H_4 + CH_3$	1.300E + 06	0.000E + 00	3.260E + 01
82	$C_2H_3 + CH_2O \rightarrow C_2H_4 + CHO$	5.400E − 03	2.810E + 00	2.450E + 01
83	$C_2H_3 + C_2H_6 \rightarrow C_2H_4 + C_2H_5$	6.000E − 04	3.300E + 00	4.390E + 01
84	$C_2H_3 \rightarrow C_2H_2 + H$	1.000E + 15	0.000E + 00	1.780E + 02
85	$C_2H_3 + HO_2 \rightarrow CH_3 + CO + OH$	3.000E + 07	0.000E + 00	0.000E + 00
86	$C_2H_3 + O_2 \rightarrow C_2H_2 + HO_2$	1.000E + 06	0.000E + 00	0.000E + 00
87	$C_2H_2 + H \rightarrow C_2H_3$	5.500E + 06	0.000E + 00	1.010E + 01
88	$C_2H_2 + O_2 \rightarrow CHCO + OH$	2.000E + 02	1.500E + 00	1.259E + 02
89	$C_2H_2 + O_2 \rightarrow CHO + CHO$	4.000E + 06	0.000E + 00	1.170E + 02
90	$C_2H_2 + H \rightarrow C_2H + H_2$	1.500E + 08	0.000E + 00	7.960E + 01
91	$C_2H_2 + OH \rightarrow C_2H + H_2O$	1.000E + 07	0.000E + 00	2.930E + 01
92	$C_2H_2 + OH \rightarrow CH_2CO + H$	3.000E + 06	0.000E + 00	4.600E + 00
93	$C_2H_2 + CH_3 \rightarrow C_2H + CH_4$	1.800E + 05	0.000E + 00	7.230E + 01
94	$C_2H + H_2 \rightarrow C_2H_2 + H$	3.500E + 06	0.000E + 00	8.800E + 00
95	$C_2H + H_2O \rightarrow C_2H_2 + OH$	4.070E + 06	0.000E + 00	5.800E + 00
96	$C_2H + CH_4 \rightarrow C_2H_2 + CH_3$	1.000E + 06	0.000E + 00	6.000E + 01
97	$C_2H + C_2H_6 \rightarrow C_2H_2 + C_2H_5$	3.600E + 06	0.000E + 00	0.000E + 00
98	$C_2H_2 + O \rightarrow CH_2 + CO$	1.750E + 07	0.000E + 00	1.330E + 01
99	$C_2H_2 + O \rightarrow CHCO + H$	9.000E + 06	0.000E + 00	1.900E + 01

Table 8-1 (*Continued*)

No.	Reaction	A	b	E_{act}
100	$C_2H + O_2 \rightarrow CHO + CO$	5.000E + 07	0.000E + 00	6.300E + 00
101	$C_2H + O \rightarrow CH + CO$	1.000E + 07	0.000E + 00	0.000E + 00
102	$CH_3CHO + H \rightarrow CH_3 + CO + H_2$	4.000E + 07	0.000E + 00	1.760E + 01
103	$CH_3CHO + O \rightarrow CH_3 + CO + OH$	5.000E + 06	0.000E + 00	7.500E + 00
104	$CH_3CHO + OH \rightarrow CH_3 + CO + H_2O$	1.000E + 07	0.000E + 00	0.000E + 00
105	$CH_3CHO \rightarrow CH_3 + CHO$	2.000E + 15	0.000E + 00	3.310E + 02
106	$CH_3CHO + HO_2 \rightarrow CH_3 + CO + H_2O_2$	1.700E + 06	0.000E + 00	4.480E + 01
107	$CH_3CHO + O_2 \rightarrow CH_3 + CO + HO_2$	2.000E + 07	5.000E − 01	1.765E + 02
108	$CH_3CHO + CH_3 \rightarrow CH_3 + CO + CH_4$	8.000E + 05	0.000E + 00	2.510E + 01
109	$CH_2CO + H \rightarrow CH_3 + CO$	7.000E + 06	0.000E + 00	1.260E + 01
110	$CH_2CO + O \rightarrow CHO + CHO$	2.000E + 07	0.000E + 00	9.600E + 00
111	$CH_2CO + OH \rightarrow CHO + CH_2O$	1.000E + 07	0.000E + 00	0.000E + 00
112	$CH_2CO + M \rightarrow CH_2 + CO + M$	1.000E + 10	0.000E + 00	2.480E + 02
113	$CHCO + H \rightarrow CO + CH_2$	3.000E + 06	0.000E + 00	0.000E + 00
114	$CHCO + O \rightarrow CO + CHO$	1.200E + 06	0.000E + 00	0.000E + 00
115	$CHCO + OH \rightarrow CHO + CO + H$	1.000E + 07	0.000E + 00	0.000E + 00
116	$CHCO + O_2 \rightarrow CO + CO + OH$	1.460E + 06	0.000E + 00	1.050E + 01
117	$C_2H_5 + CH_3 \rightarrow C_3H_8$	1.000E + 07	0.000E + 00	0.000E + 00
118	$C_2H_4 + CH_3 \rightarrow C_3H_7$	1.000E + 05	0.000E + 00	2.900E + 01
119	$C_2H_3 + CH_3 \rightarrow C_3H_6$	1.000E + 07	0.000E + 00	0.000E + 00
120	$C_2H_2 + CH_3 \rightarrow C_3H_5$	6.000E + 05	0.000E + 00	3.220E + 01
121	$C_3H_8 \rightarrow C_2H_5 + CH_3$	7.900E + 16	0.000E + 00	3.560E + 02
122	$C_3H_8 + H \rightarrow C_3H_7 + H_2$	1.300E + 08	0.000E + 00	4.060E + 01
123	$C_3H_8 + O \rightarrow C_3H_7 + OH$	3.000E + 07	0.000E + 00	2.410E + 01
124	$C_3H_8 + OH \rightarrow C_3H_7 + H_2O$	3.700E + 06	0.000E + 00	6.900E + 00
125	$C_3H_8 + CH_3 \rightarrow C_3H_7 + CH_4$	1.550E + 08	0.000E + 00	8.900E + 01
126	$C_3H_7 \rightarrow C_2H_4 + CH_3$	1.300E + 13	0.000E + 00	1.359E + 02
127	$C_3H_7 + H_2 \rightarrow C_3H_8 + H$	4.000E + 06	0.000E + 00	6.980E + 01
128	$C_3H_7 + H \rightarrow C_3H_8$	2.000E + 07	0.000E + 00	0.000E + 00
129	$C_3H_7 + HO_2 \rightarrow C_3H_8 + O_2$	1.000E + 07	0.000E + 00	0.000E + 00
130	$C_3H_7 + CH_4 \rightarrow C_3H_8 + CH_3$	7.900E + 05	0.000E + 00	6.480E + 01
131	$C_3H_7 + C_2H_6 \rightarrow C_3H_8 + C_2H_5$	1.000E + 05	0.000E + 00	4.180E + 01
132	$C_3H_7 + C_2H_4 \rightarrow C_3H_8 + C_2H_3$	1.000E + 05	0.000E + 00	6.280E + 01
133	$C_3H_7 \rightarrow C_3H_6 + H$	2.000E + 14	0.000E + 00	1.630E + 02
134	$C_3H_7 + O_2 \rightarrow C_3H_6 + HO_2$	1.000E + 06	0.000E + 00	2.090E + 01
135	$C_3H_6 + H \rightarrow C_3H_7$	4.000E + 06	0.000E + 00	1.100E + 01
136	$C_3H_6 \rightarrow C_2H_3 + CH_3$	7.000E + 16	0.000E + 00	3.600E + 02

Table 8-1 (*Continued*)

No.	Reaction	A	b	E_{act}
137	$C_3H_6 + CH_3 \rightarrow C_3H_5 + CH_4$	2.000E + 04	0.000E + 00	1.880E + 01
138	$C_3H_6 + O \rightarrow C_2H_5 + CHO$	1.510E + 06	0.000E + 00	0.000E + 00
139	$C_3H_6 + OH \rightarrow CH_3CHO + CH_3$	1.000E + 07	0.000E + 00	0.000E + 00
140	$C_3H_5 + H_2 \rightarrow C_3H_6 + H$	7.940E + 06	0.000E + 00	3.100E + 01
141	$O_2 + H \rightarrow O + OH$	1.700E + 11	−9.00E − 01	7.280E + 01
142	$O_2 + H + M \rightarrow HO_2 + M$	2.000E + 06	−8.00E + 01	0.000E + 00
143	$O + OH \rightarrow O_2 + H$	1.800E + 07	0.000E + 00	0.000E + 00
144	$O + HO_2 \rightarrow O_2 + OH$	2.000E + 07	0.000E + 00	0.000E + 00
145	$O + H_2 \rightarrow OH + H$	1.500E + 01	2.000E + 00	3.160E + 01
146	$O + H_2O \rightarrow OH + OH$	4.600E + 03	1.300E + 00	7.150E + 01
147	$O + H_2O_2 \rightarrow OH + HO_2$	6.310E + 06	0.000E + 00	2.075E + 01
148	$OH + H \rightarrow O + H_2$	6.700E + 00	2.000E + 00	2.330E + 01
149	$OH + OH \rightarrow O + H_2O$	1.500E + 03	1.140E + 00	0.000E + 00
150	$OH + H_2 \rightarrow H_2O + H$	1.000E + 02	1.600E + 00	1.380E + 01
151	$OH + H + M \rightarrow H_2O + M$	2.200E + 10	−2.00E + 00	0.000E + 00
152	$OH + HO_2 \rightarrow H_2O + O_2$	2.000E + 07	0.000E + 00	0.000E + 00
153	$OH + H_2O_2 \rightarrow H_2O + HO_2$	1.750E + 06	0.000E + 00	1.330E + 00
154	$HO_2 + M \rightarrow O_2 + H + M$	2.100E + 09	0.000E + 00	1.910E + 02
155	$HO_2 + H \rightarrow O_2 + H_2$	2.500E + 07	0.000E + 00	2.900E + 00
156	$HO_2 + HO_2 \rightarrow O_2 + OH + OH$	2.000E + 06	0.000E + 00	0.000E + 00
157	$HO_2 + H_2 \rightarrow H_2O_2 + H$	3.000E + 07	0.000E + 00	1.090E + 02
158	$H_2O_2 + H \rightarrow HO_2 + H_2$	1.700E + 06	0.000E + 00	1.570E + 01
159	$OH + OH + M \rightarrow H_2O_2 + M$	3.200E + 10	−2.00E + 00	0.000E + 00
160	$HO_2 + H \rightarrow OH + OH$	1.500E + 08	0.000E + 00	4.200E + 00
161	$H_2O_2 + M \rightarrow OH + OH + M$	3.000E + 11	0.000E + 00	1.900E + 02
162	$H_2O_2 + H \rightarrow OH + H_2O$	1.000E + 07	0.000E + 00	1.660E + 01
163	$H_2O + H \rightarrow OH + H_2$	4.600E + 02	1.600E + 00	7.770E + 01
164	$H + H + M \rightarrow H_2 + M$	1.800E + 06	−1.00E + 00	0.000E + 00

be reduced to the 164 reactions of Table 8-1 (Tsang and Hampson 1986; Gardiner 1984; Westley and Herron 1987; Warnatz 1987; Vardanyan and Nalbandyan 1985; Dagaut, Cathonnet, and Boettner 1988; Rotzoll 1986; Gesser, Hunter, and Prakash 1985; Nicholas 1976; Baulch et al. 1972; Allbright, Crynes, and Corcoran 1983) by omitting all the reactions that had no significant contribution to the product distribution. In this way, for example, the formation of methanol and reactions in which a methyl dioxygen radical is involved were skipped. This simplifies the reaction network but at the same time limits the applicability to conditions relevant for the methane oxidative coupling. The product distribution turns out to be highly sensitive toward reactions 6, 13, 15, 16, 46, 54, and 70. A reasonable fit between the model and experimental data (Elbers 1989) was obtained (see Fig. 8-5), using the kinetic constants from the literature. Only minor changes (within reported accuracy limits) were applied for

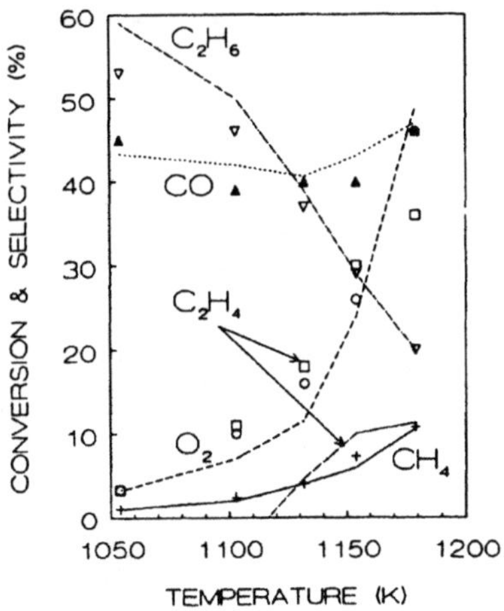

FIGURE 8-5. Comparison of experiments (symbols) with simulations (lines). Conditions: $CH_4/O_2 = 10$, $CH_4/He = 0.8$, $P_{tot} = 1$ bar.

some of the constants. Note that the ethylene selectivity predicted by the model is too low. This requires further attention.

Another powerful tool of the model is the integration of individual reaction rates over the reactor, thus demonstrating the most important reaction pathways. It appears that H, OH, HO_2, and CH_3 are the most important radicals involved in the process. After an induction period in which the radical concentrations are built up, they stay fairly constant at typical concentrations of 10^{-7}–10^{-4} mol/m^3. Figures 8-6 and 8-7 show the main reactants and reaction pathways of non-carbon-containing and of carbon-containing compounds, respectively. The thickness of the arrows corresponds to the contribution of the respective reaction and the species beside the arrows in Fig. 8-6 are the most active reactants for that reaction, in order of importance (top to bottom). Although all the reactions of the figure are reversible, it is clear, from the thickness of the arrows, that none of them is at equilibrium. Oxygen is reduced to water via HO_2 and OH mainly. Hydrogen peroxide is an intermediate product with a short lifetime, so its yield is low. The exchange of hydrogen atoms between hydrogen molecules and atoms is relatively fast. The reaction rate of hydrogen atoms with non-carbon-containing species is, however, low.

It may be obvious that the C_2-oxygenates, CH_2, CH, CH_3OH, and other omitted species are unimportant. The most important initiation reaction, no. 2 of Table 8-1, in which oxygen reacts with methane, appears

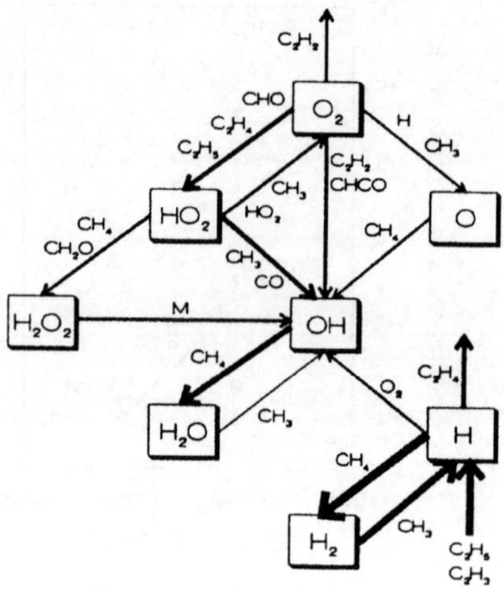

FIGURE 8-6. Reaction scheme of the non-carbon-containing components. The arrow width indicates in a logarithmic way the importance of the reaction.

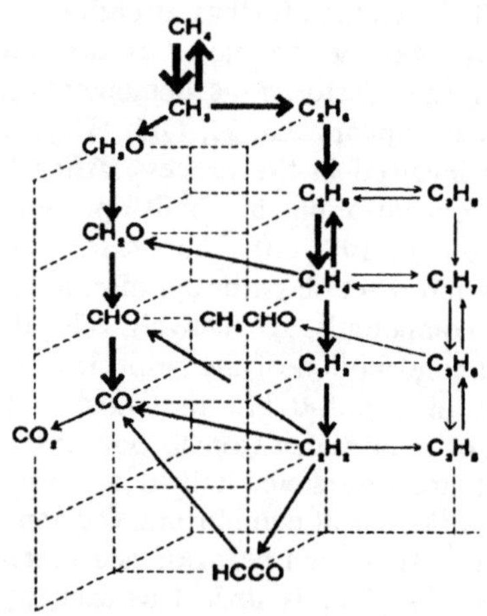

FIGURE 8-7. Reaction scheme of the non-carbon-containing components. Conditions: $T = 900°C$, $CH_4/O_2 = 6.8$, $P_{CH_4} = 0.87$, $P_{tot} = 1$ bar.

to contribute very little to the methane conversion, and the same applies to other initiation reactions. The formation of CH_3 radicals from methane appears in the scheme as a rapid system of equilibrium reactions in which various radicals are involved. In agreement with experimental results, the quantities of C_3 products are very limited, not because they are oxidized very fast, but because of their low rate of formation (see Fig. 8-7).

The C_2 species prefer to undergo hydrogen abstraction rather than coupling with methyl radicals. Ethane and ethyl radicals are hardly oxidized and the formation of CH_3 from C_2H_6 is negligibly small, which implies that an extremely high initial ethylene selectivity should be achievable in the oxidation of ethane.

Figure 8-8 shows the effect of temperature predicted by the model at a constant oxygen conversion of 80%. A comparison of this figure with the experimental results in Fig. 8-2 shows that experiments and model exhibit very similar tendencies with respect to the temperature influence on conversion and selectivities, despite slightly different reaction conditions. Although this does not fully prove the correctness of the model and the value of the parameters, it gives confidence in the principles applied. In conclusion, the model developed simulates the experiments correctly. The main carbon reaction path goes via CH_3, C_2H_6, and C_2H_5 toward C_2H_4, which is oxidized rapidly into CO. The most active reaction species in the

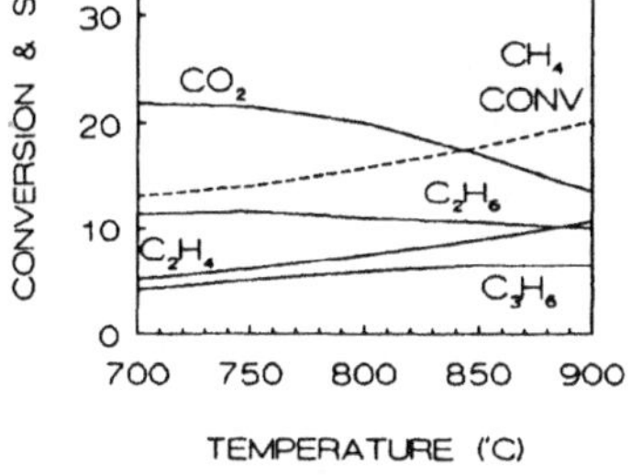

FIGURE 8-8. Simulated temperature influence on the uncatalyzed methane oxidative coupling at an oxygen conversion of 80%. Conditions: $CH_4/O_2 = 5$, $P_{CH_4} = 0.91$ bar, $P_{tot} = 1$ bar.

homogeneous gas phase are the radicals H, OH, CH_3, and OH_2. C_3 hydrocarbons are hardly produced due to their low rate of formation.

Zanthoff and Baerns (Zanthoff and Baerns 1990) also have developed a reaction model, which describes the homogeneous gas-phase reactions. Their results are very similar to those described here. McCarthy and co-workers (McCarthy, McEwen, and Quinlan 1990) and Forlani and co-workers (Forlani et al. 1990) have tried to make a model for a combination of the homogeneous and heterogeneous reactions. Their attempts have failed until now, because of the lack of accurate information on the elementary heterogeneous reaction steps.

Experiments in a Reactor with Gas Recirculation

The aim of the experiments in the gas-recirculation reactor is to investigate the kinetic influence of CO_2, water, hydrogen, and ethane on the reactions occurring in the gas phase. A gas-recirculation reactor has been chosen because it can be operated as an ideal continuous flow stirred tank reactor (CSTR) by applying a high recirculation ratio. The advantage of this type of operation is that no concentration gradients are present inside the reactor. The concentrations in the reactor can be controlled by adapting the feed gas flow or by changing feed ratios.

In all the experiments the following conditions have been applied:

Reactor temperature: 1073 or 1173 K
Reactor pressure: 1.10–1.25 bar
Reactor volume: 1.1 cm^3
Integral feed: 50–60 N cm^3/min^{-1}
Differential feed: 1700–2600 N cm^3/min^{-1}
Recycle ratio: 30–50

The Influence of Carbon Dioxide

The effect of carbon dioxide on the gas-phase reactions has not yet been reported, whereas its effect on the catalytic reactions is extensively described in literature (Korf et al. 1987). Figure 8-9 shows an experiment in which the carbon dioxide partial pressure in the reactor is varied by adding CO_2 to the other feed gases. The residence time is kept constant by diminishing the helium in the feed with the same amount as the CO_2 added. The partial pressures of ethane and ethylene in the reactor are the same. It is clear from this figure that carbon dioxide has no effect at all. No changes occur in the partial pressures inside the reactor. This is in agreement with the model developed to simulate the homogeneous gas phase reactions (Geerts et al. 1990). The formation of carbon dioxide in

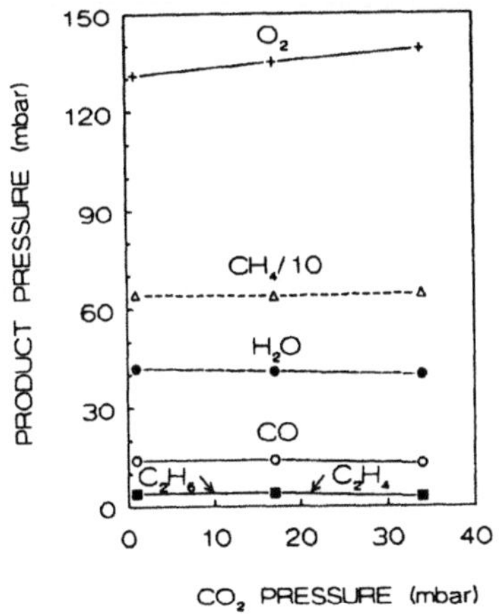

FIGURE 8-9. The influence of the carbon dioxide partial pressure on the gas-phase oxidative coupling of methane. Conditions: $T = 800°C$, $P_{tot} = 1.25$ bar, $R = 38$, $\tau_{per\ pass} = 0.032$ s.

the gas phase is rather slow, but its decomposition is still much slower (see also Fig. 8-7).

The Influence of Water

By cooling the recycle stream in a condenser kept at 273 K, the main part of the water was removed and so its effect on the oxidative coupling of methane could be observed. A higher water partial pressure increases both the methane and oxygen conversion. This is not caused by higher partial pressures of methane or oxygen, because these are equal and lower respectively (see Fig. 8-10). The net production rates of ethylene and propylene are hardly influenced by the water partial pressure. The net production rate of ethane decreases a little bit, whereas the CO_2 formation rate increases due to an increased CO concentration. The C_{2+} selectivity thus drops when the water concentration in the reactor rises. Figure 8-6 shows that the most important decomposition reaction of water is

$$H_2O + CH_3\cdot \rightarrow CH_4 + OH\cdot$$

so methyl radicals are replaced by OH· radicals. As OH· generates more secondary radicals than $CH_3\cdot$, the reaction rate increases. The C_{2+} selectivity decreases due to a relatively lower $CH_3\cdot$ concentration.

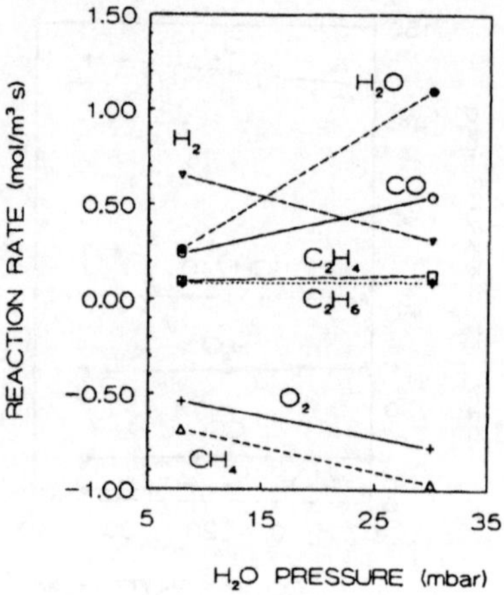

FIGURE 8-10. The influence of the water partial pressure on the overall reaction rates. Conditions: T = 900°C, P_{tot} = 1.20 bar, R = 45, $\tau_{per\ pass}$ = 0.027 s.

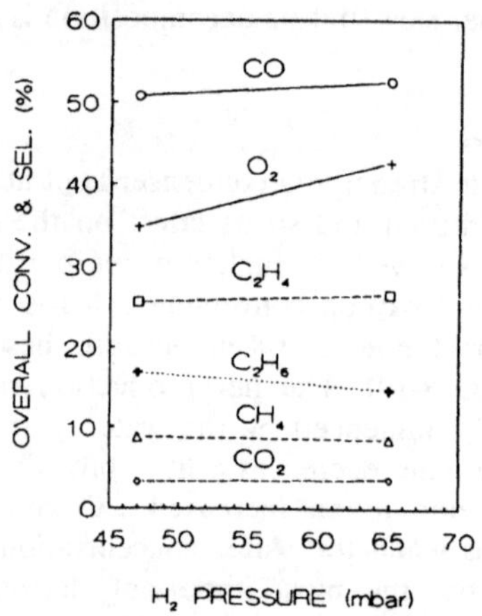

FIGURE 8-11. The influence of the hydrogen partial pressure on the gas-phase oxidative coupling of methane. Conditions: T = 900°C, P_{tot} = 1.20 bar, R = 40, $\tau_{per\ pass}$ = 0.030 s.

Table 8-2 Addition of Ethane at Constant Methane and Oxygen Partial Pressures (Conditions: $T = 800°C$, $R = 35$; $P_{tot} = 1.2$ bar)

	Reactors Partial Pressure (mbar)		Conversion Rate ($mol\ m^{-3}\ s^{-1}$)	
C_2H_6	1	14		
O_2	96	94	0.20	4.6
CH_4	603	593	0.27	3.5
H_2	1	147		
H_2O	11	106		
CO	3	94		
CO_2	0	14		
C_2H_4	1	25		
C_3H_6	0	1		

The Influence of Hydrogen

Figure 8-11 depicts the effect of the addition of hydrogen to the feed at a constant residence time. The water production and the oxygen consumption are accelerated. So, the hydrogen is burned to water, which causes a lower C_{2+} selectivity and higher CO and CO_2 selectivities (cf. Figs. 8-10 and 8-11). Note that the effects of water and hydrogen are very similar indeed.

The Influence of Ethane

Ethane has been added to the primary feed of the reactor to investigate its effect on the reaction rates. The partial pressures of methane and oxygen in the reactor are kept constant by increasing their primary feed flows. The residence time is kept constant by using helium as a balance. At the conditions chosen, $T = 1073$ K, the effect is tremendous (see Table 8-2). The conversion of methane is increased 12 times and that of oxygen 10 times. As a result, the concentrations of, especially, ethylene and, to a smaller extent, CO and CO_2 are higher, whereas the propylene concentration drops in contradiction to what is expected. This experiment proves without doubt that ethane can be dehydrogenated much more rapidly than methane. Ethane is converted into radicals relatively easily and it therefore acts as an initiator for the activation of methane.

Summary and Conclusions

The thermal conversion of methane-rich methane–oxygen mixtures ($CH_4/O_2 = 10$) in an empty tubular reactor demonstrates that even without a catalyst significant amounts of ethane and ethylene are formed, besides carbon monoxide and traces of C_3 compounds.

The initial selectivity toward ethane is very high, but the C_2 selectivity rapidly drops to very low values as the methane conversion increases: Carbon monoxide is the main product then, besides water, small amounts of hydrogen and carbon dioxide.

Remarkably, temperature variation in the range of 750–900°C has not a very large effect on the C_2 selectivity (if determined at the same oxygen conversion level). The best selectivities are found at the highest temperature level investigated.

The kinetics of the gas-phase process are adequately described by a model based on a network of elementary reactions. An initial set of over 400 reactions could be reduced to 164 reactions by successive elimination of insignificant reactions. Very minor adjustments of kinetic parameters were sufficient to achieve a very reasonable fit with experimental data. Analysis of the "reaction highways" in the network reveals that the main path for carbon is

$$CH_4 \rightarrow CH_3 \rightarrow C_2H_6 \rightarrow C_2H_5 \rightarrow C_2H_4 \rightarrow CO \rightarrow CO_2$$

However, one should realize that each step in this sequence may involve a variety of other radicals. The formation of C_3 compounds is hardly significant; model and experiment agree well on this point.

Carbon dioxide, water, hydrogen, and ethane added to the feed are of interest for an industrial catalytic process. Their influence on the gas-phase process was tested and agreed well with the prediction of the model. Carbon dioxide has no effect. Water and hydrogen slightly accelerate the reaction and slightly decrease the C_2 selectivity. Ethane addition has a dramatic accelerating effect, but the C_2 selectivity decreases if one corrects for the ethane added.

CATALYTIC OXIDATIVE COUPLING

Introduction

A variety of oxidic catalysts has been proposed to improve the selectivity of the methane oxidative coupling process (Keller and Bhasin 1982; Hinsen and Baerns 1983; Jones, Leonard, and Sofranko 1984). The most successful catalysts are strongly basic materials consisting of earth alkaline–alkali metal oxide combinations, as was shown by various investigators (Ito et al. 1985; Kolts and Lunsford 1986; Otsuka, Jinno, and Morikawa 1985; Otsuka et al. 1986; Nishiyama, Watanabe, and Aika 1990). The primary role of the catalyst is to "activate" methane, that is, to catalyze the formation of methyl radicals. However, the catalyst is also influencing many other reactions involving intermediates as well as reaction products. As a result, the oxidative coupling process is a complex network of catalytic and gas-phase reaction steps.

The experimental work described in this section mainly concerns Li/MgO catalysts operating at 800°C. The catalytic properties of Li/MgO with respect to methane activation and the formation of methyl radicals have been discovered and studied extensively by Lunsford and co-workers (Ito et al. 1985), who also developed theories regarding the mechanism. At present, Li/MgO is still the most selective material for the methane coupling, albeit that the material is not very stable and certainly is not suitable as such for industrial application. Expectedly, the mechanistic principles found for Li/MgO are applicable to many other catalysts, because the main feature of all effective catalysts is the presence of strongly basic sites and the ability to form oxygen vacancies.

First, a number of introductory experiments are discussed, by which general dependencies of the results on process conditions are clarified. Next, a systematic investigation of individual reaction steps is presented, based on experiments with a variety of feed mixtures. As a result a simplified picture of the main reaction path emerges. Finally, attention is given to the performance of the catalyst at a strongly reduced pressure (< 100 Pa). Because pressure reduction to this low level affects gas-phase reactions more than heterogeneous reactions, these experiments give additional information on the role of the catalyst.

The section concludes with a summary of the essential reaction steps and the main reaction pathways for the formation of ethane, ethylene, and carbon oxides.

Influence of Process Conditions

To determine the process conditions relevant for the catalytic oxidative coupling of methane, experiments have been carried out in which the influence of the temperature, the oxygen partial pressure, the helium partial pressure, and the size of the postcatalytic space were varied. A lithium-doped magnesium oxide catalyst was used. These experiments, which are preliminary to the investigation of the reaction mechanism and kinetics described later in this section, lead to the definition of optimal process conditions for the production of C_2 hydrocarbons. The reaction scheme that will be deduced in the next section is already used here to explain the phenomena observed.

A micro-fixed-bed reactor setup described elsewhere (Geerts 1990) was used. The catalyst was contained in a tubular reactor made of quartz.

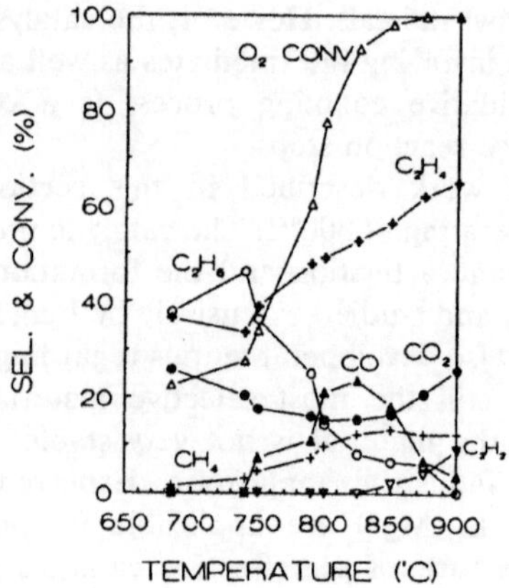

FIGURE 8-12. Influence of the reaction temperature: $P_{tot} = 1$ bar, $CH_4/O_2 = 10$, $W/F = 0.6$ g s cm^{-3}.

Reaction Temperature

The effect of the reaction temperature on the oxidative coupling of methane has been investigated at a methane–oxygen ratio of 10 (see Fig. 8-12). The conversion profiles are rather steep, which indicates that the catalytic methane activation has a high activation energy, only a little lower than that of the gas-phase process. Ethane is converted into ethylene at higher temperatures. Ethylene is, in turn, converted into acetylene, which was only detected at temperatures above 830°C. Note that the production of carbon dioxide is considerably higher than it was in the homogeneous gas-phase experiments.

Oxygen Partial Pressure

The influence of the oxygen partial pressure on the oxidative coupling of methane was determined at a total flow of 125 cm^3/min, $W/F = 0.48$ g s cm^{-3}, and a constant methane partial pressure of 0.6 bar. Figure 8-13 shows a maximum in the ethylene selectivity at an oxygen partial pressure of 0.04 bar. At low oxygen partial pressures, the carbon monoxide and carbon dioxide selectivities approach zero. This is to be expected, because in the absence of oxygen only pyrolysis reactions can occur and, hence, only ethane, ethylene, and acetylene can be formed. The reaction mechanism proposed, shown in Fig. 8-14, also suggests that at a higher oxygen partial pressure, thus at a higher conversion level, a shift from ethane to ethylene to carbon monoxide occurs.

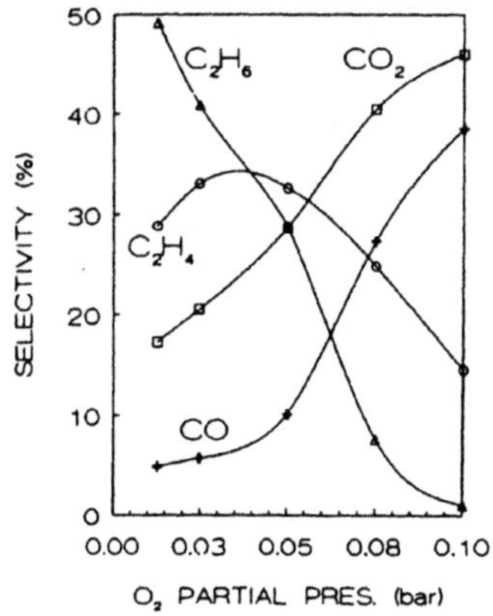

FIGURE 8-13. Selectivities as a function of the inlet oxygen partial pressure: $T = 800°C$, $W/F = 0.48$ g s ml^{-1}, $P_{tot} = 1$ bar, $P_{CH_4} = 0.6$ bar.

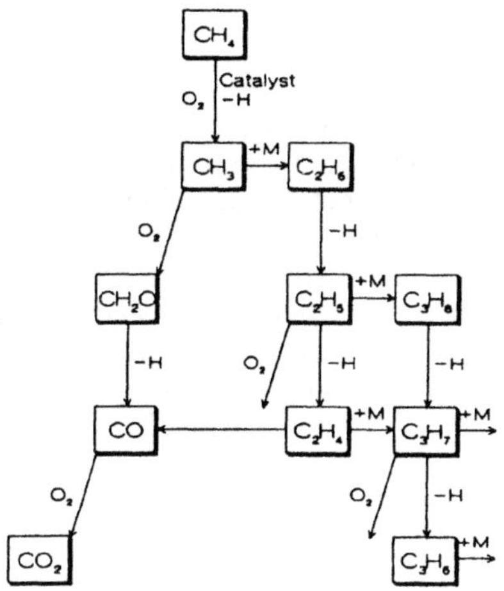

FIGURE 8-14. The proposed reaction scheme for the methane oxidative coupling.

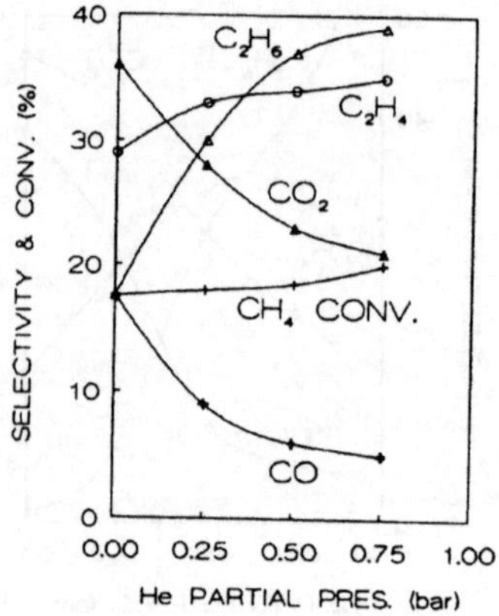

FIGURE 8-15. Selectivity and conversion as function of the helium partial pressure: $T = 800°C$, $CH_4/O_2 = 5$, $P_{tot} = 1$ bar, $W/F = 0.6$ g s ml^{-1}.

With increasing amounts of oxygen in the feed, the combustion reactions become more and more predominant. Finally, mainly CO and CO_2 are observed.

Helium Dilution

The influence of the helium partial pressure was determined at a steady total flow of 120 cm^3 min^{-1} and a constant methane–oxygen ratio of 5. From Fig. 8-15, it appears that a higher helium partial pressure, which means a dilution to a higher extent, results in higher selectivities to ethane and ethylene and lower selectivities with respect to CO and CO_2.

Otsuka and co-workers (Otsuka, Jinno, and Morikawa 1985; Otsuka et al. 1986) owe many of the high yields they obtained to the high degree of dilution used in their experiments. However, their C_2 production, in terms of absolute mole C_2H_4 per weight of catalyst per second, was very small.

This can be explained by the fact that the formation of ethylene and ethane are of a lower reaction order than the production of carbon oxides. The effect of elevated pressure has also been investigated. The results are consistent with the helium dilution results. The trend of the curves of Fig. 8-15 continues at the left side of the vertical axis, that is, at a total system pressure larger than 1. Thus the C_{2+} selectivity decreases very rapidly.

Size of the Precatalytic and Postcatalytic Space

Both thermal and catalytic activation of methane during the oxidative coupling can occur, as is clearly shown in Figs. 8-1 and 8-12. Thermal

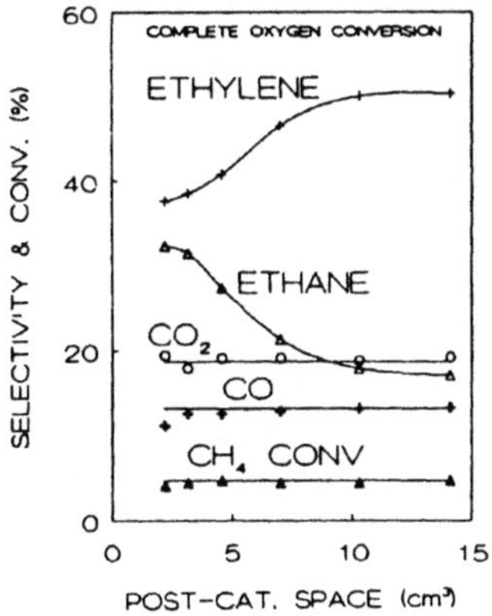

FIGURE 8-16. The dependence on the size of the postcatalytic space for complete oxygen conversion: T = 800°C, CH_4/O_2 = 10, CH_4/He = 0.8, W/F = 1.2 g s cm^{-3}, F = 67 ml min^{-1}.

activation before the reaction mixture has reached the catalyst bed, that is, reactions in the precatalytic space, can be minimized by shortening the (corresponding) residence time (Geerts, van Kasteren, and van der Wiele 1988). As it is known that additives in a methane–oxygen mixture can facilitate the thermal activation [see, e.g., Table 8-2 (the additive being ethane)], it will be very interesting to know what happens after the bed, where products are present. Figure 8-16 shows an experiment in which the space after the catalyst bed, which was heated to the same temperature as the catalyst bed itself, was varied in a way described elsewhere (van Kasteren, Geerts and van der Wiele 1988). The process conditions were carefully chosen so that the conversion in the pre-catalytic zone was negligible and that the oxygen, which was in short supply became completely converted within the catalyst bed.

It can be clearly seen in Fig. 8-16 that the methane conversion is practically constant. So, evidently, methane is not reacting in the absence of oxygen. Furthermore, selectivities with respect to the carbon oxides are hardly influenced by changing the postcatalytic space, which means that the water–gas shift reaction (water and hydrogen are present as reaction products) is rather slow compared to the cracking reactions. The most important reaction that takes place is the conversion of ethane into ethylene. This means that, in a commercial process, a large empty space after the catalyst bed (fixed or fluidized) can have the beneficial function of an ethane cracker that improves the ethylene yield and utilizes part of

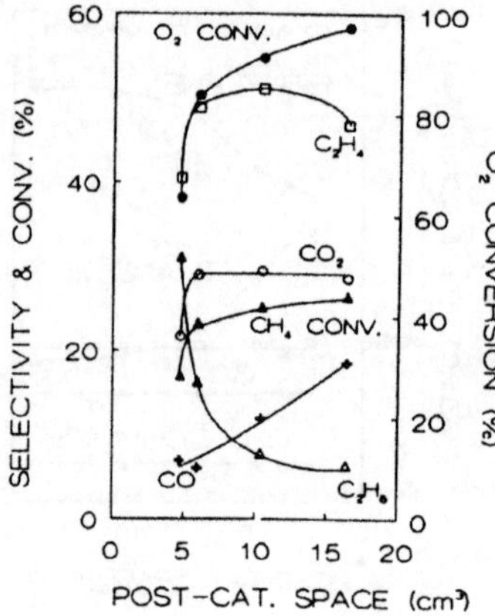

FIGURE 8-17. Selectivity and conversion as function of the postcatalytic space. Oxygen conversion $< 100\%$. $T = 770°C$, $CH_4/O_2 = 4$, $W/F = 0.6$ g s ml^{-1}.

the excess heat of the oxidation reactions. These results and ideas have been recently confirmed by Mimoun and co-workers (Mimoun et al. 1990). They added ethane in the postcatalytic space and showed that even this ethane is cracked to (mainly) ethylene in this zone.

Quite different results were obtained when similar experiments were carried out at conditions where the oxygen is only partly consumed in the catalyst bed. The remaining oxygen appears to react for a great deal with other molecules than methane, as shown in Fig. 8-17. As a result less ethylene and ethane and much more CO are produced.

Experiments with Products and Product–Methane Mixtures at Atmospheric Pressure

To elucidate the importance of various reaction steps in the oxidative conversion of methane, experiments were carried out with three reaction products: ethane, ethylene, and carbon monoxide. These products were studied separately in oxidation experiments with and without a catalyst. Moreover, the effect of admixing them to a methane—oxygen feed was investigated. All experiments were carried out in a microflow tubular quartz reactor that was either empty or filled with lithium-doped magnesium catalysts at a temperature of 800°C. The reactor setup and the preparation method of the Li/MgO catalyst are described elsewhere

(Geerts, van Kasteren, and van der Wiele 1988). Conditions similar to those used in methane oxidative coupling experiments were chosen.

Experiments with Single Feeds

First, experiments were carried out in an empty reactor. The results are shown in Figs. 8-18 and 8-19. It appears that ethane is oxidized much more easily than methane, which starts to be converted at a much higher temperature (a figure is not shown; the temperature difference is more than 150°C). The main product is ethylene. Ethylene is oxidized even faster than ethane, which results mainly in CO. The difference in oxidation rate between ethane and methane correlates with the difference in C—H bond strength in these molecules (435 ± 4 kJ mol^{-1}) for CH_4 and 410 ± 4 kJ mol^{-1} for C_2H_6 (Golden and Benson 1969) and the fact that hydrogen abstraction undoubtedly is the rate-determining step (Cant et al. 1988). The high reactivity of ethylene must be due to the fact that oxygen or oxygen-containing radicals attack the double bond. The fact that the conversion level of ethylene is lower than that of ethane does not reflect kinetics. The only reasons are that oxygen consumption is complete in both cases and that the oxidation of ethylene consumes much more oxygen per hydrocarbon molecule than methane oxidation does.

Next, similar experiments were carried out in a reactor filled with the lithium/magnesium oxide catalyst. The results are surprisingly similar to

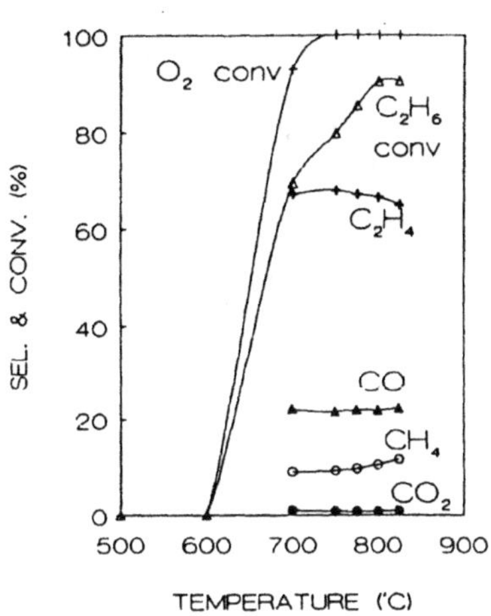

FIGURE 8-18. Homogeneous gas-phase ethane oxidation: $C_2H_6/O_2 = 2$, $He/C_2H_6 = 8.5$; $P_{tot} = 1$ bar.

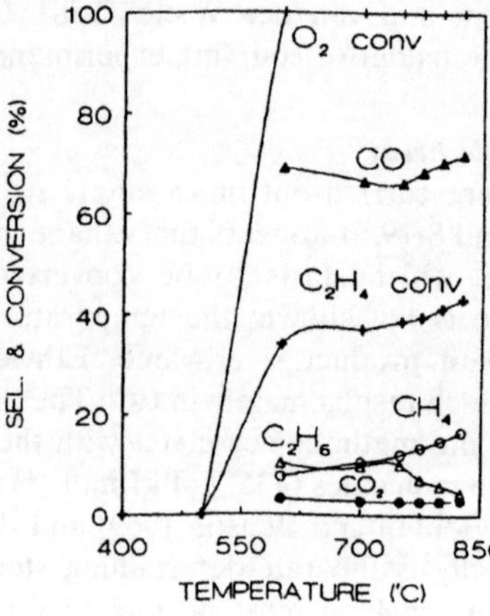

FIGURE 8-19. Homogeneous gas-phase ethylene oxidation: $C_2H_4/O_2 = 2.2$, $He/C_2H_4 = 8.5$; $P_{tot} = 1$ bar.

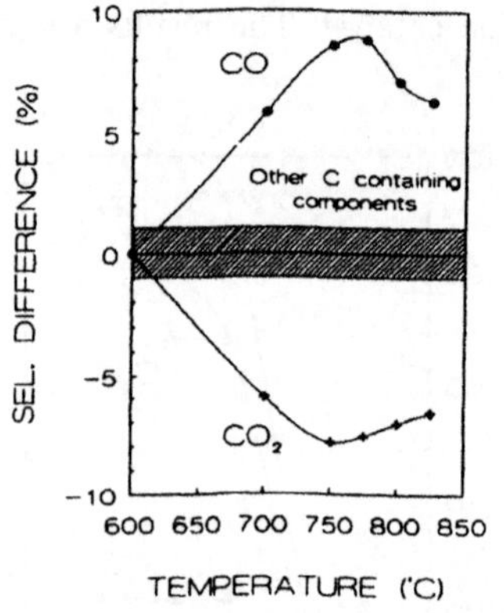

FIGURE 8-20. Differences in selectivities between gas-phase and catalytic ethane oxidation experiments: $C_2H_6/O_2 = 2.2$, $He/C_2H_6 = 8.5$; $W/F = 0.6$ g s cm^{-3}.

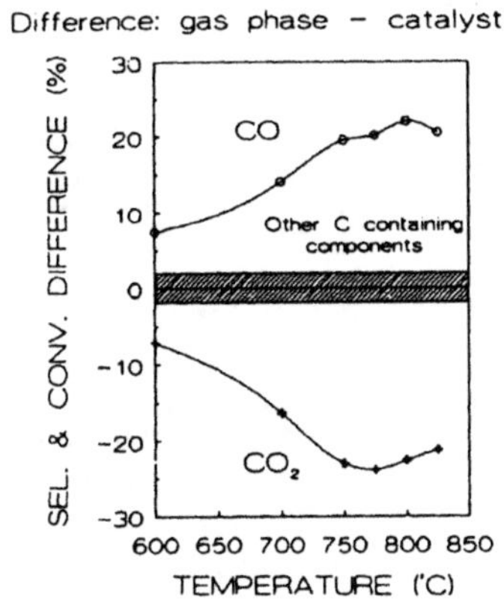

FIGURE 8-21. Differences in conversions and selectivities between gas-phase and catalytic ethylene oxidation experiments: $C_2H_4/O_2 = 2.2$, $He/C_2H_4 = 8.5$; $W/F = 0.6$ g s cm^{-3}.

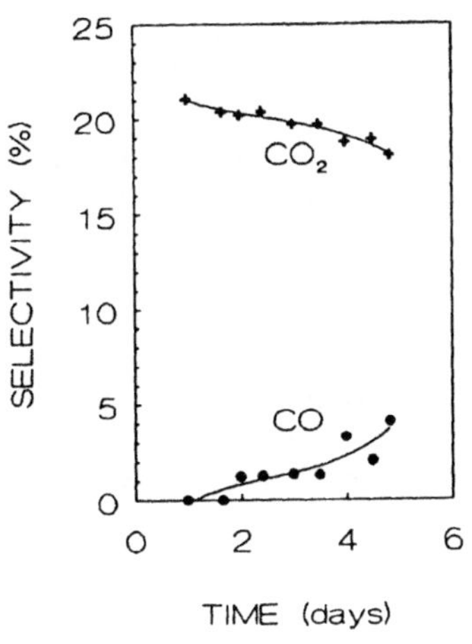

FIGURE 8-22. Catalyst deactivation. Time on stream for a Li/MgO catalyst: $T = 800$°C, $CH_4/O_2 = 10$; $W/F = 0.6$ g s cm^{-3}.

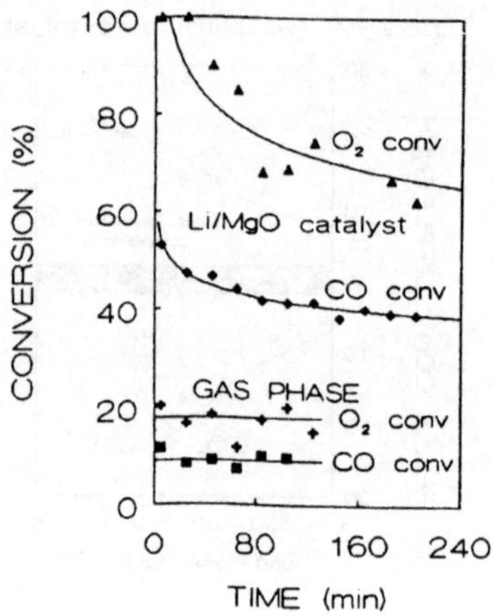

FIGURE 8-23. Gas-phase and catalytic oxidation of CO: T = 800°C, CO/O_2 = 4, CO/He = 0.06, W/F = 0.3 g s cm^{-3}.

those just described, except for a remarkable shift in the selectivities with respect to the carbon oxides, as shown in Figs. 8-20 and 8-21. Apparently, the oxidation of CO to CO_2 is strongly accelerated by the catalyst. Additional evidence stems from a similar shift that was observed during the deactivation of the catalyst in the methane oxidation experiments (Fig. 8-22).

Direct proof of the activity of the catalyst for CO oxidation is produced by separate CO oxidation experiments, as shown in Fig. 8-23. Initially, very high rates are achieved with a catalyst. Significant deactivation occurs, which may be explained by poisoning of the catalyst by CO_2, as proven by Korf and co-workers (Korf et al. 1987). However, even after catalyst deactivation, the conversion level of CO is an order of magnitude larger than in the absence of the catalyst.

Experiments with Mixed Feeds

Although the oxidation of ethane and ethylene separately fed to the reactor gives valuable information, cofeeding with methane is even more interesting. Ethane is present in natural gas, and moreover, recycling of the ethane is of interest to improve the ethylene yield. Cofeeding of ethylene is particularly useful to obtain more reliable data on its stability (its relative reactivity with respect to methane) under actual reaction conditions. The results of ethane and ethylene cofeeding (Figs. 8-24 and 8-25) show remarkable similarities. All the corresponding curves in these

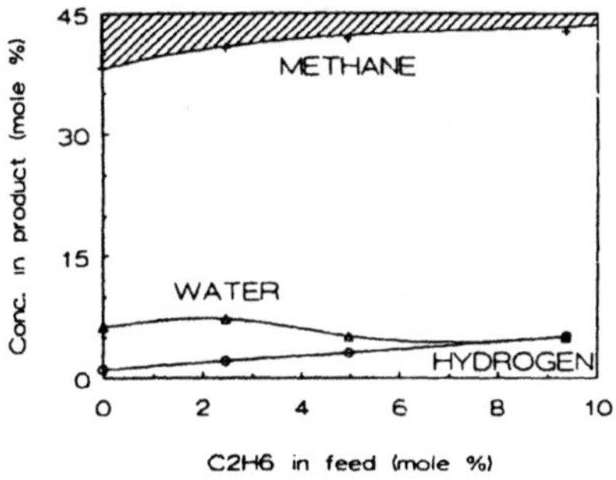

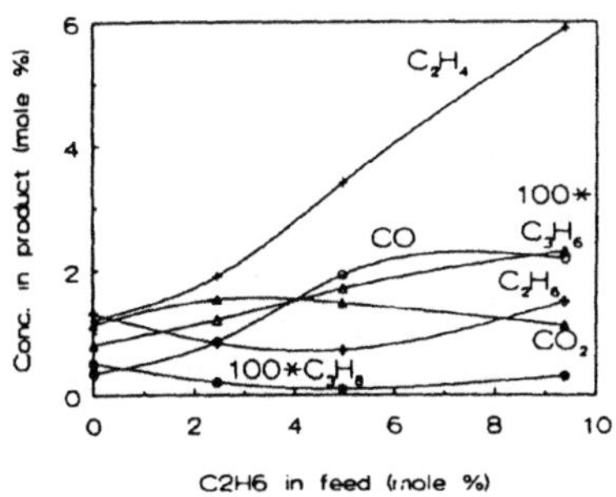

FIGURE 8-24. Admixing of C_2H_6 to a CH_4–O_2–He mixture, catalyzed by Li/MgO: $T = 800°C$, $CH_4/O_2 = 10$, $W/F = 0.6$ g s cm^{-3}.

figures have practically the same shape. It seems to make no difference whether ethane or ethylene is fed to the methane–oxygen flow. Apparently, ethane is rapidly converted into ethylene, maybe even before methane oxidation starts.

The top diagrams in Figs. 8-24 and 8-25 show the actual concentrations of methane and hydrogen, corrected for the small changes in the total volume that occur. The feed gas contains 45 mol% of methane; the height of the shaded area therefore represents the methane conversion. It is clear that the addition of C_2s to the feed stream decreases the methane conversion, because part of the oxygen available is consumed by the C_2s added. The two extra hydrogen atoms of ethane, in comparison to ethylene, cause somewhat higher hydrogen and water concentrations at the outlet of the reactor than in the case of ethylene addition. Furthermore, it is shown that the more C_2 is added, the more carbon monoxide is

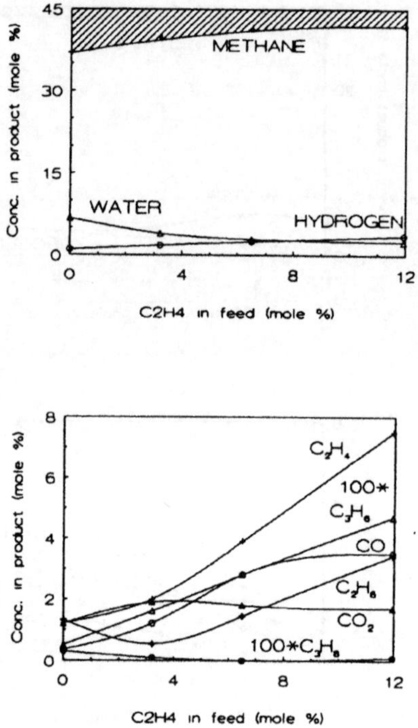

FIGURE 8-25. Addition of ethylene to a CH_4–O_2–He mixture over Li/MgO: $T = 800°C$, $CH_4/O_2 = 10$, $W/F = 0.6$ g s ml^{-1}.

produced, which is in agreement with the ethylene experiments (see Fig. 8-19). This CO is oxidized further to CO_2, the production of which also increases. However, at high added mole fractions, the oxygen is used by C_2s before CO can be oxidized, and this results in a diminished CO_2 production.

The C_3 concentrations are shown magnified a hundredfold in Figs. 8-24 and 8-25. No C_4 products could be detected. The C_3 production is enhanced by C_2 cofeeding, but the amounts formed remain relatively small. Methane oxidation thus appears to be a rather selective process for the production of ethylene, particularly if ethane is recycled. One should realize however, that the net C_2 production rapidly decreases when ethane is added. It is already zero at an added mole fraction ethane of 2% (see Fig. 8-24).

Low-Pressure Experiments

As stated in the introduction to this section, reduction of the pressure below 100 Pa (below 1 torr) will strongly affect most gas-phase reactions, probably more than heterogeneous reactions. Such experiments may therefore reveal more clearly the role of the catalyst versus that of the gas phase.

The experiments were carried out in a special low-pressure setup, described in detail by van Kasteren (van Kasteren 1990). Essential components of this equipment are an atmospheric feed gas vessel, a low-pressure section containing a quartz reactor with the catalyst, and a vacuum connection to a mass spectrometer for continuous on-line analysis. The reactor could be operated both in continuous flow and in batch mode. The latter was preferred in most experiments.

Figures 8-26 and 8-27 show the methane batch oxidation with and without the presence of the Li/MgO catalyst. Without catalyst, almost no conversion can be detected during a 1-h reaction at 800°C. With catalyst, clearly methane activation is occurring. Surprisingly, the C_{2+} product fraction is very low and the dominant products are the total oxidation products CO and CO_2. This means that coupling products are not formed in a first step or rapidly converted into total oxidation products at the catalyst surface. It is more likely, however, that at the low pressure applied the methyl radicals formed by hydrogen abstraction of methane can only react at the catalyst surface and cannot combine in the gas phase. For

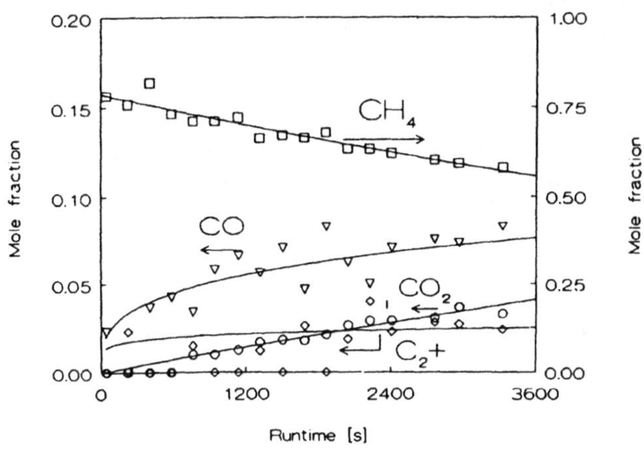

FIGURE 8-26. Methane batch oxidation over 0.25 g Li/MgO: T = 800°C, P = 66 Pa.

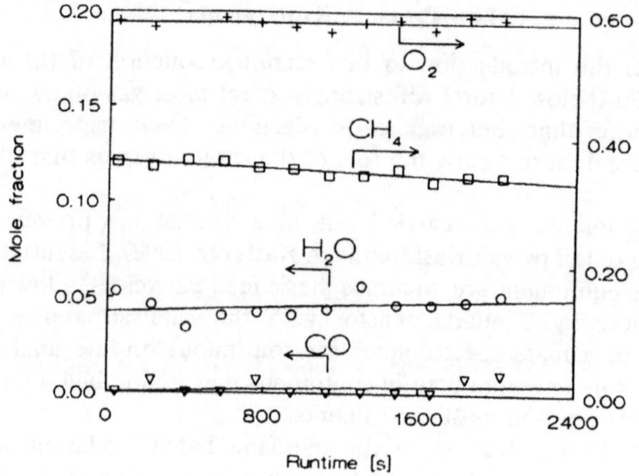

FIGURE 8-27. Methane batch oxidation without catalyst: $T = 800°C$, $P = 66$ Pa.

coupling they must loose their excess energy and this (at low pressure) is possible only at the catalyst surface. The chance of being oxidized at the catalyst surface is greater than coupling with another methyl radical. This theory is confirmed by atmospheric experiments with varying diluent gas concentration (see Fig 8-25). High concentrations of inert gas reduce the oxidation of the radicals in the gas phase and improve the radical coupling efficiency. This means that the homogeneous radical gas-phase reactions play a very important role as to the C_{2+} selectivity that can be reached. This also lays constraints on the comparison of different catalysts. A different diluent concentration can mean a large difference in the C_{2+} selectivity.

The reactivity of ethane and ethylene was investigated in separate experiments with an ethane–oxygen and an ethylene–oxygen feed, respectively. Figures 8-28 and 8-29 show the ethane oxidation with and without a catalyst as a function of the temperature for a C_2H_6–O_2–Ne (5 : 5 : 1) feed gas mixture. Gas-phase oxidation does not occur, but with a Li/MgO catalyst oxidation sets in between 600°C and 800°C. Besides total oxidation products, ethylene is formed. It can be concluded that the Li/MgO catalyzes the conversion of ethane into ethylene. These experiments clearly prove that ethylene also is produced heterogeneously from ethane over Li/MgO. The same experiment carried out with an ethylene–oxygen mixture showed that ethylene is catalytically oxidized to CO_x (Fig. 8-30).

Figure 8-31 shows the product distribution of an ethane oxygen batch experiment over Li/MgO. Clearly, the fraction of ethylene is coming up

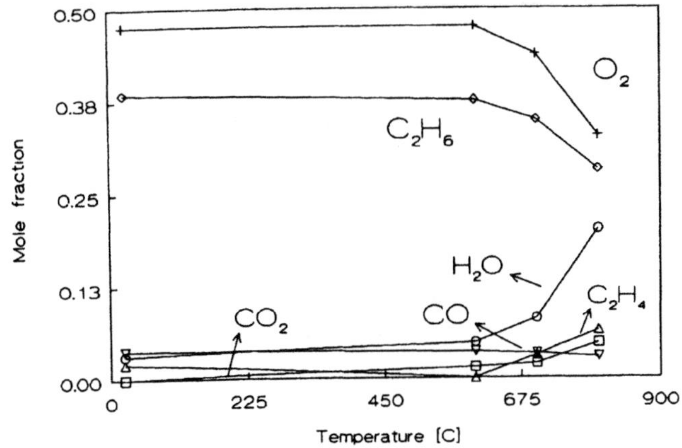

FIGURE 8-28. Ethane oxidation as function of the temperature: $P = 33$ Pa, $W = 0.5$ g Li/MgO, $F = 10^{-1}$ mol s^{-1}.

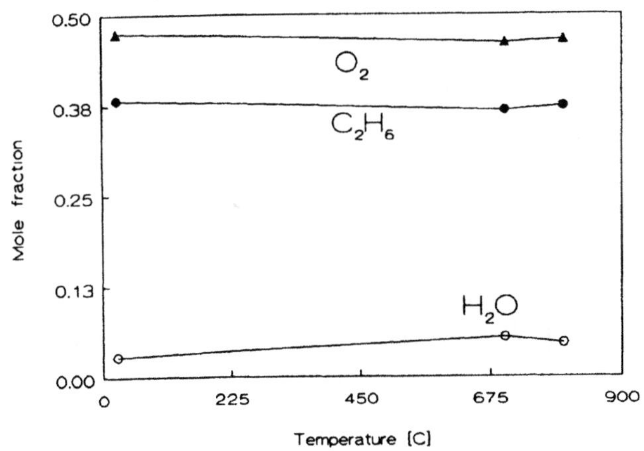

FIGURE 8-29. Ethane gas-phase oxidation as a function of the temperature: $P = 33$ Pa, $F = 10^{-7}$ mol s^{-1}.

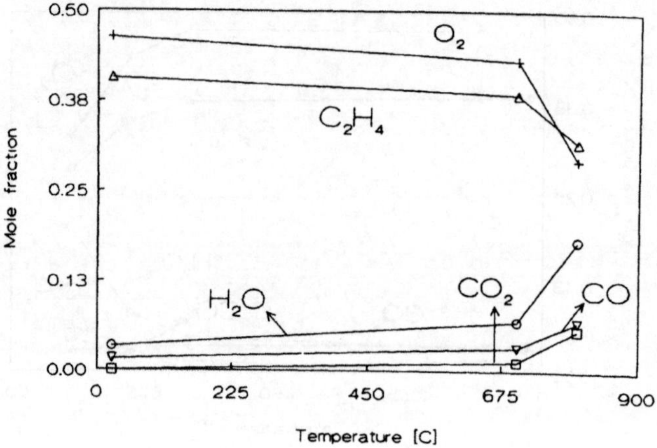

FIGURE 8-30. Ethene oxidation over Li/MgO: P = 33 Pa, W = 1 g, $F = 10^{-7}$ mol s^{-1}.

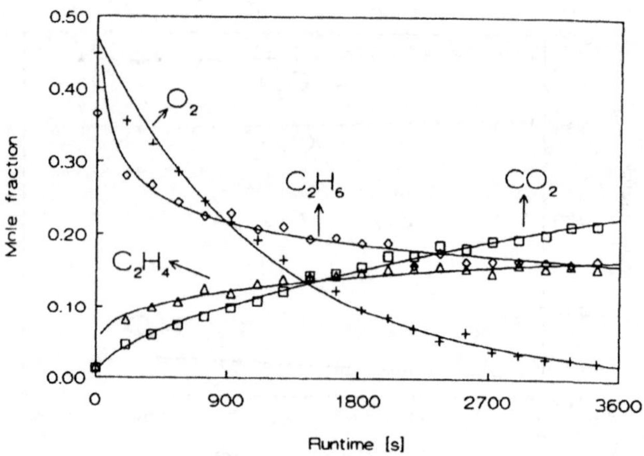

FIGURE 8-31. Ethane batch oxidation over Li/MgO at 66 Pa: T = 800°C, W = 0.25 g.

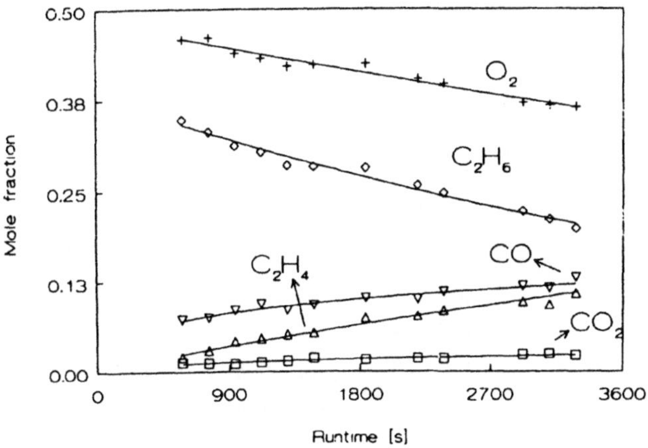

FIGURE 8-32. Ethane gas-phase batch oxidation at 66 Pa: $T = 800°C$.

first followed by CO_2. CO is hardly occurring, indicating that CO is relatively rapidly converted to CO_2. This last reaction is definitely catalyzed by the Li/MgO catalyst, because the same experiment without catalyst shows that CO and C_2H_4 are the main products (Fig. 8-32). This gas-phase batch experiment also shows that the reactions take place at a very much lower rate than in the presence of a catalyst.

These experiments have shown that the catalyst activates methane (with low coupling efficiency at very low pressure), catalyzes the ethane to ethylene conversion, and also accelerates the ethylene oxidation. Because the ratio of the rates of these reactions is essential for the C_2 selectivity that can be achieved, this was further investigated and quantified by using mixtures of CH_4–C_2H_6–O_2 and CH_4–C_2H_4–O_2.

Figures 8-33 and 8-34 show CH_4–C_2H_6–O_2 and CH_4–C_2H_4–O_2 admixing experiments carried out in the batch mode of operation at 66 Pa in the low-pressure reactor setup. As expected, ethane and ethylene react faster than methane. To make a quantitative comparison, rate constants were calculated using simple power-law rate equations of the form

$$r_i = k_i p_i p_{O_2}$$

in which i stands for CH_4, C_2H_6, or C_2H_4. Eley–Rideal type of equations are more appropriate (see below), but at these low pressures they degenerate to the simple form used here. Thus, CH_4, C_2H_6, or C_2H_4 is assumed to react independently, except for the competition for the same active oxygen on the catalyst. The amount of C_2 formed from methane in the

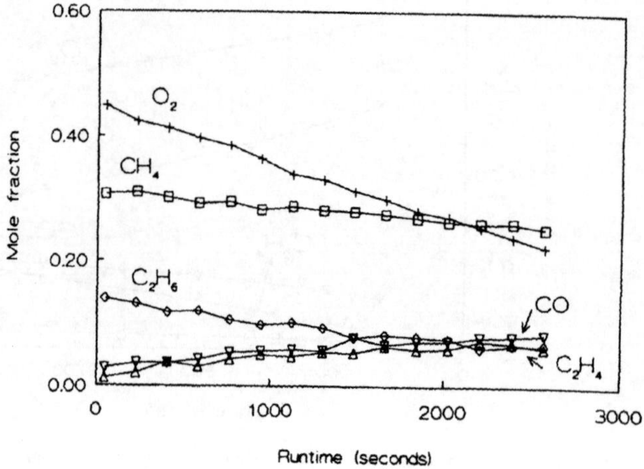

FIGURE 8-33. Admixing experiment over 0.25 g Li/MgO: T = 800°C, P = 66 Pa, CH_4–C_2H_6–O_2 mixture.

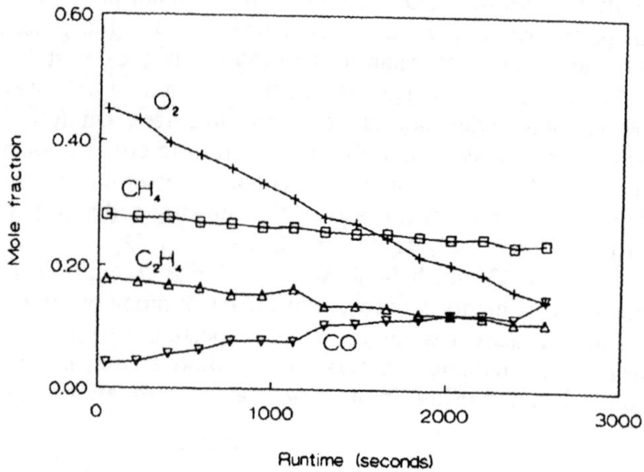

FIGURE 8-34. Admixing experiment over 0.25 g Li/MgO: T = 800°C, P = 66 Pa, CH_4–C_2H_4–O_2 mixture.

Table 8-3 ***k* Values[a] Based on Power-Law Rate Equations; Consecutive Model Applied to Methane Oxidation over 0.25 g Li / MgO at 66 Pa and 800°C**

k_{CH_4}	$k_{C_2H_6}$	$k_{C_2H_4}$	$k_{C_2H_6}/k_{CH_4}$	$k_{C_2H_4}/k_{CH_4}$
0.16	0.64	0.42	4	2.5

[a] In moles per kilogram of catalyst per second per square bar.

experiments depicted in Figs. 8-33 and 8-34 can be assumed to be negligible and it is not accounted for. Table 8-3 shows the results of the calculations. Comparison of these k values clearly shows that ethane and ethylene are activated at least a factor 2 more easily than methane. This is in agreement with the higher stability of the methane molecule compared to the C_{2+} hydrocarbons.

The results also make clear that ethane is converted a factor 1.5 faster than ethylene. This is in agreement with the results of admixing experiments carried out at atmospheric pressure (Geerts, van Kasteren, and van der Wiele 1988), although the difference is much more pronounced in that case. At atmospheric pressure it hardly makes any difference whether ethane or ethylene is fed together with methane and oxygen, the product distribution is practically the same. Apparently, gas-phase reactions contribute significantly in the ethane to ethylene reaction step.

The relatively high rate of ethylene oxidation with respect to the methane conversion restricts the maximum C_2 yield that can be reached with Li/MgO. In fact, the factor 2.5 for the ethylene oxidation is a lower limit based on catalytic reactions alone. At atmospheric pressure this factor is substantially higher due to the contribution of gas-phase reactions and accordingly the attainable yields are then lower.

Summary and Conclusions

The application of a catalyst like Li/MgO strongly accelerates the methane oxidative coupling and strongly improves the selectivity and the yield of the C_2 products: ethane and ethylene. Within the range of conditions used here the best conditions are the following:

1. a high methane–oxygen ratio (5–10);
2. relatively high temperature (800°C or higher);
3. a relatively low pressure (1 bar).

Furthermore, the precatalytic gas volume in the reactor should be minimized to avoid the start of a less-selective gas-phase process. On the other hand, the postcatalytic space can be utilized to improve the ethylene–ethane ratio by thermal cracking of ethane, provided that the oxygen is fully consumed within the catalyst bed.

The maximum C_2 yield achieved in our experiments is about 20%, corresponding with a methane selectivity of about 80% and a conversion level of about 25%. The large effect of the temperature on the reaction rate reflects a high activation energy, undoubtedly related to hydrogen abstraction from methane as the rate-determining step.

Clear evidence as to the essential reaction steps and the main reaction pathway in the oxidative coupling process has been provided by experiments with intermediates, products and mixtures thereof with methane, both at normal conditions in flow experiments and at strongly reduced pressure (< 100 Pa) in batch experiments.

The following picture emerges for the Li/MgO-catalyzed methane oxidative coupling:

CH_4

$k_{CH_4}\downarrow$ Catalytic activation–hydrogen abstraction

$CH_3\cdot$

$\downarrow$ Rapid gas-phase coupling

C_2H_6

$k_{C_2H_6}\downarrow$ Catalytic activation and activation by various radicals

$C_2H_5\cdot$

$\downarrow$ Rapid gas-phase reactions

C_2H_4

$k_{C_2H_4}\downarrow$ Catalytic activation and activation by various radicals followed by rapid consecutive gas-phase reactions

CO

$k_{CO}\downarrow$ Mainly catalytic oxidation

CO_2

- The first step is the formation of methyl radicals, definitely catalytic, and rate-limiting for the formation of ethane and ethylene.

- The second step is the relatively rapid coupling of these radicals to ethane. This is the principal reaction involving these radicals, apart from regeneration of methane. Conversion to oxygen-containing intermediates, eventually resulting in CO and CO_2, is of minor importance. This is proven by the fact that the initial selectivity to ethane approaches 100%. The coupling of methyl radicals is a gas-phase process. This follows from the fact that coupling hardly occurs at very low (total) pressure, whereas dilution of the feed with helium (at atmospheric pressure) generally improves the coupling selectivity.
- The third step is the conversion of ethane via ethyl radicals to ethylene. This process is highly selective, as is demonstrated by high initial selectivity obtained with ethane–oxygen feeds both in the absence and in the presence of a catalyst. The catalyst is definitely accelerating this process, although the "catalytic contribution" may vary according to the activity of the catalyst. Competition is clearly occurring between methane and ethane conversion. Several processes are important here: competition for the same active sites on the catalyst, competitive reaction in the gas phase (e.g., hydrogen transfer to OH radicals), and hydrogen transfer between methane, ethane, and radicals thereof. Experiments with methane–ethane mixtures clearly prove that ethane is converted 4–50 times more rapidly than methane, depending on the conditions. Again, direct conversion of ethane or ethyl radicals to oxygen-containing products and next to CO and CO_2 is insignificant.
- The next step is the conversion of ethylene to CO and CO_2. Like the conversion of ethane, the catalyst definitely accelerates this process, parallel to a variety of gas-phase reactions. Although ethylene is not converted as rapid as ethane, it is unfortunately reacting more rapidly than methane and this is the principal problem of reaching high C_2 selectivities. Even in the extremely low pressure case, where gas-phase reactions can be assumed insignificant, ethylene proved to react 2.5 times faster than methane. The ethylene conversion, whether starting at the catalyst surface or not, involves a large series of rapid consecutive gas-phase reactions and causes a "shower of radicals," the effect of which is difficult to describe but eventually leads to CO and H_2O as main products.
- The conversion of CO to CO_2, finally, is predominantly a catalytic step, particularly at the higher temperature levels. This is proven by the very pronounced effect of the presence of a catalyst and its activity on the CO_2–CO ratio.
- The production of C_3 compounds is insignificant, in agreement with accepted rates of gas-phase reactions between methyl radicals and reactive C_2 species.

In summary, the oxidative coupling of methane over Li/MgO at 800°C is a very complex process involving many catalytic and gas-phase reaction steps. The nature of the most essential reaction steps has been identified and it has been proven that the main pathway successively leads from methane via ethane to ethylene and from there to CO and CO_2. The Li/MgO catalyst is selective in the sense that "side reactions" leading to combustion products are negligible, but the selectivity is limited by the unavoidable (and rather rapid) conversion of the desired product ethylene to CO and CO_2.

KINETIC MODELING

Introduction

A kinetic model of the gas-phase oxidative coupling is presented earlier in this chapter. That model is based on elementary gas-phase reactions with the rather simple kinetics of monomolecular or bimolecular reactions. The introduction of a catalyst complicates the kinetics enormously, first of all because usually the kinetics of catalytic steps are complex in themselves. A second problem is that some catalytic reaction steps are known, but many are not: The interaction between the large "active wall" area of the catalyst and the gas phase, in which numerous radical reactions occur, is difficult to describe and is also difficult to investigate experimentally.

To reveal the mechanism of the most important catalytic step (methyl radical formation) and hence to provide a basis for a kinetic description, isotopic labeling of oxygen and methane has been applied. From these results, a clear picture emerges of the participation of oxygen that has been become part of the oxide lattice and of the limited role of adsorbed oxygen and carbon-containing species.

Next, the kinetics of the same catalytic step are investigated by differential kinetic measurements in which the methane and the oxygen partial pressure are varied independently.

Finally, overall modeling of the methane oxidative coupling process is attempted by gross simplification of the process. First a simple consecutive model is applied, assuming that catalytic reactions are rate-determining and using kinetics derived in this section. Next the usefulness of introducing a catalytic step in the gas-phase model is investigated.

Steady-State Isotopic Transient Kinetic Analysis

In the steady-state isotopic transient kinetic analysis (SSITKA) technique, labeled reactants are replaced by the normal species without disturbing the steady state of the reaction (Biloen, Nelle, and van den Berg 1983; Happel, Walter, and Lecourtier 1986).

The experiments were carried out in a special microflow setup comprising a tubular microreactor and tubing with a minimal void volume and very accurate flow control and pressure equilibration between the flows to be switched. Under typical flow conditions, with a flow rate of 50–100 ml min^{-1}, the system responded as an almost ideal plug-flow system, equivalent to a cascade of 500 ideal mixers in series. Analysis of the products was performed by a quadrupole mass spectrometer analogous to that of Miro and co-workers (Miro et al. 1990).

To investigate the interaction of both oxygen and carbon with the Li/MgO catalyst, two types of isotopic switches were applied: from CH_4–$^{18}O_2$–He to CH_4–$^{16}O_2$–He and from $^{13}CH_4$–O_2–He to $^{12}CH_4$–O_2–He. The catalyst preparation for these experiments is described elsewhere (Geerts, van Kasteren, and van der Wiele 1988).

Oxygen Transients

Figure 8-35 shows the oxygen response as function of the time after a switch from a CH_4–$^{16}O_2$–He to a CH_4–$^{18}O_2$–He feed gas mixture. Clearly, it can be seen that the $^{18}O_2$ response takes at least 50 s before it reaches its steady-state level. This means that oxygen adsorption and/or exchange with the catalyst is occurring. Also the mixed oxygen form ($^{16}O\,^{18}O$) is observed, which means that the O=O bond is broken at the catalyst

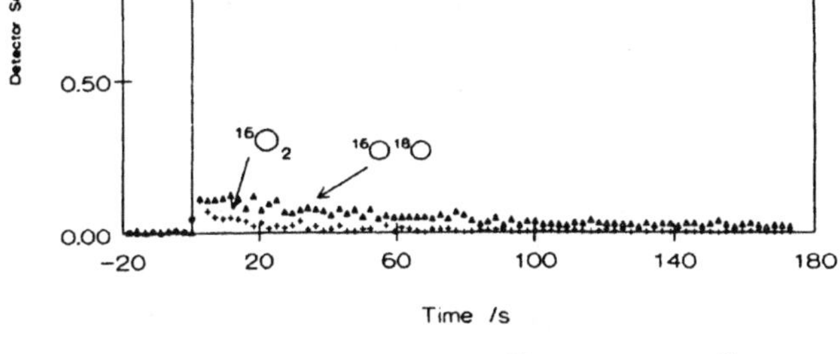

FIGURE 8-35. O_2 response after switch from CH_4–$^{16}O_2$–He to CH_4–$^{18}O_2$–He: $T = 800°C$, $CH_4/O_2 = 1.7$, $He/CH_4 = 9$, 0.25 g Li/MgO, $W/F = 0.15$ g s ml^{-1}.

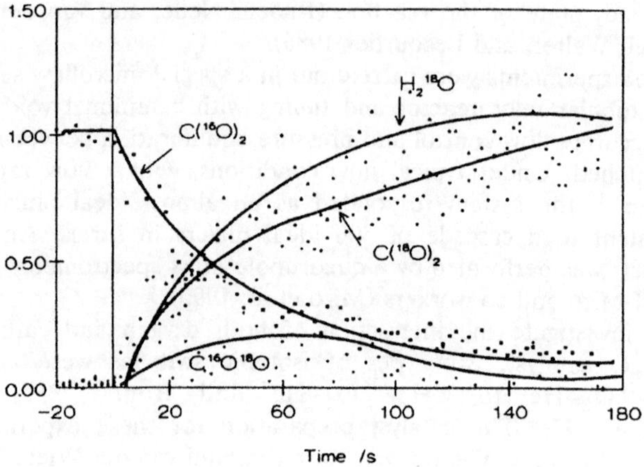

FIGURE 8-36. CO_2 and H_2O response after isotopic switch from CH_4–$^{16}O_2$–He to CH_4–$^{18}O_2$–He. Conditions identical to those of Fig. 8-35.

surface and that atomic oxygen is formed. During the course of the reaction oxygen is incorporated in water and carbon dioxide. The responses of these components were also monitored by scanning the mass numbers (m/e): 20 ($H_2{}^{18}O$), 44 ($C^{16}O_2$), 46 ($C^{16}O^{18}O$), and 48 ($C^{18}O_2$). Figure 8-36 shows the measured response curves, which result in a remarkable difference compared to the oxygen response. It takes at least 200 s for the $C^{18}O_2$ to reach its steady-state level. This can mean, for instance, that slow CO_2 desorption is occurring or that slow carbonate decomposition is taking place. The latter agrees with the idea that the decomposition of Li_2CO_3 is essential in the working principle of the Li/MgO catalyst.

Oxygen scrambling in CO_2 is occurring, because $C^{16}O^{18}O$ is clearly observed. The scrambling is not at the statistical equilibrium (the $C^{16}O^{18}O$ maximum should then be much higher). Peil and co-workers (Peil, Goodwin, and Marcelin 1989) carried out similar isotopic switching experiments with Li/MgO, but at lower reaction temperatures, where only total oxidation reactions prevail. They found oxygen scrambling in CO_2 to be at equilibrium. In our case this clearly is not the case, probably due to the higher reaction rates at the higher reaction temperature. At these temperatures carbon dioxide is formed at a much higher reaction rate and its contact time with the catalyst surface is lowered. The water ($H_2{}^{18}O$) response is also very slow compared to the oxygen response, which can

Table 8-4 Reaction Rates During Methane and after an Isotopic Coupling over Li / MgO before and after an Isotopic Switch from CH_4–$^{16}O_2$–He to CH_4–$^{18}O_2$–He

	RATE (10^{-4} mol kg-cat.$^{-1}$ s^{-1})	
Component	$^{16}O_2$	$^{18}O_2$
CH_4	10	10
C_2H_6	2.8	2.8
C_2H_4	0.5	0.5
CO_2	3	4

mean a slow water desorption from the catalyst surface or a slow rate of formation.

Table 8-4 shows the total rate of product formation before and after the isotopic switch confirming that no isotopic effects are occurring, which could alter the total reaction rate. Table 8-5 shows the total amount of ^{16}O atoms that is replaced by ^{18}O atoms. These numbers are calculated by integration of the response curves of the oxygen-containing products. The sum of these numbers is the total amount of replaced ^{16}O. This number can be related to the number of atomic layers of the catalyst that take part in the oxidation process. Assuming that the catalyst contains about 10^{19} atoms per square meter and combining this with the surface area (1 m^2 g^{-1}) and the catalyst amount in the reactor (0.25 g) it can be calculated that about 10 lattice layers of a pure MgO lattice are exchanged. This means that much more than a monolayer of oxygen is taking part in the oxidation process and thus lattice oxygen ("bulk") is participating in the oxidation process.

Table 8-5 Number of Exchangeable ^{16}O Atoms with Li / MgO Catalyst during Methane Coupling; T = 800°C

Component	Number of ^{16}O [10^{19} atoms (g-cat.)$^{-1}$]
$^{16}O_2$	3
$^{16}O^{18}O$	2.4
$C^{16}O_2$	1.6
$C^{16}O^{18}O$	0.25
$H_2{}^{16}O$	3.2
Total	10.5

Carbon Transients

The $^{13}CH_4$ switch experiments give information about the incorporation of ^{13}C into the catalyst and into the coupling products (C_2H_6 and C_2H_4). Figure 8-37 shows the CH_4 response after the isotopic switch. Also, a trace of argon was added to the feed to determine whether methane adsorption was occurring. As expected, no measurable methane adsorption was occurring, because the methane response is identical to the argon response (not adsorbing). This effect confirms our idea that methane reacts from the gas phase with adsorbed oxygen by an Eley–Rideal mechanism. Methyl radicals are released into the gas phase and couple there to form ethane. This is confirmed by the monitoring of mass (m/e) 26, which results mainly from $^{12}C_2H_4$. The $^{12}C_2H_4$ signal (not shown) appears exactly parallel to the $^{12}CH_4$ signal. This also means that the coupling products are not adsorbed at the catalyst surface. It is further interesting to examine the carbon dioxide response, which is monitored via mass (m/e) 44 ($^{12}CO_2$) and mass 45 ($^{13}CO_2$). Figure 8-38 shows the CO_2 response compared to the argon trace. The responses do not differ significantly indicating that CO_2 is hardly adsorbed at 800°C. The $^{13}CO_2$ signal reaches its steady-state level within a few seconds, indicating that the carbon atom is not incorporated into the lattice of the catalyst to a measurable extent. This effect proves that the slow $C^{18}O_2$ response (200 s) shown previously is not due to slow CO_2 desorption. The slow formation of $C^{18}O_2$ can be explained by assuming that carbon dioxide is formed

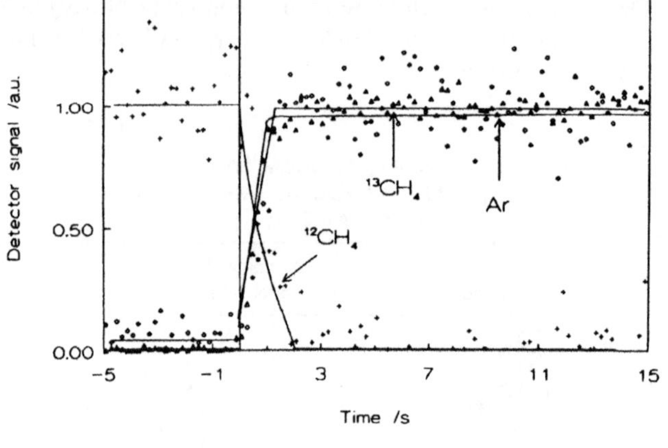

FIGURE 8-37. CH_4 response after isotopic switch from $^{12}CH_4$–O_2–He to $^{13}CH_4$–H_2–He. T = 800°C, CH_4/O_2 = 2, 0.25 g Li/MgO, W/F = 0.25 g s ml^{-1}.

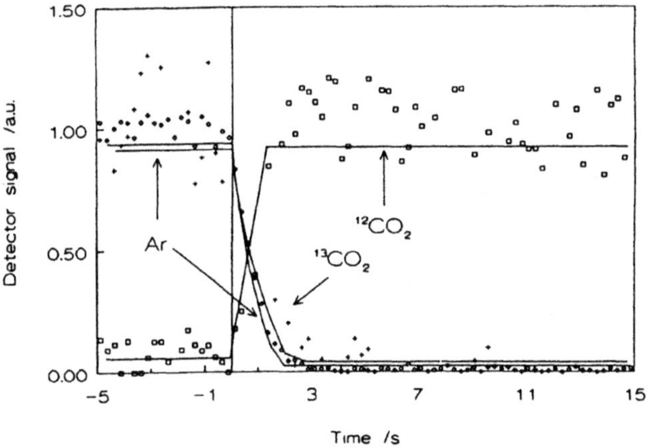

FIGURE 8-38. CO_2 response after isotopic switch from $^{13}CH_4$–O_2–He to $^{12}CH_4$–O_2–He: $T = 800°C$, $CH_4/O_2 = 2$, W = 0.25 g Li/MgO, W/F = 0.25 g s ml^{-1}.

mainly at the catalyst surface by reaction with a reactive oxygen species. This oxygen species is more reactive than the gas-phase oxygen. Before $C^{18}O_2$ can be formed, the $^{18}O_2$ must be adsorbed at the catalyst surface and the reactive oxygen species must be formed. The slow $C^{18}O_2$ response thus can be brought back to the rate of refreshment of the active oxygen species, which is present in the Li/MgO catalyst. This phenomenon resembles the work of Van Santen and De Groot (Van Santen and De Groot 1986), who investigated the mechanism of the ethylene epoxidation. They found that ethylene initially reacted more rapidly with pre-adsorbed oxygen atoms than with molecular oxygen adsorbed in the precursor state. An analogous situation seems to be the case during oxidative coupling of methane over Li/MgO.

These experiments prove that there is no large pool of a carbon-containing species (like Li_2CO_3) taking part in the catalytic cycle during steady-state operation. This rejects the theory of Korf and co-workers (Korf et al. 1987) that the active centers are involved in a continuous process of Li_2CO_3 decomposition and formation. A more likely theory is that the active centers are formed by reaction of an oxidic lithium phase, probably stabilized in the MgO matrix.

Implications with Respect to the Mechanism

Figure 8-39 gives a possible oxygen pathway that explains the observed experimental results. The oxygen is adsorbed at the catalyst surface and at

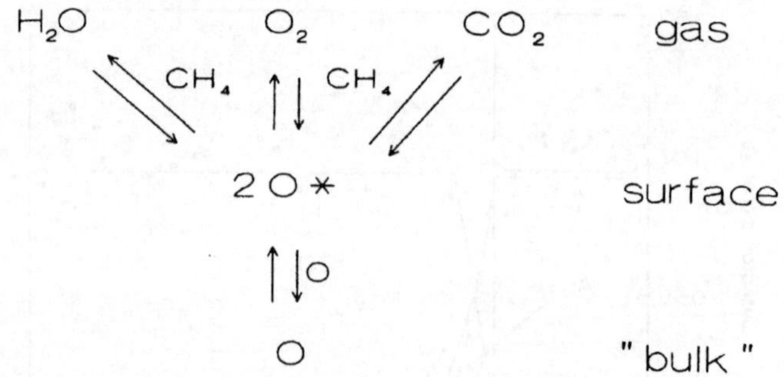

FIGURE 8-39. Oxygen pathway during methane coupling over Li/MgO.

the same time partly incorporated into the catalyst lattice. CO_2 is formed from methane or methyl radicals colliding with the active oxygen at the catalyst surface. The CO_2 formed is not strongly bound to the surface, but is released rather quickly into the gas phase. The water that is formed during the methane activation is also desorbed rapidly into the gas phase. This is concluded from on–off switching experiments. These experiments show that the water response is identical to the methane response. It can be concluded that more than a monolayer of oxygen is taking part in the catalytic cycle.

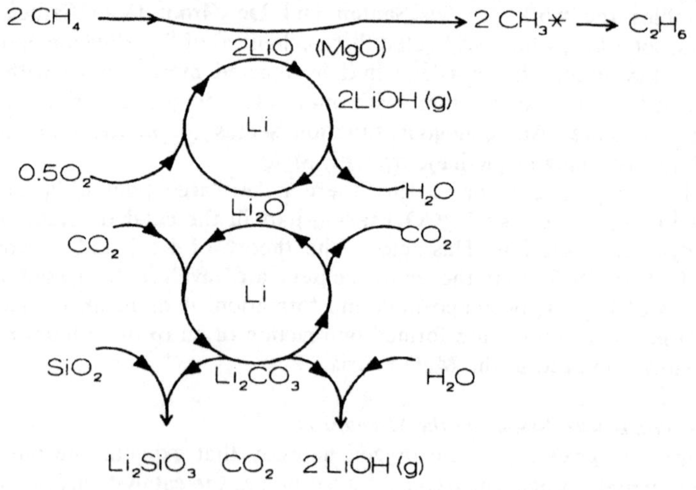

FIGURE 8-40. Working principle of Li/MgO catalyst during methane coupling.

Figure 8-40 shows the working principle of the Li/MgO catalyst during oxidative coupling of methane at 800°C. Active centers are created by uptake of oxygen by the Li/MgO catalyst. These centers react with methane from the gas phase to create methyl radicals, which combine in the gas phase to ethane. The Li/MgO catalyst possesses active centers that are very active in methane activation. The CO_2, which is formed during reaction, can react with the catalyst and form extra Li_2CO_3, although under a steady state of operation this reaction hardly takes place. The lithium carbonate in turn reacts with the water, also formed during reaction, and forms lithium hydroxide, which evaporates easily at 800°C. The lithium carbonate also reacts with the quartz reactor wall and forms lithium silicates, which are inactive in methane activation. In this way the catalyst looses its lithium gradually, which eventually results in deactivation of the catalyst.

Differential Kinetic Measurements

The experiments with isotopes have provided important facts with respect to the role of the catalyst in the reaction between methane and oxygen. It appears that oxygen is interacting so strongly with the catalyst that it can exchange with lattice oxygen at reaction conditions. Methane, on the other hand, hardly seems to adsorb, with the result that it seemingly reacts with the surface directly from the gas phase. The kinetics should reflect the preceding mechanism in the dependency of the reaction rate on the partial pressures of methane and oxygen.

Differential kinetic measurements are definitely the best way to determine these, provided that the catalyst is sufficiently stable and that transport limitations do not interfere. This was checked first. The equipment used was a microflow setup, similar to the one used in the normal flow experiments. However, a smaller tubular reactor was used, containing 0.25 g of catalyst with a particle size of 0.25–0.30 mm.

As to the stability of the catalyst, it turned out that temporary changes in the reaction conditions (variations in feed gas compositions) did not affect the catalytic activity. However, loss of lithium from the catalyst is a slow but continuing process that does cause deactivation. The kinetic experiments, therefore, were carried out after 40 h on stream, and per series of experiments within the shortest possible time range.

As to the possible interference of mass-transfer limitations, experiments and calculations were carried out. The nonoccurrence of internal diffusion limitations was demonstrated by experiments with varying catalyst particle size (0.1–0.8 mm) and was confirmed by calculations: There was no effect on conversion level or selectivity. In principle, external diffusion limitations should be absent when the pore diffusion limitation is negligible.

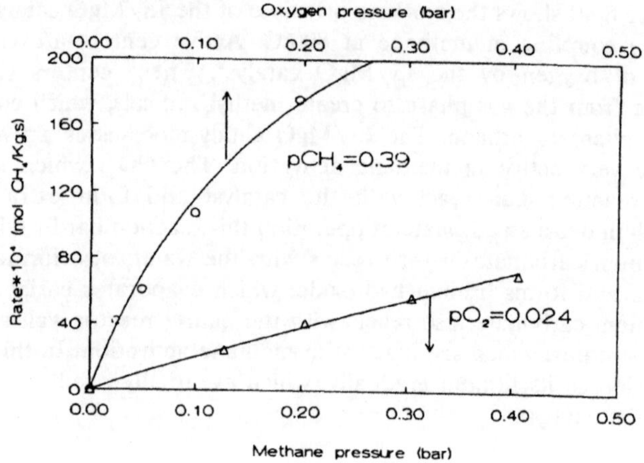

FIGURE 8-41. Methane disappearance rate during methane coupling over Li/MgO versus methane and oxygen pressure: $T = 800°C$, $W/F = 0.03$ g s ml^{-1}.

Experiments with a varying linear gas velocity at constant contact time (W/F) confirmed this.

Two series of kinetic experiments were carried out: one with varying oxygen partial pressure at constant methane, and one with varying methane and constant oxygen. The total pressure was kept at 1 bar by supplying helium.

Figure 8-41 shows the reaction rate (r_m) as a function of methane and oxygen partial pressure, respectively. Correlating this result with the simple power-law rate equation leads to

$$r_m = k_m[p_{CH_4}]^a[p_{O_2}]^b$$

where $a = 0.91 \pm 0.12$, $b = 0.62 \pm 0.13$, and $k_m = 0.145 \pm 0.015$ mol CH_4 (kg-cat. (bar)$^{a+b}$ s)$^{-1}$, with pressures in bars.

One can conclude from these data that the overall reaction rate is first-order in methane and half-order in oxygen.

The order of 0.5 with respect to oxygen can be explained as follows: Oxygen is rapidly and reversibly dissociating and reacting with anion vacancies on the Li/MgO surface according to

$$Li_2O^* + \tfrac{1}{2}O_2 \Leftrightarrow Li_2OO^*$$

The activated oxygen O* in turn reacts irreversibly with methane from the gas phase, to form surface OH and eventually water. If the fraction of the sites occupied by O* is small, the number will be proportional to the square root of the oxygen partial pressure. However, if the fraction is varying and tends to higher values, an Eley–Rideal type of kinetic equation is more appropriate. If the preceding reaction is considered irreversible, a redox type of equation results. Much more data would be required to discriminate between the kinetic models mentioned, and even then it is not sure that discrimination is possible because the assumptions underlying these models are seldom fulfilled by the reality of the chemical process.

The simplest way to describe our data quite accurately proves to be an Eley–Rideal type kinetic equation:

$$r_m = k_1[p_{CH_4}][p_{O_2}]/(1 + b_{O_2}[p_{O_2}])$$

where $k_1 = 0.45$ mol CH_4 (kg-cat. (bar)^2 s)$^{-1}$ and $b_{O_2} = 5$ (bar)$^{-1}$, with pressures in bars.

This equation reflects the strong interaction of oxygen with the catalyst, in agreement with results from the isotope experiments.

Modeling

Modeling the reaction in full detail is clearly impossible as yet. Two simplified ways of modeling the overall process will be treated here.

The first is a very simple consecutive model based on the idea that catalytic steps are rate-determining. The second is an extension of the kinetic model for the homogeneous reaction, by introducing catalytic steps.

The Consecutive Reaction Model

The consecutive reaction model is based on the simplified reaction scheme

$$CH_4 \rightarrow C_2H_6 \rightarrow C_2H_4 \rightarrow CO \rightarrow CO_2$$

and the assumption that each step is mainly determined by a catalytic reaction. The following kinetic equation, derived previously, was adopted:

$$r_i = k_i p_i p_{O_2}/(1 + b_{O_2} p_{O_2})$$

Thus the hydrocarbon molecules CH_4, C_2H_6 or C_2H_4 and CO are assumed to react from the gas phase with an adsorbed oxygen species. The contribution of parallel reactions in the gas phase is accounted for by adjusting the k values. It is therefore assumed that radical initiation reactions mainly take place at the catalyst surface. Consecutive reactions in the gas phase, leading to the next stable molecule in the series are considered infinitely fast. Thus the net production rates (R_i) of each of the components follow from the material balances:

$$R_{CH_4} = -r_{CH_4}$$

$$R_{C_2H_6} = 0.5r_{CH_4} - r_{C_2H_6}$$

$$R_{C_2H_4} = r_{C_2H_6} - r_{C_2H_4}$$

$$R_{CO} = 2r_{C_2H_4} - r_{CO}$$

$$R_{CO_2} = r_{CO}$$

Using the appropriate stoichiometry factors, R_{O_2} is easily formulated to complete these equations.

The k values were determined by fitting the model to the results of integral kinetic measurements collected in our laboratory. A nonlinear least squares estimation routine based on Marquardt's method (Marquardt 1963) was applied. The parameter b_{O_2} was fixed at a value of 5 (bar)$^{-1}$, as determined previously. Furthermore a distinction was made between data obtained with fresh catalyst samples and data from experiments with an aged catalyst. The results with respect to the k values are presented in Table 8-6.

The case of the aged catalyst is interesting as a reference, because one can assume that the catalyst is playing a minor role here. This is confirmed by the comparison of selectivity data of the aged catalyst with results of an empty reactor, both as a function of the oxygen conversion (Fig. 8-42). The results are indeed very similar, indicating that homogeneous reactions (gas-phase) predominate during the experiments with the aged catalyst. A reasonable fit of measured and calculated selectivities has been obtained both for the fresh as for the aged catalyst. Figure 8-43 shows the product selectivity versus W/F for the case of the aged catalyst (symbols) and the

Table 8-6 k Values[a] Resulting from Parameter Fitting of Integral Methane Bar; $b_{O_2} = 5$ (bar)$^{-1}$

Catalyst	k_{CH_4}	$k_{C_2H_6}$	$k_{C_2H_4}$	k_{CO}	$k_{C_2H_4}/k_{CH_4}$
Aged	0.65	37	12	1.6	19
Fresh	2.4	47	16	100	6.3

[a]In moles per kilogram of catalyst per second square bar.

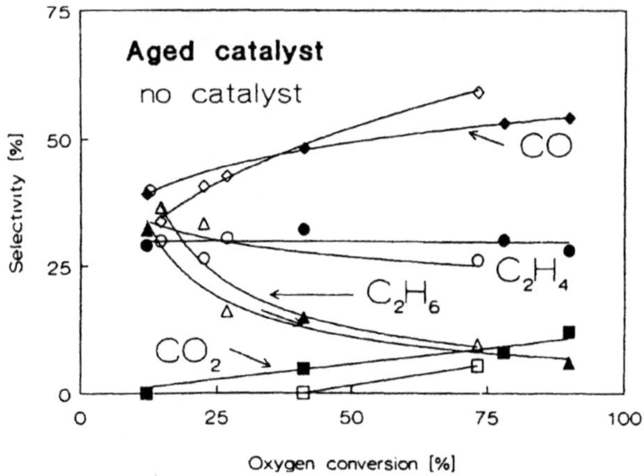

FIGURE 8-42. Product selectivity versus oxygen conversion for an aged Li/MgO catalyst and gas-phase methane coupling: $T = 800°C$, $CH_4/O_2 = 5$.

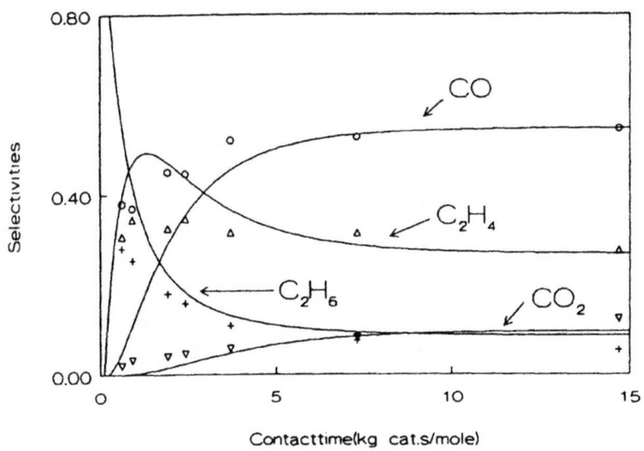

FIGURE 8-43. Product selectivity versus contact time (W/F) for experiment (symbols) and consecutive model fit: $T = 800°C$, $P = 1$ bar, $CH_4/O_2 = 5$, Li/MgO catalyst.

Table 8-7 Comparison between Experimental (Fresh Catalyst) and Consecutive Model Fitting Results

Selectivity (%)	Experiment	Fit
C_2H_6	20	20
C_2H_4	45	45
CO	7	6.7
CO_2	28	28.3
CH_4 conv. (%)	23	21
O_2 conv. (%)	90	100

calculated selectivities (lines). As can be seen, a reasonable fit has been achieved with the highest accuracy at the largest contact times and the poorest accuracy at the smallest contact times. This is not surprising, as the experimental error is the largest at the smallest W/F. Also, the fact must be considered that the model is no longer appropriate at very low conversion levels, because at very small contact times the direct oxidation of methane to carbon dioxide can no longer be neglected. For the case of the fresh catalyst it is very difficult to obtain an accurate selectivity pattern as function of the contact time, due to the fast deactivation of the Li/MgO catalyst. Therefore, no accurate selectivity fit versus W/F could be performed. On the other hand it is possible to fit the results of the optimal selectivities obtained at a certain W/F (0.3 g s ml^{-1}). This is shown in Table 8-7. By changing only the $k_{C_2H_4}/k_{CH_4}$ and the k_{CO} value, the measured selectivities of a methane oxidation experiment over a fresh Li/MgO catalyst can be fitted quite well.

Regarding the calculated k values in Table 8-6, some interesting conclusions can be drawn in connection with the low-pressure results of Table 8-3. The ratio $k_{C_2H_4}/k_{CH_4}$ is much higher for the atmospheric experiments (6.3, fresh catalyst) than for the low-pressure experiments (2.5). This increase in $k_{C_2H_4}/k_{CH_4}$ can be ascribed to the contribution of gas-phase reactions to the oxidation of ethylene at atmospheric pressure. The importance of the gas-phase reactions is confirmed by the data of the aged catalyst: Aging of the catalyst has much more effect on k_{CH_4} and k_{CO} ("catalytic steps") than on $k_{C_2H_6}$ and $k_{C_2H_4}$.

The ratio $k_{C_2H_4}/k_{CH_4}$ is the key factor in the selectivity of the methane oxidative coupling process, as demonstrated in Table 8-8 where the maximum yields are calculated for the three cases that correspond to the theoretical case (low-pressure data), the fresh catalyst, and the aged catalyst. A catalyst influences $k_{C_2H_4}/k_{CH_4}$ by increasing the methane activation rate more than the ethylene activation rate. The ethylene

Table 8-8 C_{2+} Yield as a Function of the $k_{C_2H_4}/k_{CH_4}$ Ratio for the Consecutive Model

$k_{C_2H_4}/k_{CH_4}$	C_{2+} Yield
2.6	35
6.3	14
19	6

activation rate is determined mainly by the gas-phase activation rate. Therefore, the high C_{2+} yields obtained by a catalyst are mainly due to the increase of the methane activation rate relative to the ethylene activation rate.

The Modified Gas-Phase Model

If the principal function of the catalyst is that of a radical initiator, particularly methyl radicals, one can simulate this by adapting the gas-phase model.

In Table 8-9 simulations and experiments are shown, for both catalyzed and homogeneous gas-phase reactions. The catalyst is simulated by a set of reactions that equals the overall reaction $4CH_4 + O_2 \rightarrow 4CH_3 + 2H_2O$. The same oxygen conversion level was chosen to be able to compare the selectivities and the catalytic $CH_3\cdot$ formation rate was adjusted to obtain similar selectivities.

Indeed, as shown in Table 8-9, the higher $CH_3\cdot$ radical formation rate results in a higher reaction rate and a higher ethane selectivity. The simulated reaction time, however, is much too short compared to the experimental value. Therefore, it is clear that the catalyst also acts as a radical sink. Radicals collide with the surface and react to less active

Table 8-9 Experiments and Simulations of both Catalytic and Homogeneous Gas-Phase Reactions; Reactor Made of Quartz

	HOMOGENEOUS		CATALYTIC	
	Exp.	Sim.	Exp.	Sim.
Time (s)	4.6	7.9	6.0	0.5
O_2 conv. (%)	28.2	28.2	92.8	92.8
CH_4 conv. (%)	4.1	4.2	13.4	20.2
C_{2+} sel. (%)	51.1	49.7	76.2	76.9
CO_x sel. (%)	48.8	50.0	23.8	22.8
$CH_3\cdot$ (mol m^{-3})	—	5.9E − 3		2.3E − 3

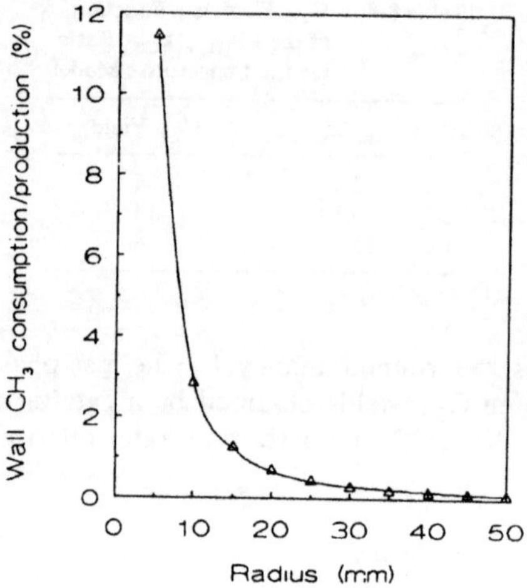

FIGURE 8-44. Fractional consumption of $CH_3\cdot$ by the wall versus pore radius. Only homogeneous gas-phase radical formation is considered.

molecules. This lowers the radical concentrations and slows down the whole process.

Iwamatsu and co-workers (Iwamatsu et al. 1987) have shown correlations of yield and specific surface area, for a large number of catalysts, that reveal the wall as an important parameter for the reaction. Tentative calculations have been performed to investigate the contribution of the pore wall to the consumption of $CH_3\cdot$ radicals formed in the homogeneous gas phase as a function of the pore radius. The basic assumptions are that the $CH_3\cdot$ production rate is only related to the homogeneous gas-phase volume, the reaction of $CH_3\cdot$ at the wall is infinitely fast (so, its concentration at the wall is zero and the process is diffusion-limited), and the diffusivity of $CH_3\cdot$ is equal to that of CH_4. Figure 8-44 clearly shows that even at pore diameters of the reactor scale the wall influence is noticeable. So, a large radical-trapping effect of the catalyst surface is indeed to be expected.

Summary and Conclusions

The kinetic modeling of the oxidative coupling process in full detail is still beyond reach. Although the homogeneous reactions in the absence of a catalyst are quite well understood and quantified as to their kinetics, there is still much uncertainty with respect to the precise role of the catalyst and the kinetic interaction with the gas-phase reactions. It has become clear, however, that a very important role of the catalyst is the conversion of methane, ethane, and ethylene into their respective radicals by abstraction of a hydrogen atom. Moreover, the catalyst strongly accelerates the conversion of carbon monoxide into carbon dioxide.

Combined evidence from experiments with isotopically labeled oxygen and methane and from differential kinetic measurements has proven that the catalyst operates as a redox system, in which oxygen is reacting with the catalyst, which in turn reacts with the hydrocarbon, abstracts hydrogen, and eventually releases water. Physically adsorbed (or chemisorbed) amounts of oxygen and hydrocarbon are negligibly small, as expected at this high temperature level.

The redox model explains the large variation in "apparent reaction orders" reported in the literature (Otsuka and Jinno 1986; Iwamatsu and Aika 1989; Hinsen, Bytyn, and Baerns 1984; Baerns, van der Wiele, and Ross 1990) and explains the strong influence of the temperature on these. The catalyst reoxidation has a much lower activation energy than the hydrogen abstraction from the hydrocarbon. Accordingly, a zero-order oxygen dependency is found at lower temperatures whereas the hydrocarbon reaction order tends to values below 1 at the higher temperatures.

Facing the inability to model the reaction kinetics in full detail, a very simple consecutive model has been adopted. This model is based on the idea that a highly active catalyst is kinetically controlling the essential reaction steps for the formation of ethane, ethylene, and carbon oxides. Formation of ethane is controlled by the catalyst because the formation of methyl radicals is the rate-determining step, followed by coupling and other gas-phase reactions. Conversion of ethane to ethylene and combustion of ethylene are similar hydrogen abstraction reactions in which the catalyst, if sufficiently active, is playing a dominating role, again being the rate-determining initiator of a series of radical reactions occurring in the gas phase.

The kinetic analysis performed with this simple model has shown that a fresh Li/MgO catalysts owes its superior selectivity to the high rate of methane activation relative to the combustion of ethylene. The less active the catalyst, the larger the contribution of gas-phase reactions in the key reaction steps. Because the influence of gas-phase reactions is much more pronounced for the attack on ethylene (with its double bond) than for methane dissociation, this means that a more active catalyst will always result in higher C_2 selectivities and yields.

The upper limit for C_2 yields and selectivities would thus be determined in particular by the ratio of catalytic reaction rates for hydrogen abstraction from methane, ethane, and ethylene. According to the results of the low-pressure experiments, this ratio is approximately 1 : 4 : 2.5, which means that an ultimate C_2 yield of 35% would be achievable with the ideal "super active" Li/MgO catalyst at 800°C. One should realize, however, that this is only true if ethane is quantitatively converted into ethylene and if decomposition of the C_2H_3 radical (eventually leading to carbon oxides) does not involve chains or branches involving ethylene molecules. This would have a multiplier effect on the catalytic hydrogen abstraction from ethylene and magnify the number 2.5.

Modeling of the oxidative coupling process by introducing "catalytic radical formation steps" into the network of gas-phase reactions has not been very successful yet, although various attempts have been made (McCarthy, McEwen, and Quinlan et al. 1990). Qualitatively, we have shown that the selectivity increases by "catalytic acceleration" of the methyl radical formation rate. However, unrealistically high overall conversion rates then result. Thus, one must conclude that the catalyst is not merely active in producing radicals, but also functions as a radical trap that retards essential gas-phase reactions.

The design of models that can cope adequately with the interaction of reactions in the gas phase and at the catalyst surface and the collection of relevant data for individual steps involving the catalyst is an extremely difficult task, but a great challenge for further understanding of high temperature, catalyzed reactions like methane oxidative coupling.

ACKNOWLEDGMENT

The financial support for this research, which was provided in part by the European Communities, under contract no. EN3C-0038-NL, and by the Netherlands Organization for Scientific Research (NWO) is gratefully acknowledged.

References

Allbright, L.F., B.L. Crynes, and W.H. Corcoran. 1983. *Pyrolysis, Theory and Industrial Practice*. New York: Academic Press.

Baulch, D.L., D.D. Drysdale, D.G. Horne, and A.C. Lloyd. 1972. *Elevated Kinetic Data for High Temperature Reactions*. London: Butterworths.

Baerns, M., K. van der Wiele, and J.R.H. Ross. 1990. Proceedings second European workshop on methane activation, 22–23 May 1989, Enschede, The Netherlands. *Catal. Today* 6(4):373–613.

Biloen, P., J.N. Nelle, and F.G.A. van den Berg. 1983. *J. Catal.* 81:450–63.

Cant, N.W., C.A. Lukey, P.F. Nelson, and R.J. Taylor. 1988. The rate determining step in the oxidative coupling of methane over a lithium-promoted magnesium oxide catalyst. *J. Chem. Soc., Chem. Commun.* 766.

Dagaut, P., M. Cathonnet, and J.C. Boettner, 1988. Experimental study and kinetic modelling of propane oxidation in a jet stirred flow reactor. *J. Phys. Chem.* 661–71.

Elbers, J. 1989. Personal communication. AKZO Salt & Basic Chemistry, Hengelo, The Netherlands.

Forlani, O., M. Lupieri, V. Picoli, S. Rossini, D. Sanfelippo, J.A. Dumesic, L.A. Aparicio, J.A. Rekoske, and A.A. Trevino. 1990. *Stud. Surf. Sci. Catal.* 55:343.

Gardiner, W.C. 1984. *Combustion Chemistry*, pp. 197–360. New York: Springer-Verlag.

Geerts, J.W.M.H. 1990. Ethylene synthesis by direct partial oxidation of methane. Ph.D. Thesis, Eindhoven University of Technology, Eindhoven, The Netherlands.

Geerts, J.W.M.H., Q. Chen, J.M.N. van Kasteren, and K. van der Wiele. 1990. Thermodynamics and kinetic modelling of the homogeneous gas phase reactions of the oxidative coupling of methane. *Catal. Today* 6:519–26.

Geerts. J.W.M.H., J.M.N. van Kasteren, and K. van der Wiele. 1988. A mechanistic study on the oxidative coupling of methane over lithium doped magnesium oxide catalysts, Proceedings of the EC. Congress: "Hydrocarbons: Source of energy," ed. G. Imarisio, M. Frias, and J.M. Bemtgen, Lyon, France, Sept. 1988, 434–40.

Gesser, H.D., N.R. Hunter, and C.B. Prakash. 1985. The direct conversion of methane to methanol by controlled oxidation. *Chem. Rev.* 85(4):235–44.

Golden, D.M., and S.W. Benson. 1969. *Chem. Rev.* 69:125.

Happel, J., E. Walter, and Y. Lecourtier. 1990. Modeling transient tracer studies in plug flow reactors. *J. Catal.* 123:12–20.

Hinsen, W., and M. Baerns. 1983. Oxidatif Kupplung von Methan zu C_2-Kohlenwasserstoffen in Gegenwart unterschiedlicher Katalysatoren. *Chem.-Ztg.* 107(7/8):223–6.

Hinsen, W., W. Bytyn, and M. Baerns. 1984. Oxidative dehydrogenation and coupling of methane. In *Proceedings of the 8th International Congress on Catalysis (ICC), Berlin*, Vol. 3, pp. 581–91.

Ito, T., J.-X. Wang, C.H. Lin, and J.H. Lunsford. 1985. Oxidative coupling of methane over a lithium-promoted magnesium oxide catalyst. *J. Am. Chem. Soc.* 107:5062–8.

Iwamatsu, E., and K. Aika. 1989. Kinetic analysis of the oxidative coupling of methane over Na^+-doped MgO. *J. Catal.* 117:416–31.

Iwamatsu, E., T. Moriyama, N. Takasaki, and K. Aika. 1988. Oxidative coupling of methane over Na^+- and Rb^+-doped MgO catalysts. *J. Catal.* 113:25–35.

Jones, L.A., J.J. Leonard, and J.A. Sofranko. 1984. US Patent, nos. 4443644 to 4443649 and no. 4444984.

Keller, G.E., and M.M. Bhasin. 1982. Synthesis of ethylene via oxidative coupling of methane. *J. Catal.* 73:9–19.

Kolts, J.H., and J.H. Lunsford. 1986. European Patent 0196541.

Korf, S.J. 1990. Catalysts for the oxidative coupling of methane, Ph.D. Thesis, University of Twente, Enschede, The Netherlands.

Korf, S.J., J.A. Roos, N.A. De Bruin, J.G. Ommen, and J.R.H. Ross. 1987. Influence of CO_2 on the oxidative coupling of methane over lithium promoted magnesium oxide catalyst. *J. Chem. Soc., Chem. Commun.* 1433–4.

Lee, J.S., and S.T. Oyama. 1988. Oxidative coupling of methane to higher hydrocarbons. *Catal. Rev.–Sci. Eng.* 30(2):249–80.

Marquardt, D.W. 1963. *J. Soc. Indust. Appl. Math.* 11(2):431–41.

McCarthy, J.G., A.B. McEwen, and M.A. Quinlan. 1990. Models of the direct catalytic partial oxidation of light alkanes. *Stud. Surf. Sci. Catal.* 55:405–15.

Mimoun, H., A. Robine, S. Bonnaudet, and C.J. Cameron. 1990. Oxypyrolysis of natural gas, *Appl. Catal.* 58:269–80.

Miro, E.E., Z. Kalenik, J. Santamaria, and E.E. Wolf. 1990. Transient studies on methane oxidative coupling over alkali-metal promoted titanate catalysts. *Catal. Today* 6:511–18.

Nicholas, J. 1976. *Chemical Kinetics: A Modern Survey of Gas Reactions*. London: Harper & Row Ltd.

Nishiyama, T., T. Watanabe, and K. Aika. 1990. Oxidative coupling of methane over CaO catalysts promoted with alkali and alkaline earth oxide. *Catal. Today* 6:391–7.

Otsuka, K., and K. Jinno. 1986. Kinetic studies on partial oxidation of methane over Sm_2O_3. *Inorg. Chim. Acta* 121:237–41.

Otsuka, K., K. Jinno, and A. Morikawa. 1985. The catalysts active and selective in oxidative coupling of methane. *Chem. Lett.* 499–500.

Otsuka, K., Q. Liu, M. Hatano, and A. Morikawa. 1986. Synthesis of ethylene by partial oxidation of methane over the oxides of transition elements with LiCl. *Chem. Lett.* 903–6.

Peil, K.P., J.G. Goodwin, Jr., and G. Marcelin. 1989. *J. Phys. Chem.* 63:5977–9.

Rotzoll, G. 1986. Mass spectrometric investigation and computer modeling of CH_4–O_2–O_3 reactions from 480 to 830 K. *J. Phys. Chem.* 661–71.

Tsang, W., and R.F. Hampson. 1986. Chemical kinetic data base for combustion chemistry. Part 1, Methane and related compounds. *J. Phys. Chem. Ref. Data* 1087–279.

van Kasteren, J.M.N. 1990. Oxidative coupling of methane over lithium promoted magnesia. Ph.D. Thesis. Eindhoven University of Technology, Eindhoven, The Nethlands.

van Kasteren, J.M.N., J.W.M.H. Geerts, and K. van der Wiele. 1988. Ethylene synthesis by catalytic oxidation of methane over Li-doped MgO catalysts: The interaction of catalytic and non-catalytic reaction steps. In *Proceedings of the 9th International Congress on Catalysis, Calgary, Alberta*, Canada, Vol. 2; ed. M.J. Phillips and M. Ternan, pp. 930–6. Ottawa: Chemical Institute of Canada.

van Kasteren, J.M.N., J.W.M.H. Geerts, and K. van der Wiele. 1989. Working principle of Li doped MgO applied for the oxidative coupling of methane. Preprint First World Congress: "New Developments in Selective Oxidation," Rimini, Italy, Sept. 18–22, 1989.

van Kasteren, J.M.N., J.W.M.H. Geerts, and K. van der Wiele. 1990. The role of heterogeneous reaction steps during the oxidative coupling of methane over Li/MgO catalysts. *Catal. Today* 6:497–502.

Van Santen, R.A., and C.P.M. De Groot. 1986. The mechanism of ethylene epoxidation. *J. Catal.* 98:530–9.

Vardanyan, I.A., and A.B. Nalbandyan. 1985. On the mechansim of thermal oxidation of methane. *Int. J. Chem. Kinet.* 17:901–24.

Warnatz, J. 1987. Hydrocarbon oxidation at high temperatures. *Ber. Bunsen-Ges. Phys. Chem.* 87:1008–22.

Westley, F., and J.T. Herron. 1987. Compilation of kinetic data for combustion chemistry. *Part 1: Nonaromatic C, H, O, N, and S containing compounds*. US Government Printing Office.

Zanthoff, H., and M. Baerns. 1990. Oxidative coupling of methane in the gas phase. Kinetic simulation and experimental verification. *Ind. Eng. Chem. Res.* 29:1–10.

9

Mechanism of Cooxidative Methane Dimerization Catalysis: Kinetic and Thermodynamic Aspects

Jon G. McCarty

Materials Research Laboratory, SRI International, Menlo Park, CA

TABLE OF CONTENTS

Introduction

A general mechanism for the direct oxidative conversion of methane is emerging from recent experimental work. Considerable evidence indicates that heterogeneous production of methyl radicals followed by gas-phase coupling to form ethane initiates the production of higher hydrocarbons during the selective catalytic oxidation of methane. The definitive experiments are those of Lunsford and co-workers (Driscoll et al. 1985; Campbell and Lunsford 1988, 1989; Driscoll, Campbell, and Lunsford 1987; Driscoll, Zhang, and Lunsford 1986), who use matrix-isolated electron paramagnetic resonance (MIEPR) measurements to determine the distribution of methyl radicals downstream of a Li/MgO catalyst bed. These results have been confirmed by other workers using mass spectrometric (Amorebieta and Colussi 1988, 1989), spectroscopic (Lee, Yu, and Lin 1990; Gulcicek, Colson, and Pfefferle 1990), and other MIEPR (Sinev, Korchak, and Krylov 1988) methods. Isotopic exchange experiments (Nelson, Lukey, and Cant 1988, 1989; Nelson and Cant 1990; Cant et al. 1988; Mims et al. 1989) also support the view that the methyl radical is the primary intermediate in the production of ethane. Several reports indicate that the reactor geometry influences the product distribution and that microreactors without catalysts (van Kasteren, Geerts, and van der Wiele 1988; Lo, Agarwal, and Marcelin 1988; Lane and Wolf 1988; Asami et al. 1988; Yates and Zlotin 1988) can selectively produce ethane and ethene. These observations affirm the significant role of homogeneous kinetics in all aspects of the reaction, including its initiation.

The essential question becomes not whether homogeneous reactions contribute to the production of hydrocarbons, but what is the relative importance of homogeneous versus heterogeneous reactions in controlling product yields. Many experimental methane oxidative dehydrodimerization studies show that the selectivity of a given catalyst consistently declines with increasing single-pass conversion (for reviews see Lane and Wolf 1988 and Jones, Leonard, and Sofranko 1987) despite wide differences in reaction conditions and choice of catalyst. Yet, for low methane conversion and low oxygen partial pressure, the hydrocarbon selectivity varies greatly (from $> 99\%$ to $< 1\%$) depending on the inherent selectivity of the catalyst. Reaction kinetic models (Geerts et al. 1990; McCarty, McEwen, and Quinlan 1990; Forlani et al. 1990; Zanthoff and Baerns 1990) show that, for high temperature and high oxygen partial pressure, the contribution of homogeneous gas-phase reactions increases with product accumulation. Domination by homogeneous reactions can lead to a leveling of selectivity, that is, the C_{2+} yields of catalysts that have relatively poor selectivity in dilute oxygen will approach the yields of the more-selective catalysts as the initial oxygen partial pressure increases. In this view, the limits to high C_{2+} yields at high pressure and high temperature (> 1000 K) involve the homogeneous reactions of O_2, whereas the intrinsic activity and selectivity of the catalyst limit C_{2+} yields at low temperature.

This chapter is intended to complement other reviews of the kinetics and mechanism of methane dimerization. In the following sections, the importance of specific heterogeneous reactions and the significance of diffusional mass transport for cooxidative (methane and oxygen cofeed) processes are discussed. Also included is an admittedly speculative discussion of the nature and thermodynamics of reactive oxygen centers on alkali and alkaline earth catalyst surfaces.

KINETICS OF METHANE OXIDATION CATALYSIS

Because the mechanism of the cooxidative methane dimerization process includes a complex set of heterogeneous and homogeneous reactions, it is desirable to construct kinetic models (Geerts et al. 1990; McCarty, McEwen, and Quinlan 1990; Forlani et al. 1990; Zanthoff and Baerns 1990). Models can help interpret and extrapolate laboratory data to conditions more appropriate for commercial processing. They also serve to evaluate the relative importance of homogeneous and heterogeneous reactions. Hydrocarbon formation in the coupling process has been attributed to the well-known gas-phase recombination kinetics of heterogeneously produced radicals. However, homogeneous hydrocarbon oxidation kinetics also include fast oxidation steps. Given a flux of methyl radicals from catalyst surfaces, it is easy to imagine a host of undesirable radical oxidation reactions as well as the desirable coupling reactions.

Application of a kinetic model to the methane coupling process shows that homogeneous oxidation reactions effectively compete with hydrocarbon formation. Our initial kinetic model (McCarty, McEwen, and Quinlan 1990; McEwen, Rossi, and McCarty 1991) consisted of 134 reversible homogeneous reactions along with a limited set of 10 reactions that involve catalyst surface sites (see Appendix 9-A for the set of rates and kinetic parameters). Most of the gas-phase reaction parameters were obtained from review compilations (Warnatz 1984, Tsang, 1986) although several rate constants were adjusted for important bimolecular steps that occur in the "falloff" region under methane coupling conditions. Hydrogen atom metathesis reaction parameters were calculated from transition-state analysis of experimental results. In addition, Rice–Ramsperger–Kassel–Marcus (RRKM) calculations were used to provide rate constants for several important bimolecular steps located in the three-body collision falloff region, under methane coupling conditions. The rate constant for methyl radical recombination is located in the falloff region at 1000 K and therefore decreases slightly at higher temperatures.

Heterogeneous reaction steps were added in a consistent manner, that is, we presumed that reactive surface oxygen species interact with all major reaction products containing C—H bonds in addition to methane. The first step in the catalytic cycle involves the activation of methane by a surface oxygen species [equation (9-1)]. We consider the rate determining step in the oxidative coupling to be methane C—H bond scission by reaction with an atomic surface oxygen species.

$$CH_4 + O_s \Rightarrow CH_3\cdot + HO_s \tag{9-1}$$

This assumption is based on a large, positive (1.5) hydrogen–deuterium isotope effect reported for Li/MgO (Nelson and Cant 1990) and reaction orders, half-order in oxygen and first-order in methane, measured (Amorebieta and Colussi 1988) for methane conversion over Li/MgO at very low pressure (10^{-6}–10^{-4} atm). These results suggest that methane reacts directly with atomic surface oxygen. It should be noted, however, that power-law kinetics and isotope effects are at best weak indicators of the reaction mechanism. In our model, the site density of oxygen centers and the activation energy for methane activation [equation (9-1)] are the only adjustable parameters. Its preëxponential factor is estimated as one-tenth the collision frequency to allow for unfavorable configurational constraints.

The heterogeneous H abstraction step can be generalized to include scission of any C—H bond by an Eley–Rideal reaction with surface oxygen (O_s) to form a gas-phase alkyl radical and a hydroxyl surface site (HO_s),

$$RH + O_s \Rightarrow R\cdot + HO_s \tag{9-2}$$

where RH = {CH_4, C_2H_6, C_2H_4, CH_3OH, or CH_2O}. We adjusted the rate constants for C_2H_6 to give the conversion rate relative to methane (Driscoll et al. 1985; Gulcicek, Colson, and Pfefferle 1990) reported for the Na/Mg/Mn catalyst at 1000 K. Rate constants for the other reactants (H—C_2H_3, H—CH_2OH, and H—CHO) were determined by fixing their frequency factors to that of methane (per H atom) and adjusting their activation energies in proportion to the differences in C—H bond dissociation energy relative to methane.

We did not explicitly define the active site in development of the model, but required only that a surface oxygen species be intimately involved in the C—H bond cleavage process. In our description of the heterogeneous chemistry of the methane coupling reaction system, we reduced the number of unknown kinetic parameters by two justifiable assumptions: (i) the heterogeneous rates of H abstraction of ethane, ethene, methanol, and formaldehyde must be included, and (ii) a high reaction probability (> 0.1) exists for the collision of methyl radicals with the reactive oxygen centers. Only the concentration and the reactivity (activation energy) of the oxygen centers remained as adjustable parameters for the methane coupling model.

The reaction of methyl radicals with surface oxygen significantly alters the selectivity predicted by the model. The oxidation of methyl radicals with the redox Na/Mg/Mn catalyst (without feed gas oxygen) was determined by kinetic analysis (Labinger and Ott 1987; Labinger et al. 1988) to be 2700 times the rate of methane activation. Yet, gas-phase methyl radicals have been observed to pass through beds of alkali, alkaline earth, and some rare earth oxide catalysts (Campbell and Lunsford 1989; Tong, Rosynek, and Lunsford 1989) under very low pressure or highly dilute conditions. The presumably very low, but highly variable surface concentration of reactive oxygen appears to govern the rate of oxidation of methyl radicals. We argue that, as in the gas phase, a surface oxygen entity reactive enough to abstract hydrogen atoms from methane must readily interact with methyl radicals. Indeed, it is easier to visualize forms of surface oxygen that rarely activate methane yet readily oxidize methyl radicals.

Therefore, the second key premise of our model is that methyl radicals irreversibly react in a nonactivated step with surface oxygen to form adsorbed methoxy complexes:

$$CH_3\cdot + O_s \Rightarrow CH_3O_s \tag{9-3}$$

We presume that CH_3O_s remains adsorbed only long enough to oxidize rapidly by reacting with another surface oxygen to produce intermediate formaldehyde, adsorbed OH, and a vacant site ($\Diamond_s$):

$$CH_3O_s + O_s \Rightarrow CH_2O + \Diamond_s + HO_s \tag{9-4}$$

As a consequence, the stationary-state concentration of adsorbed methoxy is very small and its effect on the reaction kinetics is primarily the preservation of stoichiometry.

Another important heterogeneous step is the adsorption of reactive oxygen. If atomic surface oxygen is chosen as the active center, molecular oxygen must dissociatively chemisorb to occupy dispersed vacant reactive center sites. The occupancy of surface sites by atomic oxygen is governed by the overall reversible thermodynamics of oxygen chemisorption. A critical question is the magnitude of the rate of oxygen adsorption compared with the methane conversion. Rapid adsorption and lattice oxygen exchange with $^{18}O_2$ has been reported for Sm_2O_3 (Ekstrom and Lapszewicz 1988a, 1988b). For Li/MgO (Roos et al. 1989) catalysts, oxygen adsorption or water vapor desorption were suggested as rate-determining steps rather than the abstraction step.

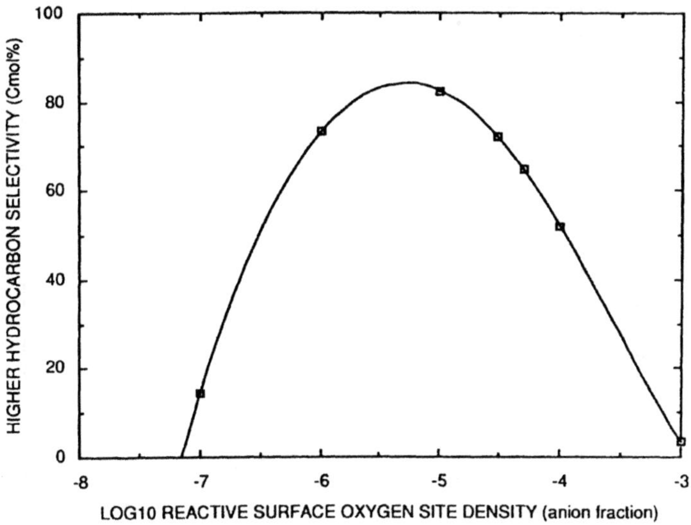

FIGURE 9-2. The effect of reactive surface oxygen site density (sites per unit surface anion) on the calculated C_{2+} selectivity at 1073 K, 0.3 atm CH_4, 0.03 atm O_2, and 8-ms reaction time.

that is, the model simulates catalytic combustion. Lower than optimal concentrations of reactive surface oxygen lead to both lower conversion and lower C_{2+} selectivity because homogeneous methyl radical oxidation processes compete with coupling under these conditions. In most of our subsequent calculations with this model, we used an optimal (but fixed) site density of 10^{10} reactive oxygen sites per square centimeter surface area.

Effect of Pressure on Higher-Hydrocarbon Selectivity

The relative importance of the gas-phase reactions varies with increased conversion. Before oxygen is depleted, the main source of methyl radicals is heterogeneous activation. Two significant gas-phase sources of methyl radicals are reactions involving the hydrogen and hydroxyl radicals. Several methyl radical oxidation pathways are important. The fastest reaction is between methyl radicals and O_2 to produce the methylperoxy radical ($OOCH_3\cdot$) in nearly equilibrated but low concentrations (approximately 4×10^{-7} atm). However, under high-temperature and low-pressure condi-

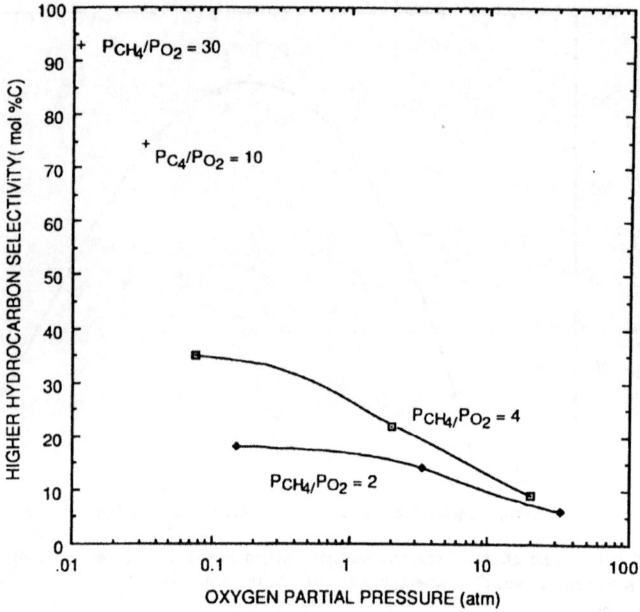

FIGURE 9-3. Calculated higher hydrocarbon selectivity as a function of oxygen partial pressure for complete oxygen conversion during cooxidative coupling of methane at 1073 K.

tions, the reaction of methyl radicals with hydrogen peroxy radicals (OOH·) is prominent. The major sources of OOH· are hydrogen abstraction reactions of unstable radicals, such as formyl (CHO·) and ethyl (C_2H_5·), with diatomic oxygen.

Using the heterogeneous–homogeneous reaction kinetics model described previously, we estimated the effect of total pressure (0.1–100 atm) on methane conversion and product selectivity for various CH_4–O_2 ratios with a fixed reactive oxygen center concentration (10^{10} cm^{-2}). The calculated selectivity for C_{2+} hydrocarbons decreased dramatically with increasing pressure (Fig. 9-3). The model predicts that homogeneous reactions increasingly dominate the kinetics at higher pressure. The most important overall reaction is the production of formaldehyde and its rapid conversion to CO (Table 9-1 lists the most rapid rates at 10 and 40% oxygen conversion calculated for 1073 K and 10 atm with $CH_4/O_2 = 4$). The decreasing importance of heterogeneous reactions is seen by the change rank of the methane activation step from initially the most rapid reaction,

Table 9-1 Calculated Reaction Rates for Methane Coupling at 1073 K and 10 atm with $CH_4/O_2 = 4$

Reactions[a] at 0.2 ms	Rate[b] ($mol\ cm^{-3}\ s^{-1}$)	Reactions[a] at 0.8 ms	Rate[b] ($mol\ cm^{-3}\ s^{-1}$)
$CH_3 + O_2 \rightleftharpoons CH_3O_2$	8.508×10^{-3}	*$CH_2OH + M \rightleftharpoons CH_2O + H + M$*	3.844×10^{-2}
$CH_3O_2 \rightleftharpoons CH_3 + O_2$	8.052×10^{-3}	*$CH_2O + H + M \rightleftharpoons CH_2OH + M$*	3.531×10^{-2}
$CH_4 + OH \rightleftharpoons CH_3 + H_2O$	2.554×10^{-3}	$CH_3 + HO_2 \rightleftharpoons CH_3O + OH$	2.112×10^{-2}
$CH_4 + HO_2 \rightleftharpoons CH_3 + H_2O_2$	1.799×10^{-3}	$CH_4 + OH \rightleftharpoons CH_3 + H_2O$	1.764×10^{-2}
$CH_3O + O_2 \rightleftharpoons CH_2O + HO_2$	1.702×10^{-3}	$CH_4 + HO_2 \rightleftharpoons CH_3 + H_2O_2$	1.479×10^{-2}
$CH_3 + HO_2 \rightleftharpoons CH_3O + OH$	1.537×10^{-3}	$CH_3 + H_2O_2 \rightleftharpoons CH_4 + HO_2$	1.326×10^{-2}
$CH_3 + H_2O_2 \rightleftharpoons CH_4 + HO_2$	1.377×10^{-3}	$CH_3O + O_2 \rightleftharpoons CH_2O + HO_2$	1.090×10^{-2}
$CH_3OH + CH_3 \rightleftharpoons CH_3O + CH_4$	6.050×10^{-4}	$HCO + O_2 \rightleftharpoons HO_2 + CO$	9.238×10^{-3}
$CH_3 + O_2 \rightleftharpoons CH_2O + OH$	5.744×10^{-4}	*$CH_3 + O_2 \rightleftharpoons CH_3O_2$*	8.691×10^{-3}
$CH_2O + H + M \rightleftharpoons CH_2OH + M$	4.321×10^{-4}	*$CH_3O_2 \rightleftharpoons CH_3 + O_2$*	7.884×10^{-3}
$CH_3O_2 + CH_3 \rightleftharpoons CH_3O + CH_3O$	4.252×10^{-4}	$CH_2O + OH \rightleftharpoons HCO + H_2O$	7.779×10^{-3}
$CH_3 + CH_3 \rightleftharpoons C_2H_6$	3.884×10^{-4}	$CH_2OH + O_2 \rightleftharpoons CH_2O + HO_2$	5.297×10^{-3}
$CH_2OH + M \rightleftharpoons CH_2O + H + M$	3.653×10^{-4}	*$H_2O_2 + M \rightleftharpoons OH + OH + M$*	3.348×10^{-3}
$CH_4 + O^* \rightarrow CH_3 + OH^*$	2.787×10^{-4}	*$HCO + M \rightleftharpoons H + CO + M$*	2.313×10^{-3}
$H_2O_2 + M \rightleftharpoons OH + OH + M$	2.733×10^{-4}	$CH_3O + CH_3O \rightleftharpoons CH_2O + CH_3OH$	2.124×10^{-3}
$HCO + O_2 \rightleftharpoons HO_2 + CO$	2.057×10^{-4}	$CH_2O + HO_2 \rightleftharpoons HCO + H_2O_2$	1.874×10^{-3}
$OH^* + OH^* \rightarrow H_2O + {}^* + O^*$	1.460×10^{-4}	$CH_3O + CO \rightleftharpoons CO_2 + CH_3$	1.791×10^{-3}
$CH_2O + OH \rightleftharpoons HCO + H_2O$	9.36×10^{-5}	$CH_3OH + OH \rightleftharpoons CH_2OH + H_2O$	1.732×10^{-3}
$CH_2O + CH_3 \rightleftharpoons HCO + CH_4$	8.27×10^{-5}	$CH_2O + CH_3 \rightleftharpoons HCO + CH_4$	1.401×10^{-3}
$CH_2OH + O_2 \rightleftharpoons CH_2O + HO_2$	8.23×10^{-5}	*$CH_3O_2H \rightleftharpoons CH_3O + OH$*	1.206×10^{-3}
${}^* + {}^* + O_2 \rightarrow O^* + O^*$	7.37×10^{-5}	$CH_3OH + OH \rightleftharpoons CH_3O + H_2O$	1.178×10^{-3}

Reactions in bold type are heterogeneous steps involving reactive oxygen (O*), surface hyrdroxyl (OH*), or vacant reactive centers (*); reactions in italics are bimolecular steps with adjustments for "falloff" behavior. Listings of reactions and kinetic parameters are included in Appendix 9-1.

A catalyst with 10-m^2 surface area per cubic centimeter gas volume, and 10^{10} reactive O* centers per square centimeter surface area has been assumed.

to 14th after 10% oxygen conversion, to virtual insignificance (33d) after 40% oxygen conversion.

Initial formation of small amounts of formaldehyde and ethane by the heterogeneous reactions quickly leads to a pool of radicals that set up a complex oxidative reaction network. By far the most effective homogeneous source for methyl ($CH_3\cdot$) radicals is H abstraction with hydroxyl ($OH\cdot$) radicals. Cooxidative methane conversion at higher oxygen partial pressure effectively removes the hydrogen atoms ($H\cdot$) that contribute to the high selectivity for coupling at low pressure. Unfortunately, the large pool of radicals (especially $OOH\cdot$) that form at elevated pressures are also very effective for the undesirable oxidation of product hydrocarbons, such as ethane, ethene, and oxygenates such as methanol.

The most significant oxidation pathways at high pressure are rapid conversion of methyl radicals to $OCH_3\cdot$ and CH_2O by reaction with hydroperoxy ($OOH\cdot$) radicals and oxygen (O_2), respectively. Reaction of methoxy ($OCH_3\cdot$) radicals with O_2 regenerates hydroperoxy and forms formaldehyde completing a chain reaction involving $OOH\cdot$. Because the fast reaction for methylperoxy ($OOCH_3\cdot$) radical formation is reversible, the conversion of methyl radicals by oxidation with $OOCH_3\cdot$ intermediates is of less significance at higher temperature. But, at lower temperature (< 1000 K), formation of methylperoxy radicals and their subsequent reaction with methyl radicals to form methoxy radicals could become another important oxidation pathway.

Highly Selective Surface Reactions

Most current kinetic models of cooxidative dimerization, including our own earlier effort, substantially overestimate the amount of by-products at elevated temperatures and pressures. Perhaps more significantly they tend to predict the wrong set of by-products: CO, CH_2O, and CH_3OH, whereas CO_2, CO, and trace CH_2O are typically observed under these conditions. Heterogeneous oxidation reactions must be responsible for conversion of product hydrocarbons into CO_2 during redox processes and may be responsible for significant oxidation of potential oxygenate products and CO and H_2 to CO_2 and H_2O during cooxidative dimerization. We have observed very high rates for conversion of methanol (Fig. 9-4) by Na/CaO catalysts. Under conditions of low pressure, low O_2 to CH_4 feed ratio, and low temperature (i.e., conditions typical for laboratory studies) heterogeneous processes may dominate the product distribution, especially for catalysts with poor selectivity. Thus, experimental evaluation of the heterogeneous oxidation reactions of major products, C_2H_6, C_2H_4, CO, and H_2 and potential by-products, CH_2O and CH_3OH, is important for a

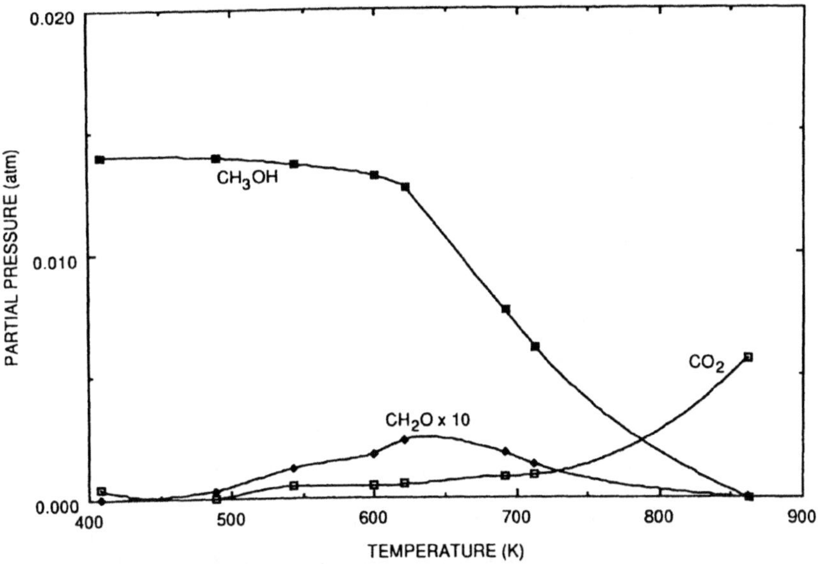

FIGURE 9-4. Catalytic partial oxidation of methanol by a Na/CaO catalyst with excess oxygen. Reactor effluent partial pressures are plotted as a function of temperature for 0.018-atm CH_3OH, 0.03-atm O_2, and balance He feed gas partial pressure at 10 $cm^3\ s^{-1}\ g^{-1}$ catalyst flow rate. Analysis of CH_2O cannot be considered quantitative because gas chromatography analysis was not quantitative for CH_2O under our experimental conditions. The partial pressure of by-product CO is not shown.

complete description of the reaction kinetics. Reactions such as the water–gas shift and methane steam reforming also may be important with typical coupling catalysts.

Effect of Transport

An important effect of increasing the reaction pressure from laboratory to commercial conditions (e.g., from < 1 to > 5 atm) is the lowered influence of heterogeneous reactions because of decreased molecular diffusion rates. The effective diffusion coefficient (D_i) decreases as the effective first-order reaction rate constant (k_i) increases roughly in proportion to pressure. The diffusion length [$\lambda_i = (D_i/k_i)^{0.5}$, i.e., the distance from a catalytically active surface into the reacting gas phase that heterogeneous reactions can significantly influence] of reactants and intermediates decreases inversely with pressure. The homogeneous reaction model can be

Table 9-2 Calculated Concentration, Reaction Lifetime, and Diffusional Length of Selected Gases and Radicals During Methane Coupling

Species	Conc. (ppm)	Conc. (nmol cm^{-3})	Rate (mmol s^{-1} cm^{-3})[b]	τ (μs)[c]	λ (μm)[d]
CH_4	731,000	83,000	23	2200	260
CH_2O	25,700	2,900	55	53	36
OOH·	242	28.	39	0.70	4.6
CH_3·	68	7.8	46	0.17	2.2
OCH_3·	52	6.0	25	0.24	2.7
OH·	1.4	0.16	30	0.0052	0.39
CH_2OH·	0.25	0.029	41	0.0007	0.12
H·	0.009	0.0010	38	0.00003	0.05

[a]Calculated for 1073 K and 10 atm pressure with $CH_4/O_2 = 4$ at 80-μs reaction time.
[b]Consumption rate, not the net reaction rate.
[c]Reciprocal of first-order time constant.
[d]Square root of the product of the estimated diffusivity at 10 atm and 1073 K and τ.

used to estimate the diffusion length for each gas-phase species (Table 9-2).

Relatively long-lived radicals, such as CH_3· and OOH·, have diffusion paths much greater than typical pore diameters, and therefore affect the homogeneous reaction rates within catalyst particles and for some distance into the gas film surrounding an individual catalyst particle. For short-lived radicals, such as OH· and CH_2OH· or CHO·, surface reactions have virtually no influence on their concentrations except very near a surface, and stationary-state conditions prevail for these species throughout the entire homogeneous phase. For reactant methane and oxygen the diffusion length is large, and therefore the assumption of uniform volumetrically averaged reaction rates is justified. However, the concentrations of moderately reactive intermediates, such as C_2H_6, CH_2O, CH_3·, and OOH·, and therefore indirectly the entire set of homogeneous reaction rates, must be modeled with heterogeneous boundary conditions for the set of nonlinear interactive reactive mass-transport equations.

A simple description of methane coupling is sufficient to show that most of the catalytic production of product ethane must occur *within* the pores of the catalyst. In this preliminary assessment, we consider uniform porosity within the catalyst particle and allow concentration gradients to form. These gradients are caused both by fast heterogeneous reaction rates and slow diffusion within the internal catalyst pores and by film diffusion in the gas phase near an external surface. Thus at a distance x away from the external catalyst particle surface,

$$D_i\, d^2C_i/dx^2 = R_i^g \qquad \text{for } x > 0$$
$$D_i^{\text{het}}\, d^2C_i/dx^2 = R_i^{\text{het}} + R_i^g \qquad \text{for } x < 0 \tag{9-6}$$

with jump conditions for C_i and the diffusion flux at $x = 0$ and boundary conditions $C_i = C_i^0$ in the gas phase at $x = 0.5d_{mp}$ and $dC_i dx = 0$ at $x = -0.5d_{mp}$, where d_{mp} is the mean catalyst particle diameter. In this calculation, the only heterogeneous reaction was net methyl radical production,

$$CH_4 \Rightarrow CH_3\cdot + \tfrac{1}{2}H_2O - \tfrac{1}{4}O_2 \tag{9-7}$$

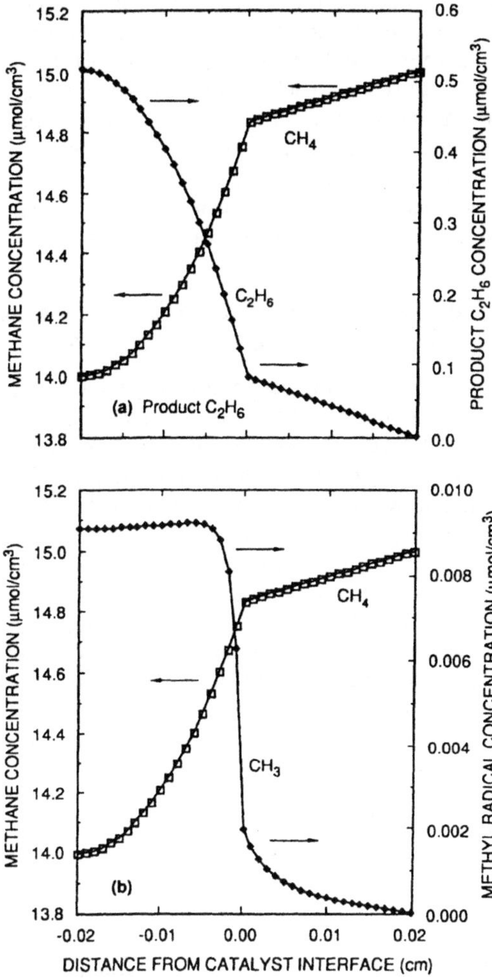

FIGURE 9-5. Calculated methane, ethane, and methyl radical concentration profiles near the surface of a catalyst particle.

whereas only coupling,

$$2CH_3\cdot \Rightarrow C_2H_6 \tag{9-8}$$

was allowed in the gas phase. Diffusion coefficients were chosen to be 5 $cm^2\ s^{-1}$ in the gas phase and 0.5 $cm^2\ s^{-1}$ within the pores (consistent with a porosity of 0.3 and tortuosity of 3.0).

The calculated concentration profiles (Fig. 9-5) show that almost all coupling takes place within the pores of the catalyst under these conditions. The calculation simulates reactive transport for a catalyst with 400-μm diameter. Larger particles would show even greater transport restriction. The inclusion of additional homogeneous and heterogeneous reactions would decrease the effective time constant for methyl radicals and again lead to greater transport restriction within the pores. This simple calculation shows that one must carefully assess reactive mass transport in the kinetic analysis of the coupling process.

NATURE OF REACTIVE CENTERS ON ALKALI AND ALKALINE EARTH CATALYSTS

Our heterogeneous and homogeneous kinetic model of catalytic oxidative methane coupling processes (McCarty, McEwen, and Quinlan 1990; McEwen, Rossi, and McCarty 1991), as described previously, suffers from several imitations. First, it overestimates the homogeneous oxidation of methane to formaldehyde and CO while it underestimates the production of CO_2 with increasing oxygen partial pressure. Inclusion of fast specific heterogeneous reactions, such as CO oxidation or fast deep formaldehyde oxidation, could remedy this limitation.

In addition, the model does not accurately describe the effect of temperature on the conversion rate nor the inhibiting effect of added by-product CO_2 on the higher-hydrocarbon selectivity of alkali catalysts. The model predicts that a very low concentration of reactive surface oxygen must be present in order to avoid rapid oxidation of intermediate methyl radicals and higher hydrocarbon products. For low-pressure cooxidative methane coupling, the optimal concentration of reactive oxygen was estimated to be about 10 parts per million anions at the surface. Such low concentrations of reactive centers can lead to unacceptably low conversion rates unless the reaction probability of methane is very high. This situation can occur only if the activation energy for the abstraction of a hydrogen atom from methane by a reactive oxygen surface atom is low. But with a low activation energy for the abstraction step, the overall apparent activation energy for methane conversion should be lower than

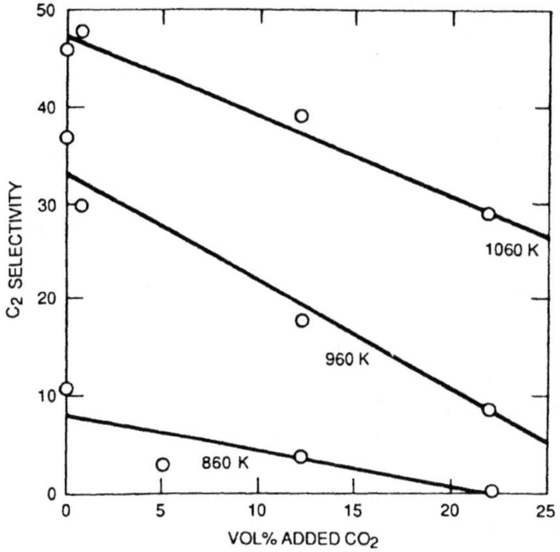

FIGURE 9-6. Effect of CO_2 partial pressure on C_{2+} selectivity for a $SrZrO_3$ catalyst with 0.3-atm CH_4 and 0.03-atm O_2 feed gas partial pressures.

typically observed, 80–120 kJ mol^{-1}. This presumes, perhaps incorrectly as discussed later, that the site density for reactive surface oxygen remains constant with temperature.

The model allows only a minor role for CO_2 as a diluent and third-body molecules to influence the methane conversion rate and product selectivity, contrary to our observations for Na/CaO and $SrZrO_3$ catalysts (Fig. 9-6) and observations (Roos et al. 1989) for Li/MgO catalysts. Both of these latter limitations of the model are overcome if we assume that the passive carbonate anions, which are formed by the adsorption of CO_2, can displace reactive oxygen species from the surfaces of strongly basic oxide catalysts.

Nature of Active Oxygen Centers

A number of reactive centers have been proposed (Lee and Oyama 1988; Otsuka 1987; Lunsford 1988) and can be considered as candidate reactive oxygen centers for the coupling process. These include the following: O^-, the radical anion (Lunsford 1987, 1988; Ito et al. 1985; Lin, Wang, and Lunsford 1988; Hutchings, Scurrell, and Woodhouse 1987); O_2^-, the superoxide anion (Lunsford 1987; Lin et al. 1989; Garrone et al. 1988; Giamello et al. 1990); O^{2-}, the oxide anion (Kimble and Kolts 1987; Che and Tenili 1982; Sokolovskii et al. 1990; Kadushin et al. 1990), which is especially reactive if it is associated with a multivalent metal cation (Che and Tenili 1982); and O_2^{2-}, the peroxide anion (Otsuka 1987; Kharas and Lunsford 1989; Otsuka et al. 1987, 1990; Korf et al. 1990). A great deal could be written about this subject, yet the nature and concentration of the reactive surface oxygen center (or centers) remains largely unknown for any methane dimerization catalyst. The nature of the methane or hydrocarbon activation step is similarly uncertain. Most authors (including ourselves) presume homolytic methane reaction (involving only the oxygen species, whereas others (Sokolovskii et al. 1990; Kadushin et al. 1990; Mehandru, Anderson, and Brazdil 1988) prefer heterolytic reaction (C—H bond stretching with CH_3^- located over the metal cation and H^+ over the oxygen anion). Much additional experimental and theoretical work remains to clarify this issue.

There are several reasons to suppose that reactive oxygen can be displaced (Peil, Goodwin, and Marcelin 1989) by adsorption and accumulation of by-product CO_2 on the surfaces of alkali-promoted, alkaline-earth, or rare-earth oxide catalysts. An examination of the thermodynamics of the Na_2CO_3/Na_2O_x system (Wagman et al. 1982; Lamoreaux, Hildenbrand, and Brewer 1987; Vol'nov 1966) shows that alkali carbonates are typically in the stable bulk phase under methane catalytic cooxidative coupling reaction conditions. Also, of the oxide anions that are possible reactive surface oxygen species capable of activating methane, O_2^- and O_2^{2-} have the greater thermochemical potential. Our temperature-programed desorption experiments (McCarty and Quinlan 1988) and other thermogravimetric (Kimble and Kolts 1987), X-ray photoelectron spectroscopic (Peng, Richardson, and Stair 1990), and isotope exchange experiments (Nelson, Lukey, and Cant 1989; Peil, Goodwin and Marcelin 1989) also point to the accumulation of carbonate layers on the working surfaces of the basic oxide catalysts.

The concentration of reactive oxygen centers on surfaces of Na-promoted catalysts may be related to the solid-state chemical potentials of competitive oxygen anions. These species could populate the surface of a sodium carbonate layer at levels that were predicted by our heterogeneous–homogeneous kinetic model as optimal for the production of higher hydrocarbons. In this view, oxygen associated with the bulk oxides of metals that are multivalent (Tong, Rosynek, and Lunsford 1989) and less basic than Na (e.g., Ce and Pr) is present at the surface in high concentration. Therefore these oxides readily reduce gas-phase methyl radicals to levels ineffective for coupling and serve as deep-oxidation catalysts.

Thermodynamics of Surface Oxygen on Alkali and Alkaline-Earth Catalysts

We postulate that reactive oxygen can form in the carbonate or oxide layers at the surfaces of selective catalysts. Bulk solid-state solutions are usually affected by steric mismatch of ions in the lattice. This constraint may be relaxed at the surface layer and more optimal cation–anion bonding distances can form. If we assume that the surface anions compose an ideal solution with the solid state, the surface carbonate salt could contain significant levels of dissolved minority species of O^{2-}, O_2^{2-}, and OH^- (see Table 9-3 for Na_2CO_3). In particular, the superoxide anion would appear to have about the optimal concentration for high C_{2+} selectivity, and both O_2^- and O_2^{2-} must be considered prospective species for the role of the surface reactive oxygen center.

The effect of CO_2 on the concentration of reactive oxygen may be substantial if a carbonate layer covers the surface. If the concentration of reactive oxygen centers on the surface of an alkali carbonate layer is governed by the thermochemical activity of superoxide and peroxide (ideally) dissolved in the carbonate layer, then the heat of formation of the centers is approximately 160 and 280 kJ mol^{-1} (Fig. 9-7), respectively, for the superoxide and peroxide with Na_2CO_3 when $P_{O_2} = P_{CO_2} = 1$ atm. We can also estimate the effect of reaction temperature and O_2 and CO_2 partial pressure on the concentration of reactive oxygen (see Fig. 9-8 for Na). As discussed previously, inclusion of rate constants for adsorption and desorption of CO_2 and O_2 on Na catalysts that are consistent with the thermochemical analysis shows the effect of O_2 and CO_2 partial pressure on the methane conversion rates by the mutual displacement of reactive oxygen and surface carbonate. We conclude from this analysis of the

Table 9-3 Thermodynamic Calculations[a] for Sodium in the Cooxidative Methane Coupling System

Component	$\Delta H_{f298°}$(kJ/mol)	$S_{f298°}$(J/mol/K)	$C_{p298°}$(J/mol)	Activity (1073 K)[b]
Na_2O	−414.22	75.06	69.12	1.3×10^{-6}
NaOH	−425.609	64.455	59.54	0.017
Na_2O_2	−510.87	95	89.24	0.5×10^{-6}
O_2	0	205.138	29.355	0.1
H_2O	−241.818	188.825	33.577	0.25
CO_2	−393.509	213.74	37.11	0.01

[a]References: Wagman et al. 1982, Lamoreaux, Hildenbrand, and Brewer 1987; and Vol'nov 1966.

[b]Thermochemical activity of (unstable) sodium oxides determined from gas-phase partial pressures and unit activity for Na_2CO_3.

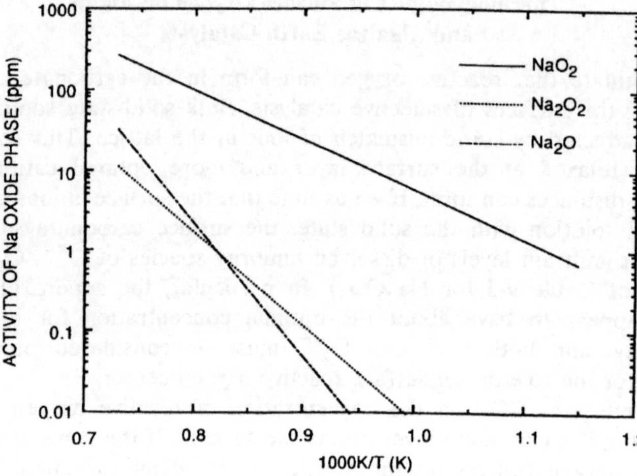

FIGURE 9-7. Thermochemical activity ($\times 10^{-6}$) of sodium superoxide, sodium peroxide, and sodium oxide versus reciprocal temperature for $P_{O_2} = P_{CO_2} = 1$ atm and $P_{H_2O} = 2.5$ atm. The stable bulk phase (and majority surface component) is sodium carbonate.

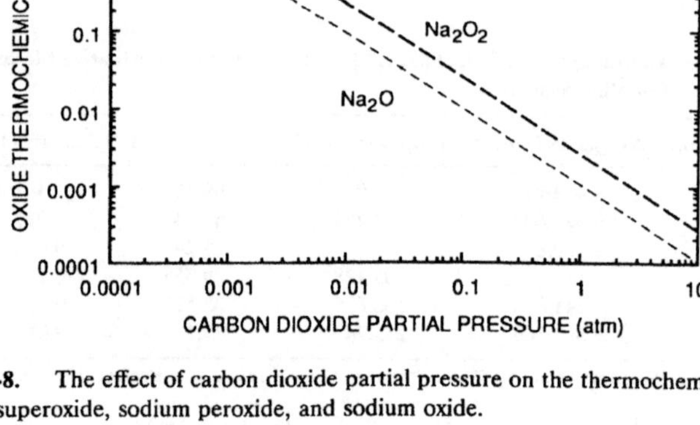

FIGURE 9-8. The effect of carbon dioxide partial pressure on the thermochemical activity of sodium superoxide, sodium peroxide, and sodium oxide.

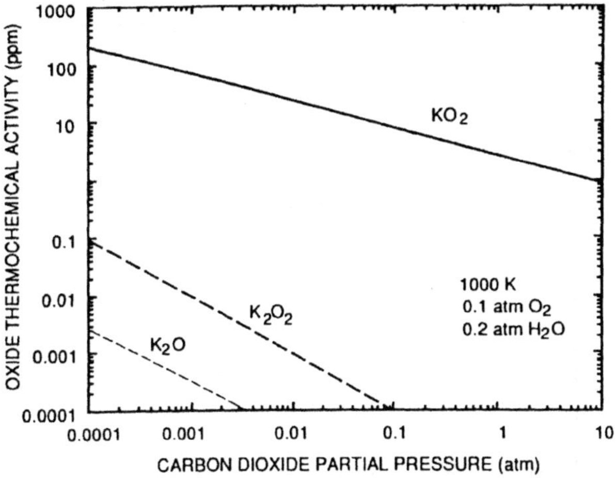

FIGURE 9-9. The effect of carbon dioxide partial pressure on the thermochemical activity of potassium superoxide, sodium peroxide, and sodium oxide.

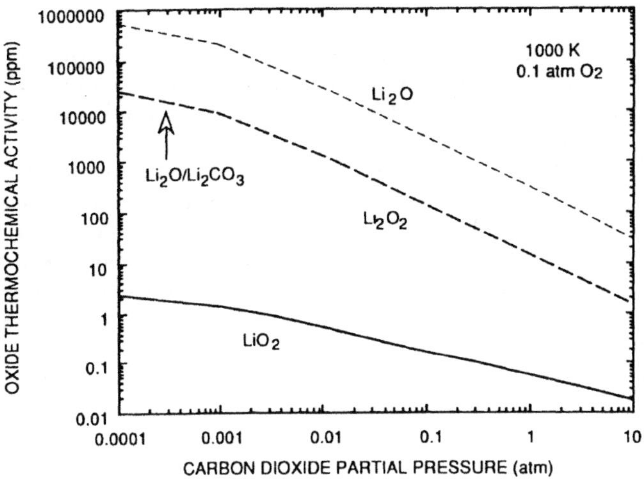

FIGURE 9-10. The effect of carbon dioxide partial pressure on the thermochemical activity of lithium superoxide, sodium peroxide, and sodium oxide. The arrow points to the decomposition point of bulk lithium carbonate.

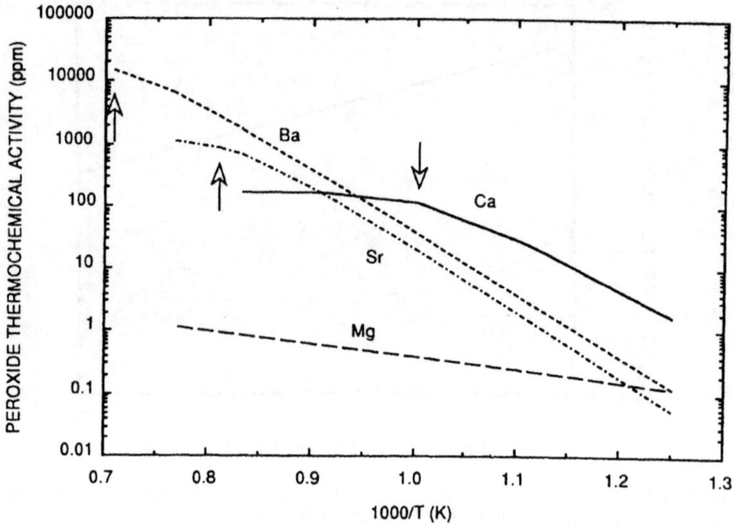

FIGURE 9-11. Thermodynamic calculation for peroxide in alkaline earth carbonates and oxides with $P_{O_2} = P_{CO_2} = 0.1$ atm and $P_{H_2O} = 0.2$ atm. Arrows refer to the carbonate decomposition points.

kinetic models and oxide thermodynamics that the surface oxygen involved in the coupling reaction on sodium catalysts is probably superoxide, O_2^-. It could be bound by about 160 kJ mol^{-1}, with a surface concentration of roughly 10 ppm of surface anion (mostly carbonate).

Similar analysis for potassium (Fig. 9-9) also shows that the superoxide ion (O_2^-) has about the surface density expected for a reactive center during selective oxidation of methane. But, for Li (Fig. 9-10) and the alkaline-earth series (Fig. 9-11), peroxide (O_2^{2-}) is the more likely reactive center, based on the estimated fractional coverage from thermochemical analysis. We can only suggest that either species may be responsible for methane activation. Identification and measurement of the actual concentration of reactive centers under reaction conditions are, of course, necessary to validate this thermochemically inspired view of the nature of the catalyst surface.

The Effect of Temperature and Gas Composition on Surface Oxygen Concentration

We can roughly assess the initial rate of methane conversion and the relative population of surface species from a set of estimated heterogeneous rates (see the following section for a discussion of recommended

Table 9-4 Effect of Gas Composition and Temperature on Calculated Initial Methane Conversion Rates for Na / CaO Dimerization Catalysts

GAS COMPOSITION (atm)				Temp.	CH_4 Rate	Fractional O_2^- Surface Conc.
CH_4	O_2	CO_2	H_2O	(K)	(μmol g^{-1} s^{-1})	(ppm)
0.6	0.2	0.03	0.1	973	0.112	0.87
0.6	0.2	0.03	0.1	1073	4.26	11
0.6	0.2	0.3	0.1	1073	1.32	3.4
0.6	0.3	0.03	0.1	1073	5.73	14
0.6	0.2	0.03	0.2	1073	4.25	11
0.5	0.2	0.03	0.1	1073	4.20	14

heterogeneous reactions and rates). Time constants for reaction of all surface species are very fast and reversible ($\tau < 10$ ms) except for carbonate accumulation ($\tau \approx 10$ ms). The effect of gas composition (CH_4, O_2, CO_2, and H_2O partial pressures) and temperature on surface population has a potentially significant effect on heterogeneous rates. This thermodynamically influenced model was used to predict methane conversion rates (Table 9-4). For high CH_4/O_2 ratio, the reaction rate became more dependent on O_2 concentration than on CH_4 concentration. Also, CO_2 strongly inhibited the conversion rate because reactive oxygen was displaced. The indicated reaction orders (power law) for the initial conversion of methane in the absence of gas-phase contributions to the rate (Table 9-4) were -0.5, 0.7, and 0.08, respectively, for CO_2, O_2, and CH_4, at 1073 K with 0.6, 0.2, 0.03, and 0.1 atm partial pressures of CH_4, O_2, CO_2, and H_2O, respectively. It is very difficult to compare these estimates with experimental data, because the experimental results have not been determined under truly differential conditions with respect to product CO_2. Also, the reaction orders for reactants have been determined for only a limited range of parameters. Reactant orders, 0.3 and 0.2 for CH_4 and O_2, respectively, were determined from experimental results for Na/CaO (60) at 898 K with 0.4 atm CH_4 and 0.04 atm O_2. The range of feed gas oxygen partial pressure was very wide in these experiments and oxygen conversions were typically very high. Therefore, the effect of CO_2 partial pressure may have offset the effect of increased O_2 partial pressure. Better agreement with the trends of Table 9-4 were measured for Li/MgO at 998 K with 0.5 atm CH_4, 0.05 atm CO_2, and 0.05 atm O_2. The power-law orders for methane conversion obtained from our analysis of data presented by Roos and co-workers (Roos et al. 1989) were -0.6, 1.0, and 0.2, respectively, for CO_2, O_2, and CH_4. The activation energy predicted from changes in oxygen superoxide surface concentration was about 310 kJ mol^{-1}, a value much greater than our model calculations with fixed reactive oxygen site density and somewhat greater than the experimental range for Na/CaO. The high estimated apparent activation energy for Na/CaO primarily results from changes in surface O_2^- concentration. Displacement of reactive oxygen, whatever its nature, may generally answer the question of why the apparent activation energy for presumably very reactive, endothermically formed surface oxygen is so high (Kadushin et al. 1990).

Recommended Heterogeneous Reaction Network

The collection of heterogeneous reactions dictates the product distribution for methane conversion, at least for low conversion. Most descriptions of the surface reaction kinetics that drive the catalytic dimerization of methane (including our own) treat them as effectively homogeneous (i.e., uniformly distributed through the reactor bed), not as boundary conditions for the set of gas-phase reactions. However, as shown previously, it appears that at high temperature and higher pressures many radical and product gases (such as CH_2O) are too reactive to be handled in this manner.

Before we can construct a more-accurate set of heterogeneous reactions, we must know something about the nature of the catalytically active sites. We currently consider the reactive oxygen species to be either surface peroxide or superoxide, O_2^{2-} or O_2^-, respectively. Adsorption of oxygen as a diatomic surface species is more easily described as fast. Based in part on reports from the literature, our previous temperature programmed desorption (TPD) results, the thermochemical analysis discussed previously, and based in part on estimated rates for fast reaction steps, a new set of general heterogeneous rates (Table 9-5) has been compiled. The fast oxidation of CH_3OH and CH_2O by Na/CaO and $SrZrO_3$ catalysts, and the likely significant reaction of peroxy radicals with vacant sites, compel us to expand our previous set of heterogeneous reactions. Estimated heterogeneous rate expressions and the heterogeneous intermediates must be made consistent with the thermodynamic properties of carbonate, oxide, peroxide, superoxide, and hydroxide compounds, as appropriate.

The activation energy for methane activation, ethane activation, ethylene activation, and CO oxidation by O(a) and O_2(a) species remain major parameters that must be determined from experimental kinetic data. The rates of conversion between O_2(a) and O(a) species is uncertain but could be important, if the latter can be more positively identified as a significant intermediate. Conversion between O_2^{2-}(a) and O_2^-(a) species may be fast and nearly equilibrated because carbonate salts are essentially semiconductors at elevated temperatures. The reaction probability for

Table 9-5 Proposed Set of Heterogeneous Reactions for the Catalytic Cooxidative Dimerization of Methane

Reaction	Rate Evaluation[a]	Comments
Adsorption–desorption		
$O_2 + \diamond(a) \Leftrightarrow O_2(a) + \diamond(a) \Leftrightarrow 2O(a)$	RDR	Superoxide or peroxide
$H_2O + \diamond(a) + O(a) \Leftrightarrow 2HO(a)$	FR	Water desorption
$CO_2 + O(a) \Leftrightarrow CO_3(a)$	FR	Carbonate formation
$CH_2O + \diamond(a) \Rightarrow CH_2O(a) + 2O(a) \Rightarrow CO + 2HO(a)$	FI	Formaldehyde adsorption and decomposition
$CH_3OH + \diamond(a) + O(a) \Leftrightarrow CH_3O(a) + HO(a)$	RDS	Methanol decomposition
Radical trapping [$R = H, CH_3$]		
$ROO \cdot + \diamond(a) \Rightarrow ROO(a)$	FI	Peroxy trapping
$RO \cdot + \diamond(a) \Rightarrow RO(a)$	FI	Fast OH and methoxy
$RO \cdot + O(a) \Leftrightarrow ROO(a)$	FI	trapping steps
$R \cdot + O_2(a) \Leftrightarrow ROO(a)$	FI	Methyl radical trapping step
Oxidation–decomposition		
$RH + O(a) \Rightarrow R \cdot + HO(a)$	RDS	Hydrocarbon activation step
$CO + O(a) \Leftrightarrow CO_2 + \diamond(a)$	RDS	CO oxidation step
$ROO(a) + \diamond(a) \Rightarrow RO(a) + O(a)$	FI	Peroxy reoxidizes surface

[a] RDS = rate-determining irreversible step—use experimental kinetic data; RDR = rate-determining reversible step—use experimental adsorption/desorption data; FI = fast irreversible step—use estimated rates; FR = fast irreversible step—assume equilibration, use thermochemical properties and estimated rates.

collision of radicals the surface and strongly exothermic adsorption of molecules with reactive surface species is taken as a high value (e.g., 0.3 or 0.1).

CONCLUSIONS

A kinetic model of the catalytic cooxidative methane coupling process shows the increased dominance of homogeneous oxidation reactions at higher pressures. The formation of small quantities of initial products CH_2O and C_2H_6 quickly leads to the establishment of a radical network that primarily oxidizes methane to CO. The initial methane activation step by Eley–Rideal H abstraction with oxygen centers of variable surface concentration was taken to be rate-determining. The volumetric concentration of reactive oxygen centers on the surfaces of a working catalyst may range between levels so high that they cause rapid heterogeneous oxidation of the methyl radicals to levels so low that they cause the coupling rate to decrease and lead to poor selectivity through the loss of methyl radicals by first-order homogeneous or secondary heterogeneous oxidation reactions. Most transition-metal oxides appear to have a surface density of reactive centers well above the limit of optimal selectivity and therefore principally operate as deep-oxidation catalysts.

An examination of the thermodynamics of the alkali and alkaline-earth carbonate, hydroxide, and oxide systems shows that superoxide and peroxide, O_2^- and O_2^{2-}, respectively, are most likely to populate the catalyst surfaces at levels optimal for the selective production of higher hydrocarbons. For base oxides, absorption or adsorption of by-product or feedstock CO_2 may displace reactive oxygen centers, whereas increasing oxygen partial pressure and temperature increases the population of reactive oxygen centers.

Reaction models help extract the complex heterogeneous and homogeneous kinetics of the catalytic methane dimerization process. Although trends for the major hydrocarbon products are predicted, the models cannot adequately describe the by-product distribution and overall stoichiometry without inclusion of a number of specific heterogeneous reactions. A one-dimensional transport description of methane activation, focusing on methyl radical intermediates, leads to the conclusion that most coupling occurs within the pores and voids of a catalyst. A great deal more needs to be learned about kinetic and thermochemical properties of dimerization catalysts before truly predictive rate models can be constructed.

ACKNOWLEDGMENT

The author gratefully acknowledges the continuing support of the Methane Reaction Science Program directed by Dr. Daniel A. Scarpiello for the Gas Research Institute and its associated Industrial Affiliate Cosponsors. The author is also grateful for the able assistance of M.A. Quinlan and E.D. Wachsman in acquiring and evaluating experimental results and A.B. McEwen and M.J. Rossi in constructing the reaction models.

APPENDIX 9-A REACTION MECHANISM AND RATE CONSTANTS

The order of the reaction steps and species symbols is slightly different than in the text. The parameters in brackets following a reaction represent enhancement factors for neutral third body molecules (M). Adsorbed species (e.g., O_s) and vacancies ($\Diamond_s$) are designated by subscript *s*. Forward irreversible reactions are designated by ($\rightarrow$) between reactants and products whereas reversible reactions are designated by ($\Leftrightarrow$). Rate expressions have the form $A(T)^n \exp(-E/RT)$, with units of mole per centimeter per second. Reverse rates are calculated from thermodynamic properties.

Reaction	Ref.	A (mol cm s)	n	E (cal mol^{-1})
$\Diamond_s + \Diamond_s + O_2 \rightarrow O_s + O_s$	*a*	5.52×10^{13}	0.0	−23,910.
$O_s + O_s \rightarrow \Diamond_s + \Diamond_s + O_2$	*a*	2.52×10^{16}	0.0	32,551.
$OH_s + OH_s \rightarrow H_2O + \Diamond_s + O_s$	*a*	6.6×10^{16}	0.0	19,119.
$CH_4 + H \Leftrightarrow CH_3 + H_2$	*b*	5.3×10^{8}	1.75	11,211.
$CH_4 + O_2 \Leftrightarrow CH_3 + HO_2$	*c*	4.8×10^{13}	0.0	56,000.
$CH_4 + OH \Leftrightarrow CH_3 + H_2O$	*c*	1.5×10^{6}	2.13	2,438.
$CH_4 + O \Leftrightarrow CH_3 + OH$	*c*	1.9×10^{9}	1.44	8,675.
$CH_4 + HO_2 \Leftrightarrow CH_3 + H_2O_2$	*c*	3.0×10^{13}	0.0	17,996.
$CH_4 + O_s \rightarrow CH_3 + OH_s$	*a*	1.73×10^{12}	0.0	15,231.
$CH_3 + O_s \rightarrow CH_3O_s$	*a*	1.78×10^{11}	0.0	0.
$CH_3O_s + O_s \rightarrow CH_2O + OH_s + \Diamond_s$	*a*	2.52×10^{16}	0.0	19,119.
$CH_3 + H \Leftrightarrow CH_2 + H_2$	*b*	6.9×10^{9}	1.30	12,576.
$CH_3 + CH_3 \Leftrightarrow C_2H_6$	*d*	1.25×10^{13}	0.0	0.
$CH_3 + CH_3 \Leftrightarrow CH_4 + CH_2$	*c*	1.7×10^{9}	0.56	12,572.
$CH_3 + CH_3 \Leftrightarrow C_2H_5 + H$	*d*	1.0×10^{10}	0.0	0.
$CH_3 + O_2 \Leftrightarrow CH_3O + O$	*e*	1.1×10^{13}	0.0	27,798.
$CH_3 + O_2 \Leftrightarrow CH_2O + OH$	*e*	3.4×10^{11}	0.0	8,941.
$CH_3 + O_2 \Leftrightarrow CH_3O_2$	*d*	7.57×10^{10}	0.0	0.
$CH_3 + OH \Leftrightarrow CH_2O + H_2$	*c*	8.0×10^{12}	0.0	0.
$CH_3 + OH \Leftrightarrow CH_2 + H_2O$	*c*	1.1×10^{6}	2.13	2,438.
$CH_3 + OH \Leftrightarrow CH_2OH + H$	*c*	1.0×10^{13}	0.0	0.
$CH_3 + O \Leftrightarrow CH_2O + H$	*c*	8.4×10^{13}	0.0	0.
$CH_3 + H \Leftrightarrow CH_4$	*d*	1.4×10^{13}	0.0	0.
$CH_3 + HO_2 \Leftrightarrow CH_3O + OH$	*c*	1.0×10^{14}	0.0	0.
$CH_3 + HCO \Leftrightarrow CH_4 + CO$	*c*	3.2×10^{11}	0.5	0.
$CH_3O_2 + CH_4 \Leftrightarrow CH_3O_2H + CH_3$	*f*	1.8×10^{11}	0.0	18,480.
$CH_3O_2 + C_2H_6 \Leftrightarrow CH_3O_2H + C_2H_5$	*f*	2.9×10^{11}	0.0	14,942.
$CH_3O_2 + H_2 \Leftrightarrow CH_3O_2H + H$	*f*	3.0×10^{13}	0.0	26,030.
$CH_3O_2 + H \Leftrightarrow CH_3O + OH$	*f*	9.6×10^{13}	0.0	0.
$CH_3O_2 + O \Leftrightarrow CH_3O + O_2$	*f*	3.6×10^{13}	0.0	0.
$CH_3O_2 + OH \Leftrightarrow CH_3OH + O_2$	*f*	6.0×10^{13}	0.0	0.
$CH_3O_2 + HO_2 \Leftrightarrow CH_3O_2H + O_2$	*f*	4.6×10^{10}	0.0	−2,583.
$CH_3O_2 + H_2O_2 \Leftrightarrow CH_3O_2H + HO_2$	*f*	2.4×10^{12}	0.0	9,935.
$CH_3O_2 + CH_2O \Leftrightarrow CH_3O_2H + HCO$	*f*	2.0×10^{12}	0.0	11,684.
$CH_3O_2 + HCO \Leftrightarrow CH_3O + H + CO_2$	*f*	3.0×10^{13}	0.0	0.
$CH_3O_2 + CH_3 \Leftrightarrow CH_3O + CH_3O$	*f*	2.4×10^{13}	0.0	0.
$CH_3O_2H \Leftrightarrow CH_3O + OH$	*e*	7.9×10^{14}	0.0	42,919.
$CH_2 + O_2 \Leftrightarrow CO + OH + H$	*c*	1.3×10^{13}	0.0	1,435.
$CH_2 + H \Leftrightarrow CH + H_2$	*b*	6.5×10^{9}	1.20	−1,200.
$CH + H \Leftrightarrow C + H_2$	*b*	4.2×10^{9}	1.34	−1,375.
$CH_3O + H \Leftrightarrow CH_2O + H_2$	*b*	1.58×10^{8}	1.76	−982.
$CH_3O + O \Leftrightarrow CH_2O + OH$	*c*	5.0×10^{12}	0.0	0.
$CH_3O + OH \Leftrightarrow CH_2O + H_2O$	*c*	5.0×10^{12}	0.0	0.

Reaction	Ref.	A (mol cm s)	n	E (cal mol^{-1})
$CH_3O + O_2 \Leftrightarrow CH_2O + HO_2$	c	5.26	3.46	574.
$CH_3O + CO \Leftrightarrow CO_2 + CH_3$	c	4.68×10^2	3.16	5,377.
$CH_3O + HO_2 \Leftrightarrow CH_2O + H_2O_2$	f	3.0×10^{11}	0.0	0.
$CH_3O + C_2H_6 \Leftrightarrow CH_3OH + C_2H_5$	f	2.4×10^{11}	0.0	7,094.
$CH_3O + C_2H_5 \Leftrightarrow CH_2O + C_2H_6$	f	2.4×10^{13}	0.0	0.
$CH_3O + C_2H_4 \Leftrightarrow CH_2O + C_2H_5$	f	1.2×10^{11}	0.0	6,756.
$CH_3O + C_2H_3 \Leftrightarrow CH_2O + C_2H_4$	f	2.4×10^{13}	0.0	0.
$CH_3O + CH_2O \Leftrightarrow CH_3OH + HCO$	f	1.0×10^{11}	0.0	2,980.
$CH_3O + HCO \Leftrightarrow CH_3OH + CO$	f	9.0×10^{13}	0.0	0.
$CH_3O + CH_3O \Leftrightarrow CH_2O + CH_3OH$	f	6.0×10^{13}	0.0	0.
$CH_3O + M \Leftrightarrow CH_2O + H + M$	c	1.0×10^{14}	0.0	25,096.
$CH_2O + O_s \rightarrow HCO + OH_s$	a	1.35×10^{12}	0.0	9,231.
$CH_2O + H \Leftrightarrow HCO + H_2$	b	5.47×10^8	1.65	3,334.
$CH_2O + CH_3 \Leftrightarrow HCO + CH_4$	c	4.1×10^{12}	0.0	8,916.
$CH_2O + OH \Leftrightarrow HCO + H_2O$	c	3.0×10^{13}	0.0	1,195.
$CH_2O + O \Leftrightarrow HCO + OH$	c	3.5×10^{13}	0.0	3,513.
$CH_2O + HO_2 \Leftrightarrow HCO + H_2O_2$	c	1.0×10^{12}	0.0	8,008.
$CH_2O + O_2 \Leftrightarrow HCO + HO_2$	c	1.0×10^{14}	0.0	40,010.
$CH_2O + M \Leftrightarrow HCO + H + M$	c	5.0×10^{16}	0.0	76,482.
$CH_2O + M \Leftrightarrow CO + H_2 + M$	c	8.2×10^{15}	0.0	69,551.
$CH_2OH + H \Leftrightarrow CH_2O + H_2$	c	2.0×10^{13}	0.0	0.
$CH_2OH + O \Leftrightarrow CH_2O + OH$	c	5.0×10^{12}	0.0	0.
$CH_2OH + OH \Leftrightarrow CH_2O + H_2O$	c	5.0×10^{12}	0.0	0.
$CH_2OH + O_2 \Leftrightarrow CH_2O + HO_2$	c	1.8×10^{13}	0.0	789.
$CH_2OH + M \Leftrightarrow CH_2O + H + M$	c	1.4×10^{13}	0.0	32,982.
$HCO + H \Leftrightarrow CO + H_2$	b	3.1×10^9	1.62	−1,105.
$HCO + O \Leftrightarrow OH + CO$	c	3.0×10^{13}	0.0	0.
$HCO + O \Leftrightarrow H + CO_2$	c	3.0×10^{13}	0.0	0.
$HCO + OH \Leftrightarrow H_2O + CO$	c	5.0×10^{13}	0.0	0.
$HCO + O_2 \Leftrightarrow HO_2 + CO$	c	4.2×10^{12}	0.0	0.
$HCO + HCO \Leftrightarrow CH_2O + CO$	c	2.0×10^{13}	0.0	0.
$HCO + HO_2 \Leftrightarrow OH + H + CO_2$	f	3.0×10^{13}	0.0	0.
$HCO + M \Leftrightarrow H + CO + M$ [H_2 = 1.87; O_2 = 0.99; H_2O = 18.5; CO = 2.1; CO_2 = 4.3]	c	1.9×10^{17}	−1.0	16,993.
$C_2H_6 + H \Leftrightarrow C_2H_5 + H_2$	b	7.0×10^8	1.70	7,827.
$C_2H_6 + CH_3 \Leftrightarrow C_2H_5 + CH_4$	b	6.1×10^6	1.74	10,444.
$C_2H_6 + OH \Leftrightarrow C_2H_5 + H_2O$	b	3.5×10^6	2.12	872.
$C_2H_6 + O \Leftrightarrow C_2H_5 + OH$	b	9.0×10^7	1.92	5,691.
$C_2H_6 + HO_2 \Leftrightarrow H_2O_2 + C_2H_5$	f	3.0×10^{11}	0.0	14,942.
$C_2H_6 + O_s \rightarrow C_2H_5 + OH_s$	a	1.35×10^{12}	0.0	13,231.
$C_2H_5 + H \Leftrightarrow C_2H_4 + H_2$	b	1.3×10^9	1.30	−695.
$C_2H_5 + H \Leftrightarrow C_2H_6$	d	3.2×10^{13}	0.0	0.
$C_2H_5 + O_2 \Leftrightarrow C_2H_4 + HO_2$	c	2.0×10^{12}	0.0	4,995.
$C_2H_5 + CH_3 \Leftrightarrow C_2H_4 + CH_4$	f	6.2×10^{11}	0.0	0.
$C_2H_5 + CH_3 \Leftrightarrow C_3H_8$	f	1.6×10^{13}	0.0	0.
$C_2H_5 + OH \Leftrightarrow C_2H_4 + H_2O$	f	2.4×10^{13}	0.0	0.
$C_2H_5 + OH \Leftrightarrow CH_3 + H + CH_2O$	f	2.4×10^{13}	0.0	0.
$C_2H_5 + O \Leftrightarrow C_2H_4 + OH$	c	3.1×10^{13}	0.0	0.

Reaction	Ref.	A (mol cm s)	n	E (cal mol^{-1})
$C_2H_5 + O \Leftrightarrow CH_2O + CH_3$	*c*	4.2×10^{13}	0.0	0.
$C_2H_5 + O \Leftrightarrow CH_3CHO + H$	*c*	5.3×10^{13}	0.0	0.
$C_2H_5 + C_2H_5 \Leftrightarrow C_4H_{10}$	*g*	1.0×10^{13}	0.0	0.
$C_2H_5 + C_2H_5 \Leftrightarrow C_2H_4 + C_2H_6$	*g*	1.4×10^{12}	0.0	0.
$C_2H_5 \Leftrightarrow C_2H_4 + H$	*g*	2.0×10^{13}	0.0	39,740.
$C_2H_5 + HO_2 \Leftrightarrow CH_3 + OH + CH_2O$	*f*	2.4×10^{13}	0.0	0.
$C_2H_4 + H \Leftrightarrow C_2H_3 + H_2$	*b*	2.9×10^{9}	1.34	7,026.
$C_2H_4 + CH_3 \Leftrightarrow C_2H_3 + CH_4$	*c*	3.9×10^{12}	0.0	13,027.
$C_2H_4 + OH \Leftrightarrow C_2H_3 + H_2O$	*c*	3.0×10^{13}	0.0	2,988.
$C_2H_4 + O \Leftrightarrow CH_2O + CH_2$	*c*	8.0×10^{8}	1.2	741.
$C_2H_4 + O \Leftrightarrow HCO + CH_3$	*c*	8.0×10^{8}	1.2	741.
$C_2H_4 + HO_2 \Leftrightarrow CH_3CHO + OH$	*f*	6.0×10^{9}	0.0	7,948.
$C_2H_4 + O_s \rightarrow C_2H_3 + OH_s$	*a*	1.35×10^{12}	0.0	16,231.
$C_2H_3 + H \Leftrightarrow C_2H_2 + H_2$	*b*	1.1×10^{8}	1.73	−1,357.
$C_2H_3 + O_2 \Leftrightarrow CH_2O + HCO$	*c*	4.0×10^{12}	0.0	−238.
$C_2H_3 + CH_3 \Leftrightarrow C_2H_2 + CH_4$	*f*	3.9×10^{11}	0.0	0.
$C_2H_3 + OH \Leftrightarrow C_2H_2 + H_2O$	*f*	3.0×10^{13}	0.0	0.
$C_2H_3 + OH \Leftrightarrow CH_3CHO$	*f*	3.0×10^{13}	0.0	0.
$C_2H_3 + HO_2 \Leftrightarrow OH + CH_3 + CO$	*f*	3.0×10^{13}	0.0	0.
$C_2H_3 + C_2H_6 \Leftrightarrow C_2H_4 + C_2H_5$	*f*	6.02×10^{2}	3.3	10,501.
$C_2H_2 + H \Leftrightarrow C_2H + H_2$	*b*	1.20×10^{9}	1.55	16,987.
$CH_3OH + H \Leftrightarrow CH_3O + H_2$	*b*	1.71×10^{8}	1.63	4,048.
$CH_3OH + H \Leftrightarrow CH_2OH + H_2$	*b*	3.93×10^{7}	1.92	5,375.
$CH_3OH + O \Leftrightarrow CH_3O + OH$	*g*	1.0×10^{13}	0.0	4,685.
$CH_3OH + O \Leftrightarrow CH_2OH + OH$	*b*	1.5×10^{13}	0.0	4,750.
$CH_3OH + OH \Leftrightarrow CH_3O + H_2O$	*g*	1.0×10^{13}	0.0	1,697.
$CH_3OH + OH \Leftrightarrow CH_2OH + H_2O$	*b*	1.5×10^{13}	0.0	1,750.
$CH_3OH + CH_3 \Leftrightarrow CH_2OH + CH_4$	*f*	3.19×10	3.17	7,171.
$CH_3OH + CH_3 \Leftrightarrow CH_3O + CH_4$	*f*	1.44×10	3.10	6,935.
$CH_3OH \Leftrightarrow CH_3 + OH$	*g*	9.4×10^{15}	0.0	89,812.
$CH_3OH + O_s \rightarrow CH_2OH + OH_s$	*a*	1.35×10^{12}	0.0	11,231.
$CO + O_2 \Leftrightarrow CO_2 + O$	*c*	2.5×10^{12}	0.0	47,807.
$CO + O + M \Leftrightarrow CO_2 + M$ [O_2 = 12.0; CO = 3.0; CO_2 = 7.0]	*c*	3.0×10^{14}	0.0	3,011.
$CO + OH \Leftrightarrow CO_2 + H$	*g*	1.5×10^{7}	1.3	−765.
$CO + HO_2 \Leftrightarrow CO_2 + OH$	*c*	1.5×10^{14}	0.0	23,586.
$HO_2 + OH \Leftrightarrow H_2O + O_2$	*c*	2.0×10^{13}	0.0	0.
$HO_2 + O \Leftrightarrow O_2 + OH$	*c*	2.0×10^{13}	0.0	0.
$HO_2 + H \Leftrightarrow H_2 + O_2$	*c*	2.5×10^{13}	0.0	693.
$HO_2 + H \Leftrightarrow OH + OH$	*c*	2.2×10^{14}	0.0	1,410.
$HO_2 + H \Leftrightarrow H_2O + O$	*c*	5.0×10^{12}	0.0	1,410.
$HO_2 + HO_2 \Leftrightarrow H_2O_2 + O_2$	*c*	1.06×10^{11}	0.0	−1,697.
$H_2O_2 + M \Leftrightarrow OH + OH + M$ [H_2 = 2.9; O_2 = 1.2; H_2O = 18.5; CO = 2.1; CO_2 = 4.3]	*c*	8.4×10^{16}	0.0	45,410.
$H_2O_2 + H \Leftrightarrow HO_2 + H_2$	*c*	1.7×10^{12}	0.0	3,752.
$H_2O_2 + H \Leftrightarrow H_2O + OH$	*c*	1.0×10^{13}	0.0	3,585.
$H_2O_2 + O \Leftrightarrow HO_2 + OH$	*c*	2.8×10^{13}	0.0	6,405.

Reaction	Ref.	A (mol cm s)	n	E (cal mol^{-1})
$H_2O_2 + OH \Leftrightarrow H_2O + HO_2$	c	7.0×10^{12}	0.0	1,434.
$H + O_2 \Leftrightarrow OH + O$	c	2.2×10^{14}	0.0	16,802.
$H + O_2 + M \Leftrightarrow HO_2 + M$	c	7.0×10^{17}	−0.8	0.
$H + H + M \Leftrightarrow H_2 + M$	c	6.4×10^{17}	−1.0	0.
$O + H_2 \Leftrightarrow OH + H$	c	1.5×10^{7}	2.0	7,552.
$OH + OH \Leftrightarrow H_2O + O$	c	2.1×10^{8}	1.4	−397.
$OH + H_2 \Leftrightarrow H_2O + H$	c	6.3×10^{6}	2.0	2,963.
$OH + H + M \Leftrightarrow M + H_2O$ [$H_2 = 1.9$; $O_2 = 1.9$; $H_2O = 9.5$; $CO = 1.9$; $CO_2 = 1.9$]	c	8.4×10^{21}	−2.0	0.

[a]Determined by thermochemical estimates and adjusted rate parameters as described in the text.
[b]Calculated by a critical review of the literature with transition state calculations (McEwen and Golden 1991).
[c]Frenklach, Wang and Rabinowitz 1991.
[d]Steward 1988.
[e]Zellner and Ewig 1988.
[f]Tsang 1986.
[g]Warnatz 1990.

References

Amorebieta, V.T., and A. Colussi. 1988. *J. Phys. Chem.* 92:4576.

Amorebieta, V.T., and A. Colussi. 1989. *J. Phys. Chem.* 93:5153.

Asami, K., K. Omata, K. Fujimoto, and H. Tominaga. 1988. *Energy and Fuels* 2:574.

Campbell, K.D., and J.H. Lunsford. 1988. *J. Phys. Chem.* 92:5792.

Campbell, K.D., and J.H. Lunsford. 1989. *J. Phys. Chem.* 93:2590.

Campbell, K.D., H. Zhang, and J.H. Lunsford. 1988. *J. Phys. Chem.* 92:750.

Cant, N.W., C.A. Lukey, P.F. Nelson, and R.J. Tyler. 1988. *J. Chem. Soc., Chem. Commun.* 766.

Che, M., and A.J. Tenili. 1982. *Adv. Catal.* 31:77.

Driscoll, D.J., K.D. Campbell, and J.H. Lunsford. 1987. *Adv. Catal.* 35:139.

Driscoll, D.J., W. Martir, J.-X. Wang, and J. H. Lunsford. 1985. *J. Am. Chem. Soc.* 107:58.

Ekstrom, A., and J.A. Lapszewicz. 1988a. *J. Am. Chem. Soc.* 110:5226.

Ekstrom, A., and J.A. Lapszewicz. 1988b. *J. Chem. Soc., Chem. Commun.* 797.

Forlani, O., M. Lupieri, V. Piccoli, S. Rossini, D. Sanfilippo, J. Dumesic, L.C. Aparico, J.A. Rekoske, and A.A. Trevino. 1990. General mechanism for the oxidative coupling of methane. In *New Developments in Selective Oxidation*, ed. G. Centi and F. Trifiro, pp. 417–25. New York: Elsevier.

Frenklach, M., Wang H., and Robinowitz, M.J., 1991 Prog. Energy Combust. Sci. (In press).

Garrone, E., E. Giamello, S. Coluccia, G. Spoto, and A. Zecchina. 1988. Mechanism of oxidation of hydrocarbons over basic oxides: Comparison between pure and Co^{++} doped MgO. In *Proceedings of the 9th International Congress of Catalysis*, Vol. 4, ed. M.J. Phillips, and M. Ternan, pp. 1577–84. The Chemical Institute of Canada.

Geerts, W.M.H., Q. Chen, J.M.N. van Kasteren, and K. van der Wiele. 1990. *Catal. Today* 6:519.

Giamello, E., E. Garrone, S. Coluccia, G. Spoto, and A. Zecchina. 1990. Oxidation of propene and toulene on CoO–MgO solid solutions: A spectroscopic study. In *New Developments in Selective Oxidation*, ed. G. Centi and F. Trifiro, pp. 405–15. New York: Elsevier.

Gulcicek, E.E., S.D. Colson, and L.D. Pfefferle. 1990. *J. Phys. Chem.* 94:7069.

Ito, T., J.-X. Wang, C.-H. Lin, and J.H. Lunsford., 1985. *J. Amer. Chem. Soc.* 107:5062.

Jones, C.A., J.J. Leonard, and J.A. Sofranko. 1987. *Energy and Fuels* 1:12.

Kadushin, A.A., O.V. Krylov, S.E. Plate, Yu.P. Tulenin, V.A. Seleznev, A.V. Bobrov, and Ya.M. Kimelfeld. 1990. SERS—In situ study of the surface species formed in methane oxidative coupling. In *New Developments in Selective Oxidation*, ed. G. Centi and F. Trifiro, pp. 447–52. New York: Elsevier.

Kimble, J.B., and J.H. Kolts. 1987. *CHEMTECH* 501.

Kharas, K.C.C., and J.H. Lunsford. 1989. *Appl. Catal.* 51:L1.

Korf, S.J., J.A. Roos, J.W.H.C. Derksen, J.A. Vreeman, J.G. van Ommen, and J.R.H. Ross. 1990. *Appl. Catal.* 59:291.

Labinger, J.A., S. Mehta, K.C. Ott, H.K. Rockstad, and S. Zoumalan. 1988. In *Catalysis 1987*, ed. J.W. Ward. New York: Elsevier.

Labinger, J.A., and K.C. Ott, 1987. *J. Phys. Chem.* 91:2682.

Lamoreaux, R.B., and D.H. Hildenbrand, 1984, *J. Phys. Chem. Ref. Data* 13:151.

Lane, G.S., and E.E. Wolf. 1988. *J. Catal.* 113:144.

Lee, J.S., and S.T. Oyama. 1988. *Catal. Rev. Sci. Eng.* 30:249.

Lee, S.-P., T. Yu, and M.C. Lin. 1990. *Int. J. Chem. Kinet.* 22:975.

Lin, C.H., K.D. Campbell, J.-X. Wang, and. J.H. Lunsford. 1989 *J. Phys. Chem.* 90:534.

Lin, C.-H., J.-X. Wang, and J.H. Lunsford. 1988. *J. Catal.* 111:302.

Lo, M.-Y., S.K. Agarwal, and G. Marcelin. 1988. *J. Catal.* 112:168.

Lunsford, J.H. 1987. *Adv. Catal.* 35:139.

Lunsford, J.H. 1988. Direct conversion of methane to methanol and higher hydrocarbons. In *Methane Conversion*, ed. D.M. Biddy, C.D. Chang, R.F. Howe, and S. Yurchak. New York: Elsevier.

McCarty, J.G., A.B. McEwen, and M.A. Quinlan. 1990. Models of direct catalytic oxidation of light alkanes. In *New Developments in Selective Oxidation*, ed. G. Centi and F. Trifiro, pp. 405–15. New York: Elsevier.

McCarty, J.G., and M.A. Quinlan. 1988. The role of alkali promoters in selective catalytic coupling of methane. *ACS Fuel Division Reprints* 33:363.

McEwen, A.B., and D.M. Golden. 1991. Unpublished.

McEwen, A.B., M.J. Rossi, and J.G. McCarty. 1991. Unpublished.

Mehandru, S.P., A.B. Anderson, and J.F. Brazdil. 1988. *J. Am. Chem. Soc.* 110:1715.

Mims, C.A., R.B. Hall, K.D. Rose, and G.R. Myers. 1989. *Catal. Lett.* 2:361.
Nelson, P.F., and N.W. Cant, 1990. *J. Phys. Chem.* 94:3756.
Nelson, P.F., C.A. Lukey, and N.W. Cant. 1988. *J. Phys. Chem.* 92:6176.
Nelson, P.F., C.A. Lukey, and N.W. Cant. 1989. *J. Catal.* 120:216.
Otsuka, K. 1987. *Sekiyu Gakkaishi* 30:385.
Otsuka, K., Y.W. Murakami, A.A. Said, and A. Morikawa. 1990. J. Catal. 121:122.
Otsuka, K., A.A. Said, K. Jinno, and T. Komatsu. 1987. *Chem. Lett.* 77.
Peil, K.P., J.G. Goodwin, and G. Marcelin. 1989. *J. Phys. Chem.* 93:5977.
Peng, X.D., D.A. Richardson, and P.C. Stair. 1990. *J. Catal.* 121:99.
Roos, J.A., S.J. Korf, R.H.J. Veehof, J.G. van Ommen, and J.R.H. Ross. 1989. *Appl. Catal.* 52:131.
Sinev, M.Yu., V.N. Korchak, and O.V. Krylov, 1988. Mechanism of catalytic oxidative methane condensation. In *Proceedings of the 9th International Congress on Catalysis*, Vol. 2, ed. M.J. Phillips and M. Ternan, pp. 968–73. Ottawa: The Chemical Institute of Canada.
Sokolovskii, V.D., O.V. Buyevskaya, S.M. Aliev, and A.A. Davydov. 1990. Heterolytic mechanism of methane activation in oxidative dehydrodimerization. In *New Developments in Selective Oxidation*, ed. G. Centi and F. Trifiro, pp. 437–46. New York: Elsevier.
Steward, P., Larson, C.W., and Golden, D.M., 1989. *Combust. Flame* 75:25
Steward, P., 1988. Private communication.
Steward, P., Smith, G.P., and Golden, D.M., 1989. *Int. J. Chem. Kinet.* 21:923.
Tong, Y., M.P. Rosynek, and J.H. Lunsford, 1989. *J. Phys. Chem.* 93:2896.
Tsang, W. 1986. *J. Phys. and Chem. Ref. Data* 15:1087.
van Kasteren, H.M.N., W.M.H. Geerts, and K. van der Wiele. 1988. Ethylene synthesis by catalytic oxidation of methane over Li-doped MgO catalysts: The interaction of catalytic and non-catalytic reaction steps. In *Proceedings of the 9th International Congress on Catalysis*, Vol. 4, M.J. Phillips and M. Ternan, pp. 930–6. Ottawa: The Chemical Institute of Canada.
Vol'nov, I.I. 1966. *Peroxides, Superoxides, and Ozonides of Alkali and Alkaline Earth Metals*, ed. A.W. Petrocelli, Chapter 2. New York: Plenum Press. (Translated by J. Woroncow.)
Wagman, D.D., W.H. Evans, V.B. Parker, R.H. Schumm, I. Halow, S.M. Bailey, K.L. Churney, and R.L. Nuttal. 1982. *J. Phys. Chem. Ref. Data* 11:Supplement No. 2.
Warnatz, J. 1984. Rate coefficients in the C/H/O system. In *Combustion Chemistry*, ed. W.C. Gardiner, Jr., Chapter 5, pp. 197–360. Berlin: Springer.
Yates, D.J.C., and N.E. Zlotin. 1988. *J. Catal.* 111:317.
Zanthoff, H., and M. Baerns. 1990. *Ind. Eng. Chem. Res.* 29:2.
Zellner, R., and F. Ewig. 1988. *J. Phys. Chem.* 92:2971.

10

Morphological Aspects of Catalysts for Oxidative Coupling of Methane

Guy A. Martin and Claude Mirodatos

Institut de Recherches sur la Catalyse, CNRS, Villeurbanne Cedex, France

TABLE OF CONTENTS

INTRODUCTION

The morphological aspects of the catalysts involved in the oxidative coupling of methane (OCM) are less considered in the literature than those related to the nature of active sites or to the mechanism. However, it can be anticipated that aging and stability problems will be considered with more attention in the future, particularly in the event of an industrial implementation. As far as we know, the lack of stability of OCM catalysts results from the severe experimental conditions that prevail, which give rise to important morphological changes due to a loss of matter from the solid (vaporization) and to severe sintering. Although sintering is a priori detrimental to productivity, it is not clear whether it is beneficial to the selectivity toward the desired products, C_2 hydrocarbons. One aim of this paper is to review data in the literature pertaining to the problem of the loss of matter from the catalyst and of sintering.

Other aspects of the morphology of OCM catalysts are related more or less directly to the nature of the active site: for single-component catalysts such as MgO or rare-earth oxides, it can be questioned whether the OCM reaction is structure-sensitive or not. Some works have already been carried out on this subject, from which a better understanding of the nature of the active site is expected. For multicomponent solids, such as alkali-"promoted" alkaline-earth or lanthanum oxides, additional questions pertaining to morphology arise. In particular, are the phases separated or do they form a stable stoichiometric compound? In the former case, does a surface enrichment occur or not, and does a synergy between both components exist?

In this chapter we will consider successively the following points: (i) loss of matter in OCM solids; (ii) surface area of OCM catalysts; (iii) structure sensitivity of OCM; (iv) phase organization in two-component OCM catalysts.

LOSS OF MATTER IN OCM CATALYSTS

With the exception of palladium-based catalysts, which still demonstrate poor yields in C_2 hydrocarbons (Chan and Wilson 1988; Thampi, Kiwi, and Grätzel 1990; Mirodatos et al. 1990), oxidative coupling of methane requires high reaction temperatures, 700–900° C. For these conditions a number of components in multiphase catalysts, such as lead, alkalis or chlorine, are susceptible to evaporation. The loss of matter is considered as one of the main causes for the deactivation, together with sintering (a point that will be taken up in another section) and to a lesser extent with carbon deposits. In this section we review the works carried out on the volatility of lead and alkali compounds associated with MgO and other supports (alkali-promoted catalysts yield greater selectivity than unpromoted alkaline-earth and rare-earth oxides). Additionally, the following questions will be considered: Is it possible to quantify the phenomenon? Is there a limit to the volatility of alkali metals and how is it possible to prevent their vaporization in order to improve stability?

Lead Vaporization

Since the early studies of oxidative coupling of methane over lead-based catalysts, significant Pb losses have been reported; traces of lead and lead oxide were detected in the cold parts of the reactor outlet in a study of Pb/Al_2O_3 and Pb/TiO_2 catalysts (Hinsen, Bytyn, and Baerns 1984). A study of the profile of lead concentration in a nonisothermal fixed-bed reactor has been carried out (Follmer, Lehmann, and Baerns 1988). The bulk lead content of PbO/Al_2O_3 catalysts (29 wt%) decreases strongly as compared to the fresh catalyst. As an example, after 270 h on stream, the concentration of lead was found to be 9.7, 4.8, 14.4, and 13.9 wt% at 0, 4, 6, and 8 cm from the outlet, respectively. It has been shown that during the calcination step (500° C) Pb–Mg–O catalysts did not lose much Pb (Agarwal, Migone, and Marcelin 1990). After reaction at 780°C the lead content in the bulk decreased significantly for catalysts above 0.8 at% Pb. Both the surface and the bulk were depleted of the Pb species. There is also a rapid replenishment of the surface Pb species from the bulk, which play the role of a reservoir toward the surface, leading to a good stability of the catalyst.

Li / MgO and other Alkali-Promoted Catalysts

The loss of Li in Li/MgO and Li/CaO catalysts has been studied (Kimble and Kolts 1986). After calcination at 850°C all the Li-containing solids have lost Li. Moreover, post-use analysis reveals that all catalysts containing more than 1 wt% Li continue to lose Li down to ca. 1%. For example, an 8% Li, Li/CaO catalyst was shown to contain 4.2% after the calcination step and 1.0% after reaction at 700°C; another one, prepared as 3% Li, declined to 1.5 and 0.9%, respectively. The possible existence of a limit on Li concentration was underlined by these authors. The authors do not speculate on its possible origin, however. It has been confirmed that Li in Li/MgO catalysts is lost during the reaction (Mirodatos et al. 1987). At variance with Kimble and Kolts, these authors did not find an important loss during the calcination step. This can be explained by the lower temperature used for calcination (750°C versus 850°C). When the reaction is run at 634°C, the lithium loss is negligible. At 750°C, however, the Li/Mg ratio decreases from 0.57 to 0.18 (2.6 wt% Li) after 60 h of reaction.

The stability of a Li/MgO catalyst at a reaction temperature of 800°C has been studied (Korf et al. 1988a, 1988b). With quartz dilution the C_2 yield decreased rapidly whereas the C_2 selectivity remained constant, indicating that the number of sites had decreased; the Li content had fallen from 3.1 wt% to ca. 0.1 wt% Li. Li had reacted with quartz to give Li_2SiO_3 and LiOH was identified downstream from the reactor. In the absence of quartz diluent the stability was better. The Li content limit, 0.1 wt%, reported in this work is obviously smaller than that observed by other authors (Kimble and Kolts 1986; Mirodatos et al. 1987). This is likely to be due to the higher operating temperatures used in the work of Korf and co-workers (800°C). An important factor is the formation of volatile LiOH, which is favored by water pressure. It has been confirmed (Kaminski et al. 1989) that diluents such as SiO_2 and Al_2O_3 can trap Li: Li leaves Li/MgO catalyst in the form of LiOH and reacts with diluents to form silicates or aluminates, which are themselves more or less active toward deactivation. It has also been confirmed (Matsukata et al. 1989) that steam addition to the gas stream results in an increase of Li loss from Li/MgO catalysts. For example, a Li content of 7.09 wt% drops to 3.02 and 1.19 wt% after 5 h of reaction at 725°C without or with steam, respectively. The calcination temperature of Li/Ce/Mg catalysts changes the Li loading, the surface area, and the C_2 yield (Bartsch and Hofman 1990). However, the surface area decreases more rapidly than the Li content and the C_2 yield, suggesting that deactivation is strongly affected by Li losses. The initial weight fraction of Li (0.77 wt%) decreases with time on stream to 0.09 wt%. After 15 h on stream, the Li depletion is more important at the reactor inlet; this is explained by the higher temperature prevailing at the inlet leading to a faster LiOH evaporation. Remarkably, the step in Li concentration profile disappears after 66 h on stream. The Li vaporization at the inlet and its redeposition on the other parts of the catalyst bed can explain that deactivation proceeds more slowly if the amount of catalyst used is increased (Bartsch and Hofman 1990; Korf et al. 1987).

It has also been observed that sodium in Na/Sm_2O_3 was lost during the reaction (Korf et al. 1989). The content has fallen from 55 to 20 mol%.

In the development of our model, we quickly found that the rate of single-step O_2 dissociative chemisorption at adjacent vacant sites ($\Diamond_s$) was rate-limiting for active site concentrations low enough to give good higher-hydrocarbon selectivity. The calculated rate of oxygen chemisorption had to be enhanced by assuming either that oxygen was diatomic or that adsorption of atomic oxygen required two distinct steps: the adsorption of diatomic O_2 on a vacant surface site ($\Diamond_s$) and its subsequent rate-determining dissociation to form two active atomic oxygen centers [equation (9-5)].

$$\Diamond_s + O_2 \Leftrightarrow (O_2)_s + \Diamond_s \Leftrightarrow 2O_s \tag{9-5}$$

The oxygen adsorption process was assumed to be exothermic for both steps with no activation barriers. The frequency factors were taken as again one-tenth of the collision frequency for $(O_2)_s$ and $10^{12}\ s^{-1}$ for reaction of $(O_2)_s$ with a vacant site ($\Diamond_s$). Subsequent rates calculated with these parameters showed that typically the reactive oxygen sites were much more than half-occupied even at high temperature and low oxygen partial pressure. The density of reactive oxygen sites was found (McEwen, Rossi, and McCarty 1991) to be more important than the fractional occupancy level. Therefore, the practical effect of including Equation 9.5 in the reaction model is that the heterogeneous conversion of methane stops when gaseous oxygen is consumed. The assumption that oxygen adsorption is fast relative to the rate of methane conversion is not obvious, however, and is discussed further later in this chapter.

We used the Chemkin kinetic modeling program (kindly provided by the Department of Energy, Lawrence Livermore Laboratory) to solve the set of nonlinear differential equations. All calculations were performed as batch reactions rather than as flow reactions. This should be an accurate simplification for low flow velocities and a very small density of surface centers, because the time required for the intermediate surface concentrations to adjust to changes in gas composition was typically a few tens of microseconds. This assumption is not valid for carbonate accumulation. As in all other models that we know about, mass transport is presumed fast relative to the reaction rates. Therefore we ignore temperature and concentration gradients. Finally, we optimized the (two adjustable) heterogeneous rate parameters involving surface sites to fit our experimental results for methane conversion over Na/CaO at 1073 K.

We first used this model (McCarty, McEwen, and Quinlan 1990; McEwen, Rossi, and McCarty 1991) for pressures ranging from 0.1 to 100 atm and for temperatures between 873 and 1223 K with various methane and oxygen concentrations. Calculations for reaction at 1073 K with CH_4

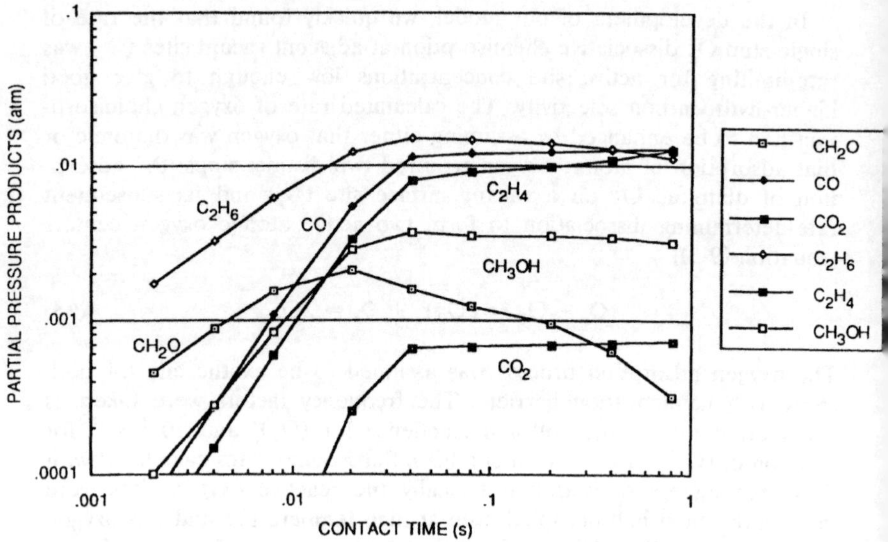

FIGURE 9-1. Calculated product distribution as a function of reaction time for methane cooxidative coupling at 1073 K with 0.3-atm CH_4 and 0.03-atm O_2 feed gas partial pressure.

and O_2 concentrations of 0.3 and 0.03 atm, respectively, showed that ethane was the major carbon-containing product, as expected, followed by CO and ethylene (Fig. 9-1). Other significant products, methanol and formaldehyde, decreased in relative importance with increased reaction time. These findings are in general agreement with those of a similar kinetic model (Geerts et al. 1990).

Effect of Reactive Site Density on Higher-Hydrocarbon Selectivity

The most significant finding of our model was that the selectivity of direct catalytic methane oxidation in both redox and cooxidative coupling processes could be related to the concentration of reactive oxygen on catalyst surfaces. Product selectivity was investigated as a function of the active site density (i.e., the concentration of surface oxygen centers normalized to the gas specific volume of the catalyst bed) at 1073 K and methane-to-oxygen feed gas ratio equal to 10 (Fig. 9-2) using the reaction model. Higher than optimal surface concentrations of reactive oxygen result in greater methane conversion rates, but with very low coupling selectivity caused by rapid heterogeneous oxidation of intermediate methyl radicals,

Table 10-1 Lithium Losses from Alkali-Promoted Catalysts for Oxidative Coupling of Methane in Reaction Conditions

Catalysts	ALKALI WT% Before Reaction	ALKALI WT% After Reaction	Reaction Temperature (deg)	Timeon Stream (h)	Reference
Li/MgO		4.8	634	128	Mirodatos et al. 1987
Li/MgO	7.09	3.02	725	5	Matsukata et al. 1989
Li/MgO	6.6	4.4	750	1.33	Mirodatos et al. 1987
	—	3.8	750	16	Mirodatos and Martin 1988
	—	2.6	750	60	Mirodatos and Martin 1988
Li/MgO +quartz	3.1	0.1	800	—	Korf et al. 1988a, 1988b
Li/CaO	4.2	1.0	700	—	Kimble and Kolts 1986
	1.5	0.9	700	—	Kimble and Kolts 1986
$Li/CeO_x/MgO$	0.77	0.3	750	15	Bartsch and Hofman 1990
	—	0.09	750	66	Bartsch and Hofman 1990

A white deposit (NaOH) was observed downstream from the reactor; it was also found that Na had reacted with the quartz reactor to form $NaSiO_3$. In the same way, the Rb^+ and Na^+ content of Rb/MgO and Na/MgO catalysts decreases after calcination (800°C) and reaction (750°C) (Iwamatsu et al. 1988). The loss is more important for Rb/MgO.

The question of the lithium content limit deserves further comments. An inspection of Table 10-1 reveals that if this limit exists, it could be of the order of magnitude of 0.1 wt%, the lower figure obtained on Li/MgO. The Li solubility in MgO cannot account for this value: As a matter of fact, electrical conductivity measurements on Li/MgO with varying Li contents show that at 700°C in CO_2-rich oxidizing atmosphere, the Li solubility in MgO is of the order of magnitude of 0.1 mol% (0.017 wt%) (Anderson and Norby 1990).

It can also be speculated that this limit results from the formation of a Li_2CO_3 monolayer on the MgO surface. It is not possible to check with accuracy the validity of this hypothesis because the BET surface areas are not always available. However, if one assumes a 1 m^2/g^{-1} surface area, which is a plausible order of magnitude, one can estimate the Li content limit in the following way: Lithium carbonate is monoclinic ($A = 0.8359$ nm, $B = 0.49767$ nm, $C = 0.6194$ nm, $b = 114° 71$) and the unit cell is tetramolecular (eight Li atoms). The average surface area of the unit cell is 0.3982 nm^2. If a surface of 1 m^2 g^{-1} MgO is covered with a lithium carbonate monolayer, the calculated Li content is found to be 0.023 wt% (Li). This figure is far from the lowest Li content ever observed after reaction. This suggests that in actual catalysts lithium carbonate is in large excess and that in all cases the present amount is sufficient to ensure a full coverage of the MgO surface.

It can be questioned if LiOH vapors, which are permanently present in the gas phase, can play a role in the oxidative coupling of methane, which, as is now well established, involves gas-phase homogeneous reactions. As a matter of fact, it can be remarked that steam addition to the gas stream results in an increase of activity (Matsukata et al. 1989) and of Li loss, indicating that more Li is present in the gas phase. Conversely, it has been shown that CO_2 addition decreases both activity and Li vaporization (Korf et al. 1988; Kaminsky et al. 1989).

Influence of the Nature of Supports on Alkali Vaporization

In the course of a study of potassium addition to Fe/MgO, Ni/MgO, and Ni/SiO_2 carried out in our laboratory (Moral, Praliaud, and Martin 1987), a loss of potassium from magnesia-supported catalyst was observed at reduction temperatures in excess of 500°C under flowing hydrogen. In contrast, silica-supported catalysts yielded a different behavior: Silica was found to inhibit alkali volatilization, probably because of the formation of a stable K silicate hindering K loss. This cannot occur on MgO, where Li, which cannot combine with MgO, is in the form of more volatile Li_2O or LiOH. A leaching technique based on the difference of solubility of K silicates and K_2O (or KOH) in water and alcohols was used to confirm our assumptions. We took advantage of the fact that K_2O and KOH are soluble in both solvents whereas silicates are soluble only in water. The quantity of potassium extracted from MgO-based catalysts by water is equal to that extracted by ethanol. For silica-based catalysts most of the potassium is extracted by water while a small fraction is obtained from ethanol leaching, revealing the formation of an alkali silicate in agreement with our assumptions.

Owing to the relevance of these results to the oxidative coupling catalysts, a more systematic study of the volatility of alkali compounds deposited on various supports has been undertaken (Perrichon and

Durupty 1988). Different alkali metal promoted supports were prepared by impregnating SiO_2, Al_2O_3, Cr_2O_3, and MgO with solutions of lithium, sodium, or potassium nitrate on carbonate. After calcination under an air flow at increasing temperatures, it was found that the deposited alkali metals are fairly stable on Cr_2O_3 or Al_2O_3 and slightly less on SiO_2. With MgO, the important loss of alkali metal already observed (Moral, Praliaud, and Martin 1987) was confirmed at calcination temperatures in excess of 500°C. This effect increases when going from Li to K.

Inhibition of Alkali Vaporization

Two possibilities have been considered to prevent alkali losses: (i) modifications of solids and (ii) changes in the reaction conditions.

In the patent literature, addition of phosphorus (as phosphate or pyrophosphate) is claimed to improve the stability of Ru, Mn, and alkaline-earth metal catalysts promoted with alkalis (Withers 1984; Jones and Sofranko 1985). In the patents, however, it is not clearly stated whether the role of phosphates consists in stabilizing the alkali metal promoter, as for the case of silica alumina and chromia (Perrichon and Durupty 1988), or in being themselves active species.

The stability improvement induced by phosphorus addition to alkali-promoted Mn/SiO_2 (and rare-earth oxides) is thought to be related to enhanced retention of surface area (Jones, Leonard, and Sofranko 1987; Gaffney et al. 1988) and of alkali as well. Silica addition to Li/MgO catalyst can have a beneficial effect on Li loss (Perrichon and Durupty 1988). This idea was put to the test in our laboratory by preparing basic magnesium silicates. Their structure is derived from that of brucite, $Mg(OH)_2$, where OH groups are replaced by silica tetrahedra according to the following formulas:

$(OH)_3$	$OHSi_2O_5$	$OHSi_2O_5$
/	/	/
Mg_3	Mg_3	Mg_3
\	\	\
$(OH)_3$	$(OH)_3$	$OHSi_2O_5$
brucite	chrysotile	talc
	(lizardite, antigorite)	

Figure 10-1 shows that in calcination experiments Li is stabilized on chrysotile, at variance with MgO, and that its stability is similar to that observed on silica (Martin et al. 1991). Under reaction conditions it has

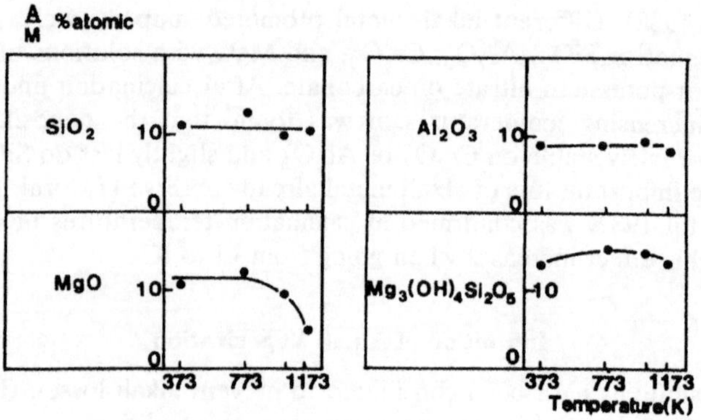

FIGURE 10-1. Li stability as a function of calcination temperature: Li/MgO, Li/SiO_2, Li/Al_2O_3, Li/chrysotile [$Mg_3(OH)_4Si_2O_5$].

been confirmed that Li vaporization is hindered in Li/chrysotile, which appears to be more active, more stable, and more selective toward ethylene than Li/MgO catalysts (the selectivity toward C_{2+} hydrocarbons is nearly similar) (Martin et al. 1990). It should be noted, however, that the chrysotile phase is destroyed after reaction and that a local segregation of Mg and Si phases occurs.

It is also possible to inhibit Li loss and to stabilize Li/MgO catalysts against deactivation by the addition of CO_2 to the gas feed (Korf et al. 1988, 1987). As a matter of fact, the formation of volatile LiOH can be minimized by decreasing the H_2O/CO_2 ratio by the addition of CO_2 to the feed (conversely, as already seen, water addition is very detrimental to the stability of the catalyst, by promoting LiOH formation). It has been confirmed (Kaminsky et al. 1989) that CO_2 addition in the gas stream stabilizes the Li/MgO catalyst against deactivation (at the expense of a lower activity, however) whereas H_2O decreases the activity maintenance (with the benefit of an improved conversion). They also attribute this behavior to the volatility of LiOH, the concentration of which increases in the presence of water and decreases with CO_2 in the gas stream. In addition, these authors have investigated the response of Li/MgO performance to H_2S addition. H_2S was shown to increase deactivation and Li losses. It was assumed that H_2S is oxidized into SO_2, which reacts with Li_2O to form sulfite or bisulfites (which are even more volatile than LiOH).

To minimize Li loss, another possibility can be considered, based on the fact that the lower the temperature, the higher the Li stability. Efforts to decrease the reaction temperature can be promising. From this viewpoint, the work carried out by a Dutch group (Korf et al. 1989) is of special interest; they have shown that the addition of small amounts of additive (SnO_2, TiO_2, Dy_2O_3, and Tb_4O_7) resulted in a noticeable decrease in the temperature required to give optimal yields of C_2 products. Li/Sn/MgO gives optimal yields 70°C lower than for the unpromoted Li/MgO, thus improving its stability.

SURFACE AREA OF OCM CATALYSTS

A recurring idea in oxidation catalysis is that too-large surface areas favor consecutive reactions, leading to deep oxidation and to a loss of hydrocarbon selectivity. It also can be anticipated that for oxidative coupling of methane the surface area can play in itself an important role, independently of structure-sensitivity effects. In this review, we present some data available in the literature dealing with this question without considering problems of porosity and transport limitation (Follmer, Lehmann, and Baerns 1989).

It is the experience of laboratories working on the OCM catalysts that dramatic sintering occurs during reactions conducted at high temperature and in the presence of water vapor. We therefore also examine the sintering process in calcination and reaction conditions, the influence of alkalis, and the possible methods to stabilize surface areas.

Influence of Surface Area on the Catalytic Pattern

In the first paper reporting on this subject (Moriyama et al. 1986), various metal-doped MgO samples were examined. The specific surface area after heat treatment at 800°C for 2 h was shown to decrease continuously from 70 $m^2\ g^{-1}$ for unpromoted MgO down to 1 $m^2\ g^{-1}$ for Na^+ mole content of 20%. The selectivity toward C_2 hydrocarbons increases in a parallel way from ca. 10 to 55% when the sodium content does not exceed 15 mol%. These preliminary results were confirmed (Iwamatsu et al. 1987) by examining another series of catalysts based on MgO. Addition of alkali results in sintering (loss of area) and in an improvement of C_2 selectivity. It was speculated that the specific rate (per unit weight) of methyl radical formation increases with the surface area, leading to higher specific rate of reaction. However, the rate of collision of radicals with the surface is increased. This would lead to total oxidation into CO_2, thus decreasing the selectivity into C_2 hydrocarbons. The possible relationship between specific surface area and the reaction was detailed in another paper. In the following reaction scheme, steps 1, 2, 3, (5), and 6 are supposed to be heterogeneous whereas steps 4 and 5 are homogeneous.

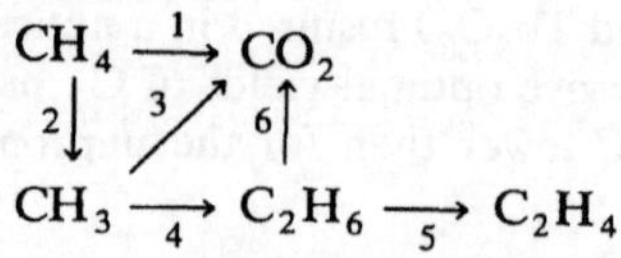

Methyl radicals could be oxidized to CO_2 on the surface (step 3) or undergo recombination in the gas phase (step 4). Step 4 should be independent of the surface area whereas steps 2 and 3 should be improved with increasing the surface area. This leads to the idea that an optimal surface area will give a maximum C_2 yield. Figures 10-2 and 10-3 show the

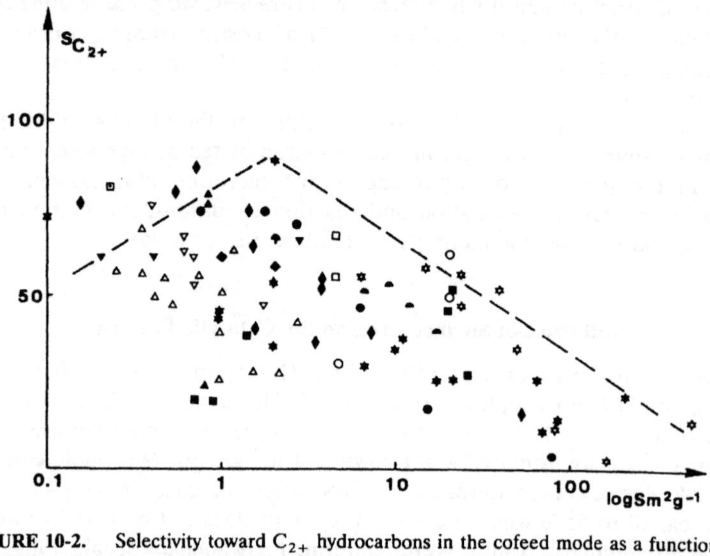

FIGURE 10-2. Selectivity toward C_{2+} hydrocarbons in the cofeed mode as a function of the BET surface area measured after use for various catalysts in different experimental conditions: ●, Martin et al. 1991; ▲, Lunsford et al. 1989; ▼, Andorf and Baerns 1990; ■, Aagarwal, Migone, and Marcelin 1990; ◆, Ito et al. 1985; ♠, Lin et al. 1987; ▬, Perrichon and Durupty 1988 (the surface area determined before reaction after calcination at 850°C); ★, Moriyama et al. 1986; Iwamatsu 1987 (S measured after calcination at 800°C or 1000°C before use). The following experimental points refer to surface area measurements that are carried out before use after calcination at a temperature lower or comparable to the reaction temperature, or in conditions that are not specified: O, Hargreaves, Hutchings, and Joyner 1990; □, Sokolovskii 1990; ◆, Burch, Squire, and Tsang 1988; ▽, Anshits et al. 1989; △, Matsuura et al. 1989; ☆, Bytyn and Baerns 1986; Carreiro, and Baerns 1987; Hinsen, Bytyn, and Baerns 1984; ⊡ , Otsuka, Hatano, and Komatsu 1989.

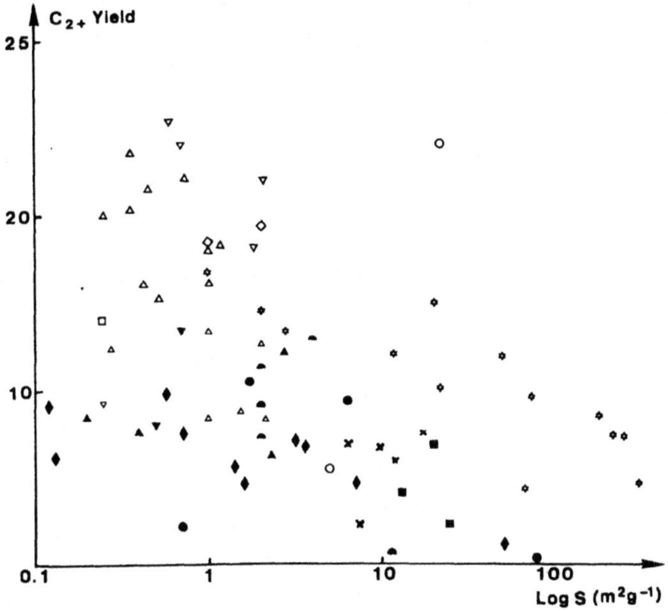

FIGURE 10-3. C_2 yield versus BET surface area. The symbols have the same significance as in Fig. 10-2.

variations of the selectivity toward C_{2+} hydrocarbons and the yield, respectively, with the specific surface area *S* of various solids reported during OCM studied in the cofeed mode at various reaction conditions; the data obtained in the redox mode (Jones, Leonard, and Sofranko 1987) are not shown. In some works, the BET surface area was measured after calcination before catalytic use. When the calcination temperature was higher than the reaction temperature (Moriyama et al. 1986; Iwamatsu et al. 1987; Korf et al. 1989), we have considered that catalytic run did not lead to significant additional sintering. If we consider these points together with points corresponding to surface areas measured after use (black symbols in Figs. 10-2 and 10-3), one can see that in the $S_{C_{2+}}$–log S plot experimental points are situated below a volcano-shaped envelope with a maximum at ca. 2 m^2 g^{-1}. The situation is less clear if one plots the C_{2+} yield versus log S. This seems to substantiate the idea that too large a surface area is detrimental to a high selectivity toward C_{2+} hydrocarbons (Moriyama et al. 1986).

This concept, however, can be criticized on the following basis: (i) in this kind of experiment the physical factor (specific surface area) is changed together with the chemical factor. As an example, the results dealing with the addition of Na to MgO, which leads to a decrease of the specific surface area and an increase of C_{2+} hydrocarbons selectivity, can be explained alternatively in terms of variable sodium coverage of MgO, lower coverage (corresponding to lower content) giving rise to bad selectivities. (ii) The reaction scheme on which the relationship between surface area and selectivity is based is not universally accepted.

In particular, the gas-phase nature of the recombination of methyl radicals is still a subject of discussion (Baerns and Ross 1990): One cannot exclude that methyl radicals formed on the surface combine on the same surface. Furthermore, some data indicating that the specific surface area is a negative factor have already been published. As an example, among three Li/MgO catalysts prepared by different techniques, the most selective toward C_2 hydrocarbons has the greatest specific surface area (Lunsford 1990; Lunsford et al. 1989). The authors, however, do not contend that their results do not preclude the concept of an optimum surface area, but they do not rule out a general inverse relationship between surface area and C_2 yield, either. The same trend is observed for Pb/MgO catalysts (Agarwal, Migone, and Marcelin 1990). In our laboratory (Mirodatos et al. 1989) we have varied the specific surface area of MgO from 70 to 28 $m^2\ g^{-1}$ by sintering at 750°C in the reacting mixture. Afterwards the selectivity toward C_2 hydrocarbons decreases from 6.2 to 3.8%. It does not fit well with the idea that surface area is detrimental to selectivity. It cannot be overlooked, however, that heat treatments may alter the surface morphology. Because oxidative coupling of methane is probably a structure-sensitive reaction (see below), sintering experiments may not be conclusive. It can be concluded, therefore, that much work is needed to substantiate the concept of optimal surface area in oxidative coupling of methane, particularly by comparing catalysts of different dispersion, the chemical factor being kept unchanged.

Sintering of Alkali Metals Deposited on Oxides during Calcination

Some studies have been done in our laboratory (Perrichon and Durupty 1988; Mirodatos and Martin 1988) aimed at clarifying the influence of the calcination temperature on the surface area of alkali-promoted oxide catalysts relevant to the oxidative coupling reaction catalysts: SiO_2, Al_2O_3, Cr_2O_3, and MgO were impregnated with solutions of lithium, sodium, and potassium nitrates or carbonates. It has been observed that, whatever the alkali nature and the calcination temperature, the addition of alkali results in a loss of BET-specific surface area. From a quantitative point of view, this phenomenon depends on the nature of the alkali and of the oxide. The alkali-induced sintering effect varies according to the following sequences:

$$MgO > SiO_2 > Cr_2O_3 > Al_2O_3 \tag{10-1}$$

$$K, Na \gg Li \tag{10-2}$$

Lithium, added as nitrate or carbonate, achieves the most interesting behavior with respect to catalyst resistance to sintering, whereas MgO suffers the most significant sintering in the presence of an alkali. To illustrate this feature, it can be underlined that the surface area of a Li/MgO solid calcined at 750°C decreases from 70 to 0.4 $m^2\ g^{-1}$ whereas an unpromoted MgO is hardly sintered (from 94 to 54 $m^2\ g^{-1}$). Al_2O_3 appears to be a very stable support: Its specific surface area is only slightly altered in the presence of alkalis at temperatures as high as 930°C. In the same way (but to a lesser extent), Li/SiO_2 catalyst is relatively stable: The BET surface area decreases from 138 to 91 $m^2\ g^{-1}$ when it is calcined at 750°C. It cannot be overlooked, however, that Li/Al_2O_3 and Li/SiO_2 have low selectivity for the oxidative coupling of methane.

The better selectivity of SiO_2 can be advantageous to try to promote Li/MgO stability. This has been done in our laboratory by using hydrosilicates of magnesium impregnated with Li (Perrichon and Durupty 1988). The BET surface area of these solids after reaction at 750°C increases from 0.5 to 3 $m^2\ g^{-1}$ when the 3Si/Mg ratio of the precursor varies from 0 to 12. The selectivity toward C_2 hydrocarbons is little altered and the conversion increases by a factor of 5. The rate of deactivation during the first 2 h of the catalytic run is 5 times smaller. Thus silica addition appears to be very beneficial by improving productivity and stability. This can be partly related to the stabilization of the dispersion due to SiO_2, which can be considered as a texture promoter (another beneficial effect of silica is to reduce lithium volatilization). Similar experiments could be carried out with Al_2O_3, which is probably a better stabilizer than SiO_2. In particular, the utilization of hydroaluminate of magnesium can be considered.

Only a few papers have been published reporting on the influence of the quantity of added promoter on the BET surface area after calcination at a given temperature. It has been shown (Moriyama et al. 1986) that S equals 70, 21, 11, 2, and 1 $m^2\ g^{-1}$ after calcination at 800°C, when the Na^+ content in Na/MgO is 0, 2, 5, 10, and 20 mol%, respectively. CaO with different Li content were prepared (Anshits et al. 1989); when the Li content varies from 0.1 to 1.8 at%, the surface area after calcination at

950°C does not change significantly (ca. 0.4 m^2 g^{-1}). For the same conditions the surface area of pure CaO is 2.8 m^2 g^{-1}. $Pb/\gamma\text{-}Al_2O_3$ catalysts promoted with increasing amounts of sodium display the same surface area (ca. 44 m^2 g^{-1}) after calcination in standard conditions (Hinsen, Bytyn, and Baerns 1984). The surface area of Pb/MgO calcined at 800°C decreases continuously from 36 to 11 m^2 g^{-1} when the lead content increases from 0 to 4.4% (Agarwal et al. 1990).

Sintering during Reaction

Demonstrative experiments were carried out with catalysts that were calcined at moderate temperature before use (Mirodatos et al. 1989; Mirodatos and Martin 1988). Some results obtained in our laboratory (Table 10-2) illustrate the sintering of Li/MgO in the course of reaction.

From Table 10-2 it can be seen that at 634°C the admission of the reacting mixture gives rise to an important decrease in surface area from 50 to 4.3 m^2 g^{-1} although at this temperature calcination does not lead to sintering. The nature of the reaction atmosphere plays an important role in surface area loss. In the presence of the reacting mixture, an additional sintering occurs, probably via the formation of reaction products such as H_2O or CO_2. When the catalyst is heated up to 750°C, either in flowing O_2 or in the reacting mixture, a further decrease of the surface area is observed that is independent of the nature of gases present in the catalyst bed. This second sintering step is very beneficial for catalytic performance because it corresponds to an improvement of the selectivity toward C_{2+} hydrocarbons. It corresponds also to the embedding of MgO by Li_2CO_3, which is probably liquid at this temperature, and to pore plugging (see the

Table 10-2 Sintering of Li / MgO Catalysts during Reaction

TREATMENT				
Gas	Temp. (° C)	Time on Stream (mn)	Surface Area (m^2 g^{-1})	BJH Pore Size (nm)
O_2	500	240	60	6.7–9.0
	630	—	70	7.7–10.3
	750	—	0.5	No porosity
$CH_4 + O_2$	634	1080	4.3	No porosity
	750	80	0.57	No porosity
	750	960	0.56	No porosity
	750	3600	0.44	No porosity

section on phase organization in two-component catalysts). The degree of agglomeration of powders seems also to be an important parameter for sintering. For instance, after calcination at 650°C in CH_4–O_2 atmosphere, the surface area of NaOH/CaO is 2.2 and 1.3 $m^2\ g^{-1}$ for powder (0.25 and 0.35 mm) and pellets (d = 10 mm, h = 4.6 mm), respectively, and 0.8 and 0.5 $m^2\ g^{-1}$, respectively, after reaction at 700–800°C. (Andorf and Baerns 1990). Pb–Mg–O catalysts are remarkably dispersed after reaction at 780°C (Agarwal, Migone, and Marcelin 1990). The surface area loss after reaction is less than 30% (before reaction, solids were calcined at 800°C).

STRUCTURE SENSITIVITY OF OCM

Few studies have attempted to investigate the relationship between the structure of solids and their catalytic behavior in the oxidative coupling of methane. The concept of structure sensitivity was introduced (Boudart et al. 1966) in order to characterize a catalytic reaction for which the activity and/or the selectivity toward the formation of the products depend on the structure of the catalytic site. This concept initially proposed for metal catalysts seems also to be applicable to selective oxidation reactions over oxides (Volta et al. 1984; Volta and Portefaix 1985; Hicks et al. 1990). This is not completely unexpected because electron-donor properties and chemical reactivity of surface oxygen ions depend on their coordination (Che and Tench 1982). The case of O^{2-} in MgO is well-documented. As far as oxidative coupling of methane involves surface-assisted steps, it can be questioned whether this reaction is also structure sensitive, and if so, what is the reason for this behavior.

MgO and MgO-Based Catalysts

A comparative study of the structure and catalytic properties of two MgO samples has been carried out (Hargreaves, Hutchings, and Joyner 1990): MgO from decomposition of basic carbonate is found to give higher selectivities to C_2 hydrocarbons (60–70% in typical conditions) compared to MgO obtained from the oxidation of metal ribbons in air (30%). On the basis of unit surface area (21 and 5.1 $m^2\ g^{-1}$, respectively) the total activity of the latter is twice that of the former. However, the intrinsic activity for the production of C_2 hydrocarbons is three times larger. It is concluded that MgO ex basic carbonate has a higher relative concentration of the active sites for the coupling reaction. It can be remarked that the catalyst with the higher surface area displays the higher selectivity. Thus, the variation of selectivity cannot be explained in terms of a possible detrimental effect of surface area on selectivity (Moriyama et al. 1986; Iwamatsu et al. 1988).

As seen from electron microscopy, the latter catalyst consists of agglomerated MgO crystallites whereas the least selective solid consists of single MgO crystallites with regular cubic structure. It is concluded that the morphology of MgO is an important factor in the oxidative coupling of methane and that high-index planes with higher concentration of low-coordinate oxide ions are more selective for C_{2+} hydrocarbon production. This observation could be rationalized in terms of a heterolytic adsorption mechanism of methane ($CH_4 \rightarrow CH_3^- + H^+$, where CH_3^- and H^+ are associated to Mg^{2+} and O^{2-}, respectively) that could occur on a pair of three coordinated Mg and O atoms on the surface that are present on a high-index plane of MgO such as (111) (Ito et al. 1983, 1987). This can be understood easily because the charges of the oxygen ions are retained less by the smaller number of Mg^{2+} ions of low coordination sites. The electron-donor properties of O^{2+} are thus increased, making the heterolytic adsorption easier. A detailed picture of the interaction of CH_4 with the three coordination sites of a Mg_4O_4 model cluster has been investigated (Kobayashi and Yamagushi 1989) by means of an ab initio molecular orbital (MO) method. It is shown that only the heterolytic adsorption ($OH/MgCH_3$) leads to stabilization, and the activation energy is found to be nearly equal to 7 kcal mol^{-1}. This value compares well with the activation energy of the CD_4–CH_4 equilibration reaction over MgO, 8.9 kcal mol^{-1} (Quanzi and Amenomiya 1986), which proceeds via the heterolytic adsorption mechanism.

An interpretation of the promoting effect of Li addition to MgO based on the structure sensitivity of the OCM reaction has been proposed (Anpo et al. 1989; Matsuura, Utsumi, and Doi 1989). Two luminescence bands were observed on Li/Mg (not during reaction). The first one, at 220–270 nm, is associated to the charge-transfer processes on the coordinatively unsaturated surface site with a coordination number of 4 (Coluccia and Tench 1981):

$$Mg^{2+}O^{2-} \underset{h\nu}{\overset{h\nu}{\rightleftharpoons}} Mg^{+}O^{-} \tag{10-3}$$

A second band is observed at ca. 450 nm that develops as Li is added: A good relationship between the intensity of this newly observed band and the catalytic activity is found. Transmission electron microscopy (TEM) shows that the hexahedral crystal of MgO has a well-defined outline and presents the same plane indexed as (100). Upon adding Li to MgO, sintering occurs and the regular hexahedral shape transforms into a sphere, which is supposed to be a 26-sided polyhedron. Various surface planes, such as (111), (011), (110), and (001), appear as seen from the interference fringe. It is assumed that the photoluminescence band at 450 nm would correspond to a charge-transfer process [equation (10-3)] on a lower-coordinative surface site, perhaps 3, located on the high-index planes of MgO. This surface site, produced by incorporating lithium in MgO would play a significant role in this reaction.

It can be concluded that convincing evidence of the structure sensitivity of the OCM reaction over pure MgO has been provided and that high-index planes associated to low-coordinate oxide ions could be required for this reaction. This observation parallels the assumption that the heterolytic adsorption of CH_4 on MgO occurs also on such low-coordinate ions. This poses the problem of the relationship between the heterolytic adsorption of methane and the OCM reaction, which can be dealt with on experimental grounds by comparing the rate of the reaction between CH_4 and CD_4 in to CH_3D and CD_3H, and the rate of the OCM reaction by reacting a mixture of CH_4 and CD_4 with O_2 over the OCM catalyst (Mirodatos et al. 1990). The structure sensitivity of this reaction over MgO provides a possible explanation for the promoting effect of lithium. It will be seen later, however, that alternative hypotheses can also be formulated.

Sm_2O_3 and Sm_2O_3-Based Catalysts

Rare-earth sesquioxides Ln_2O_3 exhibit three crystallographic forms: hexagonal, monoclinic, and cubic. The monoclinic arrangement derives from the hexagonal one by a small shift of $(MO)^{n+}{}_n$ planes in a direction perpendicular to the C axis of the hexagonal structure. The structure depends on the size of the cation: It is hexagonal for large cations such as La and Ce and cubic for smaller cations. A temperature-dependent polymorphism has also been observed; for example, Sm_2O_3 is cubic below 850–900°C (Rosynek 1977). This metastable form changes irreversibly into the stable monoclinic structure at higher temperatures. This provides a good opportunity to examine the sensitivity to the *bulk* structure of the OCM reaction. This has been done independently by two laboratories and the results are in good agreement. It has been shown (Korf et al. 1989) that the cubic form is not only more active but that it is also much more selective than the monoclinic form; the yield obtained on the cubic form is three times that of the monoclinic one. Addition of Li to Sm_2O_3 has a detrimental effect on the catalytic performance whereas Ca and Na have a beneficial effect. It was shown by X-ray diffraction that Li has a destabilizing effect on the resultant catalyst, favoring the cubic–monoclinic transformation. The order of stabilizing effect towards the cubic phase of additives to Sm_2O_3 is (Korf et al. 1988a, 1988b)

$$Ca > Mg > La$$

Furthermore Na, as Li induces a decrease in stability. The explanation of these effects seems to lie in the ionic radii of the different ions and the structure of the pure promoter phase that would be present under these conditions. CaO and MgO are cubic and have ionic radii (0.099 and 0.066 nm, respectively) that are of the same order or smaller than that of Sm^{3+} (0.096 nm); the ions can thus fit into the surface of Sm_2O_3 without stress. La^{3+} has a larger radius (0.102 nm) and a smaller stabilizing effect. On the other hand, Na_2CO_3, Li_2CO_3, and LiOH have a monoclinic structure, encouraging the phase transition of Sm_2O_3 from cubic to monoclinic. In the same vein, the cubic phase has been prepared (Lo, Kamat, and Shrader 1988) by calcining $Sm(NO_3)_3 \cdot 6H_2O$ at 900° C for 1 h and the monoclinic form by calcining Sm_2O_3 at 1100° C for 22 h. It has been shown that the cubic phase is more active and more selective toward C_2 hydrocarbons than the monoclinic phase. In addition, these authors have prepared by hydrothermal treatments of Sm_2O_3 a mixture of monoclinic and cubic phases that yields a still better selectivity toward C_2 hydrocarbons. A study of Ca/Sm_2O_3 catalysts has confirmed that, when the cubic and monoclinic forms coexist, the solid is more active and more selective toward C_2 hydrocarbons than the cubic or monoclinic structure (Sokolovskii et al. 1990). The best performance of the cubic–monoclinic mixed phases was tenatively attributed to the highest mobility of oxygen vacancies.

It can be concluded that the sensitivity of the oxidative coupling of methane to the bulk structure of Sm_2O_3 has been confirmed by two groups. In the same way, the better performances of solids where two phases coexist seem well-established. So far, the reasons for this behavior do not appear clearly. A plausible hypothesis, relating the catalytic activity to the presence of defects in the solid has been considered and would deserve further investigations: In particular, it can be questioned whether in a metastable form the density of defects is not larger than in the high-temperature stable form and if the interface between the cubic and the monoclinic phase does not create additional defects. It is also interesting to note that, as for the case of Li/MgO, the effect of additives to Sm_2O_3 (Li, Na, Ca, Mg, and La) are interpreted in terms of structural modification.

La_2O_3

In a paper recently submitted for publication (Le Van et al. 1990) provides evidence of the structure sensitivity of this reaction over La_2O_3 and compares the catalytic performance and the morphology of La_2O_3 prepared by decomposition of nitrate at low (La_2O_3 LT) and high (La_2O_3 HT) temperature. The thin-plate morphology, revealing the presence of (001) planes (which is observed for La_2O_3 LT) gives rise to a better selectivity toward C_2 hydrocarbons and to a smaller intrinsic activity. This is typical of a structure-sensitive behavior. It is also observed that the oxycarbonate is thermally less stable on La_2O_3 LT. It is suggested that the structure sensitivity is related to this stability, the deep oxidation of methane into CO_2 being favored by this stability. This work, however, can be criticized on the following basis: The most selective catalyst has the smaller surface area, in contrast with the case of MgO catalytic data (Hargreaves, Hutchings, and Joyner 1990). As far as the role of surface area on selectivity (see below) will not be clearly elucidated, this work cannot be considered as definitively conclusive and it will require further investigation, particularly by varying separately the two parameters, surface area and structure.

Other Systems

It has been suggested that the bulk structure in itself (Matsuura et al. 1989) is an important factor for oxidative coupling of methane. This assumption is based on the fact that complex oxides such as $LiNiO_2$, $LiTiO_2$, and $LiYO_2$ (Otsuka and Hatano 1987; Lane and Wolf 1988), which have a NaCl structure (bcc, fluorite type), can act as catalysts of high selectivity. In the same way, alkaline-earth metal have a typical NaCl-type structure. $A^+M^{3+}O_2$ and $A^+{}_2M^{4+}O_3$ (with $A^+=$ Li, Na, or K, $M^{3+}=$ Mn, Fe, Co, Ni, or In, $M^{4+}=$ Ti, Ge, Zr, Sn, or Hf) with NaCl-type structure were prepared. These compounds are active and relatively good selective catalysts. It can be argued, however, that solids with other structures have been shown to be reasonably selective catalysts. This is the case, for instance, for La_2O_3, which has a hexagonal structure (Lin et al. 1986).

PHASE ORGANIZATION IN TWO-COMPONENT CATALYSTS

Most active and selective OCM catalysts contain at least two components, A and B. In this kind of binary system a number of questions have been raised, the main issue being about what is the active phase—A, B, or AB. To answer this question, morphological information is needed. This can be drawn from ion scattering spectroscopy (ISS), X-ray diffraction (XRD),

X-ray photoelectron spectroscopy (XPS), and electron microscopy, via measurements carried out on precursors, on solids after reaction, and, preferably, on catalysts during reaction. The morphology studies addressed the following questions: Do A and B form stable binary compounds or are they separated? Does A or B cover the other phase (surface enrichment)? A good description of the phase organization between A and B is expected to shed some light on the question of a possible synergy between A and B and on the nature of the active phase. We will first examine catalysts prepared by impregnation techniques (Li/MgO, La_2O_3; LiCl/NiO; Na/Ln_xO_y) and, as a second step, well-defined binary oxides.

Li / MgO Catalysts

The bulk and surface composition of Li/MgO catalysts from chemical analysis and XPS measurements have been compared before and after reaction at 750°C (Mirodatos et al. 1989). After reaction the XPS Mg–Li ratio decreases from 2.8 to 0.18 while the bulk ratio increases from 1.6 to 3.6 due to lithium losses. This is interpreted as an evidence of a spread of the lithium phase (LiOH and Li_2CO_3 as shown by XRD and XPS) over the MgO phase, leaving a small fraction of MgO surface accessible to the gas. However, the area of the accessible fraction cannot be estimated with accuracy. XPS data together with TEM and chemical analysis clearly show that two different morphologies have to be considered: (i) Before reaction at 750°C the catalyst consists of flakes of MgO and separated Li_2CO_3 (ii) After reaction the magnesia is almost completely covered with Li. It can be questioned if the catalytic behavior of Li/MgO catalysts is that of the lithium phase and if MgO acts only as a support of the Li-containing active phase. It should be noted that this hypothesis is at variance with that of Matsuura and co-workers, who suggest that the presence of Li in Li/MgO induces structural changes of MgO favoring low-coordination sites (high-index planes) yielding higher selectivity; it is also at variance with that of Lunsford's group (Ito et al. 1985), who propose that the active site of oxidative coupling is O^-, resulting from the substitution of Mg^{2+} by Li^+ in the MgO lattice (both hypotheses can be termed synergetic). This has led us (Mirodatos et al. 1989) to compare the "primary intrinsic" activity of MgO, Li/MgO, and pure Li_2CO_3 toward CO_2 and C_2H_6, the main products observed at the relatively low reaction temperature (630°C). By "primary" we understand at low conversion, that is, in the absence of secondary reactions, and "intrinsic" means per unit area. Table 10-3 shows that the catalytic properties of unsupported Li_2CO_3 are comparable to those of Li/MgO: the selectivities toward C_2 are of the same order of magnitude (41% versus 38–60%, respectively). The intrinsic overall activ-

Table 10-3 Catalytic Data[a] for the Oxidative Methane Coupling over MgO, Li/MgO, and Unsupported Li (ex Li_2CO_3) at 630°C (Mirodatos et al. 1989)

Primary Intrinsic Activity (mmol h^{-1} m^2)	CATALYSTS				
	MgO		Li/MgO		
	Before Treatment	After Treatment[b]	Before Treatment	After Treatment[b]	Unsupported Li
Into CO_2	0.03	0.05	0.74	0.64	0.25
Into C_2H_6	0.002	0.002	0.46	0.96	0.17
Into $CO_2 + C_2H_6$	0.032	0.052	1.20	1.60	0.42
C_2 selectivity (%)	6.2	3.8	38.3	60	41

[a]Pressure CH_4, O_2, 7.8 and 4.6 kPa; flow rate, 3.6 l h^{-1}.
[b]The catalyst is heated at a reaction temperature of 750°C before cooling down at 630°C.

ity of Li_2CO_3 is ca. four times smaller than that of Li/MgO. If it is considered that the oxidative coupling of methane is a structure-sensitive reaction, the factor 4 observed between the intrinsic activity of unsupported Li_2CO_3 and Li/MgO is not large enough to rule out the hypothesis of MgO merely playing the role of a support towards Li, the active phase.

In a recent paper (Peng, Richards, and Stair 1990) it has been confirmed that MgO was completely encapsulated by lithium, by comparing the C_{1s} and O_{1s} XPS spectra of pure Li_2CO_3 and Li/MgO catalyst after reaction. Thus, it seems well-established that MgO is embedded in the Li-containing phase. This result may lead one to consider the Li^+O^- hypothesis with more criticism, because these sites, which were actually detected by ESR, should be located in the MgO phase, probably unattainable by gas molecules.

Li/ZnO and Li/TiO_2

Similar qualitative conclusions have been reached for the case of Li/ZnO (Giusti et al. 1990) after calcination in air at 750°C, XPS indicates that Li covers ZnO by forming Li_2CO_3. The morphology of Li/TiO_2 has been studied extensively (Lane, Miro, and Wolf 1989). After calcination, only Li_2CO_3 is detected by XPS although many bulk phases are present. At this stage it can be stressed that in many cases Li_2CO_3 is detected in ex situ experiments such as XPS. This does not mean that this is necessarily representative of the catalyst in reaction conditions and that Li_2CO_3 is the active phase.

Li / La_2O_3

The influence of the reaction temperature, space velocity, and methane and oxygen pressures for the oxidative coupling of methane over Li/La_2O_3 and Li/MgO has been investigated (Kooh, Mimoun, and Cameron 1989). Product selectivities extrapolated at zero percent methane conversion are similar, suggesting that both rare earth and alkaline earth based catalysts involve similar mechanisms. In this paper, however, it is not stated that similar conclusions can be reached regarding the nature of the active site. This would fit with the idea that La_2O_3 is covered with the Li-containing phase, which would impose its catalytic pattern to the Li/La_2O_3 system as for the Na/La_2O_3 and Li/MgO catalysts.

Li / TiO_2, Li / $La_2O_3 \cdot 2TiO_2$, and Li / $NiTiO_3$

A transient method to assess the role of the different oxygen species involved in oxidative coupling of methane (gas-phase, adsorbed, or lattice) is applied to Li/TiO_2, Li/$La_2O_3 \cdot 2TiO_2$, and Li/$NiTiO_3$ (Miro et al. 1990). Despite the fact that all the catalysts studied have the same alkali-metal promoter and nearly the same catalytic pattern, their behavior in the transient experiments is clearly different: For Li/$NiTiO_3$, the oxygen stored in the lattice of the Li–Ni mixed oxide is responsible for the selective activation of methane whereas for the Li/TiO_2 the C_2 yield is linked to the presence of gas-phase oxygen, which presumably leads to weakly adsorbed oxygen. Thus, for these cases the simple picture where the Li-containing active phase imposes its catalytic pattern, covering a support that would not participate in the catalytic reaction, would not be appropriate.

LiCl / NiO

The case of LiCl/NiO (Otsuka, Hatano, and Komatsu 1989) is of special interest because at first sight it seems to be a well-documented case of synergy between the phases of a two-component catalyst. It is first speculated that most of LiCl must be present on the surface without strong interaction with NiO. This assessment is based on the following two considerations: First, in contrast with $LiNO_3$, Li_2CO_3, or LiOH-added NiO, where solid solutions between Li and NiO were observed by XRD (Hatano and Otsuka 1988), no mixed compounds were detected; furthermore, at reaction temperature, LiCl is liquid (the melting point of LiCl being 606° C). It is also noted that the intrinsic activity of a LiCl/SiO_2 (surface area, 600 m^2 g^{-1}) is two or three orders of magnitude lower than that of the LiCl/NiO solid, indicating, according to the authors, that the catalytic activity of LiCl without NiO is negligible. From these observations it was deduced that a new active species at the LiCl/NiO interface is created, which was tenatively ascribed to Ni^{3+} or $Ni^{(2+\delta)+}$ on the basis of XPS data.

The experiment carried out on the LiCl/SiO_2 solid, however, does not seem conclusive to us. In fact, we have performed similar experiments with the Li/SiO_2 system to compare it with Li/MgO, which demonstrates a catalytic behavior comparable to that of unsupported Li (Mirodatos et al. 1989). The selectivity toward C_2 hydrocarbons and the intrinsic activity (per unit area measured after reaction) are, respectively, 7% and 0.008 mmol h^{-1} m^{-2} for Li/SiO_2 and 72% and 4 mmol h^{-1} m^{-2} for Li/MgO. This means that if the Li-containing active phase is deposited on SiO_2, for some still unclear reasons (silicate formation?), the catalytic behavior of the composite catalyst has nothing to do with that of the corresponding unsupported catalyst. This situation could also occur for the case of LiCl-based catalysts. We conclude that the synergetic hypothesis proposed for LiCl/NiO still needs further work to be accepted.

Our experiments carried out on Li/SiO_2 also cast some doubts on conclusions drawn from other SiO_2-supported catalyst experiments. For example, the fact that $MgCl_2$/SiO_2 catalysts exhibit poor activity and poor selectivity for the C_2 formation has led to the conclusion that $MgCl_2$ in $MgCl_2$/CaO is not the active phase (Fujimoto et al. 1988).

Na / Rare-Earth Oxides (La_2O_3, CeO_x, Yb_2O_3)

Unsupported Na catalysts ex Na_2CO_3 (denoted in what follows as Na_2CO_3) have been compared (see Table 10-4) with Na/La_2O_3, Na/CeO_x and Na/Yb_2O_3 (Lunsford and Tong 1989). It was first shown that the catalyst was completely covered with an alkali-containing compound, as was first observed for the case of Li/MgO catalysts (Mirodatos et al. 1989), suggesting that the surface-phase enrichment into alkali is a general feature in alkali-promoted oxides. In spite of the large difference in catalytic performance among the pure lanthanide oxides, the sodium-promoted catalysts display the same catalytic behavior. Moreover, on the basis of unit area, Na_2CO_3 demonstrates a very similar catalytic pattern. For example, the intrinsic activity (per unit area) of Na_2CO_3 is only 2.2 times smaller than that of Na/Y_2O_3. Once again this situation is quite comparable to that of unsupported Li and Li/MgO. It is concluded that the active site centers on Na/Ln_xO_y are similar to those on Na_2CO_3 and that lanthanum oxides are serving as supports for the active sodium-containing phase. The active site is supposed to be located on the Na_2O phase in

Table 10-4 Catalytic Behavior of Na/Ln_xO_y compared to that of Na_2CO_3 (Lunsford and Tong 1989)

Catalysts	4% Na/ La_2O_3	4% Na/ CeO_2	4% Na/ Yb_2O_3	10% Na CeO_2	Na_2CO_3
CH_4% conv.[a]	6.1	2.5	8.7	3.5	0.2
C_{2+} prod.[b]	0.7	0.23	0.95	0.28	0.02
Surface area ($m^2\ g^{-1}$)	1.5	0.84	1.3	1.2	0.08
Intrinsic activity (per unit area)	4.1	3.0	6.5	2.9	3
Intrinsic C_{2+} prod.[c] (per unit area)	0.5	0.27	0.73	0.23	0.25

[a] $P_{CH_4} = 0.5$ atm; $CH_4:O_2 = 5:1$; T + 730°C, atmospheric pressure, 0.05-g catalysts. GHSV = 72,000 h^{-1}.
[b] C_{2+} productivity in g(CH_4 to C_{2+})/g(catalyst) · h (specific productivity).
[c] C_{2+} productivity in g(CH_4 to C_{2+})/m^2 (catalyst) · h (intrinsic productivity).

equilibrium with Na_2CO_3 according to the equation

$$Na_2CO_3 = Na_2O + CO_2$$

It is also suggested that the lanthanide oxides do not alter this equilibrium significantly. [Are the factors 2.2. and 4 observed between specific activities of unsupported Na (Li) carbonates and Na/Ln_xO_y (Li/MgO), respectively, attributable to some change in this equilibrium induced by the presence of the support?] In this respect, the series of papers indicating that alkali and alkaline-earth peroxides are active in oxidative coupling of methane (Sinev, Korchak, and Krylov 1986; Otsuka et al. 1987) are particularly interesting because they suggest that the active site could be the peroxide ion O_2^{2-}, highly unstable in reaction conditions, which can disproportionate with the formation of O^{2-} and atomic oxygen actively convert CH_4 into methyl radicals (Anshits et al. 1989a). If the peroxide ion is really the active site of this reaction, there is no need to invoke any synergetic effects between the alkali-containing phase and the lanthanide or alkaline earth oxide phase.

Na/Pr_xO_y

The case of this catalyst (Gaffney et al. 1988) is also interesting from the viewpoint of the synergy between the alkali-containing phase and lanthanide oxides. Based on the fact that pure Pr_xO_y leads selectively to CO_2 and Na/Pr_xO_y to C_2 hydrocarbons and on other available experimental data such as the capacity of Na_2O_2 to dimerize CH_4 (Otsuka et al. 1987),

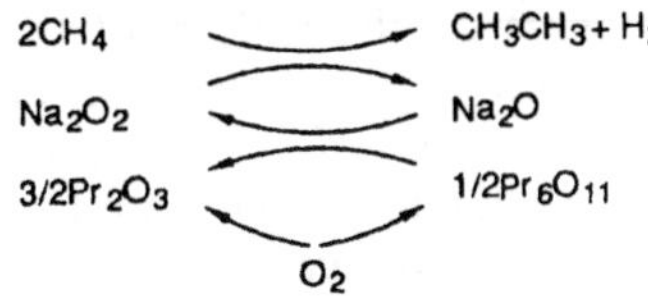

FIGURE 10-4. The reaction scheme proposed by he group of ARCO for the oxidative coupling of methane over Na/Pr_xO_y catalysts in cyclic or cofeed (catalytic) modes (Gaffney et al. 1988).

a reaction scheme has been proposed where the active phase is the sodium-containing compound. Methane is activated by a peroxide anion O_2^{2-} to yield the primary product ethane via methyl formation and O^{2-} anion. Sodium oxide is regenerated into sodium peroxide by praseodymium oxide, which goes from Pr_6O_7 to Pr_2O_3, giving up one oxygen atom to the sodium-containing phase. Molecular oxygen regenerates Pr_2O_3, which is oxidized into Pr_6O_{11}. The praseodymium oxide phase appears as a kind of cocatalyst: In itself it is not directly active toward the reaction, but participates in the regeneration of the active phase (Fig. 10-4).

This scheme, however, raises two difficulties. The first is that the scheme implies that the molecular gaseous oxygen has access to Pr_xO_y. This suggests that the latter oxide is not fully covered with the sodium-containing phase and that a certain fraction of its surface is free. If this were so, it is difficult to understand why such catalysts are so selective toward C_2 hydrocarbons, whereas pure Pr_xO_y heads only to CO_2. It can be argued, however, that O_2 sorption does not require a large fraction of free Pr_xO_y surface. It can also be speculated that the Na-containing phase covers completely the Pr_xO_y solid and that it forms a selective molecular sieve that selectively allows the access of O_2 to Pr_xO_y and inhibits the access of CH_4. This occurrence would prevent CH_4 reaction over Pr_xO_y into CO_2. This hypothesis has already been proposed for the case of potassium-promoted Ni/SiO_2 catalysts in our laboratory (Martin and Praliaud 1991) to account for the good accessibility to Ni of small molecules (H_2, CO, O_2) and the inhibition of the adsorption of larger ones such as H_2S and C_2H_6.

The other difficulty of this reaction scheme is the following: If Pr_xO_y really acts as a cocatalyst promoting the regeneration of Na_2O into Na_2O_2, one expects that Na/Pr_xO_y is more active than pure unsupported Na catalysts. No experimental data are available in this respect. However, one can speculate that the situation should be similar with Na and Na/Ce_xO_y catalysts due to the similarity between both oxides. The experimental results of Lunsford and Tong show clearly that this is not the case. This casts some doubts on the validity of this interesting scheme. Obviously, further work involving the comparison of intrinsic activities of Na_2CO_3 and Na/Pr_xO_y and the surface composition are needed to be more conclusive. In our laboratory some studies are under way on Li/Ce_xO_y catalyst: We take advantage of the fact that Ce^{4+} is diamagnetic whereas Ce^{3+} is paramagnetic to follow in situ in a Faraday balance the redox properties of Ce_xO_y and Li/Ce_xO_y in the presence of reacting gases, from which we expect information on the reaction scheme, the nature of active sites, and the role of the redox support.

Well-Defined Binary Oxides

Binary oxides, Na_2PrO_3 and Li_2PrO_3, have been reported to be selective in the redox oxidative coupling of methane (Gaffney et al. 1988). In the cofeed mode, (Kiennemann et al. 1990) has shown that $LnLiO_2$ (Ln = Sm, Nd, and La) are more selective toward C_2 hydrocarbons than Ln_2O_3. For all the $LnLiO_2$ catalysts the XRD patterns before and after reaction are identical, showing that the structure is retained.

The same selectivity improvement is observed when comparing Ln_2O_3 and $LnNaO_2$ (Hatano and Otsuka 1989). The structure, however, is less well conserved after reaction. Stoichiometric $LiNiO_2$ are also selective and active catalysts for the oxidative coupling of methane (Lane, Kalenik, and Wolf 1989). Complex oxides with the rock salt structure, $LiNiO_2$, $LiYO_2$, $LiTiO_3$, $NaFeO_2$..., were prepared (Ungar, Zhang, and Lambert 1988). Among them, $LiNiO_2$ and $LiYO_2$ appear promising catalysts for the oxidative coupling of methane. $LaAlO_3$ prepared by the "mist decomposition" technique (Imai and Tagawa 1986; Tagawa and Imai 1988) seems to be a highly active selective and stable catalyst. Tin-containing rare-earth pyrochlores, $Ln_2Sn_2O_7$ have also been reported to be active and selective in oxidative coupling of methane (Ashcroft et al. 1989), with an enhanced conversion to ethylene for those lanthanides that exhibit mixed valence behavior (Pr, Sm, Eu, Tb, Tm, and Yb). Titanate catalysts have been compared (Lane, Kalenik, and Wolf 1989). Many of them have poor selectivities toward C_2 hydrocarbons (Zn_2TiO_4, $CaTiO_3$, and $Na_2Ti_3O_7$). The more selective is lanthanum titanate, $MgTiO_3$, $NiTiO_2$, and K_2TiO_3 being intermediate. XRD indicates that the solids used are a complex mixture in which the main titanate predominates but other phases are also present. This poses the problem of the morphology definition of this type of binary well-defined oxides. It is not clearly established whether the binary phase per se is active or whether the surface is not altered and more or less selectively covered with one component, which would impose its catalytic pattern. In this respect XPS and ISS experiments that give information on the A-to-B ratio are particularly helpful.

SUMMARY

Under reaction conditions, lead and alkali metal vaporize, leading to deactivation. For the case of lithium-based catalysts, Li vaporizes under the form of volatile LiOH and can recombine with diluents such as SiO_2 or Al_2O_3. The question of the Li content limit has been raised but not definitively answered. So far the role of LiOH vapor permanently present in the gas phase has not been assessed. The rate of vaporization of supported alkali metal depends on the nature of the support: It is larger on MgO than on SiO_2, Al_2O_3, and Cr_2O_3. It seems possible to inhibit alkali vaporization, either by modifying the solid (e.g., SiO_2 addition to Li/MgO) or by changing the reaction condition. However, in the perspective of an implementation, where very stable catalysts are needed, alkaline-earth compounds associated with rare-earth oxides could be preferred to lead or Li-based catalysts.

It has been suggested that because OCM involves many consecutive steps, too large a surface area could be detrimental to selectivity and that an optimal area could exist. However, much work is needed to substantiate this concept, particularly by comparing catalysts of different dispersion, the chemical factor being kept unchanged. The sintering of OCM catalysts depends on many parameters: It varies with the nature of the oxide and of the promoter. The loss of surface area increases when T increases, when powders are agglomerated under the form of pellets, and in the presence of the reacting mixture. It is also enhanced when a component of composite solids melts, as for the case of Li_2CO_3 in Li/MgO catalysts.

In the event of an industrial implementation, the problem of OCM catalyst stability and of sintering will be considered more carefully. Solutions to improve stability similar to that we used for Li/MgO catalysts then should emerge.

The structure sensitivity of the oxidative coupling of methane reaction is apparently well-established when the reaction is run over MgO, Sm_2O_3, and, probably, La_2O_3. This can be taken as a proof that the reaction involves at least one surface-assisted step and that this latter is rate-determining. This step could be the hydrogen abstraction from methane by the surface, as suggested by isotopic effect experiments carried out by several groups (Cant et al. 1988; Mirodatos et al. 1990). We can anticipate that, in the future, structure-sensitivity investigations will develop. It can be hoped that the surface area and the structure of solids will be studied independently. Interesting ideas are emerging to account for the variations of activity and/or selectivity with the structure: low coordination oxide ions, oxygen vacancies, selective formation of carbonates on certain crystallographic planes, etc. Structure-sensitivity studies will help us in a better description of the nature of the active sites.

From this review of experimental data on phase organization in OCM two-component catalysts it can be concluded that for Li/MgO and Na/Ln_xO_y (La_2O_3, CeO_2, Yb_2O_3) converging experimental evidence of a more or less complete encapsulation of MgO and Ln_xO_y by alkali has been provided. Comparison of the catalytic performance of supported and supported alkali carbonates suggests strongly in both cases that MgO or Ln_xO_y is serving as a carrier for the alkali-containing phase where activity sites are located. This gives some support to the idea that the active site for these catalysts could be O_2^{2-} located on the alkali phase and that no important synergy exists between the alkali and the carrier. This does not preclude that one given carrier can be a bad or a good carrier, favoring or not the alkali spreading on its surface.

Conversely, according to us, if one excepts the case of $Li/NiTiO_3$ (where the oxygen of the support lattice was shown to play an important role), no convincing evidence of synergy between components of impregnated catalysts has so far been provided. For the case of binary oxides, well-defined before reaction, too little information concerning the composition and the structure of the surface during or after reaction is available to conclude with confidence that the binary phase is, per se, the active phase and that no surface enrichment in one component occurs.

References

Agarwal, S.K., R.A. Migone, and G. Marcelin. 1990. *J. Catal.* 121:110.

Anderson, A.G. and T. Norby. 1990. *Catal. Today* 6:575.

Andorf, R., and M. Baerns. 1990. *Catal. Today* 6:445.

Anpo, M., Y. Yamada, T. Doi, I. Matsuura, S. Coluccia, A. Zechina, and M. Che. 1989. In *Structure and Reactivity of Surfaces*, ed. C. Morterra, A. Zechina, and G. Costa, p. 1. Amsterdam: Elsevier.

Anshits, A.G., N.P. Kirik, V.G. Roguleva, A.N. Shigapov, and G.E. Selyutin. 1989. *Catal. Today* 4:399.

Anshits, A.G., E.N. Voskresenskaya, and A.N. Shigapov. 1989. Paper 88, presented at Pacifichem 89, Honolulu, Hawaii, Dec. 17–20.

Ashcroft, A.T., A.K. Cheetham, M.L.H. Green, P.C. Grey, and P.D.F. Vernon. 1989. *J. Chem. Soc., Chem. Commun.* 1667.

Baerns, M., and J.R.H. Ross. (1990). Unpublished.

Bartsch, S., and H. Hofman. 1990. *Catal. Today* 6:527.

Boudart, M., A.W. Aldag, J.E. Benson, N.A. Dougharty, and J. Girvin-Harkins. 1966. *J. Catal.* 6:92.
Burch, R., G.D. Squire, and S.C. Tsang. 1988. *Appl. Catal.* 43:105.
Bytyn, W., and M. Baerns. 1986. *Appl. Catal.* 28:199.
Cant, N.W., C.A. Luckey, P.F. Nelson and R.J. Tyler. 1988. *J. Chem. Soc., Chem. Commun.* 766.
Carreiro, J.A.S.P., and M. Baerns. 1987. *React. Kinet. Catal. Lett.* 35 (1-2):349.
Chan, Y.W., and R.B. Wilson. 1988. Preprint 33, no. 3, Fuel Chem. Division, Am. Chem. Soc., Sept. 25–30, p. 453.
Che, M., and A.J. Tench. 1982. *Adv. Catal.* 31:77.
Coluccia, S., and A.J. Tench. 1981. In *Proceedings of the Seventh International Congress on Catalysis, Tokyo, 1980*, ed. T. Seiyawa and K. Tanabe, p. 1154. Tokyo/Amsterdam: Kodansha/Elsevier.
Ekstrom, A., and J.A. Lapszewicz. 1989. *J. Phys. Chem.* 93:5230.
Follmer, G., L. Lehmann, and M. Baerns. 1988. Preprint, Petroleum Chemistry (Inc.) Division, Am. Chem. Soc., Sept. 25–30, p. 453.
Follmer, G., L. Lehmann, and M. Baerns. 1989. *Catal. Today* 4:323.
Fujimoto, K., S. Hashimoto, K. Asami, K. Omata, and H. Tominaga. 1988. Preprint, Bicentenary Catalysis Conference, Sept. 1–2, Sydney, Australia, 1988, p. 60.
Gaffney, A.M., C.A. Jones, J.J. Leonard, and J.H. Sofranko. 1988. *J. Catal.* 114:422.
Giusti, A., S. Contarini, C. Flego, S. Palmery, and G. Piro. 1990. *Catal. Today* 6:463.
Hargreaves, J.S.H., G.S. Hutchings, and R.W. Joyner. 1990. *Catal. Today* 6:481.
Hatano, M., and K. Otsuka. 1988. *Inorg. Chim. Acta* 146:243.
Hatano, M., and K. Otsuka. 1989. *J. Chem. Soc., Faraday Trans. 1* 85(2):199.
Hicks, R.F., H. Qi, M.L. Young, and R.G. Lee. 1990. *J. Catal.* 122:280.
Hinsen, W., W. Bytyn, and M. Baerns. 1984. In *Proceedings of the Eighth International Congress on Catalysis*, Part III, p. 581. Berlin: Verlag Chemie.
Imai, H., and T. Tagawa. 1986. *J. Chem. Soc., Chem. Commun.* 52.
Ito, K., M. Kuramato, M. Yoshioka, and T. Tokuda. 1983. *J. Phys. Chem.* 87:4411.
Ito, T., T. Tashiro, T. Watanabe, K. Toi, and I. Ikemoto. 1987. *Chem. Lett.* 1723.
Ito, T., J.X. Wang, C.H. Lin, and J.H. Lunsford. 1985. *J. Am. Chem. Soc.* 107:5062.
Iwamatsu, E., T. Moriyama, N. Takasaki, and K.I. Aika. 1987. *J. Chem. Soc., Chem. Commun.* 19.
Iwamatsu, E., T. Moriyama, N. Takasaki, and K.I. Aika. 1988. *J. Catal.* 113:25.
Jones, C.A., J.J. Leonard, and J.A. Sofranko. 1987. *J. Catal.* 103:311.
Jones, C.A., and J.A. Sofranko. 1985. US patents 4,495,374 (Jan. 22) and 4,449,322 (Feb. 12).
Kaminsky, M.P., G.A. Huff, M.J. Spangler, J.Z. Shyn, and T.P. Kobylinski. 1989. Preprint 84, Pacifichem 89, Honolulu, Hawaii, Dec. 17–20.
Kiennemann, A., R. Kieffer, A. Kaddouri, P. Poix, and J.L. Rehspringer. 1990. *Catal. Today* 6:409.

Kimble, J.B., and J.H. Kolts. 1986. *Energy Process*. 6:226.
Kobayashi, H., and M. Yamagushi. 1989. Paper 114, presented at Pacifichem 89, Honolulu, Hawaii, Dec. 17–20.
Kooh, A., H. Mimoun, and C.J. Cameron. 1989. *Catal. Today* 4:333.
Korf, S.J., J.A. Roos, N.A. de Bruijn, J.G. van Ommen, and J.R.H., Ross. 1987. *J. Chem. Soc., Chem. Commun*. 1433.
Korf, S.J., J.A. Roos, N.A. de Bruijn, J.G. van Ommen, and J.R.H. Ross. 1988a. *Catal. Today* 2:535.
Korf, S.J., J.A. Roos, J.M. Diphoorn, R.H.J. Veekof, J.G. van Ommen, and J.R.H. Ross. 1988b. Preprint, 196th ACS National Meeting, Vol. 33, p. 437.
Korf, S.J., J.A. Roos, J.M. Diphoorn, R.H.J. Veehof, J.G. van Ommen, and R.J.H. Ross. 1989. *Catal. Today* 4:279.
Korf, S.J., J.A. Roos, J.A. Vreeman, J.W.H.C. Derksen, and J.G. van Ommen. 1989. Paper 29, Pacifichem 89, Honolulu, Hawaii, Dec. 17–20.
Lane, G.S., Z. Kalenik, and E.E. Wolf. 1989. *Appl. Catal*. 53:183.
Lane, G.S., E. Miro, and E.E. Wolf. 1989. *J. Catal*. 119:161.
Lane, G.S., and E. Wolf. 1988. In *Proceedings of the Ninth International Congress on Catalysis*, ed. M.J. Phillips and M. Ternan, Vol. 2, p. 944. Ottawa: Chem. Inst. Canada.
Lane, G.S., and E.E. Wolf. 1989. *J. Catal*. 119:161.
Le Van, T., M. Che, M. Kermarec, C. Louis, and J.M. Tatibouët. 1990. *Catal. Lett*. 6:395.
Lin, C.H., D.C. Campbell, J.X. Wang, and J.H. Lunsford. 1986. *J. Phys. Chem*. 90:534.
Lin, C.H., T. Ito, J.X. Wang, and J. Lunsford. 1987. *J. Am. Chem. Soc*. 109:4808.
Lo, M.Y., S.N. Kamat, and G.L. Schrader. 1988. Preprints, 196th ACS National Meeting, Vol. 33, p. 378.
Lunsford, J.H. 1990. *Catal. Today* 6 (3):235.
Lunsford, J.H., M.D. Cisneros, P.G. Hinson, Y. Tong, and H. Zhang. 1989. *Faraday Discuss. Chem. Soc*. 87:212.
Lunsford, J.H., and Y. Tong. 1989. Paper 46, Pacifichem 89, Honolulu, Hawaii, Dec. 17–20.
Martin, G.A., M.C. Durupty, C. Mirodatos, N. Mouaddib, and V. Perrichon. 1991. In Fifth International Symposium on Scientific Bases for the Preparation of Heterogeneous Catalysts, Louvain-la-Neuve, Belgium, Sept. 3–6. Poncelet et al. Eds. Elsevier Science Pub. Amsterdam, 269.
Martin, G.A., and H. Praliaud. 1991. *Catal. Lett*. 1991, 9:751.
Martin, G.A., P. Turlier, V. Ducarme, C. Mirodatos, and M. Pinabiau. 1990b. *Catal. Today* 6:373.
Matsukata, M., E. Okanari, K. Komori, T. Matsuda, E. Kikuchi, and Y. Morita. 1989. *Pacifichem 89*, Honolulu, Hawaii, Dec. 17–20.
Matsuura, I., Y. Utsumi, and T. Doi. 1989. *Appl. Catal*. 47:299.
Matsuura, I., Y. Utsumi, T. Doi, and K. Ohyachi. 1989. Paper 92, Pacifichem 89, Honolulu, Hawaii, Dec. 17–20.
Mims, C.A., R.B. Hall, K.D. Rose, and G.R. Myers. 1989. *Catal. Lett*. 2:361.

Miro, E.E., Z. Kalenik, J. Santamaria, and E.E. Wolf. 1990. *Catal. Today* 6:511.
Mirodatos, C., V. Ducarme, H. Mozzanega, A. Holmen, J. Sanchez-Marcano, Q. Wu, and G.A. Martin. 1990. *Natural Gas Conversion Symposium*, Oslo.
Mirodatos, C., M.C. Durupty, P. Moral, and V. Perrichon. 1987. In *Catalyst Deactivation 1987*, ed. B. Deluron and G.F. Froment, p. 153. Amsterdam: Elsevier Science Publications.
Mirodatos, C., A. Holmen, R. Mariscal, and G.A. Martin. 1990. *Catal. Today* 6:601.
Mirodatos, C., and G.A. Martin. 1988. In *Proceedings of the Ninth International Congress of Catalysis*, Vol. 2, ed. M.J. Phillips and M. Ternan, p. 899. Ottawa: Chemical Institute of Canada.
Mirodatos, C., G.A. Martin, J.C. Bertolini, and J. Saint-Just. 1989. *Catal. Today*, 4:301.
Moral, P., H. Praliaud and G.A. Martin. 1987. *React. Kinet. Catal. Lett.* 34:1.
Moriyama, T., N. Takasaki, E. Iwamatsu, and K.I. Aika. 1986. *Chem. Lett.* 1156.
Otsuka, K., and M. Hatano. 1987. *Shokubai* 29:46.
Otsuka, K., M. Hatano, and T. Komatsu. 1989. *Catal. Today* 4:409.
Otsuka, K., M. Inaida, Y. Wada, T. Komatsu, and A. Morikawa. 1989. *Chem. Lett.* 1531.
Otsuka, K., A.A. Said, K. Jinno, and T. Komatsu. 1987. *Chem. Lett.* 77.
Peng, X.D., D.A. Richards, and P.C. Stair. 1990. *Appl. Catal.* 121:99.
Perrichon, V., and M.C. Durupty. 1988. *Appl. Catal.* 42:217.
Quanzi, Li, Y. Amenomiya. 1986. *Appl. Catal.* 23:173.
Rosynek, M.P. 1977. *Catal. Rev.—Sci. Eng.* 16:111.
Sinev, M.Y., U.N. Korchak, and O.V. Krylov. 1986. *Kim. Catal.* 27:1274.
Sokolovskii, V.D., O.V. Buyevskaya, L.M. Plyasova, G.S. Litvak, and N.Ph. Uvarov. 1990. *Catal. Today* 6:489.
Tagawa, T., and I. Imai. 1988. *J. Chem. Soc., Faraday Trans. 1* 84 (4):923.
Thampi, K., J. Kiwi, and M. Grätzel. 1990. *Catal. Lett.* 4:49.
Ungar, R.K., X. Zhang, and M. Lambert. 1988. *Appl. Catal.* 42:L1.
Volta, J.C., and J.L. Portefaix. 1985. *Appl. Catal.* 18:1.
Volta, J.C., J.M. Tatibouët, C. Phichitkul, and J.E. Germain. 1984. In *Proceedings of the Eighth International Congress on Catalysis*, Vol. 4, p. 451. Berlin: Verlag Chimie.
Withers, H.P. 1984. US patent 4,489, 215 (Dec. 18).

11

Basic Solids as Catalysts for the Oxidative Coupling of Methane

Manfred Baerns

Ruhr-Universität Bochum, Bochum, Germany

TABLE OF CONTENTS

INTRODUCTION

In the oxidative coupling of methane to ethane, ethylene, and minor amounts of higher hydrocarbons (i.e., C_{2+} hydrocarbons) a large variety of solid metal oxide materials has been applied as catalysts; they may be loosely categorized as reducible and nonreducible ones. Only the former solids can be applied in the redox type of operation: Methane and oxygen are intermittently passed over the solid; first, methane reacts with the reducible oxide ("reagent"), which also catalyzes the desired reaction, and then the reduced metal oxide is reoxidized by gas-phase oxygen (cf., e.g., Keller and Bhasin 1982). Both types of catalysts, however, can be applied in the cofeed method in which the oxidant (most commonly oxygen) and methane flow simultaneously through the bed of solid material acting as the catalyst (Hinsen and Baerns 1982, 1983, 1986). Already in the early stages of the research on oxidative methane coupling it became apparent that doping reducible metal oxides, for example, alumina-supported PbO, with alkali compounds, or applying alkali and alkaline-earths compounds of high basicity as well as their mixtures as catalysts, or using supports of different acidity and basicity, respectively, affect the selectivity of the oxidative coupling reaction (Hinsen, Bytyn, and Baerns 1985; Baerns and Bytyn 1984; Ito and Lunsford 1985; Ito et al. 1985; Otsuka, Jinno, and Morikawa 1985; Baerns, Carreiro, and Bytyn 1985; Carreiro and Baerns 1989a, 1989b). As a general rule it has been observed that decreasing acidity and/or increasing basicity of the catalyst improves C_{2+} selectivity and reduces the formation of the undesired carbon oxides. The present contribution deals with these experimental findings, their interpretations, and some preliminary conclusions that might be derived from these results with respect to the catalysis of the reaction and to further catalyst improvement.

RESULTS ON THE EFFECT OF ACIDITY AND BASICITY ON C_{2+} SELECTIVITY

The results presented in this section are mainly based on the observations made in the author's laboratory when applying catalysts in the cofeed procedure of the oxidative methane coupling reaction. Nevertheless, reference is made to related work where it appears appropriate; however, it is not intended to review completely all research that has been done at other places that has been partly cited elsewhere (Hutchings, Scurrell, and Woodhouse 1989; Baerns and Ross 1991). The presentation is structured along the following lines:

1. PbO-based catalysts.
   a. Different support materials.
   b. Doping of γ-alumina–supported PbO catalysts with sodium oxide.
2. Alkali–alkaline-earths metal compounds as catalysts.
3. La_2O_3/CaO catalysts.

Table 11-1 **C_n Selectivity ($n \geq 2$) of PbO Catalysts when Using Supports of Different Acidity or Basicity[a]**

	SELECTIVITIES (%)			
Catalyst (wt%)	C_2	C_3	C_4	ΣC_i
$PbO(20)/Al_2O_3 \cdot xSiO_2$	20	—	—	20
$PbO(31)/\alpha\text{-}Al_2O_3$	53	1	—	54
$PbO(05)/\gamma\text{-}Al_2O_3$	56	2	1	59
$PbO(08)/SiO_2$	56	9	—	65
$PbO(31)/MgO$	65	3	1	69

[a] Experimental conditions: catalyst size, 3 × 2 mm, $T = 740°C$; $P_{CH_4}/P_{O_2} = 8$; $P_{CH_4} = 0.65$ bar.

PbO-Based Catalysts

Different Support Materials for PbO-Based Catalysts

It is well known that acidity and basicity of different solid surfaces vary over a wide range. Although alumina silicate, $Al_2O_3 \cdot xSiO_2$, exhibits strong acidity, magnesia MgO is of basic nature. The acidity of various supports decreases toward basicity in the following order: $Al_2O_3 \cdot SiO_2 > Al_2O_3 > \alpha\text{-}Al_2O_3 > SiO_2 > MgO$. Lead oxide, being a catalytically active compound for C_{2+} hydrocarbons formation by oxidative coupling of methane, was impregnated on these supports. The C_{2+} selectivities obtained by such catalysts are shown in Table 11-1 (Baerns 1988). It is obvious that decreasing acidity of the support material results in an increase of C_{2+} selectivity. The C_{2+} selectivity of the catalyst supported on silica that is almost neutral (see Table 11-1) is close to that of

Table 11-2 **Effect of Na_2O Doping of an Alumina Silicate Supported PbO Catalyst on Selectivity $S(i)$**

Support Material	$Al_2O_3 \cdot xSiO_2$				
PbO content (%)	0	8.3	8.2	20	18.5
Na_2O content (%)	0	0	4.6	0	4.1
$S(C_2H_4)$ (%)	0.1	2.1	5.7	5	8.0
$S(C_2H_6)$ (%)	0.4	4.7	19.0	14.5	33.1
$S(\Sigma C_i)$ (%)	0.5	6.8	24.8	19.5	41.1
W/F (g-cat s ml^{-1})	0.392	1.03	1.03	1.03	1.03
X_{O_2} (%)	90	94.5	50	58.4	25.3
X_{CH_4} (%)	4.9	5.9	3.5	3.9	2.1

unsupported PbO (approximately 60%) within the limits of experimental accuracy. From these findings it can be concluded that surface acidity of the support material favors nonselective oxidation of methane to CO_x whereas surface basicity contributes to better selectivity of the catalytic compound by a synergistic effect that is, however, still not well-defined yet. The nonselective oxidation may be tentatively explained by assuming that Bronsted- or Lewis-type acidic sites cause the formation of a carbonium ion, which is easily attacked by a negatively charged adsorbed oxygen species (e.g., O^- or O_2^-) leading finally to total combustion.

The proposition that surface acidity enhances CO_x formation and that basicity favors C_{2+} selectivity is supported by two further observations, which are described later.

Alkali Oxide Doping of Supported PbO Catalysts

When applying an acidic $Al_2O_3 \cdot SiO_2$ support the addition of sodium oxide, which is achieved by impregnation with an aqueous sodium nitrate

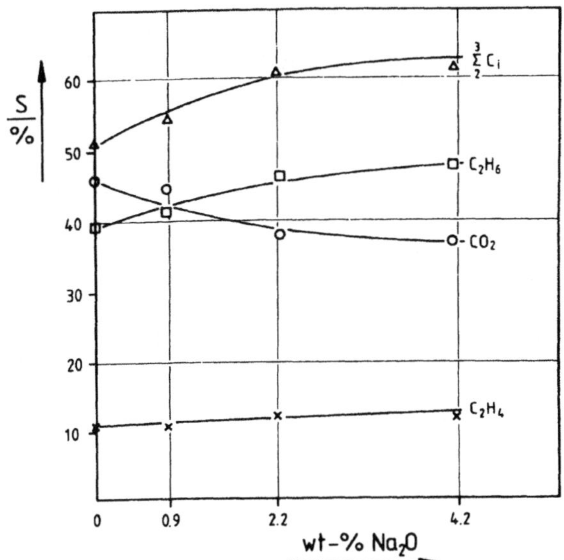

FIGURE 11-1. Effect of Na_2O doping of a PbO(22 wt%)/γ-Al_2O_3 catalyst on selectivity (experimental conditions: $T = 750°C$; $P^0_{CH_4} = 0.7$ bar; $P^0_{O_2} = 0.07$ bar; $W/F = 0.046$ g s ml^{-1}).

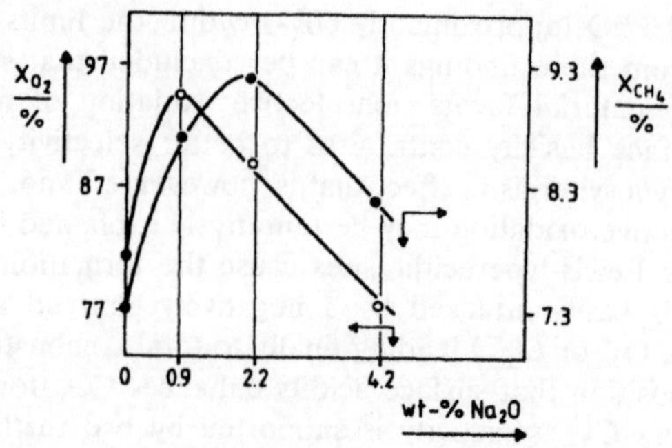

FIGURE 11-2. Effect of Na_2O doping of a PbO(22 wt%)/γ-Al_2O_3 catalyst on its activity expressed by oxygen and methane conversion, respectively (experimental conditions: T = 750°C; P_{CH_4} = 0.7 bar; P_{O_2} = 0.07 bar; W/F = 0.046 g s ml^{-1}).

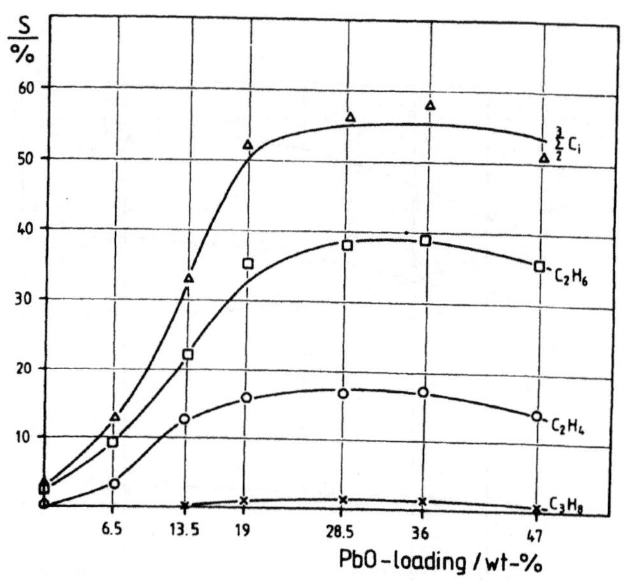

FIGURE 11-3. Effect of PbO loading on C_n ($n \geq 2$) selectivity of γ-alumina-supported catalysts (experimental conditions: T = 740°C; P_{CH_4} = 0.7 bar; P_{O_2} = 0.07 bar; X_{O_2} = 95–99%).

solution and subsequent decomposition of the nitrate, improves C_{2+} selectivity for various PbO loadings of the catalyst as is shown in Table 11-2. A similar effect was observed for a γ-alumina–supported PbO catalyst (see Fig. 11-1); again, doping with Na_2O led to an increase in C_{2+} selectivity. These results may be interpretated in such a way that strong acidic sites being favorable for CO_x formation are neutralized by alkali addition. It should also be mentioned that addition of small amounts of alkali increases the activity of the catalyst but if a certain alkali amount is exceeded, it is reduced as shown in Fig. 11-2; however, this should not be considered important compared to the gain in selectivity. A detailed account of this work has been given elsewhere (Hinsen, Bytyn, and Baerns 1985; Baerns and Bytyn 1984, 1987; Bytyn and Baerns 1986).

Amount of PbO Loading of γ-Alumina Supports

As explained previously, the selectivity of a γ-alumina–supported PbO catalyst is improved by presumably neutralizing the acidic sites. Another way of inhibiting their undesired properties is to cover them up with the catalytic compound. Experimental results obtained by applying this procedure are shown in Figs. 11-3 and 11-4 (similar results also may be derived from Table 11-2 for the $Al_2O_3 \cdot xSiO_2$-supported catalysts). It is obvious that C_{2+} selectivity is improved up to a maximum value when the lead

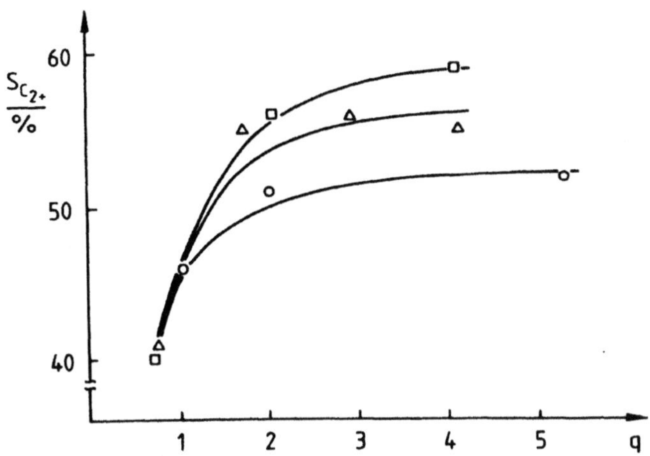

FIGURE 11-4. Effect of PbO loading on C_n ($n \geq 2$) selectivity for various γ-alumina (Rhône Poulenc) supported catalysts (experimental conditions: $T = 750°C$; $P_{CH_4} = 0.7$ bar; $P_{O_2} = 0.07$ bar; $W/F = 0.046$ g s ml^{-1}; $X_{O_2} = 95–99\%$).

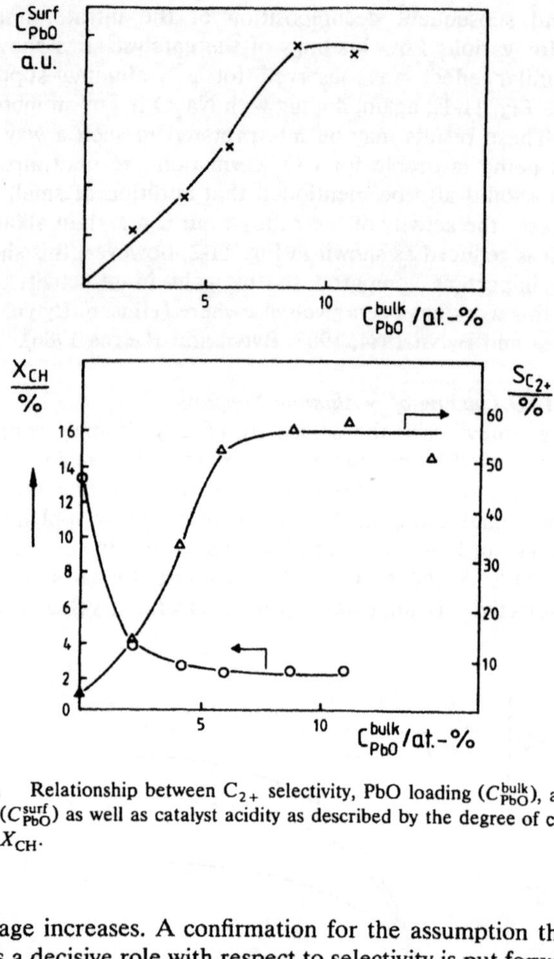

FIGURE 11-5. Relationship between C_{2+} selectivity, PbO loading (C_{PbO}^{bulk}), and surface concentration (C_{PbO}^{surf}) as well as catalyst acidity as described by the degree of cyclohexene isomerization X_{CH}.

oxide coverage increases. A confirmation for the assumption that surface acidity plays a decisive role with respect to selectivity is put forward by Fig. 11-5. In the upper diagram, C_{2+} selectivity and the degree of isomerization of cyclohexene to 1-methyl-cyclopentene as a measure of acidity (Bytyn and Baerns 1986; Pines and Haag 1960) is plotted versus lead oxide loading. From the pattern of this dependency it becomes clear that C_{2+} selectivity is improved as a consequence of decreasing surface acidity, which is related to the decrease in cylcohexene isomerization. Furthermore, the lower diagram shows the dependency between surface concentration and loading of lead; when also considering the upper diagram it may be concluded that a defined surface coverage with PbO is required to eliminate the nonselective catalytic activity of the support material.

Conclusions

From the preceding it can be concluded that C_{2+} selectivity of supported reducible metal oxide catalysts may be positively affected by the choice of the carrier material but also by the addition of alkali metal oxides. Although these relationships have been elucidated only in the present contribution for PbO as the catalytic species (see also Asami et al. 1987), similar results have been reported for other catalytic systems, for example, manganese oxide, nickel oxide, praseodymiun oxide, and titanium oxide (Sofranko, Leonard, and Jones 1987a, 1987b; Otsuka, Suga, and Yamanaka 1988; Hatano and Otsuka 1988; Jones, Leonard, and Sofranko 1987; Lane, Miro, and Wolf 1989). Some of the authors do not solely ascribe the effect of alkali addition on C_{2+} selectivity to a change in surface acidity but also to the formation of a specific solid. In which way the support material and/or the dopant may also contribute to the type of surface oxygen participating in the oxidative coupling reaction is still an unresolved question that needs further attention.

Alkaline-Earths and Alkali–Alkaline-Earths Metal Compound Catalysts

Alkaline-Earths Metal Compounds

When analyzing the relationships presented in the previous section, it becomes obvious that a suitable support for achieving high C_{2+} selectivities should not have any acidic but instead basic properties. Because it was also observed that the support material itself possesses catalytic activity, various basic oxides (i.e., the alkaline-earths oxides) were studied for their catalytic performance in the oxidative methane coupling reaction (Baerns, Carreiro, and Bytyn 1985; Carreiro and Baerns 1989a, 1989b, 1987; Carreiro et al. 1988); the lattice oxygen, that is, O^{2-}, of these compounds is of highly basic nature, increasing from BeO to BaO. C_{2+} as well as C_{3+} selectivity was lowest for BeO and highest for SrO and BaO (see Table 11-3). The activity of these materials increased from BeO to CaO (SrO and BaO could not be included in this pattern due to hydroxide formation and loss of surface area by sintering) (Baerns, Carreiro, and Bytyn 1985; Carreiro and Baerns 1989a, 1989b). As a proof that the O^{2-} anion plays an important role in the selective oxidative coupling reaction, the selectivity results obtained when using different calcium compounds should be considered (see Table 11-4). These data show a marked effect of the

Table 11-3 **C_{2+} Hydrocarbon Selectivity of Alkaline-Earths Oxides (Carreiro and Baerns 1989a)[a]**

Oxide	BeO	MgO		CaO		SrO	BaO
S_{BET} ($m^2\ g^{-1}$)	Not determined	24[b]	37[c]	7.6[d]	6.3[e]	0.1	Not determined
$S_{C_{2+}}$ (%)	22	47	22	47	55	72	70
$S_{C_{3+}}$ (%)	0.2	1.8	0.1	1.2	3.2	7.6	7.2

[a]Experimental conditions: $T = 740°C$; $P_{CH_4} = 660$ mbar; $P_{O_2} = 80$ mbar; $P_{N_2} = 260$ mbar; $X_{O_2} \geq 80$–90%.
[b]Prepared from $Mg_5(OH)_2(CO_3)_4 \cdot 4H_2O$.
[c]Prepared from $Mg(OH)_2$.
[d]Prepared from $CaCO_3$.
[e]Prepared from $Ca(OH)_2$.

Table 11-4 **C_{2+} Hydrocarbon Selectivity of Various Calcium Compounds (Carreiro and Baerns 1989a)[a]**

Compound	CaO	$CaSO_4$	Ca_2SiO_4	$CaAl_2O_4$	CaF_2	$Ca_3(PO_4)_2$
$S_{C_{2+}}$ (%)	47–55	53	43	18	14	11

[a]Experimental conditions: see Table 11-3.

anion; the fact that $CaSO_4$ compared to the other compounds exhibits a rather high selectivity similar to CaO may be explained by assuming that this catalyst contains at least some O^{2-} anions due to decomposition of $CaSO_4$ at reaction temperature.

Summarizing the selectivity results of the alkaline-earths metal oxides, it can be concluded that only those oxides that are highly basic (i.e., SrO and BaO) show higher C_{2+} selectivities than the various PbO-based catalysts. However, these compounds appear not to be applicable in practice without further catalyst modifications because their interaction with water vapor results partly in liquid hydroxides. Materials of high basicity similar to SrO and BaO without having the aforementioned shortcomings can be prepared by the incorporation of alkali compounds into the surface of stable alkaline-earth oxides like MgO and CaO (which is dealt with in the next section).

Composite Catalysts of Basic Solids

Catalysts consisting of the Li–Mg–O and Na–Ca–O types have been extensively applied in the oxidative coupling of methane. It has been shown (Ito and Lunsford 1985; Ito et al. 1985) that especially lithium-doped magnesium oxide is an active and selective catalyst; subsequently, many researchers have turned to this catalyst and have used it as a reference. Our own approach as outlined in the preceding section was initiated by Japanese workers (Tanabe 1981; Matsuda et al. 1986): According to the acid–base theory the electronic charge distribution on metal oxide surfaces may be severely perturbed by doping with metal ions of different valence. This has been well-investigated, particularly for acid catalysts (Imelik et al. 1985). There existed evidence (Matsuda et al. 1986) indicating that a similar approach may also be applied to basic catalysts; it was shown that the basicity of MgO increases when doped with sodium compounds. Based on the experimental findings on the effect of acidity and basicity, respectively, on C_{2+} selectivity as reported previously, we assumed that the concepts described for the formation and stabilization of surface basicity (Tanabe 1981; Matsuda et al. 1986; Imelik et al. 1985) are more general and ought to be applied for oxidative-methane-coupling catalysts.

Against this background the hypothesis was put forward and successfully proven (Baerns, Carreiro, and Bytyn 1985; Carreiro and Baerns 1989a, 1989b) that the use of catalysts exhibiting high surface basicity achieved by dispersing alkali compounds on other basic materials (i.e., alkaline-earths oxides) would significantly improve C_{2+} selectivity. For illustration of this work, some essential results are presented that have

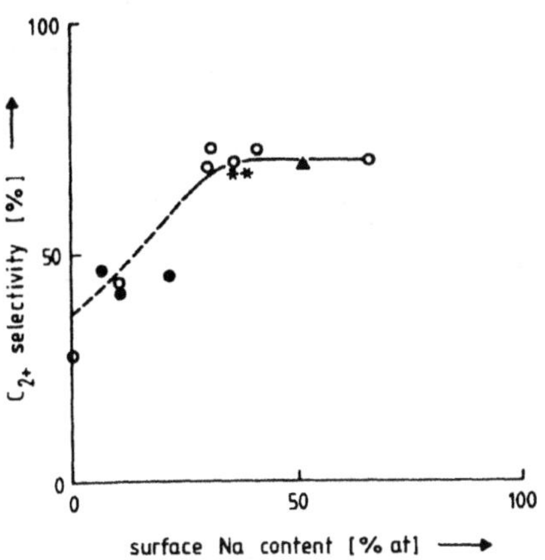

FIGURE 11-6. The dependence of C_{2+} selectivity on the concentration of sodium dispersed on CaO (experimental conditions: see Table 11-3; $X_{O_2} \geq 75\%$); $CH_4/O_2 = 6.5$ (•), 7.1 (▲), 7.3 (∗), and 8.5 (○).

Table 11-5 Activity (as Oxygen Conversion X Related to Modified Reaction Time W/F or as Pseudo-First-Order Rate Constant k) and Selectivity of Na_2CO_3 and K_2CO_3 Catalysts[a]

	ACTIVITY			SELECTIVITY			
Catalytic Material	W/F (g s ml^{-1})	X_{O_2} (%)	k (ml g^{-1} s^{-1})	C_2H_6 (%)	C_2H_4 (%)	C_{3+} (%)	ΣC_nH_m (%)
Na_2CO_3	46.4	71	0.03	19	27.8	4.2	51
K_2CO_3	4.1	97	0.86	22.8	10.8	0.4	34

[a]Experimental conditions: see Table 11-3.

been obtained more recently (Grzybek et al. 1990). C_{2+} selectivity increases with the amount of NaOH that is dispersed on CaO until a final value is reached (see Fig. 11-6). The maximum selectivity of approximately 70% is significantly above the selectivity achieved for CaO alone, which was around 50% (see Table 11-3). This result indicates that a minimum surface concentration of the alkali compound is necessary for establishing optimum catalytic conditions. It should be mentioned that the increase in C_{2+} selectivity must not be attributed to the alkali compound itself but to its interaction with CaO. This becomes evident when considering the catalytic performance of sodium carbonate and potassium carbonate catalysts as shown in Table 11-5 and of pure CaO (see Table 11-3).

Further results on the effect of alkali doping of alkaline-earths oxides are presented in Table 11-6. Over a wide range of such catalytic materials, remarkable C_{2+} selectivities have been obtained, which have been reported earlier (Baerns, Carreiro, and Bytyn 1985; Carreiro and Baerns 1989a, 1989b). It can, however, not be excluded that some of the C_{2+} selectivities listed might be too high by about 5–10% due to pore-diffusional effects within the catalyst pellets (3 × 4 mm) used in these experiments; such an effect has been reported elsewhere (Follmer, Lehmann, and Baerns 1989).

Conclusions

From the experimental findings presented in this section it becomes apparent that basicity plays an important role for the oxidative coupling reaction. One would assume that high surface basicity results in high C_{2+} selectivity. In fact, this has been ascertained by determining the pK_A values (where $pK_A = 7$ is neutral, >7 is basic, and <7 is acidic) of the catalysts by suitable indicators adsorbed on the surface of the solids dispersed in a nonaqeous solvent (Bytyn and Baerns 1986) and relating them to C_{2+} selectivity (Baerns 1988) as shown in Table 11-7.

La_2O_3/CaO Catalysts

Based on the preceding results it was to be expected that La_2O_3/CaO catalysts, being of highly basic nature due to their composition, should yield high C_{2+} selectivities in the oxidative coupling reaction. An advantage of rare-earth oxide catalysts is their long-time stability with respect to selectivity and activity as compared to most of the alkali–alkaline-earths compound catalysts. For these reasons La_2O_3–CaO catalysts of various compositions were applied to identify the optimum composition for maximum C_{2+} selectivity (Becker and Baerns 1991). The catalytic materials were characterized by bulk-phase and surface composition as well as by their surface basicity using the isopropyl alcohol conversion to acetone and/or propene as a probe reaction (Sofranko, Leonard, and Jones 1987a, 1987b; Bytyn 1987). In the following the essential relationships between catalytic results and surface properties are summarized.

Catalyst activity is increased when adding La_2O_3 to CaO; this can be derived from the contact time ($m_{cat}/\dot{V}$) required for obtaining an oxygen conversion of approximately 95%:

CaO—0.11 g s ml^{-1}
La_2O_3 (27%)/CaO—ca. 0.02 g s ml^{-1}
La_2O_3 (50%)/CaO—ca. 0.02 g s ml^{-1}

(under the following experimental conditions: $P_{CH_4} = 93$ kPa; $P_{O_2} =$ 7 kPa; $T =$ ca. 1000 K). C_{2+} selectivity also was affected by the addition of La_2O_3 to CaO. There was, however, no clear effect of the bulk composition. A clear relationship could be visualized only when selectivity was plotted versus the surface concentration of La as shown in Fig. 11-7. Lanthanum was always enriched on the surface; the degree of enrichment was influenced by the catalyst precursors and the methods of preparation as demonstrated in Fig. 11-8, which contains these data without specifying all conditions in detail. Although there is some scatter in the data of Fig. 11-7 it is nevertheless rather obvious that C_{2+} selectivity passes through a maximum. The assumption that the maximum in C_{2+} selectivity would coincide with the maximum of basicity which was based on the conclusions drawn in the preceding sections, however, could not be confirmed. This is illustrated by the relationships shown in Fig. 11-9: The ratio of acetone to propene, which is plotted versus lanthanum concentration of the surface, can be considered as a qualitative measure of basicity, that is, the larger this ratio, the higher the basicity; its maximum is apart from the maximum of selectivity. It appears, however, that there exists a slight dependency of selectivity on the absolute rate of propene formation catalyzed by acidic

Table 11-6 Activity and Selectivity of Alkali-Compound-on-CaO Catalysts[a]

Alkali Compound (mole%)	ACTIVITY W/F (g s ml^{-1})	X_{O_2} (%)	X_{CH_4} (%)	k (ml g^{-1} s^{-1})	SELECTIVITY C_2H_6 (%)	C_2H_4 (%)	ΣC_2 (%)	C_3H_8 (%)	C_3H_6 (%)	ΣC_3 (%)	ΣC_4 (%)	ΣC_nH_m (%)	CO_x (%)
1.2% Li_2CO_3	0.37	92	15	6.8	27	40	67	2.3	2.5	4.8	4.2	76	24
3.0% Li_2CO_3	0.36	91	16	6.7	27	41	68	2.3	2.5	4.8	4.0	77	23
5.8% Li_2CO_3[b]	0.44	93	16	6.0	30	38	68	2.6	2.6	5.2	4.6	77	23
1.2% Na_2CO_3	0.14	86	16	14.0	24	44	68	—	4.6	4.6	5.4	78	22
3.5% Na_2CO_3	0.14	83	15	12.7	25	42	67	—	4.6	4.6	5.3	78	22
6.2% Na_2CO_3[b]	0.81	91	12	3.0	29	33	62	2.6	1.2	3.8	2.9	70	30
1.5% K_2CO_3	0.36	98	15	10.9	22	41	63	—	3.6	3.6	5.6	72	28
3.7% K_2CO_3	0.14	91	14	17.2	20	42	62	—	3.3	3.3	5.0	72	28
7.1% K_2CO_3[c]	0.38	96	13	8.5	22	43	65	—	3.6	3.6	1.8	70	30
2.3% LiOH	0.62	87	15	3.3	30	39	69	3.3	2.1	5.4	2.7	77	23
5.7% LiOH	0.63	100	16	> 11	31	36	67	3.4	1.7	5.1	4.5	77	23
11.0% LiOH[b]	1.37	96	16	2.3	34	29	63	5.9	—	5.9	6.6	76	24
2.6% NaOH	2.59	100	16	> 3	29	34	63	3.5	1.7	5.2	4.8	74	26
6.0% NaOH	0.15	90	14	15.4	24	43	67	—	4.5	4.5	2.5	74	26
11.9% NaOH[b]	0.96	93	15	2.8	30	36	66	5.3	—	5.3	4.1	75	25
2.6% KOH	0.26	95	15	11.5	22	43	65	—	3.8	3.8	3.5	73	27
5.5% KOH	0.26	89	13	8.5	20	44	64	—	3.5	3.5	2.3	70	30

11.3% KOH[c]	1.93	96	14	1.7	30	30	60	4.8	—	4.8	7.5	72	28
1.4% Li_2O	0.37	92	14	6.8	28	39	67	2.5	2.3	4.8	3.4	75	25
3.0% Li_2O	0.38	97	14	9.2	24	40	64	—	4.0	4.0	3.4	72	28
6.0% Li_2O[b]	0.36	99	14	12.8	24	41	65	—	3.9	3.9	3.4	72	28
1.3% Na_2O	0.36	97	17	9.7	27	40	67	2.7	2.8	5.5	5.5	78	22
3.0% Na_2O	0.36	95	16	8.3	25	40	65	—	4.8	4.8	5.6	72	28
8.7% Na_2O[b]	0.72	96	14	4.5	29	35	64	4.1	1.4	5.5	6.7	73	23
1.2% K_2O	0.25	93	15	10.6	24	42	66	—	3.9	3.9	2.5	74	26
2.9% K_2O	0.25	83	14	7.1	23	42	65	—	4.2	4.2	2.9	72	28
6.0% K_2O[c]	0.43	94	14	6.5	21	42	63	—	3.7	3.7	2.3	69	31
1.4% Li_2SO_4	0.91	100	16	>7.6	35	31	66	5.8	—	5.8	4.3	76	24
3.4% Li_2SO_4	1.40	99	16	3.3	33	32	65	5.3	—	5.3	5.1	76	24
6.5% Li_2SO_4[d]	0.95	93	14	2.8	30	38	68	3.6	6.0	4.6	2.2	74	26
1.4% Na_2SO_4	0.24	95	14	12.5	24	40	64	—	3.7	3.7	4.3	73	27
3.5% Na_2SO_4	0.19	92	16	13.3	26	45	71	—	4.8	4.8	2.2	78	22
5.2% Na_2SO_4[e]	0.20	82	15	8.6	27	42	69	—	4.9	4.9	4.0	78	22
1.2% K_2SO_4	0.22	85	15	8.6	27	42	69	2.5	2.3	4.8	2.4	76	24
3.4% K_2SO_4	0.18	80	14	8.9	24	46	70	—	4.2	4.2	2.0	76	24
5.2% K_2SO_4[f]	0.23	92	15	11.0	27	42	69	—	4.2	4.2	2.8	76	24

[a]Experimental conditions: see Table 3. Catalyst preparation: drying of precursor, that is, alkali salt on $Ca(OH_2)$, for 2 h at 130°C; subsequent pelletization to 3 × 4 mm; calcination under reaction conditions.

[b-f]Phase composition by X-ray diffraction: [b]CaO, $Ca(OH)_2$; [c]CaO, $Ca(OH)_2$, $K_2Ca(CO)_3$; [d]CaO, Li_2SO_4-I; [e]CaO, Na_2SO_4-II; [f]CaO, K_2SO_4 (arcanite).

Table 11-7 Selectivity and Acid–Base Properties of Various Catalytic Oxides

Catalytic Oxide (wt%)	$S_{C_{2+}}$ (%)	pK_A
1.5% Li_2O/CaO	73	+12.2
3.0% Na_2O/CaO	67	+9.3
4.5% K_2O/CaO	67	+9.3
SiO_2, *kieselgel*	15	+6.8
γ-Al_2O_3	3	+3.3/+4.0
$Al_2O_3 \times SiO_2$	1	−5.6

sites. The increase of C_{2+} selectivity goes along with an increase of this rate of formation; its maximum, however, is shifted to higher lanthanum concentrations than that of C_{2+} selectivity. From this finding one might tentatively conclude that not only the basicity of the catalyst surface, but also its acidity affects C_{2+} selectivity; such a cooperative effect involving acidic sites (i.e., in the present case La^{3+} and/or Ca^{2+}) and basic O^{2-} sites also has been postulated for other reactions [side-chain alkylation of toluene by methanol (Miyamoto et al. 1989) and hydrogenation of 1,3-butadiene (Tanabe 1981)]. If a heterolytic splitting of the methane molecule is assumed,

$$CH_4 \nearrow CH_3^{+} + H^{-}$$
$$CH_4 \searrow CH_3^{-} + H^{+}$$

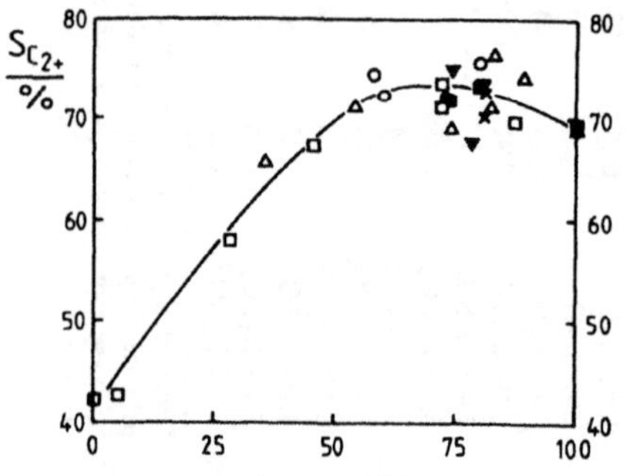

FIGURE 11-7. Effect of surface concentration of lanthanum on C_{2+} selectivity for differently prepared La_2O_3/CaO catalysts (experimental conditions: P_{CH_4} = 93 kPa; P_{O_2} = 7 kPa; T = 1013 K; $X_{O_2} \geq 90\%$; d_p = 0.25–0.36 mm).

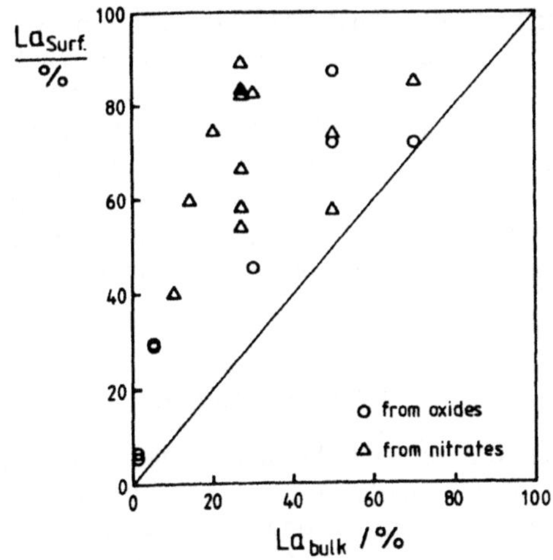

FIGURE 11-8. Enrichment of La on the surface as compared to bulk composition.

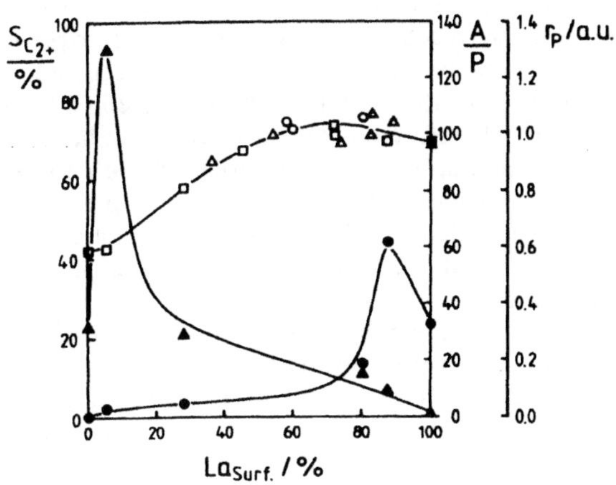

FIGURE 11-9. C_{2+} selectivity, relative basicity (acetone–propene), and rate of propene formation as a function of surface concentration of lanthanum (experimental conditions: see Fig. 11-7).

it could be imagined either that CH_3^- is stabilized by a Lewis center to which its electron is then transfered and that such CH_3 species either recombine on the surface if closely located to each other or desorb, or that CH_3^+ interacts with negatively charged oxygen existing as O^- or O_2^- on the surface, leading to nonselective oxidation. Accepting this hypothesis could lead to the conclusion that electronic as well as geometric surface properties affect the oxidative coupling reaction and that optimum conditions with respect to these properties are required to maximize C_{2+} selectivity.

CONCLUSIONS

The results reported here generally suggest that basicity and acidity of the catalyst surface play a major role for the selectivity of the oxidative coupling reaction of methane. Strong acidity favors total oxidation of the hydrocarbons, that is, methane and C_{2+} hydrocarbons formed. In contrast, high basicity results in high C_{2+} selectivity; this might be attributed to (1) suppression of the nonselective pathways and/or (2) promotion of the selective routes. Total oxidation has been ascribed to interaction of negatively charged oxygen either with a protonated methane molecule or CH_3^+ species, or to reaction between radical-like oxygen and methyl species. However, none of these routes, which are discussed in more detail elsewhere (Baerns and Ross 1991), have been definitely confirmed or rejected up to now. Therefore no final conclusions can be drawn on how to design a selective catalyst based on the various elementary reaction steps. Thus, the design of such catalysts has to rely on the preceding experimental findings, suggesting that high basicity of the catalyst surface is required although some additional acidity appears to be favorable as observed for a La_2O_3/CaO catalyst.

In the preceding reasoning, basicity and acidity were claimed to be the decisive factors for C_{2+} selectivity. These properties may, of course, also be related to the electronegativity of the catalytic solids. There exists a direct relationship between acidity or basicity of binary metal oxides and their electronegativity, that is, the lower the electronegativity, the higher the basicity and vice versa with respect to acidity (Smith 1987; Bratsch 1988). The electronegativity x of a binary oxide can be calculated as an average of the two metal cations using Pauling's scale of the electronegativities x of metals (Shibata et al. 1973; Tanabe and Ozaki 1967):

$$x_{\text{cations}} = (1 + z)x_{\text{metal}}$$

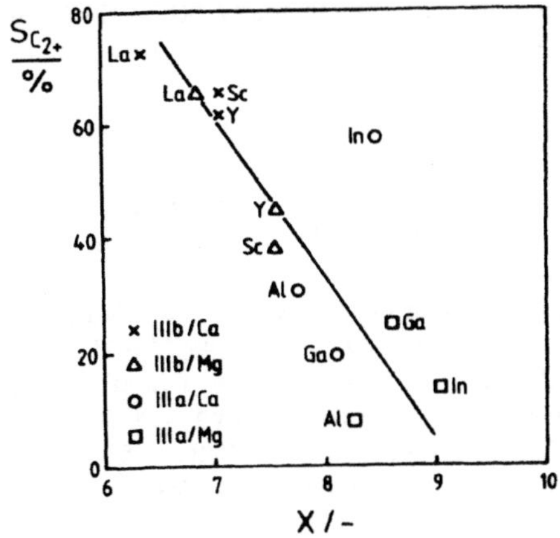

FIGURE 11-10. Relationship between C_{2+} selectivity and the electronegativity of various binary metal oxides (experimental conditions: similar to those of Fig. 11-7).

with z being the valence of the metal cation;

$$x = 0.5(x_{\text{cation A}} + x_{\text{cation B}})$$

For some selected binary oxides used as catalysts, the C_{2+} selectivities that had been experimentally determined are plotted as function of the respective electronegativity values in Fig. 11-10 (Becker 1990). The trend indicated by this plot is in agreement with the suggested dependence of C_{2+} selectivity on basicity–acidity and electronegativity, respectively, although the latter relationship $S_{C_{2+}}(x)$ still needs some further support by analyzing the available experimental data.

The final conclusion from the present contribution and the extensive work of others on the catalysis of the oxidative coupling of methane may be summarized by stating that significant progress has been made during the last decade in designing more-selective catalysts but that a comprehensive fundamental understanding of the surface processes and their dependence on the surface properties is still lacking.

ACKNOWLEDGMENT

The support of several collaborators who provided the experimental material incorporated in this contribution and who assisted in various ways in finalizing the manuscript during many other time-consuming events in the author's present daily work is most gratefully acknowledged.

References

Asami, K., S. Hashimoto, T. Shikada, K. Fujimoto, and H. Tominaga. 1987. Selective oxidative coupling of methane to ethane and ethylene over supported lead oxide catalysts. *Ind. Eng. Chem. Res.* 26:1485–88.

Baerns, M. 1988. Development of catalysts for the oxidative coupling reaction. In *Successful Design of Catalysts*, ed. T. Inui, pp. 333–41. Amsterdam: Elsevier.

Baerns, M., and W. Bytyn. 1984. German Patent Application 3.406.751A1 (24.02.1984). [Corresponding Canadian Patent 458,797 (1987).]

Baerns, M., and W. Bytyn. 1987. Canadian Patent 458,797. [Corresponding German Patent Application 3.406.751A1 (24.02.1984).]

Baerns, M., J. Carreiro, and W. Bytyn. 1985. German Patent Application 3.534.530A1 (27.09.1985).

Baerns, M., and J.R.H. Ross. 1991. Catalytic chemistry of methane conversion. In IUPAC Monograph, *Perspectives in Catalysis*, ed. J.M. Thomas and K.I. Zamaraev. In press.

Becker, S. 1990. Doctoral Dissertation, Ruhr-Universität Bochum.

Becker, S., and M. Baerns. 1991. Oxidative coupling of methane over La_2O_3–CaO catalysts—Effect of bulk and surface properties of the solids on catalytic performance. *J. Catal.* 128:512.

Bratsch, S.G. 1988. Electronegativity and the acid–base character of binary oxides. *J. Chem. Educ.* 65:877–8.

Bytyn, W. 1987. Doctoral Dissertation, Ruhr-Universität Bochum.

Bytyn, W., and M. Baerns. 1986. Supported PbO catalysts for the oxidative coupling of methane—The effect of surface acidity of the support on C_{2+} selectivity. *Appl. Catal.* 28:199–207.

Carreiro, J., and M. Baerns. 1987. Catalytic conversion of methane by oxidative coupling to C_{2+} hydrocarbons. *React. Kinet. Catal. Lett.* 35:349–60.

Carreiro, J., and M. Baerns. 1989a. Oxidative coupling of methane—Part I: Alkaline earths compound catalysts. *J. Catal.* 117:258–65.

Carreiro, J., and M. Baerns. 1989b. Oxidative coupling of methane—Part II: Composite catalysts of basic materials. *J. Catal.* 117:396–403.

Carreiro, J., G. Follmer, L. Lehmann, and M. Baerns. 1988. Oxidative coupling of methane to C_{2+} hydrocarbons—Considerations on the performance of lead compound and alkali/alkaline earth compound catalysts. In *Proceedings of the 9th International Congress on Catalysis*, Vol. 2, ed. M.J. Philips and M. Ternan, pp. 891–8. Ottawa: The Chemical Institute of Canada.

Follmer, G., L. Lehmann, and M. Baerns. 1989. Effect of transport limitation on C_{2+} selectivity in the oxidative methane coupling reaction using a NaOH/CaO catalyst. *Catal. Today* 4:323–32.

Grzybek, T., G.C. Maiti, D. Scholz, and M. Baerns. 1990. Unpublished.

Hatano, M., and K. Otsuka. 1988. Alkali-metal doped transition metal oxides active for oxidative coupling of methane. *Inorg. Chim. Acta* 146:243.

Hinsen, W., and M. Baerns. 1982. German Patent Application 3.237.079A (07.10.1982). [Corresponding US Patent 4,608,449 (26.08.1986).]

Hinsen, W., and M. Baerns. 1983. Oxidative coupling of methane to C_2 hydrocarbons in the presence of different catalysts. *Chem. Ztg.* 107:223–6.

Hinsen, W., W. Bytyn, and M. Baerns. 1985. Oxidative dehydrogenation and coupling of methane. In *Proceedings of the 8th International Congress on Catalysis, Berlin (West) 1984*, Vol. III, pp. 581–92. Weinheim: Verlag Chemie.

Hinsen, W., and M. Baerns. 1986. US Patent 4,608,449 (26.08.1986). [Corresponding German Patent Application 3.237.079A (07.10.1982).]

Hutchings, G.J., M.S. Scurrell, and J.R. Woodhouse. 1989. Oxidative coupling of methane using oxide catalysts. *Chem. Soc. Rev.* 18:251–83.

Imelik, B., C. Naccache, G. Coudurier, Y. Ben Taarit, and J.C. Vedrine (eds.). 1985. *Catalysis by Acids and Bases*. Amsterdam: Elsevier.

Ito, T., and J.H. Lunsford. 1985. Synthesis of ethylene and ethane by partial oxidation of methane over lithium-doped magnesium oxide. *Nature* 314:721–2.

Ito, T., J.-X. Wang, C.-H. Lin, and J.H. Lunsford. 1985. Oxidative dimerisation of methane over a lithium-promoted magnesium oxide catalyst. *J. Am. Chem. Soc.* 107:5062–8.

Jones, C.A., J.J. Leonard, and J.A. Sofranko. 1987. Fuels for the future: Remote gas conversion. *J. Energy and Fuels* 1:12.

Keller, G.E., and M.M. Bhasin. 1982. Synthesis of ethylene via oxidative coupling of methane—I. Determination of active catalysts. *J. Catal.* 73:9–19.

Lane, G.S., E, Miro, and E.E. Wolf. 1989. Methane oxidative coupling—II. A study of lithium-titania-catalyzed reactions of methane. *J. Catal.* 119:161–78.

Matsuda, T., Z. Minami, Y. Shibata, S. Nagano, H. Miura, and K. Sugiyama. 1986. Catalytic properties of magnesium oxides doped with sodium compounds. *J. Chem. Soc., Faraday Trans 1* 82:1357–66.

Miyamoto, A., S. Iwamoto, K. Agusa, and T. Inui. 1989. Acid–base cooperative catalysis in the side-chain alkylation of toluene with methanol as visualized by the computer-graphics. In *Acid–Base Catalysis*, ed. K. Tanabe et al., pp. 497–504. Weinheim: Verlag Chemie.

Otsuka, K., K. Jinno, and A. Morikawa. 1985. The catalysts active and selective in oxidative coupling of methane. *Chem. Lett.* 499–500.

Otsuka, K., K. Suga, and I. Yamanaka. 1988. Electrochemical control for oxidative coupling of methane over $LiNiO_2$ using electrolytes. *Chem. Lett.* 317–18.

Pines, H., and W.O. Haag. 1960. Alumina: Catalyst and support—I. Alumina, its intrinsic acidity and catalytic activity. *J. Am. Chem. Soc.* 82:2471–83.

Shibata, K., T. Kiyoura, J. Kitagawa, T. Sumiyoshi, and K. Tanabe. 1973. Acidic properties of binary metal oxides. *Bull. Chem. Soc. Jpn.* 46:2985–8.

Smith, D.W. 1987. An acidity scale for binary oxides. *J. Chem. Educ.* 64:480–1.

Sofranko, J.A., J.J. Leonard, and C.A. Jones. 1987a. The oxidative conversion of methane to higher hydrocarbons. *J. Catal.* 103:302–10.

Sofranko, J.A., J.J. Leonard, and C.A. Jones. 1987b. The oxidative conversion of methane to higher hydrocarbons over alkali-promoted Mn/SiO_2. *J. Catal.* 103:311–19.

Tanabe, K. 1981. Solid acid and base catalysts. In *Catalysis—Science and Technology*, Vol. 2, ed. J.R. Anderson and M. Boudart, pp. 231. Berlin: Springer-Verlag.

Tanabe, K., and A. Ozaki. 1967. Acid–base properties and catalytic activity of solid surfaces. *J. Catal.* 8:1–7.

12

The Direct Conversion of Methane to Methanol (DMTM)

H.D. Gesser and N.R. Hunter

Chemistry Department, The University of Manitoba, Winnipeg, MB, Canada

TABLE OF CONTENTS

INTRODUCTION

Methanol is an important industrial chemical, ranked 22d (Chang and Tikkamen 1988) in the top 50 products in the United States. It is also an alternate automotive fuel that, if adopted even partially, will require a multifold increase in production to meet the higher demand.

Table 12-1 Uses of Methanol

1.	Formaldehyde	31%
2.	Acetic acid	12%
3.	Solvents	11%
4.	Chlorinated hydrocarbons	9%
5.	Methyl esters	8%
6.	Amines	4%
7.	Miscellaneous	25%

As an industrial chemical it is used extensively in the manufacture of many other chemicals, some of which are listed in Table 12-1. The most important of these is formaldehyde, which is ranked 25th in industrial production.

The partial oxidation of CH_4 has been the subject of four recent independent reviews (Garcia and Loffler 1984; Gesser, Hunter, and Prakash 1985; Foster 1985; Pitchai and Klier 1986) and the references given there will not be discussed in detail here.

When considering the partial oxidation reaction of CH_4, a distinction will be made between (a) oxidative coupling in which oxygenated products other than CO and CO_2 are not formed and (b) where CH_3OH and HCHO are produced. The treatment of oxidative coupling has been reviewed recently (Kuo 1987; Hutchings, Scurrell, and Woodhouse 1989) and a bibliography reported (Baerns, van der Wiele, and Ross 1989). This review will only be concerned with the partial oxidation reactions in which CH_3OH and HCHO are formed. The activation of the methane molecule involves the $CH_3—H$ bond and is common to both processes and will be briefly discussed.

THERMODYNAMIC CHARACTERISTICS

The thermodynamic characteristics of some relevant reactions are given in Table 12-2 and show that all the reactions are thermodynamically feasible. However, because the reactions are known not to occur spontaneously

Table 12-2 Thermochemical Values of Some CH_4 Reactions

	$\Delta H°$	$\Delta G°$	$\Delta S°$
Reaction	Kilojoules per Mole of CH_4		
$CH_4 + 0.5O_2 \rightarrow CH_3OH$	−127.1	−111.2	−48.9
$CH_4 + O_2 \rightarrow HCHO + H_2O$	−276	−290.8	16.2
$CH_4 + 2O_2 \rightarrow CO_2 + 2H_2O$	−519	−543.5	81.3
$CH_4 + 1.5O_2 \rightarrow CO + 2H_2O$	−802	−800.7	−5.1
$2CH_4 + O_2 \rightarrow CH_3OCH_3 + H_2O$	−139	−120.6	−61.1

under ambient conditions, it is worth examining the reaction under (1) higher temperatures and (2) higher pressures. The effect of temperature on the free energy of a reaction at constant pressure is given by

$$\delta \Delta G/\delta T = -\Delta S$$

Because the entropy changes are small, increasing the temperature results in relatively small changes in the free energy of the reactions. This confirms the greater importance of the kinetic factors than the thermodynamic parameters on the reaction. The effect of pressure on the free energy is given by

$$\delta \Delta G/\delta P = \bar{V}$$

Again, a simple calculation would show that the effect of changing the pressure from 1 to 100 atm is too small to be significant.

ACTIVATION OF THE C—H BOND

The C—H bond in methane is stronger than in any other molecule. Thus, there exists a misconception that the activation of CH_4 means that the conditions are such that the products that contain C—H bonds will be unstable. This is not correct and, indeed, it is possible to create conditions where the CH_4 molecule can react so that its partial oxidation products, HCHO and CH_3OH, are stable enough to survive. Several activation methods have been used.

Photochemical

The photolysis of CH_4 can be carried out at wavelengths less than 147 nm (Calvert and Pitts 1966). However, in the presence of oxygen, the competitive reaction of oxygen atoms would occur. The photosensitized oxidation of CH_4 by Hg ($^3P_1$) (λ = 254 nm) has been reported (Bell and McDowell 1961). However, the quenching cross section of O_2 is 19.9 Å^2 relative to 0.084 Å^2 for CH_4. Hence, only at very low O_2 concentrations will the CH_3—H bond be broken.

Ogura and co-workers (Ogura and Kataoka 1988; Ogura, Migita, and Fujita 1988; Ogura, Migita, and Yamada 1990) studied the photochemical conversion of CH_4–H_2O mixtures at temperatures between 60 and 90°C using the Hg resonance lines at 185 and 254 nm. Of the approximately 20 products detected, the major ones had selectivities of methanol 70%, formic acid 11%, ethanol 5%, formaldehyde 5%, acetone 4%, and acetic acid 3%.

The use of laser excitation and sensitization was reported by Baughcum and co-workers (Baughcum et al. 1986) who added acetone to the CH_4–O_2 mixture. At 193 nm the acetone decomposed to form $CH_3\cdot$ radicals, which initiated the reaction. The authors showed both a thermal as well as a photochemical sensitization to form methanol.

The photochemical activation of methane to form $CH_3\cdot$ on an irradiated surface has been studied extensively. Volodin (Volodin and Cherkashin 1981) has shown that ZnO when irradiated at $\lambda < 410$ nm formed O^-, which reacted with CH_4 at 77 K to form $CH_3\cdot$ radicals.

Kubokawa and co-workers (Yun et al. 1980) studied the photocatalytic oxidation of CH_4 (and other hydrocarbons) on MgO with N_2O as the oxidant. The products C_2H_4 and C_2H_6 were obtained at room temperature and C_3H_6 was also found at temperatures up to 673 K. Although the reaction was attributed to O^-, no oxygenates were reported as products.

Gerasimov (Gerasimov 1988) studied the photooxidation of CH_4 on V/SiO_2 with $\lambda < 340$ nm. Using $^{18}O_2$ as a tracer, it was argued that CH_3O^- was formed on the surface and was converted to HCHO, which oxidized to CO_2.

The effects of O_2, N_2O, and H_2O on the photoinduced reaction of CH_4 on Mo/SiO_2 surfaces showed coupling but no oxygenated products (Hill, Shelimov, and Karansky 1987).

The photoactivation of CH_4 on $(SiW_{12}O_{40})^{4-}/TiO_2$ and TiO_2 at room temperature was shown to occur (Gratzel, Thampi, and Kiwi 1989). Although CH_3, CH_3O^-, and CH_3OH were proposed as intermediates, only CO and CO_2 products were detected and analyzed.

The laser ($\lambda = 380$ nm) excitation of Li_2O on various supports (MgO, TiO_2, SiO_2) showed primary conversion of CH_4 to C_2 products and carbon (Sayyed and Stair 1990).

The laser induced ($\lambda = 193$ nm) photooxidation of CH_4 using N_2O was shown (Oshima et al. 1988) to produce $O(^1D)$, which is known to insert into the C—H bond (Yamazaki and Cvetanovic 1964). The products at 200–300°C were CO, CO_2, C_2H_6, C_2H_4, C_3H_8, CH_3OH, C_2H_5OH, $(CH_3)_2O$, CH_3CHO and $HCOOCH_3$. At such high temperatures abstraction as well as insertion reactions occur. The low-temperature (−70°C) photolysis ($\lambda = 185$ nm) of N_2O–CH_4 mixtures showed the primary product to be methanol, indicating the absence of abstraction reactions (Gesser 1989).

The activation of CH_4 by radiolysis in also possible (Ausloos, Rebbert, and Lias 1965), but not much has been reported on the oxidative conversion.

Electrochemical

The electrochemical activation of CH_4 has been reported (Seimonides and Stoukides 1986) using the apparatus shown in Fig. 12-1. At 700°C and 1 atm pressure for a feed ratio of $O_2 : CH_4 = 0.02$, the products were CO, CO_2, C_2H_4, and C_2H_6 (neither formaldehyde nor formic acid was detected). When the current was applied to the cell, the CO_2 formation increased whereas the C_2H_4 decreased. Reversing the current flow showed the effect of oxygen pumping.

A description has been published of the pumping of O_2 through an yttria-stabilized zirconia (YSZ) cell ($Po_2{}^1, M_1 \setminus YSZ \setminus M_2, Po_2{}^{11}$, where M_i = electrodes). The electrodes were Ag and Bi_2O_3–Ag and the O_2 pumped system showed greater activity than the nonelectrochemical system (Otsuka, Yokoyama, and Morikawa 1985).

A more dramatic illustration of O_2 flow was reported by Anshits and co-workers (Anshits et al. 1990), who allowed oxygen to diffuse through a thin (0.13 mm) silver membrane and react with CH_4 on the other side. Because the oxygen migrates through the silver as atomic oxygen (Perkins 1973), it is conceivable that the initial reaction at the silver surface is

$$CH_4 + O \rightarrow CH_3\cdot + OH\cdot$$

If sufficient O_2 is available to react with the $CH_3\cdot$ radicals, then the chain

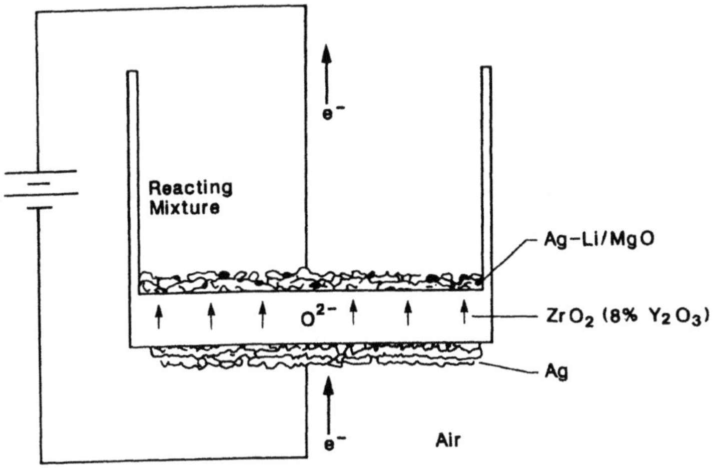

FIGURE 12-1. Apparatus for electrochemical oxygen pumping (Seimonides and Stoukides, 1986).

reaction to form methanol would occur instead of the coupling reaction of C_1 to C_2.

Eng and Stoukides (Eng and Stoukides 1988) activated CH_4 in an oxygen-conducting solid electrolyte cell with Ag/BiO_2 catalyst at 700–900°C.

The use of an oxide membrane (PbO/MgO) to transport oxygen was described (Omata et al. 1989). At 750°C, it was shown that oxidative coupling to C_2 occurred with a selectivity of over 97%.

Alkane activation in an electrochemical cell using iron-porphyrin deposited on a graphite cathode was described (Khenkin and Shilov 1987). Although methane was not studied, it was shown that cyclohexane was converted to cyclohexanol and cyclohexanone. This topic has been reviewed (Khenkin and Shilov 1989) but the system shows a rather low reactivity with CH_4.

A photochemical–electrochemical cell was described (Ogura, Migita, and Ito 1990) for the partial oxidation of methane. The process consisted of the electrochemical oxidation of Cl^- to Cl atoms and the subsequent generation of $CH_3\cdot$ and formation of CH_3Cl, which was hydrolyzed to CH_3OH. The selectivities for the products were CH_3Cl 70–90%, CH_3OH 1–10%, CH_2Cl_2 1–20%, $CHCl_3$ 0.2–7%, and CCl_4 ~ 0.1% with regular potentiostatic electrolysis and current efficiency from 3 to 14%. When a pulsed rectangular wave was used the selectivity of CH_3OH was close to 100% with a current efficiency of 100% at low concentrations of Cl^- ($[Cl^-] = 0.05$ M). The light source used was a low-pressure mercury resonance lamp emitting light primarily at 185 and 254 nm.

Catalytic Processes for Activation of Methane

The activation of alkanes by rhodium complexes was described (Schwartz 1985, 1987). A review of the mechanism of C—H bond activation by transition-metal complexes was reported (Bergman 1984; Ryabov 1990). A molecular orbital study of CH_4 activation on Fe, Ni, and Pt surfaces has been described (Anderson and Maloney 1987).

The partial oxidation of CH_4 by porphyrin and phthalocyanine complexes in zeolites was shown (Chan and Wilson 1987) to yield CO_2 and H_2O with some CH_3OH at 37°C, 50 psig and $CH_4 : O_2 = 4$.

The bioconversion of CH_4 to CH_3OH was reported (Patras and Tang 1987). The low yields of methanol obtained suggested the need for genetic modification for the optimization of methanol production. A similar bioconversion process was described (Corder et al. 1987), where the rate of production of methanol was about 1 g l^{-1} for a 24-h period.

FIGURE 12-2. Acid–base pair interaction on a catalytic surface (Sokolovskii 1987).

A computer-aided molecular design of alkane-activation catalysts was described (Shelnutt, Stohl, and Granoff 1987). A method has been suggested (Sokolovskii 1987) whereby catalysts can be chosen for conversion of organic compounds at C—H bonds. Activation of the C—H bond was believed to occur by the interaction with an acid–base pair on the catalyst surface (Fig. 12-2). Partial oxidation occurred when a cation fragment (a high-valent element) adds lattice oxygen. The scheme was supported by various reaction systems but the production of methanol was not included.

The formation of HCHO by the partial oxidation of CH_4 on Vycor glass, quartz glass, SiO_2, MgO, and Al_2O_3 was investigated (Kastanas, Tsigdinos, and Schwank 1988a, 1988b). Selectivities as high as 81% were reported for very low conversion (~ 0.26%) that decreased to 2.7% selectivity at about 4% conversion at 620–720°C. The activation energy for HCHO formation (about 130 kJ mol^{-1}) was independent of the catalytic surface, implying formation in the gas phase.

A study of modified natural mordenites as a catalyst for the partial oxidation of CH_4 used air ($CH_4:O_2 = 2$). The C_2, CH_3OH, and HCHO products were determined by the extent of decationization of the zeolite and the resulting acidity (Il'chenko et al. 1989). A comparison between the heterogeneous and homogeneous oxidation of CH_4 has been reported (Meltser et al. 1989). The addition of HCl showed a decrease in the induction period, but the HCHO yield remained unchanged because of the poisoning effect of HCl on the active centers on the SiO_2 surface. The formation of HCHO from CH_4 over MoO_3 supported on aerosil and other supports has been reported (Mac Giolla Coda et al. 1990) using N_2O as the oxidant. The reaction of N_2O with CH_4 over various metallic oxide catalysts was studied (Il'chenko et al. 1987, 1988; Il'chenko, Bostan, and Golodets 1988; Il'chenko and Golodets 1988) and both CH_3OH and HCHO were shown to be formed at 773 K. Although such studies may give some information regarding the mechanism, the use of N_2O has no practical value because of its high cost.

Some thermodynamic consideration has been given to the C—H activation by iridium (Stoutland et al. 1988). However, in general most studies have been conducted at atmospheric pressures even though it is realized that higher pressures would be more favorable. A collection of alkane oxidation papers (Hill 1989a) presents a rather wide variety of activation methods. The first chapter is a historical review of the homogeneous alkane activation system (Shilov 1989). A similar collection of papers in the *New Journal of Chemistry* (Hill 1989b) merits attention.

The homogeneous activation of CH_4 by O_2 has been studied extensively and the previous reviews (Garcia and Loffler 1984; Gesser, Hunter, and Prakash 1985; Foster 1985; Pitchai and Klier 1986) have referenced these studies. Theoretical treatment (Hamilton and Schaefer 1989) of the initiation reaction

$$CH_4 + O_2 \rightarrow CH_3\cdot + HO_2\cdot \quad (12\text{-}1)$$

has been compared with the alternate reaction

$$CH_4 + O_2 \rightarrow CH_2 + H_2O_2 \quad (12\text{-}2)$$

and quantum chemical modeling was shown that reaction (12-1) is more favored both by enthalpy and activation energy calculations. A more recent study (Hamilton and Schaefer 1989) has confirmed the calculated values with the experimental activation energy of 57 kcal mol^{-1} (238 kJ mol^{-1}).

Olah has studied the conversion of methane by electrophilic reagents (Olah 1987a) and strong acids (Olah 1987b). The reactivity of metal atoms with methane has also had some success (Klabunde and Tanaka 1983; Jeong and Klabunde 1986).

A recent report (Hunter et al. 1990) on the heterogeneous catalytic conversion of CH_4 to CH_3OH has shown that, in all but one case, the noncatalytic homogeneous reaction gave higher CH_3OH selectivities than the catalytic reaction. Only SnO_2 showed promise as a catalyst by initiating the reaction at lower temperatures. However, the effect did not persist and the catalytic activity could not be revived.

Anthony (Helton and Anthony 1989) reported on the effect of ethane (to 4%) on the catalytic oxidation of CH_4 by SnO_2. Methanol formation was favored by high pressure (50 vs. 20 atm) and lower O_2 concentrations (4.8 vs. 10.1%) with formaldehyde selectivities between 2 and 20% reported for carbon conversions equal to or less than 4%. The addition of ethane lowered the temperature by 40°C for equivalent reactivity. This confirmed the earlier work by others (Wiezevich and Frolich 1934; Hunter et al. 1985; and Kastanas, Tsigdinos, and Schwank 1988a). A comparison of the SnO_2 catalytic effect with the noncatalytic system showed a 20°C difference in temperature for comparable reactivity.

In a comparable study (Fukuoka, Omata, and Fujimoto 1989) it was shown that the addition of propane (2.7 mol%) to CH_4 decreased the onset of the CH_4 oxidation reaction from 420 to 380°C and increased the methanol yield from about 0.5 to 1%.

A homogeneous sensitization study was reported (Hunter et al. 1990) and, although many of the sensitizers showed positive effects, the concentration required for effectiveness is so high that they are of little practical value. The sensitizing effect of C_{2+} hydrocarbons in natural gas are equally effective and as little as 0.5% C_2H_6 in methane is sufficiently influential to lower the temperature for the onset of the DMTM reaction (Gesser, Hunter, and Morton 1990).

It has also been shown (Yarlagadda et al. 1990) that, at O_2 concentrations greater than 5%, temperature spikes occur in the reactor. This has been attributed to cool flames that propagate through the reactor. Evidence indicated that it might be possible to "tune" the reaction for optimum yield of methanol by adjusting the temperature in terms of the observable spikes.

It has been clearly indicated in the past that only at higher pressures methanol formation will become favorable. This has been demonstrated recently (Zhu et al. 1989) for various oxides that convert CH_4 to CH_3OH and HCHO at 5 MPa pressure and between 450 and 500°C. The results reported are given in Table 12-3. It was shown that the selectivity for methanol and formaldehyde was favored by low surface area on the V_2O_5/SiO_2 catalyst (Table 12-4). What was not reported was the noncatalytic conversion for comparison. Gesser and co-workers (Hunter et al. 1990) showed that in several cases the catalysts did not significantly improve upon the noncatalytic results and usually reduced the alcohol and aldehyde selectivities in comparison to the homogeneous reaction.

The dependence of the rate of methane oxidation (to C_2) on the surface area of boride catalysts has been reported (Il'chenko, et al. 1988).

Table 12-3 The Activity and Selectivity of Several Catalysts (Zhu et al. 1989)[a]

Catalysts	Conversion of Methane (%)	SELECTIVITY (%)	
		CH_3OH	CH_2O
$MoO_3\ Cr_2O_3/SiO_2$	5.75	50.28	16.4
$Cr_2O_3\ MoO_3/SiO_2$	3.43	17.59	51.7
Cr_2O_3/SiO_2	3.18	17.76	2.68
V_2O_5/SiO_2	3.89	36.49	7.83
$Fe_xMo_yO_z$	6.47	64.48	11.31

[a]Reaction conditions: temperature 475°C; pressure 5 MPa; GHSV = 6000. Feed ratio (vol%): 58.5% CH_4, 3.81% O_2, 37.38% N_2.

Table 12-4 The Effect of Surface Area of Supports (Zhu et al. 1989)[a]

Catalyst	Surface Area of Silica Gel ($m^2 g^{-1}$)	Conversion of CH_4 (%)	SELECTIVITY (%)			
			CH_3OH	CH_2O	CO	CH_2
V_2O_5/SiO_2	220	3.3	0	0	52.60	27.20
	120	3.89	36.49	7.83	37.56	18.12

[a]Reaction conditions: temperature 475°C; pressure 5 Mpa; GHSV = 6000 l (kg-cat)$^{-1}$ h^{-1}. Feed ratio (vol%): 58.5% CH_4, 3.81% O_2, 37.38% N_2.

With both O_2 and N_2O the rate increased with decreasing surface area. A positive catalytic effect with an iron/sodalite catalyst was achieved at 815 psia using a CH_4–air ratio of 3 at 400°C (Durante et al. 1989). Methanol selectivity of over 70% was reported with 6.5% methane conversion. A reaction scheme (Fig. 12-3) was proposed to illustrate the steps leading to CH_3OH and HCHO.

Electrical Activation

The activation of methane by electrical fields was reviewed 50 years ago (Glockler and Lind 1939) and more recently (Venugopolon 1971; Gould 1959; Holloban and Bell 1974). Recent reports on the electrical activation of methane show (Meubus and Jean 1987) the oligomerization by a hollow cathode. Using He instead of Ar reduced carbon formation and increased the rate of polymerization. Sliepcevich and coworkers (Mallinson, Sliepcevich, and Rusek 1987) examined the partial oxidation of CH_4 under the influence of an alternating field (O_3 generator). With a CH_4–O_2 ratio of 3.1, a flow of 566 ml min^{-1} and 8 kV RMS at 200 Hz the conversions were 15% for CH_4 and 24% for O_2 with the fractional yields being 0.07 for CH_3OH and 0.33 for H_2O (moles per mole of CH_4 reacted). Other products detected were formaldehyde, ethanol, methyl formate, and formic acid. At lower O_2 concentrations (~ 1.2%) the yield of methanol remained unchanged (0.073) but water formation was substantially reduced and ethane formation increased. The effect of reactor configuration was also reported.

In an analogous work (Gesser and Foster 1986) the effect of a spark on the initiation of the oxidative conversion of CH_4 was studied. Under conditions where normally no reaction was observed, it was shown that a spark (Tesla coil) caused the O_2 to be consumed with the accompanying formation of HCHO. The effect was more pronounced at higher pressures (5 atm) and higher O_2 concentrations (3.5%). It is believed that at 30–50

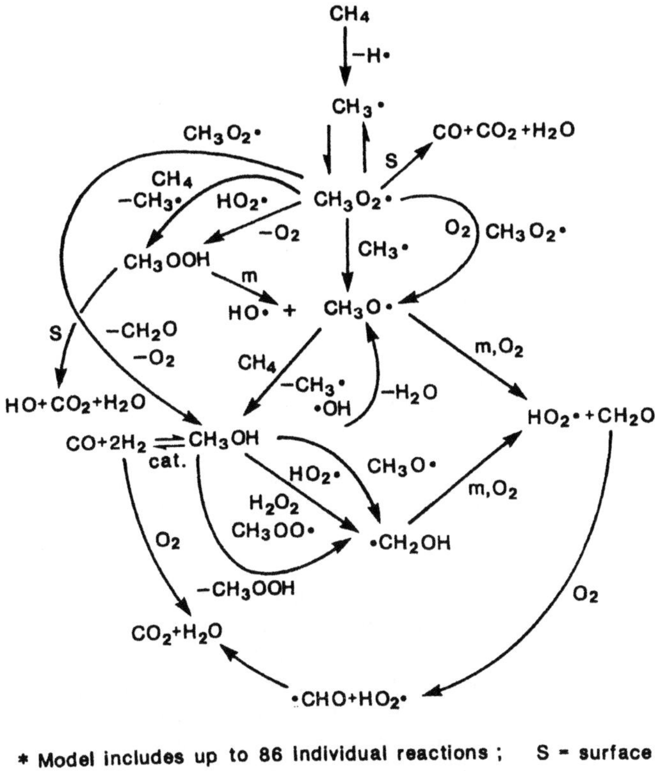

FIGURE 12-3. Schematic reaction mechanisms (Durante et al. 1989).

atm the methanol formation would be favored over the formaldehyde if complete consumption of the oxygen could be realized.

RECENT WORK ON DMTM

Wall effects and other reaction parameters were studied (Burch, Squire, and Tsang 1989) for the high-pressure (to 50 atm) partial oxidation of CH_4 to CH_3OH. The effect of reactor surfaces on methanol selectivity is shown in Fig. 12-4 for $CH_4:O_2:N_2 = 20:1:19$. The temperature for optimum conversion decreased from about 500 to 400°C as the pressure increased from 5 to 50 atm. The effect of residence time (τ) on CH_3OH selectivity was significant only at $\tau < 100$ s, whereas yields were optimum at about

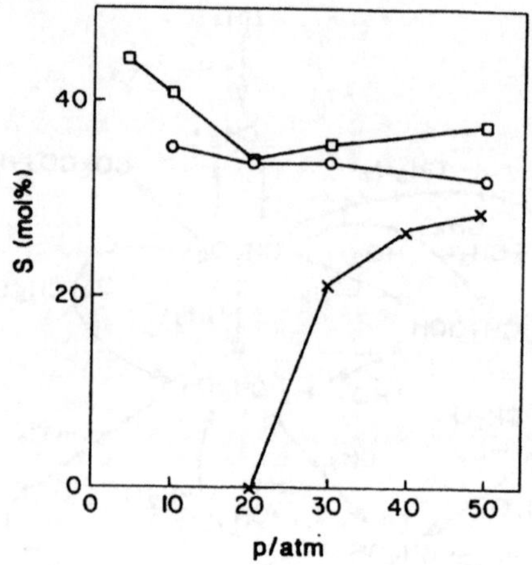

FIGURE 12-4. Effect of reactor type on methanol selectivity at 450°C in the pressure range 0–50 bar: ×, 316 stainless steel; ○, quartz; □, Pyrex (Burch, Squire, and Tsang 1989).

200 s. Methanol selectivity was also optimum at about 5% O_2. One aspect of this work that differs from that of Gesser and co-workers (Yarlagadda et al. 1986) is that Burch reported no effect on the methanol selectivity when the inert gas N_2 was removed. Gesser and Hunter have shown that CO_2 at 20% in the CH_4–O_2 mixture can reduce the methanol selectivity from 70 to 63%. Thus, some inconsistency exists between the two groups besides the differences in the maximum selectivities obtained for methanol.

The high-pressure (to 1000 atm) oxidation of CH_4 in a static system has been studied by Sliepcevich and his students (Lott 1965; Hardwicke 1966; Bauerle 1964) and several reports have appeared (Lott and Sliepcevich 1967; Hardwicke, Lott, and Sliepcevich 1969, Bauerle, Lott, and Sliepcevich 1974a, 1974b). At very high pressures (to 600 atm) the normal combustion mechanism no longer occurs. Methanol and formaldehyde become important primary products. The mechanism proposed is shown in Fig. 12-5.

The importance of the radical recombination reaction [A14] was stressed. The occurrence of this reaction can account for the higher yields of methanol relative to water. It has been suggested (Dowdy 1988) that the stoichiometry of the methanol formation would be optimum for the overall reactions

$$3CH_4 + 2.5O_2 \rightarrow 2CH_3OH + 2H_2O + CO$$

A0	CH_4	+	O_2	$\longrightarrow$	$\cdot CH_3$	+	$HO_2\cdot$
A1	$\cdot CH_3$	+	O_2	$\longrightarrow$	$CH_3OO\cdot$		
A2	$CH_3OO\cdot$			$\longrightarrow$	$HCHO$	+	$HO\cdot$
A3	$CH_3OO\cdot$	+	CH_4	$\longrightarrow$	CH_3OOH	+	$\cdot CH_3$
A4	CH_3OOH			$\longrightarrow$	$CH_3O\cdot$	+	$\cdot OH$
A5	$CH_3O\cdot$	+	CH_4	$\longrightarrow$	CH_3OH	+	$\cdot CH_3$
A6	CH_4	+	$\cdot OH$	$\longrightarrow$	$\cdot CH_3$	+	H_2O
A7	CH_4	+	$HOO\cdot$	$\longrightarrow$	$\cdot CH_3$	+	$HOOH$
A8	$HCHO$	+	$\cdot OH$	$\longrightarrow$	$H\dot{C}O$	+	H_2O
A9	$HCHO$	+	$HOO\cdot$	$\longrightarrow$	$H\dot{C}O$	+	H_2O_2
A10	$HCHO$	+	O_2	$\longrightarrow$	$H\dot{C}O$	+	$HOO\cdot$
A11	$H\dot{C}O$	+	O_2	$\longrightarrow$	CO	+	$HOO\cdot$
A12	CO	+	$HOO\cdot$	$\longrightarrow$	CO_2	+	$\cdot OH$
A13	$2\,H_2O_2$			$\longrightarrow$	$2\,H_2O$	+	O_2
A14	$\cdot CH_3$	+	$\cdot OH$	$\longrightarrow$	CH_3OH		
A15	$H\dot{C}O$	+	$\cdot OH$	$\longrightarrow$	$HCOOH$		
A16	$H\dot{C}O$	+	$CH_3O\cdot$	$\longrightarrow$	$HCOOCH_3$		
A17	$CH_3OO\cdot$			$\longrightarrow$	termination		
A18	$\cdot CH_3$			$\longrightarrow$	termination		
A19	$HOO\cdot$			$\longrightarrow$	termination		
A20	$\cdot OH$			$\longrightarrow$	termination		
A21	$H\dot{C}O$			$\longrightarrow$	termination		

FIGURE 12-5. Mechanism of partial oxidation of CH_4 at high pressure (Bauerle, Lott, and Sliepcevich 1974b).

or

$$3CH_4 + 3O_2 \rightarrow 2CH_3OH + 2H_2O + CO_2$$

giving a maximum selectivity of 67%. However, the higher selectivities obtained by Gesser and co-workers (Yarlagadda et al. 1988) and by others (Durante et al. 1989) could be accounted for by the importance of reaction [A14] at high pressures, as well as by the reaction

$$CH_3\cdot + CH_3OO\cdot \rightarrow 2CH_3O\cdot$$

B1	CH_4	+	O_2	$\longrightarrow$	$\cdot CH_3$	+	$HO_2\cdot$
B2	$\cdot CH_3$	+	O_2	$\longrightarrow$	$CH_3OO\cdot$		
B3	$CH_3OO\cdot$			$\longrightarrow$	HCHO	+	$\cdot OH$
B4	$CH_3OO\cdot$	+	CH_4	$\longrightarrow$	CH_3O_2H	+	$\cdot CH_3$
B5	CH_3O_2H			$\longrightarrow$	$CH_3O\cdot$	+	$\cdot OH$
B6	$CH_3O\cdot$			$\longrightarrow$	HCHO	+	$H\cdot$
B7	$CH_3O\cdot$	+	CH_4	$\longrightarrow$	CH_3OH	+	$\cdot CH_3$
B8	$\cdot OH$	+	CH_4	$\longrightarrow$	$\cdot CH_3$	+	H_2O
B9	$\cdot CH_3$	+	$\cdot CH_3$	$\longrightarrow$	C_2H_6		
B10	$\cdot CH_3$	+	$\cdot OH$	$\longrightarrow$	CH_3OH		
B11	$\cdot CH_3$	+	$CH_3O\cdot$	$\longrightarrow$	CH_3OCH_3		
B12	CH_2O	+	$\cdot CH_3$	$\longrightarrow$	$\cdot CHO$	+	CH_4
B13	$\cdot CHO$	+	O_2	$\longrightarrow$	CO	+	$HO_2\cdot$
B14	CH_2O	+	$CH_3O\cdot$	$\longrightarrow$	CH_3OH	+	$\cdot CHO$
B15	$\cdot CHO$	+	$\cdot CH_3$	$\longrightarrow$	CO	+	CH_4

etc.

FIGURE 12-6. Plug-flow mechanism for the partial oxidation of CH_4 at high pressure (Yarlagadda et al. 1988).

A question remains as to why there appears to be a falloff in methanol selectivity at pressures above 180 atm as determined by Boomer (Boomer and Thomas 1937) or about 100 atm as found by Gesser and co-workers (Morton, Hunter, and Gesser 1984). One possible explanation must be the competitive first-order second-order reactions of the $CH_3OO\cdot$ and $CH_3O\cdot$ radicals. These small radicals may in fact show first-order to second-order transition at these high pressures. It has been shown by Troe and co-workers (Borrell et al. 1985) that such high-pressure falloff values are common for small molecules and radicals.

This is illustrated by the comparable mechanism proposed by Gesser and co-workers (Yarlagadda et al. 1988) shown in Fig. 12-6, where reactions [B3] and [B4] and reactions [B6] and [B7] compete. Thus, at high pressures, reactions [B3] and [B6] are first-order, but at lower pressure they may become pseudo-second-order and their relative importance with the truly second-order reactions [B4] and [B7] may be diminished. Thus, the methanol selectivity should show a maximum with increasing pressure and this maximum could be dependent on the surface–volume ratio of the reactor.

Several attempts have been made to model the reaction and to obtain some insight into the sensitivity of the various reactions to the overall mechanism. Early work by Russian workers (Karmilova, Enikolopyan, and Nalbandyan 1960, 1961; Vardanyan, Yan, and Nalbandyan 1981) was limited to ambient pressure and static conditions. Sensitivity analysis of the mechanism of methane oxidation was reported for atmospheric condition (Boni and Penner 1977). The conclusion concerning the four most important reactions clearly showed the difference between low- and high-pressure modeling.

A critical collection of kinetic parameters for CH_4 related reactions has been prepared by the National Bureau of Standards (Tsang and Hampson 1986). Other data sets (Kondrat'ev 1971; Westbrook and Dryer 1984) have been used for methanol and related modeling systems (Zanthoff and Baerns 1990).

The activation energies for O_2 disappearance at 25–30 atm were determined (Yarlagadda et al. 1986) using first-order kinetics in O_2 and zero-order for CH_4. The values were 38.8 kcal/mol^{-1} for 5.5% O_2 and 41.1 kcal mol^{-1} for 10.1% O_2. These values are lower than the 57 kcal mol^{-1} value accepted as the energy for the homogeneous initiation reaction

$$CH_4 + O_2 \rightarrow CH_3\cdot + HO_2\cdot$$

and implies either a heterogeneous initiation step or a chain reaction or both. We tend to favor the chain reaction.

More recently, Vedeneev and co-workers have reported on a study of the high-pressure model system suitable for flow systems (Vedeneev et al. 1987, 1988). The agreement between calculated and experimental values was considered good and the model gave a reasonable basis for further refinement. However, the model did not predict a maximum in methanol selectivity with increase in pressure.

Onsager and co-workers have also reported their results of a comparison with experimental results and their model for the high-pressure partial oxidation of CH_4 (Onsager, Soraker, and Lodeng 1989). A refinement (Onsager 1990) of the mechanism showed the reaction

$$CH_3OOH + M \rightarrow CH_3O\cdot + OH\cdot + M$$

to be important for medium conversion of oxygen.

ECONOMIC EVALUATION OF THE DMTM PROCESS

The results of an economic evaluation of the direct process compared to the syngas route has shown varied results. The direct process was predicted to become favorable only when the methanol selectivity is over 77% (Edwards and Foster 1988). However, Edwards and Foster assumed that a catalyst was involved and included the expenses associated with its cost and recycling of the catalyst.

Other comparisons (Cran 1988) show that the direct oxidative process can produce methanol about 20–30% cheaper than the present methods. Two more recent studies (Fox, Chen, and Degen 1990; Geerts, Hoebind, and van der Wiele 1990) may be worth noting. However, the first study is incomplete because the comparison is made with gasoline (i.e., the methanol is converted to gasoline) and the second study seems to use unrealistic figures for plant costs. Nevertheless, real valid comparison between the direct and conventional methods will only be possible when pilot plant studies on the direct process have been done.

The practical implementation of the direct methane to methanol process has been studied in the Soviet Union (Vedeneev 1990), where a 100-tonne-per-year plant has been in operation for several years. The once-through CH_4–air system will soon be followed by a CH_4–O_2 recycling system with an estimated 5000 tonnes per year throughput.

CONCLUSIONS

The methanol market is uncertain and fluctuating (Crocco 1990). The future depends on many things (Morton, Hunter, and Gesser 1990) and, with the present politically unstable Middle East to factor into the equation, any prediction is bound to be unreliable.

What is certain is that methanol will continue to be a world commodity and that use will be made of the small gas wells that are, at present, uneconomical for conversion to methanol via the syngas process. This is where it is believed that the direct process will have a major advantage. The smaller capital investment and smaller physical size may make mobile plants worth considering.

The use of fluidized-bed reactors, which appear to be highly effective in oxidative coupling (Edwards and Tyler 1989), must be considered as a promising approach to the DMTM process if high pressures can be incorporated. The fluidized-bed reaction at atmospheric conditions has been shown to produce formaldehyde (McConkey and Wilkinson 1967) and it is believed that only at higher pressures will it be possible to form methanol.

An interesting aspect of the DMTM process has been extended (Yarlagadda et al. 1987) by the addition of ZSM-5 catalyst to the reactor and converting the methanol formed into gasoline—all in the one reactor. This approach eliminated the methanol–water separation and the expensive reforming process to produce the syngas required for the conventional methanol to gasoline process. The methanol–water separation that is necessary in the DMTM process may require a new approach such as pervaporation, which is being extensively studied for the ethanol–water separation from fermentation mixtures. The integration of the oxygen plant, which is necessary, with the low temperatures required for the removal of the methanol–water mixture from the product stream is another example of integrating the DMTM system that must be examined as new technology rather than as an assembly of processes, some of which are already available. Normally, one is tempted to incorporate these as separate units.

ACKNOWLEDGMENT

This work was partially supported by the Natural Sciences and Engineering Research Council of Canada in the form of a Strategic Grant.

References

Anderson, A.B., and J.J. Maloney. 1987. Activation of methane on iron, nickel and platinum surfaces. A molecular orbital study. American Chemical Society Preprints, Div. Fuel Chem. 194th National Meeting New Orleans, Aug. 31–Sept. 4, 1987 Vol. 32/3, p. 291.

Anshits, A.G., A.N. Shigapov, S.N. Vereshchagin, and V.N. Shevin. 1990. C_2 hydrocarbon formation from methane on silver membrane. *Catal. Today*, 6:593.

Ausloos, P., R.E. Rebbert, and S.G. Lias. 1965. Direct and Inert-Gas-Sensitized Radiolysis and Photolysis of Methane in the Solid Phase. *J. Chem. Phys.* 42:540.

Baerns, M., K. van der Wiele, and J.R.H. Ross. 1989. Methane Activation—A Bibliography *Catal. Today* 4:471.

Bauerle, G.L. 1964. Surface and catalytic effects in high pressure oxidation of methane. Ph.D. Dissertation, University of Oklahoma, Norman, OK.

Bauerle, G.L., J.L. Lott, and C.M. Sliepcevich. 1974a. Oxidation of methane at elevated pressures I. Ignition delay. *J. Fire & Flammability* 5:136.

Bauerle, G.L., J.L. Lott, and C.M. Sliepcevich. 1974b. Oxidation of methane at elevated pressures II. A reaction mechanism. *J. Fire & Flammability* 5:190.

Baughcum, S.L., R.C. Oldenborg, W.C. Danen, G.E. Streit, and C. Rofer. 1986. Laser-initiated chain reactions in the partial oxidation of methane. In *SPIE Proceedings*, Vol. 669, Laser Appl. Chem, June.

Bell, K.M., and C.A. McDowell. 1961. The mercury photosensitized oxidation of hydrocarbons. *Can. J. Chem.* 39:1419.

Bergman, R.G. 1984. Selective oxidative of methane. *Science* 223:902.

Boni, A.A., and R.C. Penner. 1977. Sensitivity analysis of a mechanism for methane oxidation kinetics. *Combust. Sci. Technol.* 15:99.

Boomer, E.H., and V. Thomas. 1937. The oxidation of methane at high pressures. III. Experiments using pure methane and principally copper catalysts. *Can. J. Res. Sect. B.* 15:414.

Borrell, P., C.J. Cobos, A.E. Croce de Cobos, H. Hippler, K. Luther, A.R. Ravishankara, and J. Troe, 1985. Radical association reactions in gases at high pressures. *Ber. Bunsenges. Phys. Chem.* 89:337.

Burch, R., G.D. Squire, and S.C. Tsang. 1989. Direct conversion of methane to methanol. *J. Chem. Soc., Faraday Trans. 1* 85:3561.

Calvert, J.G., and J.N. Pitts., Jr. 1966. *Photochemistry*. New York: J. Wiley & Sons.

Chan, Y.W., and R.B. Wilson, Jr. 1987. Partial oxidation of methane using supported porphyrin and phthalocyanine complexes. American Chemical Society Preprints, Div. Fuel. Chem. 194th National Meeting, New Orleans, Aug. 31–Sept. 4, 1987, Vol. 33/3, p. 453.

Chang, R., and W. Tikkanen. 1988. Kinetics of the photochemical reaction of gas mixtures of methane and water and the effects of added oxygen. In *The Top Fifty Industrial Chemicals*. New York: Random House.

Corder, R.E., E.R. Johnson, J.L. Vega, E.C. Clausen, and J.L. Gaddy. 1987. Biological production of methanol from methane. American Chemical Society Preprints, Div. Fuel Chem. 194th National Meeting, New Orleans, Aug. 31–Sept. 4, 1987, Vol. 33/3, p. 469.

Cran, J. 1988. Alta Plan, Calgary, Private communication.

Crocco, J. 1990. Methanol yesterday, today and tommorrow. *Chem. Ind. (London)* 97.

Dowdy, D. 1988. Private communication.

Durante, V.A., J.E. Lyons, and D.W. Walker. 1989. Vapor phase catalytic hydroxylation of methane. Preprints, Methane Activation Symposium, Pacifichem 89, Honolulu, Hawaii, Dec. 17–20, 1989, p. 23.

Edwards, J.H., and N.R. Foster. 1988. The potential for methanol production from natural gas by direct catalytic partial oxidation. *Fuel Sci. Technol. Intl.* 48:365.

Edwards, J.H., and R.J. Tyler. 1989. The oxidative coupling of methane in a fluidized-bed-reactor. *Catal. Today* 4:365.

Eng, D., and M. Stoukides. 1988. Partial oxidation of methane in a solid electrolyte cell. In *Proceedings of the 9*th *International Congress on Catalysis, Calgary, 1988*, Vol. 2, p. 975.

Foster, N.R. 1985. Direct Catalytic oxidation of methane to methanol—A review. *Appl. Catal.* 19:1.

Fox, J.M., III, T.P. Chen, and B.D. Degen. 1990. An evaluation of direct methane conversion processes. *Chem. Eng. Prog.* Apr.:43.

Fukuoka, N., K. Omata, and K. Fujimoto. 1989. Effect of additives on partial oxidation of methane. Preprints, Methane Activation Symposium, Pacifichem 89, Honolulu, Hawaii, Dec. 17–20, 1989, p. 106.

Garcia, E.Y., and D.A. Loffler. 1984. Partial oxidation of methane. *Lat. Am. J. Chem. Eng. Appl. Chem.* 14:267.

Geerts, J.W.M.H., J.H.B.J. Hoebind, and K. van der Wiele. 1990. Methanol from natural gas. Proven and new technologies. *Catal. Today* 6:613.

Gerasimov, S.P. 1988. Mechanism of methane photoxidation on V/SiO_2 catalyst. *Kinet. Catal. Lett.* 32:57.

Gesser, H.D. 1989. Unpublished results.

Gesser, H.D., and N.R. Foster. 1986. Unpublished results.

Gesser, H.D., N.R. Hunter, and L.A. Morton. 1990. Unpublished results.

Gesser, H.D., N.R. Hunter, and C.B. Prakash. 1985. The direct conversion of methane to methanol by controlled oxidation. *Chem. Rev.* 85:235.

Glockler, G., and S.C. Lind. 1939. *The Electrochemistry of Gasses and Other Dielectrics*. New York: J. Wiley & Sons.

Gould, R.F. (ed.). 1959. *Chemical Reactions in Electrical Discharges*. Adv. Chem. Series 80 Washington, DC: American Chemical Society.

Gratzel, M., K.R. Thampi, and J. Kiwi. 1989. Methane oxidation at room temperature and atmospheric pressure activated by light via polytungstate dispersion on titania. *J. Phys. Chem.* 93:4128.

Hamilton, T.P., and H.F. Schaefer III. 1989. The reaction of methanol with molecular oxygen: A semiquantitative estimate of the activation energy. *J. Chem. Phys.* 90:6391.

Hardwicke, N.L. 1966. The mechanism of partial oxidation of methane at high pressures. Ph.D. Dissertation, University of Oklahoma, Norman, OK.

Hardwicke N.L., J.L. Lott, and C.M. Sliepcevich. 1969. Oxidation of methane at high pressures. *Ind. Eng. Chem., Proc. Des. Dev.* 8:133.

Helton, T.E., and R.G. Anthony. 1989. Methanol and formaldehyde production from natural gas. Preprints, Methane Activation Symposium, Pacifichem 89, Honolulu, Hawaii, Dec. 17–20, p. 143.

Hill, C.L. (ed.). 1989a. *Activation and Functionalization of Alkanes*. New York: J. Wiley & Sons.

Hill, C.L. (guest ed.). 1989b. Alkane activation and functionalization. *New J. Chem.* 10/11:645–750.

Hill, W., B.N. Shelimov, and V.B. Karansky. 1987. Photoinduced reactions of methane with molybdena supported on silica. *J. Chem. Soc., Faraday Trans. 1* 83:2381.

Holloban, J.R., and A.T. Bell (eds.). 1974. *Techniques and Applications of Plasma Chemistry*. New York: Wiley.

Hunter, N.R., H.D. Gesser, L.A. Morton, and D.P.C. Fung. 1985. The direct conversion of natural gas to methanol by controlled oxidation at high pressures. In *Proceedings of the 35th Canadian Chemical Engineering Conference, Calgary, Oct. 6–9,* Vol. 2, p. 486.

Hunter, N.R., H.D. Gesser, L.A. Morton, and P.S. Yarlagadda. 1990. Methanol formation at high pressure by the catalyzed oxidation of natural gas and by the sensitized oxidation of methane. *Appl. Catal.* 57:45.

Hutchings, G.J., M.S. Scurrell, and J.R. Woodhouse. 1989. Oxidative coupling of methane using oxide catalysts. *Chem. Soc. Rev.* 18:251.

Il'chenko, N.I., A.I. Bostan, L.Yu. Dulgikhnd, and G.I. Golodets. 1987. A partial oxidation and oxidative condensation of methane with participation of nitrous oxide on a vanaduim pentoxide/silica catalyst. *Theoret. Experim. Chem. USSR* 23:641.

Il'chenko, N.I., A.I. Bostan, and G.I. Golodets. 1988. Effect of the Chemical nature of Supported Metal Ions on/heir Catalyst Properties in Methane Oxidation by Nitrous Oxide. *Theoret. Experim. Chem. USSR* 24:455.

Il'chenko, N.I., and G.I. Golodets. 1988. Causes of the higher selectivity of methane partial oxidation catalysts using nitrous oxide as an oxidizing agent. *Theoret. Experim. Chem.* 24:727.

Il'chenko, N.I., L.N. Raevskaya, A.I. Bostan, and G.I. Golodets. 1988. Oxidative condensation of methane on metal-like compounds. *Theoret. Experim. Chem. USSR* 24:638.

Il'chenko, N.I., V.G. Ilyine, L.N. Raevskaya, N.V. Turtutina, A.D. Onishchenko, and A.I. Bostan. 1989. Oxidative condensation of methane on modified natural mordenites. *React. Kinet. Catal. Lett.* 38:141.

Jeong, G., and K.J. Klabunde. 1986. Metal atom reactions with methane. Boron, aluminum, gallium and indium atoms and dimers. *J. Am. Chem. Soc.* 108:7103.

Karmilova, L.A., N.S. Enikolopyan, and A.B. Nalbandyan. 1960. The kinetics and mechanism of the oxidation of methane. *Zh. Fiz. Khim.* 34:550, 990, 1176.

Karmilova, L.A., N.S. Enikolopyan, and A.B. Nalbandyan. 1961. The kinetics and mechanism of methane oxidation. *Zh. Fiz. Khim.* 35:1046, 1435, 1458.

Kastanas, G.N., G.A. Tsigdinos, and J. Schwank. 1988a. Effect of small amounts of ethane on the selective oxidation of methane over silicic acid and quartz glass surfaces. *J. Chem. Soc., Chem. Commun.* 1298.

Kastanas, G.N., G.A. Tsigdinos, and J. Schwank. 1988b. Selective oxidation of methane over Vycor glass, quartz glass, and various silica, magnesium and alumina surfaces. *Appl Catal.* 44:33.

Khenkin, A.M., and A.E. Shilov. 1987. Electrochemical model of alkane oxidation by cytochrome P-450. *React. Kinet. Catal. Lett.* 33:125.

Khenkin, A.M., and A.E. Shilov. 1989. Biomimetic alkane oxidation in the presence of iron complexes. *New J. Chem.* 13:659.

Klabunde, K.J., and Y. Tanaka. 1983. Activation of methane with metal atoms at 10K without photolysis. *J. Am. Chem. Soc.* 105:3544.

Kondrat'ev, V.N. 1971. *Rate Constants of Gas-Phase Reactions*. Nauka: Moscow.

Kuo, J.C.W. 1987. Evaluation of direct methane conversion to higher hydrocarbons and oxygenates. Final Report Contract No. DE-AC22-86PC90009, US Department of Energy.

Lott, J.L. 1965. The selective oxidation of methane at high pressures. Ph.D. Dissertation, University of Oklahoma, Norman, OK.

Lott, J.L., and C.M. Sliepcevich. 1967. Partial oxidation of methane at high pressures. *Ind. Eng. Chem., Proc. Des. Dev.* 6:67.

MacGiolla Coda, E., M. Kennedy, J.B. McMonagle, and B.K. Hodnett. 1990. Oxidation of methane to formaldehyde over supported molybdena catalysts at ambient pressure. Isolation of the selective oxidation product. *Catal. Today* 6:559.

Mallinson, R.G., C.M. Sliepcevich, and S. Rusek. 1987. Methane partial oxidation in alternating electric fields. American Chemical Society Preprints, Div. Fuel Chem. 194th National Meeting, New Orleans, Aug. 31–Sept. 4, 1987, Vol. 32/3, p. 266.

McConkey, B.H., and P.R. Wilkinson. 1967. Oxidation of methane to formaldehyde in a fluidized bed reactor. *Ind. Eng. Chem. Proc. Des. Dev.*, 6:436.

Meltser, L.Z., T.A. Garibyan, R.R. Grigoryan, A.A. Muradyan, and L.N. Kurina. 1989. Effect of Homogeneous Initiator on Heterogeneous Homogeneous oxidation of Methane. *React. Kinet. Catal. Lett.* 38:229.

Meubus, P., and G. Jean. 1987. Methane polymerization usings hollow cathode. American Chemical Society Preprints, Div. Fuel Chem. 194th National Meeting, New Orleans, Aug. 31–Sept. 4, 1987, Vol. 32/3, p. 260.

Morton, L.A., N.R. Hunter, and H.D. Gesser. 1984. The production of methanol by the controlled oxidation of methane at high pressure. Final Report, CANMET, Ottawa, June 1984. CANM-87:000414, EDB-88-025858, Report No. MU/C-8406; Microlog-87-02229.

Morton, L., N. Hunter and H. Gesser. 1990. Methanol: A fuel for today and tomorrow. *Chem. Ind. (London)* 457.

Ogura, K., and M. Kataoka. 1988. Photochemical conversion of methane. *J. Mol. Catal.* 43:371.

Ogura, K., C.T. Migita, and M. Fujita. 1988. Conversion of methane to oxygen-containing compounds by the photochemical reactions. *Ind. Eng. Chem. Res.* 27:1387.

Ogura, K., C.T. Migita, and Y. Ito. 1990. Combined photochemical and electrochemical oxidation of methane. *J. Electrochem. Soc.* 137:500.

Ogura, K., C.T. Migita, and T. Yamada. 1990. Kinetics of the photochemical reaction of gas mixtures of methane and water and the effects of added oxygen. *J. Photochem., Photobiol. A, Chem.* 52:241.

Olah, G.A. 1987a. Electrophilic methane conversion. *Acc. Chem. Res.* 20:422.

Olah, G.A. 1987b. Electrophilic methane conversion. *Makromol. Chem., Macromol. Symp.* 13/14:1.

Omata, K., S. Hashimoto, H. Tominaga, and K. Fujimoto. 1989. Oxidative coupling of methane using a membrane reactor. *Appl. Catal.* 52:L1.

Onsager, O.T. 1990. Private communication.

Onsager, O.T., P. Soraker, and R. Lodeng. 1989. Experimental investigation and computer simulation of homogeneous gas phase oxidation of methane to methanol. Preprints, Methane Activation Symposium, Pacifichem 89, Honolulu, Hawaii, Dec. 17–20, 1989, p. 113.

Oshima, Y., M. Saito, S. Koda, and H. Tominaga. 1988. Partial oxidation of methane by laser-initiated chain reaction. *Chem. Lett.* 203.

Otsuka, K., S. Yokoyama, and A. Morikawa. 1985. Catalytic activity and selectivity control for oxidative coupling of methane by oxygen-pumping through yttria-stabilized zirconia. *Chem. Lett.* 319.

Patras, L.E., and A. Tang. 1987. Bioconversion of methane to methanol by *Methylobacterium organophilum*. American Chemical Society Preprints, Div. Fuel Chem. 194th National Meeting, New Orleans, Aug. 31–Sept. 4, 1987, Vol. 33/3, p. 462.

Perkins, W.G. 1973. Permeation and outgassing of vacuum materials. *J. Vac. Sci. Technol.* 10:543.

Pitchai, R., and K. Klier. 1986. Partial oxidation of methane. *Catal. Rev.—Sci. Eng.* 28:13.

Ryabov, A.D. 1990. Mechanism of intermolecular activation of C—H bonds in transition-metal complexes. *Chem. Rev.* 90:403.

Sayyed, B.A., and P.C. Stair. 1990. Laser-induced activation of methane at oxide surfaces: A probe of radical–surface interactions. *J. Phys. Chem.* 94:409.

Schwartz, J. 1985. Alkane activation by oxide-bound organorhodium complexes. *Acc. Chem. Res.* 18:302.

Schwartz, J. 1987. Alkane activation by oxide-bound organorhodium complexes. American Chemical Society Preprints, Div. Fuel Chem. 194th National Meeting, New Orleans, Aug. 31–Sept. 4, 1987, Vol. 32/3, p. 286.

Seimonides, S. and M. Stoukides. 1986. Electrochemical modification of Ag–MgO catalyst electrodes during methane oxidation. *J. Electrochem. Soc.* 133:1535.

Shelnutt, J.A., F.V. Stohl, and B. Granoff. 1987. Computer-aided molecular design of alkane-activation catalysts. American Chemical Society Preprints, Div. Fuel Chem., 194th National Meeting, New Orleans, Aug. 31–Sept. 4, 1987, Vol. 33/3, p. 479.

Shilov, A.E. 1989. In *Activation and Functionalization of Alkanes*, ed. C.L. Hill, P.I. New York: Wiley.

Sokolovskii, V.D. 1987. Some principles of choosing catalysts for selective conversions of organic compounds at C—H bonds. *React. Kinet. Catal. Lett.* 35:337.

Stoutland, P.O., R.G. Bergman, S.P. Nolan, and C.D. Hoff. 1988. The thermodynamic driving force for C—H activation at iridium. *Polyhedron* 7:1429.

Tsang, W., and R.F. Hampson. 1986. Chemical kinetic data base for combustion chemistry. *Phys. Chem. Ref. Data* 15:1087.

Urban, J., V. Klimo, and J. Tino. 1984. Quantum chemical modelling of reactions in hydrocarbon combination. Study of the $CH_4 + O_2$ reactions. *Collect. Czech. Chem. Commun.* 49:1440.

Vardanyan, I.A., S. Yan, and A.B. Nalbandyan. 1981. Mechanism of the thermal oxidation of methane. *Kinet. Katal.* 22:845.

Vedeneev, V.I. 1990. Private communication.

Vedeneev, V.I., M.Y. Gol'denberg, N.I. Gorban, and M.A. Teitel'boim. 1987. Mechanism of the initial stage of the oxidation of methane at high pressures. *Chem. Phys. USSR* 6:626.

Vedeneev, V.I., M.Y. Gol'denberg, N.I. Gorban, and M.A. Teitel'boim. 1988. Quantitative model of oxidation of methane at high pressures. Parts I, II, III, and IV. *Kinet. Katal.* 29:7, 14, 1291, and 1297.

Venugopolon, M. (ed.). 1971. *Reactions under Plasma Conditions*. New York: Wiley-Interscience.

Volodin, A.M., and A.E. Cherkashin. 1981. ESR spectrum of methyl radicals on ZnO surface. *React. Kinet. Catal. Lett.* 18:243.

Westbrook, C.K., and F.L. Dryer. 1984. Chemical kinetic modeling of hydrocarbon combustion. *Prog. Energy Combust. Sci.* 10:1.

Wiezevich, P.K., and P.K. Frolich. 1934. Direct oxidation of saturated hydrocarbons at high pressures. *Ind. Eng. Chem.* 26:267.

Yamazaki, H., and R.J. Cvetanovic. 1964. Collisional deactivation of the excited singlet oxygen atoms and their insertion into the C—H bonds of programs. *J. Chem. Phys.* 41:3703.

Yarlagadda, P.S., L.A. Morton, N.R. Hunter, and H.D. Gesser. 1986. Product optimization of the direct conversion process of natural gas to methanol by controlled oxidation. Final Report, Sept. 1986, CANMET, Ottawa. CANM-87:000386, EDB-86:025860 Report No. MU/C-8609; Microlog-87-02221.

Yarlagadda, P.S., L.A. Morton, N.R. Hunter, and H.D. Gesser. 1987. Direct catalytic conversion of methane to higher hydrocarbons. *Fuel Sci. Technol. Intl.* 5:169.

Yarlagadda, P.S., L.A. Morton, N.R. Hunter, and H.D. Gesser. 1988. Direct conversion of methane to methanol in a flow reactor. *Ind. Eng. Chem. Res.* 27:252.

Yarlagadda, P.S., L.A. Morton, N.R. Hunter, and H.D. Gesser. 1990. Temperature oscillations during the high pressure partial oxidation of methane in a tubular flow reactor. *Combust. & Flame* 79:216.

Yun, C., M. Anpo, Y. Mizokoshi, and Y. Kubokawa. 1980. Oxidation of alkanes and alkenes by N_2O over UV irradiated MgO. *Chem. Lett.* 799.

Zanthoff, H., and M. Baerns. 1990. *Ind. Eng. Chem. Res.* 29:2.

Zhu, Q., M. Zhang, Y. Xiao, and X. Zhao. 1989. Studies of the controlled oxidation of methane to formaldehyde and methanol. Preprints, Methane Activation Symposium, Pacifichem 89, Honolulu, Hawaii, Dec. 17–20, p. 117.

II

Engineering Aspects

13

The OXCO Process: The Direct Conversion of Natural Gas to Olefins and Liquid Fuels Using Fluidized-Bed Technology

J.H. Edwards, K.T. Do, and R.J. Tyler

CSIRO *Division of Coal and Energy Technology, North Ryde, Australia*

TABLE OF CONTENTS

INTRODUCTION

In Australia CSIRO, in collaboration with The Broken Hill Proprietary Company, Limited (BHP), is conducting a nationally coordinated research program to evaluate the technical and economic feasibility of using methane oxidative coupling as the basis of an alternative route to conventional synthesis-gas–based processes for producing olefins and transport fuels from natural gas.

The research program is comprehensive and has focused research effort on areas identified in preliminary technical evaluations as being crucial to the successful development of a new process. Topics studied include catalyst development, process chemistry (Tyler and Lukey 1987; Edwards and Tyler 1988), reactor development (Edwards and Tyler 1989; Edwards, Do, and Tyler 1990; Edwards, Tyler, and White 1990), reaction mechanisms (Cant et al. 1988; Nelson, Lukey, and Cant 1988; Nelson and Cant 1990; Ekstrom 1990), catalyst characterization (Maitra 1990a), gas-phase reactions (Mackie et al. 1990), ethylene oligomerization (Chin et al. 1988; Masters and Brown 1990), and the evaluation of conceptual process economics (Jackson, Thompson, and Whitehead 1989).

This research activity has led to the identification of a new natural-gas conversion process, known as the OXCO process, in which a single fluidized-bed reactor is used to combine the methane oxidative coupling step with the pyrolysis of ethane and higher alkane components present in natural gas to provide an efficient method for the total conversion of natural gas to olefins.

This chapter covers work on reaction engineering research at the North Ryde laboratory of the CSIRO Division of Coal and Energy Technology, which is focused on the development of methane coupling catalysts for use in fluidized beds and on their detailed performance assessment in laboratory-scale fluidized-bed reaction facilities.

The OXCO conceptual natural-gas conversion process is described and results obtained from CSIRO's fluidized-bed reactor studies on methane oxidative coupling for a variety of operating conditions are presented. These results include the successful demonstration of the key technical feature of the OXCO process, namely, that methane coupling and alkane pyrolysis can be effectively combined within a single fluidized-bed reactor. The experimental data are then used to calculate the overall process mass balance and product yields that are potentially achievable for the OXCO process.

CSIRO'S METHANE COUPLING REACTION ENGINEERING RESEARCH STRATEGY

The oxidative coupling of methane involves reacting methane and oxygen over suitable catalysts at 750–900°C to promote the following overall reactions:

$$2CH_4 + \tfrac{1}{2}O_2 \rightarrow C_2H_6 + H_2O(g) \qquad (\Delta H_{25°C} = -177 \text{ kJ mol}^{-1}) \tag{13-1}$$

$$C_2H_6 + \tfrac{1}{2}O_2 \rightarrow C_2H_4 + H_2O(g) \qquad (\Delta H_{25°C} = -105 \text{ kJ mol}^{-1}) \tag{13-2}$$

$$C_2H_6 \rightarrow C_2H_4 + H_2 \qquad (\Delta H_{25°C} = +137 \text{ kJ mol}^{-1}) \tag{13-3}$$

The undesired oxidation of methane to carbon oxides and water also occurs:

$$CH_4 + \tfrac{3}{2}O_2 \rightarrow CO + 2H_2O(g) \qquad (\Delta H_{25°C} = -520 \text{ kJ mol}^{-1}) \tag{13-4}$$

$$CH_4 + 2O_2 \rightarrow CO_2 + 2H_2O(g) \qquad (\Delta H_{25°C} = -803 \text{ kJ mol}^{-1}) \tag{13-5}$$

The water–gas shift reaction also plays a role in determining the overall product spectrum:

$$CO + H_2O(g) \Leftrightarrow CO_2 + H_2 \qquad (\Delta H_{25°C} = -41.4 \text{ kJ mol}^{-1}) \tag{13-6}$$

Depending on the reaction conditions employed, smaller amounts of acetylene, C_{3+} olefin and aromatics are produced via the pyrolysis of ethane, which is generally regarded as the primary product of the oxidative coupling step (Ito et al. 1985; Nelson, Lukey, and Cant 1988).

Because reactions (13-4) and (13-5) will inevitably proceed to some extent, the methane coupling process is strongly exothermic. Even on a laboratory scale this exothermicity places severe restrictions on the maximum methane conversion levels that can be studied in a single-stage fixed-bed reactor, using methane–oxygen mixtures without diluents, before excessive temperature gradients across the catalyst bed make data interpretation impossible. Also, any practical reactor conducting methane coupling on a large scale must be able to cope with the high rates of heat release in a safe and controlled manner.

The fluidized-bed reaction mode has the ability to maintain the catalyst bed at essentially isothermal conditions due to good solids backmixing and excellent heat-transfer characteristics. This allows methane conversions of practical significance to be obtained in a single reactor without the need for staging oxygen addition and costly interstage cooling of the process gas. The fluidized-bed also has the feature that the heat evolved by the oxidative coupling reactions can be used to preheat the feed gas directly within the bed via the recirculating catalyst particles. This minimizes the need for the costly and potentially dangerous external preheating of the feed gas.

The fluidized-bed reaction mode has a further advantage in that it is capable of being operated with continuous addition and withdrawal of catalyst from the reaction zone, so that catalysts that either have a relatively short operating life or require continuous regeneration can be used without having to shut down the reactor at regular intervals.

Fluidized-bed systems allow a wide range of gas–solids contacting modes and residence times to be achieved through the selection of appropriate operating conditions within any one of the well-known bubbling, turbulent, and fast or circulating fluidized-bed regimes.

Apart from the work of Follmer and co-workers (Follmer, Lehmann, and Baerns 1988), Andorf and Baerns (Andorf and Baerns 1990), and the Atlantic Richfield Company (ARCO) (Sofranko and Jubin 1989), little work has been reported in the literature on the use of fluidized-bed reactors for methane coupling. However, for the reasons outlined previously, CSIRO's reaction engineering research has focused on studying methane coupling in small-scale fluidized-bed reactors operating under practical conditions using oxygen (rather than air) as oxidant and in the absence of any diluents.

DESCRIPTION OF THE OXCO PROCESS CONCEPT FOR NATURAL-GAS CONVERSION

The research program is focused on the development of a process concept that is based on methane coupling but that also utilizes the noncondensate higher alkanes present in the natural gas, as would most likely be required for the conversion of gas from reserves in remote locations. Table 13-1 gives a typical composition of natural gas from Australia's North West Shelf region and shows that ethane and higher alkanes represent a substantial portion of the carbon content of the gas (36% in this case). These nonmethane components must be utilized efficiently in any practical process.

A conceptual flow diagram of the OXCO process is shown in Fig. 13-1. As well as the methane coupling reactor it includes facilities for feed gas pretreatment (to remove H_2O, CO_2, sulphur compounds, and part or all of the ethane and higher alkanes), reaction heat recovery, product gas compression and treatment to remove the H_2O and CO_2 formed by the reaction, and recovery of the C_2 and higher hydrocarbons by cryogenic technology, followed by oligomerization of the olefins to transport fuels. In order to minimize production of carbon oxides, the per-pass conversion of methane probably will be restricted to 25–35%. The unconverted methane must therefore be recycled to the reactor and consequently oxygen, rather than air, will be required, to avoid diluting the recycle methane with nitrogen.

An important feature of the process is the pyrolytic conversion of the higher alkanes (i.e., ethane produced by the process plus any higher alkanes that are separated from the feed gas) to ethylene within a fluidized-bed reactor using heat generated by the methane coupling reactions. This is achieved by injecting the alkanes directly into the oxygen-free zone of the fluidized bed so that the pyrolysis is carried out under substantially isothermal conditions using heat that is transferred continuously from the methane coupling zone to the pyrolysis zone via the internal recirculation of the catalyst particles. In this manner it should be

Table 13-1 Composition of Natural Gas from Australia's Northwest Shelf (Goodwyn Field) after the Condensate Fraction Has Been Removed

	COMPONENT[a]							
	CH_4	C_2H_6	C_3H_8	C_4H_{10}	C_5H_{12}	C_6H_{14}	N_2	CO_2
Percentage (v/v)	81.2	7.9	3.8	1.9	0.9	1.2	1.7	1.4

[a]Carbon in C_{2+} components as a percentage of total carbon: 36.4%.

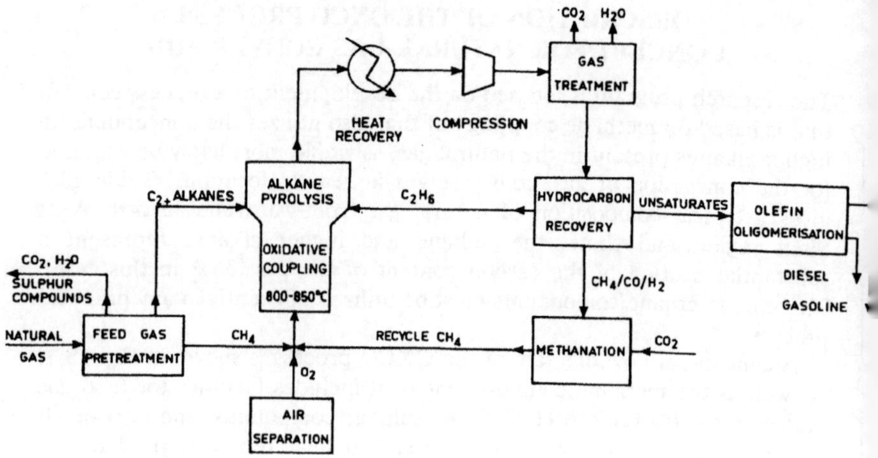

FIGURE 13-1. The OXCO conceptual process for natural gas conversion.

possible to conduct the pyrolytic conversion step with high selectivities to olefins and this would have a very beneficial effect on the overall process performance and efficiency, particularly when the process is employed for the total utilization of natural gas (rather than for only the methane component).

The carbon utilization efficiency of the process may be enhanced by including a catalytic methanation step in which hydrogen, produced by the methane coupling and alkane pyrolysis processes, is reacted with part of the carbon oxides to produce methane for recycling back to the reactor. The utilization of the by-product carbon oxides and hydrogen, however, is dictated by the overall process energy balance and the feasibility of methanating significant amounts of CO_2.

EXPERIMENTAL DETAILS

Fluidized-bed reactors of 30- and 60-mm diameter were constructed from quartz glass in order to prevent the undesired cracking of hydrocarbons being catalyzed by hot metal surfaces. The gas distributors were quartz frits with a nominal pore size 120 μm. In each case the quartz thermowell was connected to the top seal of the reactor to allow the axial temperature profile to be measured simply by raising and lowering the thermocouple within the thermowell. The reactors were electrically heated and a three-zone furnace was used for the 60-mm-diameter reactor so that the temper-

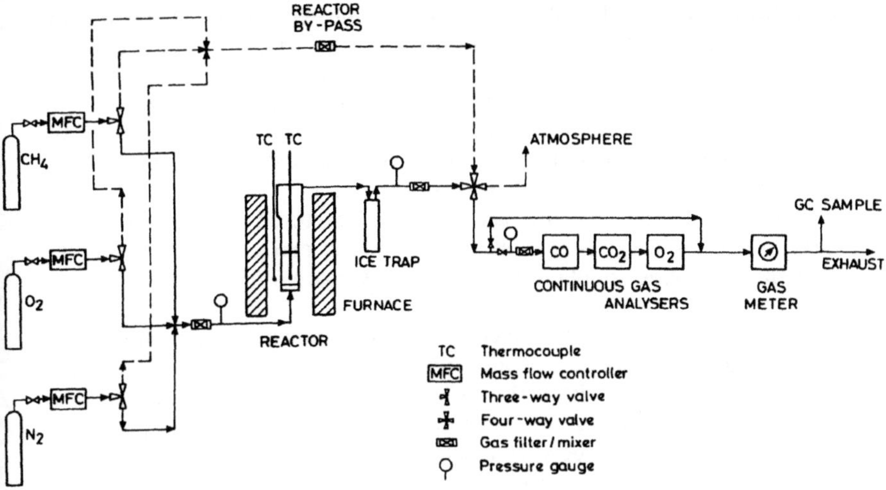

FIGURE 13-2. Flow diagram of CSIRO's small-scale fluidized-bed reactor facility (Edwards and Tyler 1989).

atures of the gas preheating zone, catalyst bed, and the disengaging zone above the bed could be controlled separately to the desired values.

Both reactors used the feed and product analysis system shown schematically in Fig. 13-2. Each run was commenced by heating the reactor to reaction temperature with the catalyst being fluidized by nitrogen. Methane and oxygen flows were set to the desired values using electronic mass flow controllers and directed to the gas analysis system via the reactor bypass line for measurement of the feed gas oxygen level. When the reactor was at the required temperature the nitrogen flow was stopped and feed gas switched over to the reactor to commence operation. Reactor product gas was cooled, dehumidified in an ice bath, and then passed to continuous gas analyzers for measurement of CO and CO_2 (Horiba PIR2000, nondispersive infrared analyzers) and oxygen (Servomex Model 1100 paramagnetic analyzer). Total product gas flow rates were measured by positive displacement flowmeters and samples were taken intermittently for C_1–C_6 hydrocarbon and hydrogen analyses by gas chromatography.

The 60-mm-diameter reactor was equipped with a quartz sample probe that enabled gas samples to be taken axially through the reactor. The probe was fitted with a quartz frit at the tip to avoid catalyst being sucked out with the gas sample while the gas samples for product analyses were withdrawn from the fluidized-bed via a diaphragm pump. Adjusting the position of the sample probe within the reactor during a run enabled axial gas concentration profiles to be determined through both the catalyst bed and the disengaging zone above the bed. The sampling system was designed so that the residence time of the gas in the probe was less than 5% of the gas residence time in the fluidized-bed when about 10% of the total gas throughput was withdrawn through the probe.

The same probe was also used to inject metered streams of ethane directly into the oxygen-free zone of the catalyst bed in order to study the combined methane-coupling–ethane-pyrolyis process in a single fluidized-bed reactor.

Reactor performance was calculated from a knowledge of both the flow rate and detailed analyses of all input and output streams. Carbon balances were generally within the range 98–102%. Methane conversion is defined as the percentage of input methane converted to total measured carbon-containing products and selectivity as the amount of methane converted to a particular product expressed as a percentage of the total methane converted. The yield of a component is then simply (methane conversion × component selectivity)/100.

In most cases catalyst particles in the size range $-250 + 150$ μm were used in both reactors. Some limited data are also reported for catalysts in the size ranges of $-425 + 250$ μm and $-150 + 53$ μm to provide an indication of the effect of particle size on reactor performance. The reactors were operated in the bubbling-bed regime (Kunii and Levenspiel 1977) with superficial gas velocities ranging from 5 to 20 times the minimum fluidization velocity. Details of the catalysts cannot be disclosed because they are currently the subject of patent applications. Results have been presented here for these specific catalyst formulations, which have been designated as catalysts A, B, and C.

EXPERIMENTAL RESULTS AND DISCUSSION

Methane Oxidative Coupling Performance Data

The Influence of Temperature on Hydrocarbon Selectivity

It is a general feature of the methane coupling catalysts developed by CSIRO that they exhibit maximum selectivity to hydrocarbons at 800–870°C when they are operated at the relatively short contact times employed in both the 30- and 60-mm-diameter fluidized-bed reactors. This is illustrated for catalyst A in Fig. 13-3, which shows, for both reactors, the influence of reaction temperature on hydrocarbon selectivity for feed

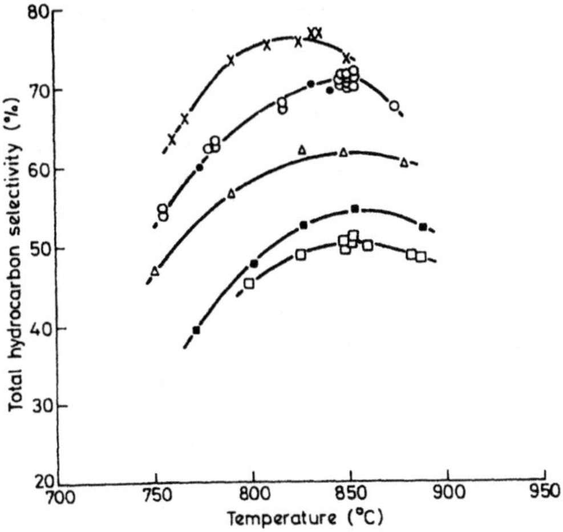

FIGURE 13-3. The influence of temperature on the hydrocarbon selectivity for catalyst A for various feed gas oxygen levels in the 30-mm-diameter [10 (○) and 25% (□) v/v O_2) and 60-mm-diameter [5 (×), 10 (●), 13 (△) and 19% (■) v/v O_2) fluidized-bed reactors at nominal contact times of 0.1–0.2 g s ml^{-1}.

gases containing oxygen levels ranging from 5 to 25% v/v. The data were obtained at nominal contact times, W/F [defined here as the weight of catalyst (in grams) divided by the feed gas flow-rate in milliliters per second at operating conditions] of 0.1–0.2 g s ml^{-1}. It is clear that the maximum hydrocarbon selectivity was obtained for this catalyst within the temperature range of about 830–850°C for all feed gas oxygen levels investigated, and the selectivity declined rapidly on either side of this fairly narrow temperature range. Other catalysts exhibited similar behavior in the fluidized-bed reactors although in some cases the maximum hydrocarbon selectivity was maintained over a somewhat broader temperature range.

The attainment of maximum selectivity to hydrocarbons at relatively high temperatures is a highly desirable result in the context of the OXCO process concept because temperatures in excess of 800°C are necessary in order to obtain a reasonable per-pass pyrolytic conversion of the non-methane alkanes when they are injected into the oxygen-free zone of the fluidized bed. In view of these considerations most of the experimental results reported here have been obtained in the temperature range 800–870°C.

The Influence of Methane Conversion on Product Selectivities

The relationship between methane conversion and the product selectivities in the fluidized-bed reactors is important in determining factors such as the amount of feedstock consumed per unit of product and the size of the recycle stream in the process concept. An example of such a relationship for one particular catalyst (catalyst B) in the 30-mm-diameter reactor is given in Figs. 13-4 and 13-5. Figure 13-4 shows the selectivity to total hydrocarbons as a function of methane conversion at 800–850°C and contact times in the range 0.1–0.4 g s ml^{-1}. At all times the reactor operated at total oxygen consumption and the methane conversion was increased by progressively raising the feed gas oxygen level from 5 to 22% v/v. The results, which were obtained using four separate batches of catalyst, showed that the hydrocarbon selectivity declined steadily from 85 to about 62% as the methane conversion was increased from 9.5 to 28%. Most of the data in Fig. 13-4 was obtained using catalyst particles in the size range −250 + 150 μm. However, a limited number of results were obtained at the higher methane conversion level with both finer (−150 + 53 μm) and coarser (−425 + 250 μm) particles and it can be seen that the hydrocarbon selectivities were similar in all cases. The ability to use coarser particles in fluidized-bed reactors without reducing the methane coupling performance is a significant result because it enables higher gas velocities and hence higher specific throughputs to be employed in large-scale reactors.

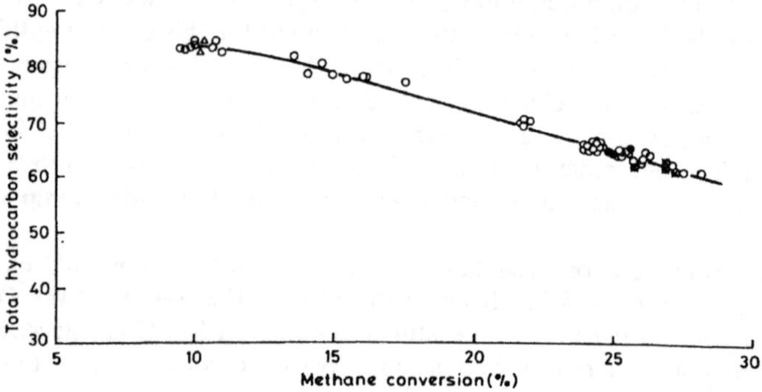

FIGURE 13-4. The influence of methane conversion on the hydrocarbon selectivity for catalyst B in the 30-mm-diameter fluidized-bed reactor [temperature 800–850°C; total oxygen consumption; nominal contact time 0.1–0.4 g s ml^{-1}; catalyst particle size ranges −150 + 53 μm (⊗), −250 + 150 μm (○), and −425 + 250 μm (●)]. Fixed-bed microreactor data (△) (temperature 850°C; nominal contact time 0.02 g s ml^{-1}).

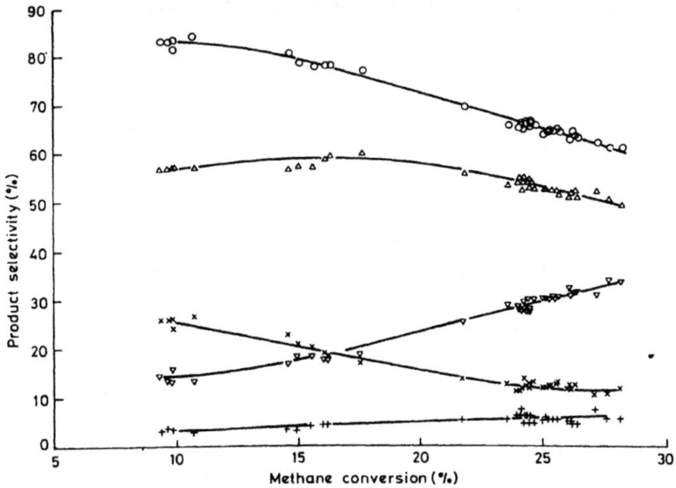

FIGURE 13-5. The influence of methane conversion on the product selectivities [total hydrocarbons (○), total unsaturates (△), ethane (×), CO (+), and CO_2 (▽)] for catalyst B in the 30-mm-diameter fluidized-bed reactor (temperature 800–850°C; total oxygen consumption; nominal contact time 0.1–0.4 g s ml^{-1}).

CSIRO's catalyst development program includes the detailed evaluation of the performance of catalysts in fixed-bed microreactors. However, due to the development of excessive temperature gradients over the catalyst bed, it is not practical to operate the microreactors at methane conversions above the 10–15% level. Although this does not allow the comparison of fixed- and fluidized-bed reactor performances at the higher methane conversion levels, it can be seen from Fig. 13-4 that the performance of catalyst B has been successfully translated from the fixed-bed to the fluidized-bed reaction mode at the 10% conversion level. This is in contrast with the previously reported failure to translate the performance of a lithium-promoted magnesium oxide catalyst (Li/MgO) from fixed- to fluidized-bed reactors (Edwards, Tyler, and White 1990). In the latter case the fixed-bed reactor had a significantly better performance, although this was considered to be at least partly due to the difficulties in preparing individual batches of the Li/MgO catalyst with reproducible properties.

In Fig. 13-5 the major component selectivities are shown as a function of methane conversion. Due largely to the relatively high reaction temperature, the unsaturated hydrocarbons (mainly ethylene along with smaller amounts of acetylene, C_3 and C_4 olefins, and benzene) were a high proportion of the total hydrocarbons, constituting some 69 and 83% of the total at the low and high methane conversions, respectively. The decline in total hydrocarbon selectivity with increasing methane conversion was due largely to the ethane selectivity decreasing from 26 to 10% over the range of methane conversions investigated. In common with most methane coupling catalysts, CO_2 was the major carbon oxide produced, with its selectivity being 4.7–5.7 times that of CO.

The measured hydrogen levels in the product gas were used to calculate the percentage of the "available hydrogen" (the hydrogen equivalent to that contained in the converted methane minus the hydrogen contained in the C_{2+} hydrocarbons) that is present in the product as H_2 rather than water. Although this parameter varied from catalyst to catalyst, depending on their total oxidation capability, it was generally found that between 13 and 30% of the available hydrogen appeared as H_2 when the fluidized-bed reactors were operated within the temperature range 800–870°C. For example, with catalyst B the values were about 22% at the low methane conversion level, declining to 13% at the high methane conversion level. It was also found that those catalysts that gave relatively high H_2 yields also produced higher CO yields. This feature would have a significant impact in a large-scale process because the higher the H_2 and CO yields, the lower are both the consumption of oxygen and the heat released in the methane coupling reactor.

The Influence of Methane Conversion on Hydrocarbon Yields and Consumption of Specific Reactants

In evaluating the relevance of the reactor performance to a large-scale process, it is useful to consider the influence of the per-pass methane conversion on both the hydrocarbon yields [i.e., (conversion × hydrocarbon selectivity)/100] and the specific consumption rates of reactants (defined as the weights of methane and oxygen consumed per unit weight of hydrocarbon produced). These parameters have been calculated for the results in Fig. 13-5 and are presented in Fig. 13-6. It can be seen that, although the rate of increase in hydrocarbon yield is slowing down as the methane conversion increases, it has by no means reached a maximum value, indicating that increasing the methane conversion to higher levels will produce additional hydrocarbons and not just more carbon oxides. However, both the specific methane and oxygen consumption rates (important cost factors in a large-scale process) rose steadily with methane conversion, increasing from 1.4 to 2.0 $g\ g^{-1}$ and from 1.5 to 4.0 $g\ g^{-1}$, respectively, as the methane conversion increased from 9.5 to 28%. The optimum per-pass methane conversion in a large-scale process will depend on, among other things, balancing the savings in capital and operating

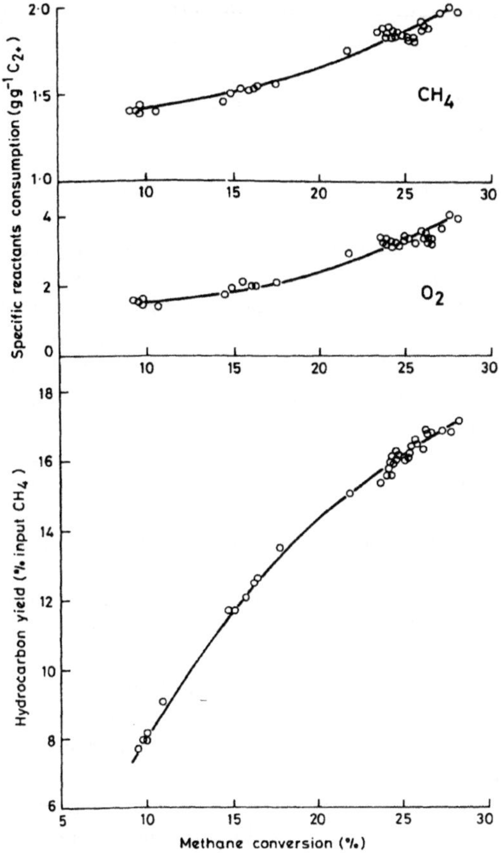

FIGURE 13-6. The influence of methane conversion on the hydrocarbon yield and the specific reactants consumption for catalyst B in the 30-mm-diameter fluidized-bed reactor (temperature 800–850°C; total oxygen consumption; nominal contact time 0.1–0.4 g s ml^{-1}).

costs brought about by reducing the methane recycle load, that is, by increasing the per-pass methane conversion, with the resulting increase in feedstock cost due to the increased consumption rates of specific reactants.

The results in Fig. 13-6 can be compared with similar data reported previously (Edwards, Tyler, and White 1990) for a lithium-promoted magnesium oxide catalyst. In that case, there was no significant increase in hydrocarbon yield at methane conversions above about 16%, at which point the hydrocarbon yield was only 11% of the input methane. In contrast, the present catalyst B gave a significantly improved performance, with hydrocarbon yields in excess of 17% of input methane achieved at the highest methane conversion levels investigated, with no evidence still of a plateau being reached.

The Influence of Methane Conversion on the Adiabatic Temperature Rise

The highly exothermic nature of the methane coupling reactions has been mentioned previously. However, temperature profile measurements through the bed of the 30-mm-diameter fluidized-bed reactor showed it to be operating nearly isothermally (5–10°C gradient) even under the most severe conditions investigated, where the methane conversion was in excess of 40% and the selectivity to carbon oxides was high (approximately 60%). This clearly demonstrated the ability of the fluidized-bed reaction mode to conduct the coupling reactions in a safe and controlled manner using methane–oxygen mixtures (without diluents) that were well inside the explosive limits.

In terms of conducting methane coupling in large-scale reactors, the adiabatic temperature rise (the calculated difference in temperature between the product and reactant gases when the product stream absorbs the total heat of reaction adiabatically) is a good indicator of the engineering problems associated with exothermic reactions because large-scale reactors approach adiabatic behavior if no allowance is made in reactor design for heat removal. The adiabatic temperature rise calculated for the data in Fig. 13-4 is shown as a function of methane conversion in Fig. 13-7. The temperature rise is independent not only on methane conversion but also on the CO and CO_2 selectivities, on the fraction of available hydrogen remaining as H_2 and on the product alkene–alkane ratio. To separate the effects of methane conversion and product selectivities on the results in Fig. 13-7, the adiabatic temperature rise has also been calculated as a function of methane conversion assuming a constant selectivity equal to that obtained at the 9.5% conversion level (i.e., 83% to hydrocarbons, 3% to CO, and 14% to CO_2). These results are included as the lower line in Fig. 13-7. It is clear that the temperature rise is very significant and that if the selectivity had remained constant, increasing the methane conversion from 9.5 to 28% would have raised the adiabatic temperature rise from 200 to about 590°C. However, the actual temperature rise increased from 200 to 1060°C; the additional rise of 470°C being largely due to the increase in carbon oxides selectivities.

The magnitude of the adiabatic temperature rise is such that it would not be possible to conduct the reaction on a large-scale in a single-staged fixed-bed reactor. It would be necessary to stage the oxygen addition in a

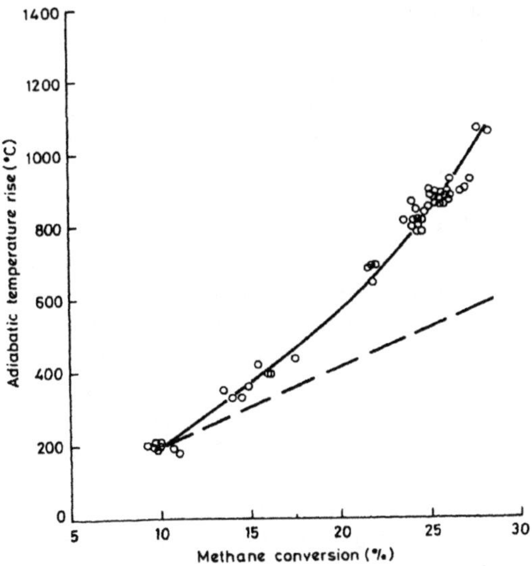

FIGURE 13-7. The influence of methane conversion on the adiabatic temperature rise for catalyst B for the results given in Fig. 13-4.

multibed reactor (to limit the temperature gradient across each bed) with costly interstage cooling of the gas and extensive preheating of the feed. On the other hand, fluidized-bed reactors are well-suited for this process due to their excellent heat-transfer characteristics, which result from the backmixing of the solid phase. In addition, unlike fixed-bed reactors, fluidized-bed reactors are capable of using the reaction heat to preheat the reactants directly within the bed, thus minimizing the need for costly indirect heat exchangers for feed gas preheating. For example, if the desired reaction temperature were 850°C, the results in Fig. 13-7 imply that catalyst B could be operated at about 25% per-pass methane conversion *without* the need for either feed gas preheating or heat removal from the fluidized-bed.

Comparison of Single- and Two-Staged Fluidized-Bed Reactor Performances

The previously discussed results clearly showed that the fluidized-bed reactor is capable of conducting the oxidative coupling process at methane conversion levels of practical significance in a single-stage reaction step. However, in a large-scale methane coupling reactor staging the oxygen addition in a multistage fluidized-bed system is one potential option for achieving the desired overall per-pass conversion of methane while maintaining the oxygen level outside the flammability limits. This can be considered practical only if the staging of the oxygen addition does not result in excessive selectivity losses due to the subsequent oxidation of C_{2+} hydrocarbons produced in previous stages. In order to check the effect of staging the oxygen addition, experiments have been conducted in a single-stage 30-mm-diameter reactor and in two 30-mm-diameter reactors installed in series. This allowed the performance of the single-stage reactor to be compared directly with a two-stage reactor in which the same total amount of oxygen was consumed as in the single reactor but in which half the oxygen was added to the cooled gas after the first reactor. Experiments have been conducted in this manner with catalysts A, B, and C at nominally 850°C and with total oxygen consumption and the results are given in Table 13-2. With catalysts A and B the same total amount of oxygen has been used in the single- and two-stage reactors. With catalyst C slightly more oxygen was used in the two-stage reactor experiment. Taking this into account all catalysts gave a remarkably similar performance (in terms of conversion, hydrocarbon selectivity, and product distributions) in the two-stage reactor compared with the single-stage reactor. The results also indicate that there was no benefit to be gained, in terms of overall reactor performance, from staging the oxygen addition. Nonetheless the data are important because they indicate that a large-scale reactor could conceivably be operated with staged oxygen addition (an inherently safer mode of operation) *without* loss of overall performance.

Comparison of 30- and 60-mm-Diameter Fluidized-Bed Reactor Performances

Although much of the catalyst evaluation work was conducted in the 30-mm-diameter fluidized-bed reactor, a 60-mm-diameter reactor has been used to study the performance of selected catalysts over a wider range of contact times than was possible in the 30-mm-diameter reactor. The 60-mm-diameter reactor has also been used to gain insights into the mechanisms governing catalyst performance and to study the pyrolysis behavior of ethane when it is injected into the oxygen-free zone of the catalyst bed.

When the two reactors were operated under similar conditions it was found that the 60-mm-diameter reactor had at least as good and in some cases a better performance than the 30-mm-diameter reactor. This is illustrated in Fig. 13-8 and Table 13-3. Figure 13-8 shows the total hydrocarbon selectivity as a function of methane conversion for catalyst C at 780–860°C and at total oxygen consumption. The results include data from both reactors over a fairly wide range of contact times (0.1–0.8 $gsml^{-1}$) and show that in all cases the two reactors gave a similar

Table 13-2 Comparative Performance Data for Single- and Two-Stage Fluidized-Bed Reactors

	NUMBER OF STAGES							
	CATALYST A		CATALYST B				CATALYST C	
	1	2	1	2	1	2	1	2
Feed O_2 level (% v/v)	14.6	2 × 7.3	10	2 × 5	18.4	2 × 9.2	9.6	2 × 5.1
Temperature (°C)	849	846	844	849	849	844	844	846
CH_4 conversion (%)	21.8	21.7	15.9	15.2	24.6	24.2	14.2	15.4
O_2 consumption (%)	100	100	100	100	100	100	100	100
Selectivity (%):								
C_2H_6	17.3	16.8	19.6	18.9	13.2	13.5	25.2	21.2
Unsaturates	53.7	53.8	59.0	59.3	55.0	54.8	48.1	51.9
Total HC	71.0	70.6	78.6	78.2	68.2	68.3	73.3	73.1
CO	4.7	5.2	1.9	1.8	3.2	2.9	5.1	5.6
CO_2	24.3	24.2	19.5	20.0	28.6	28.8	21.6	21.3

performance. Table 13-3 shows detailed reactor performance data for each reactor for catalysts B and C. For catalyst B it can be seen that the 60-mm-diameter reactor gave somewhat better results, with a total hydrocarbon selectivity of 70% at the 24% methane conversion level compared with 66.5% from the 30-mm-diameter reactor. The reduced hydrocarbon selectivity in the 30-mm-diameter reactor was largely due to an increase in selectivity to CO (6.0% cf. 3.7%). The relative distribution in selectivities

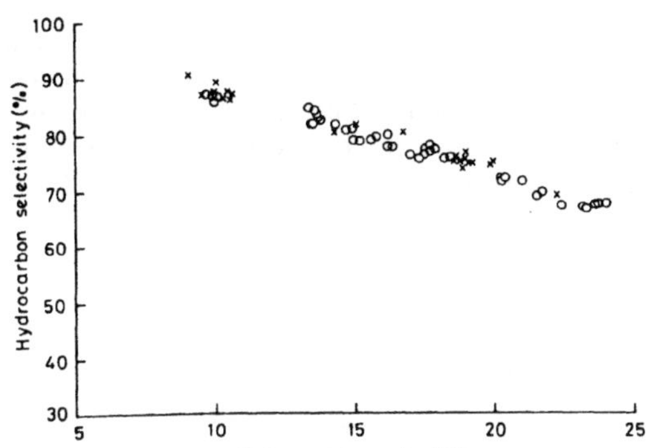

FIGURE 13-8. The influence of methane conversion on the hydrocarbon selectivity for catalyst C in the 30-mm-diameter (O) and 60-mm-diameter (×) fluidized-bed reactors (temperature 780–860°C; total oxygen consumption; nominal contact time 0.1–0.8 g s ml^{-1}).

Table 13-3 Comparison of 30- and 60-mm-Diameter Fluidized-Bed Reactor Performance

	CATALYST B		CATALYST C			
	"HIGH" CONVERSION		"LOW" CONVERSION		"HIGH" CONVERSION	
Reactor	30 mm	60 mm	30 mm	60 mm	30 mm	60 mm
Temperature (°C)	852	843	830	850	861	854
Feed O_2 level (%v/v)	17.3	17.0	5.0	5.0	16.8	16.8
Contact time (g s ml^{-1})	0.16	0.14	0.25	0.55	0.22	0.42
Methane conversion (%)	24.4	24.0	10.0	10.6	21.5	22.2
Product selectivity (%):						
C_2H_6	12.5	13.2	26.3	15.4	12.7	13.7
C_2H_4	44.7	46.6	48.0	54.0	44.5	43.0
C_3H_8	0.3	0.3	0.5	0.2	0.3	0.4
C_3H_6	4.3	4.7	6.7	9.2	5.9	6.5
Other unsaturates[a]	4.7	5.2	3.8	7.2	5.2	5.6
Total hydrocarbons	66.5	70.0	85.3	86.0	68.6	69.2
CO	6.0	3.7	1.5	1.1	3.3	3.0
CO_2	27.5	26.3	13.2	12.9	28.1	27.8
Percentage of HC as unsaturates	80.8	80.7	68.6	81.9	81.0	79.7

[a]Mainly C_4H_6 and C_6H_6 with smaller amounts of C_2H_2 and C_4H_8.

to the hydrocarbon components from the two reactors were similar, presumably reflecting the fact that both the contact times and temperatures were similar in the two reactors and hence the pyrolysis of ethane formed by the methane coupling reaction proceeded to the same extent.

Data for catalyst C at the "low" (10%) and high (22%) methane conversion levels showed that the two reactors had similar performance in terms of conversion and total hydrocarbon selectivity. The product distributions at the high conversion level were essentially the same but at the low conversion level substantially more pyrolysis had occurred (i.e., lower ethane selectivity and higher selectivity to unsaturates) in the 60-mm-diameter reactor as a result of both the higher temperature and longer contact time employed in this reactor.

Identification of Carbon Formation as a Mechanism Influencing Reactor Performance

The results in Figs. 13-4 and 13-8 showed that the performances of both catalysts B and C were relatively independent of contact time over the range of values investigated (up to 0.4 g s ml^{-1} for catalyst B and up to 0.8 g s ml^{-1} for catalyst C). By contrast, the performance of some catalysts,

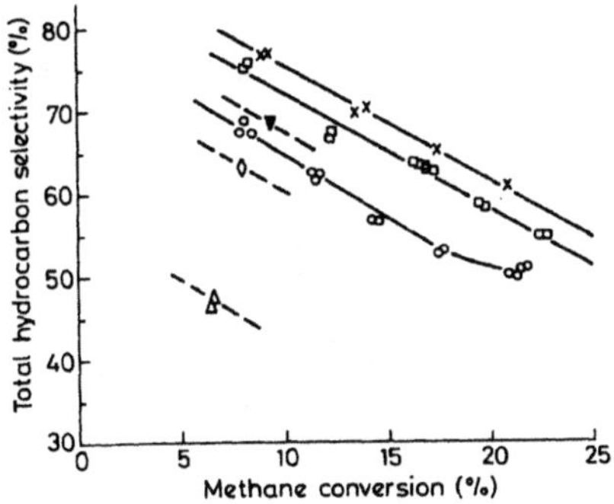

FIGURE 13-9. The influence of methane conversion on the hydrocarbon selectivity for catalyst A in the 60-mm-diameter fluidized-bed reactor at 850°C and for various nominal contact times [0.11 (×), 0.22 (□), 0.35 (▼), 0.45 (○), 0.65 (◇), and 1.40 (△) g s ml^{-1}] (Edwards, Do, and Tyler 1990).

such as catalyst A, was found to be extremely dependent on contact time. This is illustrated in Fig. 13-9, which shows for catalyst A the effect of methane conversion, at 850°C and at total oxygen consumption, on the selectivity to hydrocarbons with contact time as a parameter. For a given methane conversion the hydrocarbon selectivity declined strongly with increasing contact time. For example, at a methane conversion of 15% the hydrocarbon selectivity declined from 69 to 54% as the contact time was increased from 0.1 to 0.45 $gsml^{-1}$. At a contact time of 1.4 $gsml^{-1}$ the hydrocarbon selectivity was only 47% at a methane conversion of 6%.

The effect of contact time on hydrocarbon selectivity was also strongly influenced by temperature. This is shown in Fig. 13-10, where selectivity is plotted against temperature, for various contact times, for a feed gas containing 5% v/v oxygen. The curve obtained with the same catalyst in the fixed-bed microreactor (Maitra 1990b) operating at a low contact time of 0.013 g s ml^{-1} is also included in Fig. 13-10 for comparison with the 60-mm-diameter reactor results. It can be seen that contact time had its greatest effect on hydrocarbon selectivity in the 60-mm-diameter reactor at temperatures above 800°C. The effect on selectivity declined as the temperature was reduced to the point when, at 730–750°C, the hydrocarbon selectivity was essentially independent of contact time and approached the value obtained in the fixed-bed reactor at this temperature.

The axial gas concentration profile through the fluidized-bed was measured in an attempt to elucidate the mechanism governing the influence of contact time on hydrocarbon selectivity. An example of the type of data

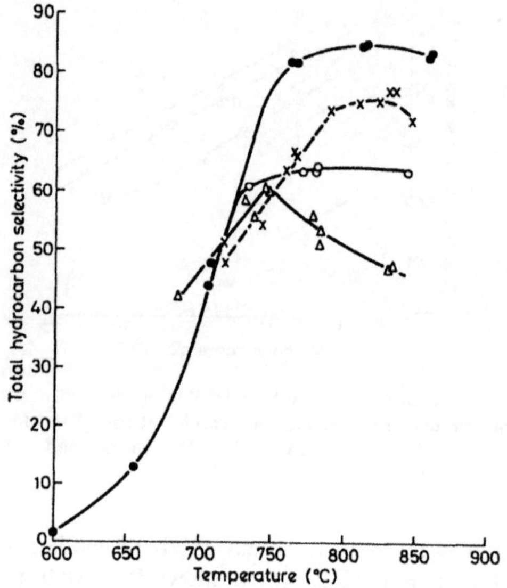

FIGURE 13-10. The influence of temperature on the hydrocarbon selectivity for catalyst A (feed gas composition 5% O_2, 95% CH_4) at various contact times [fixed-bed microreactor, 0.013 (●) g s ml^{-1}; 60-mm-diameter fluidized-bed reactor, 0.12 (×), 0.7 (O), and 1.4 (△) g s ml^{-1}] (Edwards, Do, and Tyler 1990).

obtained for one set of conditions (850°C, 13% v/v oxygen in feed, a contact time of 0.21 g s ml^{-1}, a methane conversion of 16.3%, and a total hydrocarbon selectivity of 61.3%) is shown in Fig. 13-11. The data presumably represent an average of the values in the bubble and emulsion phases. Up to 0.2 cm above the gas distributor very little reaction had occurred, indicating that no substantial reaction had taken place either prior to the gas entering the catalyst bed or in the gas sample probe. However, between 0.2 and 0.3 cm above the distributor a very rapid reaction occurred, resulting in complete consumption of oxygen and formation of products. Beyond this point the ethane level declined from 2.9 to 1.5% v/v at 5 cm above the distributor, effectively the top of the fluidized bed. Surprisingly, there was no significant increase in the ethylene concentration over this section of the bed, as might have been expected from ethane cracking, although the hydrogen increased from 3.3 to almost 5% v/v. However, ethane dehydrogenation to ethylene and hydrogen occurred in the gas zone above the bed as shown by comparing the composition of the product gas exiting the reactor with that of the gas at 5 cm above the

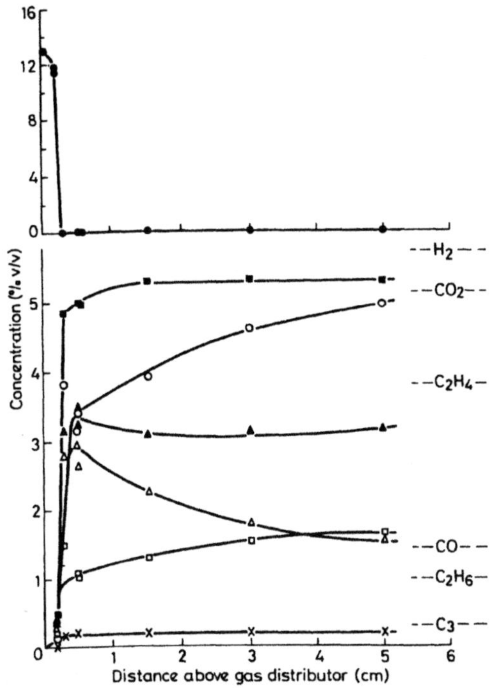

FIGURE 13-11. Variation of gas concentration [O_2 (●), H_2 (○), CO (□), CO_2 (■), C_2H_6 (△), C_2H_4 (▲), and C_{3+} (×)] with distance above the gas distributor for catalyst A in the 60-mm-diameter fluidized-bed reactor (feed gas composition 13% O_2, 87% CH_4; nominal contact time 0.21 g s ml^{-1}). Dashed lines are component levels in the reactor product gas (Edwards Do, and Tyler 1990).

distributor. There was also a small increase in CO concentration, and to a lesser extent CO_2, after the oxygen was consumed. However, both the amount and distribution of these components throughout the bed were also likely to be affected by the steam reforming and water–gas shift reactions.

Overall, the most important observation from the data in Fig. 13-11 was that, after the oxygen was rapidly consumed, there was a loss of carbon in the form of C_{2+} hydrocarbons as the gas passed through the oxygen-free zone of the bed. This loss could not be accounted for by an equivalent increase in carbon oxides. As shown in Fig. 13-12, the net carbon loss from the gas phase was equivalent to about 18% of the carbon in the C_{2+} hydrocarbons measured at the point in the bed where the oxygen was

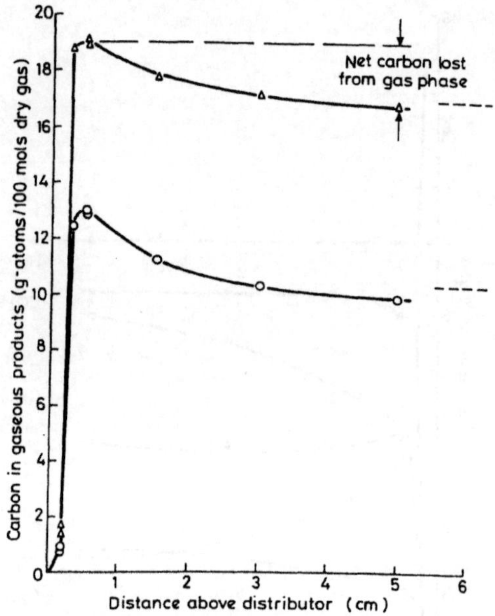

FIGURE 13-12. Variation of carbon in products [C_{2+} hydrocarbons (○) and total products (△)] with distance above the gas distributor for the results given in Fig. 13-11. Dashed lines are values for the reactor product gas (Edwards, Do, and Tyler 1990).

totally consumed. Further, the results in Fig. 13-12 also show that the carbon loss effectively ceased when the gas left the top of the bed because both the total carbon in measured products and the carbon in the hydrocarbons were essentially the same at this point as the corresponding values in the reactor product gas. This was despite the fact that, as noted previously, a substantial pyrolytic conversion of ethane to ethylene had occurred in the hot gas zone above the bed.

On the basis of these results, it is postulated that the loss of carbon from the gas phase in the fluidized bed was due to the cracking of hydrocarbons to form carbon. The fact that the carbon loss ceased once the gas left the catalyst bed suggested that the cracking process was catalyzed by the coupling catalyst. This mechanism explains why the hydrocarbon selectivity declined as both the contact time (i.e., the gas residence time in the oxygen-free zone of the bed) and temperature were increased, because increasing either of these parameters should favor carbon formation.

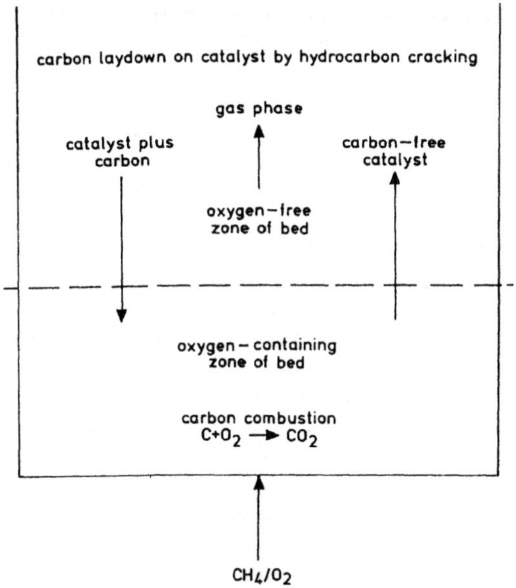

FIGURE 13-13. Mechanism for carbon formation and its removal from the catalyst in the fluidized-bed reactor (Edwards, Do, and Tyler 1990).

At the end of the run, carbon deposits were observed on the reactor wall and sample probe above the point where the oxygen was consumed. Although inspection of the catalyst revealed only minor amounts of carbon in the bed, it is further postulated that the carbon formed on the catalyst was continuously recirculated within the fluidized bed to the oxygen-containing zone of the reactor, where it was combusted. This mechanism of selectivity loss is depicted schematically in Fig. 13-13. This phenomenon is doubly detrimental to overall reactor performance because not only did it result in a significant loss of hydrocarbon but it also resulted in a wasteful consumption of oxygen that would otherwise be available for the coupling reaction.

It should be noted that the results in Figs. 13-11 and 13-12 were obtained at the relatively short contact time of 0.21 $gsml^{-1}$, but even at this condition the hydrocarbon loss was significant. Thus it is not surprising that the loss in hydrocarbon selectivity at the contact time of 1.4 $gsml^{-1}$ was as dramatic as shown in Fig. 13-9.

The degree to which the carbon laydown phenomenon was affecting reactor performance can be estimated in the following manner. It is assumed that the carbon formed by hydrocarbon cracking is recirculated to the oxygen-containing zone of the bed, where it is combusted to CO_2 (rather than CO). This quantity of CO_2 is then deducted from the CO_2 level measured at the point in the bed where the hydrocarbons were at their maximum level and the methane conversion and product selectivity are then recalculated using this corrected CO_2 level. The oxygen saved by not having to combust the carbon deposit is assumed to convert additional methane at the aforementioned selectivity. When this is done the methane conversion/hydrocarbon selectivity performance improves very significantly from 16.3%/61.3% to 18%/79%.

Although the isothermal operation of the reactor is good evidence that the solid phase was well-backmixed, it is clear from Fig. 13-11 (which shows substantial variations in the axial concentrations of ethane and hydrogen) that the gas phase was far from being fully backmixed. This reduces the possibility of secondary oxidation of hydrocarbons via their recirculation to the oxygen-containing zone, and it is a crucial result if alkane pryolysis is to be conducted in the oxygen-free zone of the bed without excessive production of carbon oxides.

Similar axial probing studies conducted with catalysts B and C showed that these catalysts were also extremely active, with oxygen being totally consumed within several millimeters above the gas distributor plate. However, unlike catalyst A, both these catalysts had a very much lower propensity to crack hydrocarbons to coke in the oxygen-free zone of the fluidized bed and this at least partly explains why they gave a better overall methane coupling performance, which was also less dependent on contact time.

Utilization of Ethane and Higher Alkanes

The effective utilization of ethane and higher alkane components of the feed gas, as well as the ethane produced by methane coupling, is an essential part of any natural-gas conversion process based on this route. This problem has been addressed by Institut Français du Petrole (IFP) (Mimoun et al. 1990) by separating the higher alkanes from the natural gas prior to conducting methane coupling in a fixed-bed reactor. These alkanes, together with the ethane produced by methane coupling, are pyrolyzed using the sensible heat of the reactor product gas. On the other hand the ARCO process (Sofranko and Jubin 1989) cofeeds ethane to the coupling reactor within the methane.

The OXCO process takes advantage of the fact that the methane coupling catalysts developed by CSIRO are very active and consequently most of the fluidized bed is in an oxygen-free environment. This enables the pyrolysis of ethane and higher alkanes to be conducted by injection into the fluidized bed, where the endothermic pyrolysis reactions proceed under essentially isothermal conditions using heat generated from the oxidative coupling reactions. In this manner the total exothermic heat generated by the coupling reaction is potentially available for alkane pyrolysis. By contrast, if the alkanes were simply injected into the hot gas zone above the catalyst bed, as proposed by IFP, the only heat available for pyrolysis is the sensible heat in the product gas. This severely restricts the amount of pyrolysis that could be accomplished because the gas temperature would rapidly decrease to a level that was unfavorable with respect to both the equilibrium and kinetics of the pyrolysis reactions.

Although it is envisaged that propane and higher hydrocarbons would be separated from the feed gas with essentially 100% efficiency, there may be instances where complete recovery of the ethane is neither practical nor economical. Under these circumstances the OXCO process would have to tolerate at least part of the ethane being cofed with the methane and oxygen to the oxidative coupling section of the reactor.

Experiments have been conducted to assess the performance of the fluidized-bed reactor for both the ethane pyrolysis and cofeed options. Results for each case are now presented and discussed.

Ethane Pyrolysis Option

Table 13-4 presents data obtained with catalyst B at 850°C in the 60-mm-diameter fluidized-bed reactor with and without ethane injection into the oxygen-free zone of the bed, for a feed gas to the oxidative coupling zone containing 83% v/v methane and 17% v/v oxygen. Gas samples extracted axially from the fluidized bed showed that the oxygen had been consumed within 2–3 mm above the gas distributor and that the rest of the bed was oxygen-free. After methane coupling measurements were made (see the first column in Table 13-4) ethane was injected via a quartz probe into the oxygen-free zone of the bed at the rate of 1.24 g-mol h^{-1} [i.e., 2.48 (g-atom C) h^{-1}]. Reactor performance measurements made in this combined methane-coupling–ethane-pyrolysis mode were used with the methane coupling data to determine the performance of the ethane pyrolysis reaction (i.e., the second column in Table 13-4).

The results show that ethane pyrolysis was achieved within the fluidized bed *without* affecting the methane coupling performance (e.g., by processes such as backmixing of the injected ethane into the oxygen-containing zone). This is seen by the fact that the total carbon in reactor products during ethane pyrolysis [i.e., 4.87 (g-atom C) h^{-1} including the methane formed by pyrolysis] was equal, within experimental error, to the products from oxidative coupling alone *plus* the amount of added ethane [i.e.,

Table 13-4 Performance Data for Catalyst B in the 60-mm-Diameter Fluidized-Bed Reactor at 850°C With and Without Ethane Injection into the Oxygen-Free Zone of the Fluidized-Bed (Feed Gas to Methane Coupling Zone 83% v / v CH_4, 17% v / v O_2)

	Without Ethane Injection	With Ethane Injection
Methane feed rate [(g-atom C) h^{-1}]	10.0	10.0
Methane conversion (%)	24.5	24.5
Hydrocarbon selectivity for methane coupling (%)	65.7	65.7
Ethane injection rate [(g-atom C) h^{-1}]	—	2.48
Products [(g-atom C) h^{-1}]:		
CH_4	7.56	7.67
C_2H_6	0.25	0.96
C_2H_4	1.16	2.60
C_{3+} unsaturates	0.20	0.36
Total C_{2+} hydrocarbons	1.61	3.92
CO	0.15	0.20
CO_2	0.69	0.64
Total CO_x	0.84	0.84
Total C_{2+} hydrocarbons plus CO_x [(g-atom C) h^{-1}]	2.45	4.87
Per-pass pyrolytic conversion of ethane (%)	—	71.0
Selectivity of ethane conversion (%):		
CH_4	—	7
C_2H_4	—	84
C_{3+} unsaturates	—	9
Total unsaturates	—	93
CO_2	—	Nil

2.45 + 2.48 = 4.93 (g-atom C) h^{-1}]. Further, there was no measurable increase in the total production rate of carbon oxides upon injection of the ethane, confirming the absence of ethane backmixing or significant coking of the catalyst. Any carbon laid down on the catalyst in this manner would have been recirculated to the oxygen-containing zone and combusted to carbon oxides, resulting in an increase in these components and a reduction in methane coupling performance. Inspection of the catalyst and reactor surfaces on completion of the experiments showed no evidence of carbon deposits, confirming that pyrolytic carbon formation was not a problem.

The per-pass pyrolytic conversion of the injected ethane was 71% based on the measured inputs and outputs of ethane and allowing for ethane produced by methane coupling. The carbon selectivity of ethane conversion was 7% to methane and 93% to unsaturated hydrocarbons, of which more than 90% was ethylene

The combined methane-coupling–ethane-pyrolysis experiments have been conducted with ethane injection rates, relative to the methane fed to the coupling zone, that are far in excess of those pertaining to a natural-gas feedstock. Figure 13-14 shows the per-pass pyrolytic conversion of ethane and the selectivity of this conversion to unsaturates, methane, and carbon oxides as a function of the molar ratio of ethane to methane fed to the reactor. The results were obtained when the reactor was operating with catalyst B and with the same methane coupling performance as that for the data in Table 13-4, that is, 24.5% conversion and 65.7% selectivity to

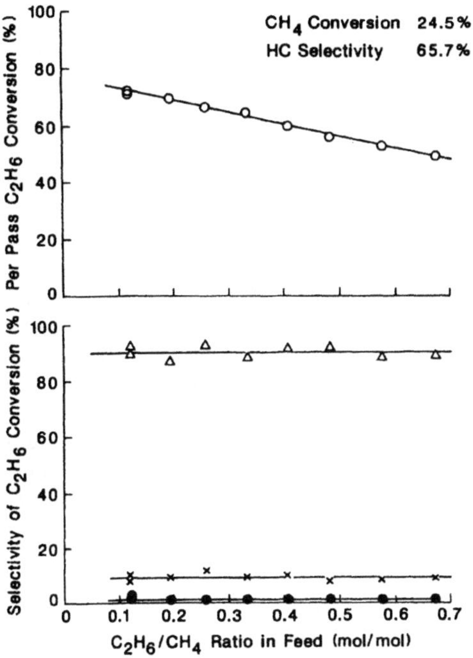

FIGURE 13-14. The influence of the C_2H_6–CH_4 molar ratio in the reactor feed gases on the pyrolytic C_2H_6 conversion and the selectivity of C_2H_6 conversion to products [unsaturates (△), CH_4 (×), and CO_x (●)] for catalyst B in the 60-mm-diameter fluidized-bed reactor at 850°C.

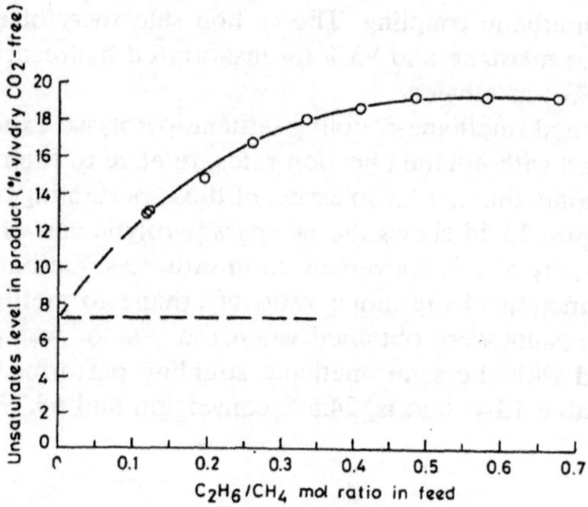

FIGURE 13-15. The influence of the C_2H_6–CH_4 molar ratio in the reactor feed gases on the unsaturates level in the product gas for the results given in Fig. 13-14. The dashed line is the unsaturates level in the gas coming from the oxidative coupling zone.

hydrocarbons. Figure 13-14 shows that the per-pass ethane conversion declined linearly from 72 to 49% as the ethane-to-methane molar ratio in the feed was increased from 0.125 to 0.675. The selectivity to unsaturates remained essentially constant at around 90% as did those to methane (approximately 9%) and CO_x (approximately 1%). It must be remembered that in a full-scale process any methane formed by the pyrolysis of ethane and higher alkanes would ultimately be recycled back to the oxidative coupling zone of the reactor, where it would be converted to unsaturates with the same selectivity as the methane feed. For the conditions pertaining in Table 13-4 and Fig. 13-14, this results in an overall selectivity of ethane conversion to unsaturates in excess of 95% for all levels of ethane injection investigated. It should also be noted that with an ethane-to-methane feed ratio of 0.675 the reactor product contains in excess of 19% v/v unsaturates, more than 90% of which is ethylene (see Fig. 13-15). This level of unsaturates is more than 2.6 times that which is coming from the oxidative coupling zone alone.

These results successfully demonstrate the OXCO combined methane-coupling–hydrocarbon-pyrolysis process concept. The ethane-to-methane feed ratios studied were far in excess of those pertaining to a natural-gas feedstock, thus demonstrating the flexibility of the process to handle a wider range of feedstocks than natural gas alone.

Table 13-5 Performance Data for Catalyst C in the 30-mm-Diameter Fluidized-Bed Reactor at 850°C and Total Oxygen Consumption With and Without Ethane in the Feed Gas

	Without Ethane	With Ethane
Feed composition (% v/v):		
CH_4	81.7	77.9
C_2H_6	—	4.7
O_2	18.3	17.4
Carbon conversion (%)	18.5	20.9
Selectivity (%)		
Unsaturates	52.4	59.7
CO	3.6	3.9
CO_2	44.0	36.4
CO_x production rate [(g-atom C) h^{-1}]	0.246	0.262
Unsaturates yield (percentage input C)	9.7	12.5

Ethane Cofeed Option

When ethane is cofed with the methane and oxygen to the oxidative coupling zone of the reactor, it is not possible to distinguish the products from ethane from those from methane. Also, in this case the carbon in ethane can be given no greater value than that in methane. It is thus necessary to redefine the carbon conversion as being the carbon present in the unsaturates and carbon oxides as a percentage of the total carbon in the feed gas, and to define a selectivity to unsaturates that is equal to the carbon in the unsaturates as a percentage of the carbon in unsaturates plus carbon oxides. Such definitions mean that ethane produced by the coupling reaction is *not* included in the product spectrum.

Table 13-5 shows that when these revised definitions of conversion and selectivity are used the fluidized-bed reactor performance using catalyst C improved significantly with 4.7% v/v ethane in the feed compared with that obtained for a methane–oxygen feed alone. The carbon conversion increased from 18.5 to 20.9% and the selectivity to unsaturates increased from 52.4 to 59.7%, resulting in the unsaturates yield being raised from 9.7 to 12.5% of input carbon. The CO_x production rate increased only slightly, from 0.246 to 0.262 (g-atom C) h^{-1}. Similar results with lithium-promoted magnesium oxide catalysts, demonstrating the improvement in reactor performance with ethane in the feed gas, have been reported previously (Edwards, Tyler, and White 1990).

Of the two options for ethane utilization, the pyrolysis option is preferred in terms of overall reactor performance. This is seen by calculating reactor performance for the results in Table 13-5 except that, rather than cofeeding the ethane with the methane, the same quantity of ethane is injected into the oxygen-free zone of the fluidized bed, where it is pyrolyzed with the same conversion and selectivity as shown in Table 13-4. In this case the per-pass carbon conversion to unsaturates and carbon oxides is 28% and the unsaturates selectivity is 70%. The corresponding figures for the cofeed case are 20.9% conversion and 59.7% selectivity. The main reason for the improvement is that in the pyrolysis option the added ethane is not competing with the methane for the available oxygen and hence the methane coupling reaction remains unaffected.

OVERALL PERFORMANCE OF THE CONCEPTUAL OXCO PROCESS

The experimental results outlined in the previous sections of this chapter have, among other things, demonstrated the technical feasibility of the OXCO process concept, which combines methane oxidative coupling with the pyrolysis of higher alkanes in a single fluidized-bed reactor. A computerized mass balance simulation model of the OXCO process has been developed. This mass balance model has been run using experimental data generated to date, producing a preliminary assessment of the overall performance of the OXCO process for natural-gas conversion.

Overall mass balances have been calculated for the conceptual process of Fig. 13-1 using the high-conversion methane coupling data in Table 13-3 and the ethane pyrolysis data in Table 13-4. The calculations are based on the feed gas composition in Table 13-1 assuming the complete separation of the higher alkanes from the methane prior to the conversion reactor and also making the simplified assumption that the propane and heavier components have the same pyrolysis performance as ethane. The mass balance simulation also includes a methanation step where it is assumed that the hydrogen produced from the methane coupling and hydrocarbon pyrolysis steps can be consumed by converting the CO and part of the CO_2 to methane for recycling back to the reactor.

The results are summarized in Fig. 13-16, which shows that the conversion reactor operates with a per-pass conversion of the carbon input (i.e., the feed and recycle methane *plus* the higher alkanes separated from the feed *plus* the ethane recycled from the product separation section) of 29.5% and the overall carbon selectivity of the process to unsaturates is 86%. The product gas, after removal of CO_2 and water, contains 16.8% v/v C_{2+} hydrocarbons, of which almost 75% are unsaturates.

The inclusion of the methanation step allows more than 25% of the carbon oxides produced by methane coupling to be converted back to

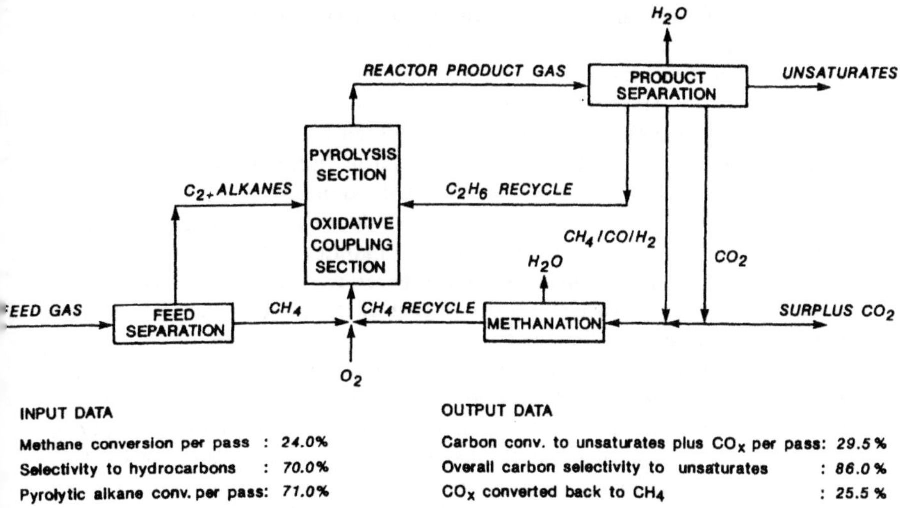

FIGURE 13-16. Overall performance simulation for the OXCO process.

methane for recycling to the coversion reactor. The effect this has on process performance can be seen by the fact that if methanation is not included, the overall unsaturates selectivity declines to 80.5% whereas the per-pass carbon conversion increases slightly to 30%.

Whether these process performance figures can be achieved in practice will obviously depend on the detailed process design and overall process energy balances. The design and optimization of the process in Fig. 13-1 are an important ongoing part of the CSIRO–BHP oxidative coupling research program.

CONCLUDING REMARKS

The CSIRO–BHP methane oxidative coupling research program has identified a new process for the direct conversion of natural gas into liquid fuels and chemicals. This process, known as the OXCO process, combines methane coupling with the pyrolysis of ethane and higher alkanes in a single fluidized-bed reactor to provide an efficient method for utilizing all of the hydrocarbon components present in natural gas. The OXCO process would thus appear to be particularly suitable for the upgrading of remotely located natural-gas reserves.

As part of its catalyst research program, CSIRO has developed a range of active and selective methane coupling catalysts suitable for use in fluidized-bed reactors and these catalysts have been studied extensively in both 30- and 60-mm-diameter reactors operating in the bubbling fluidized-bed regime. Methane conversions and hydrocarbon selectivities of practical significance have been obtained in these reactors, which have at all times operated essentially isothermally, demonstrating the suitability of the fluidized-bed reaction mode for conducting the highly exothermic coupling reactions in a safe and controlled manner. The methane coupling performance of the 60-mm-diameter reactor as found to be at least as good as and in some cases better than that of the 30-mm-diameter reactor when operated under similar conditions.

Axial probing studies in the 60-mm-diameter reactor confirmed the very high activity of the catalysts developed by CSIRO, as evident by the fact that the oxygen was totally consumed within a very short distance above the feed gas distributor plate. Similar experiments have also identified an important mechanism governing the performance of the fluidized-bed reactors, namely, the strong propensity of some of the catalysts to crack C_{2+} hydrocarbons to produce coke, which is deposited on the catalyst and is subsequently combusted to carbon oxides when the catalyst is recirculated to the oxygen-containing zone of the fluidized bed. Improved catalysts having substantially reduced coking propensities have now been developed. Operation of the reactors with these catalysts has successfully demonstrated the technical feasibility of the two major options for ethane utilization, these being (a) pyrolysis in the oxygen-free zone of the bed and (b) cofeeding the ethane with methane and oxygen to the oxidative coupling zone of the reactor.

Idealized mass balance simulations of the OXCO process, using experimental results generated to date, indicate that it has the potential for being a highly selective process for the total utilization of natural gas. Ongoing technical–economic evaluations by the research program's industrial partner, BHP, have confirmed the commercial potential of the OXCO process and have identified areas where further research and development are required. The BHP evaluations have clearly shown that the two factors having the greatest influence on process costs are, first, the need to develop catalysts that are capable of higher methane conversions for a given hydrocarbon selectivity and, second, the need to operate the primary conversion reactor at as high a pressure as possible to minimize the gas compression requirements prior to the cryogenic product recovery section of the process.

Current work at CSIRO is focused both on the further development of inherently improved methane coupling catalysts and on the optimization of their performance in the fluidized-bed reactors. The technical–economic evaluation of the OXCO process has identified pressure as a key variable having a major effect on the process economics. The effect of pressure is being studied in a 30-mm-diameter fluidized-bed reactor that is being operated at pressures up to 1.0 MPa.

ACKNOWLEDGMENTS

The authors gratefully acknowledge contributions to the overall research program by colleagues at other CSIRO divisions, at universities, and The Broken Hill Proprietary Company, Limited. Particular thanks are due to Dr. A.M. Maitra for catalyst development and preparation and to Mr. O. Skrybin for assistance in reactor operation. The CSIRO–BHP oxidative coupling research program is partly funded by the National Energy Research, Development and Demonstration Program, which is administered by the Commonwealth Department of Primary Industries and Energy.

References

Andorf, R., and M. Baerns. 1990. Oxidative coupling of methane in fluidized- and packed-fluidized-bed reactors. *Catal. Today* 6:445–52.

Cant, N.W., C.A. Lukey, P.F. Nelson, and R.J. Tyler. 1988. The rate controlling step in the oxidative coupling of methane over a lithium-promoted magnesium oxide catalyst. *J. Chem. Soc., Chem. Commun.* 766–8.

Chin, V.W.L., A.F. Masters, M. Vender, and R.J. Tyler. 1988. The oligomerization of ethylene produced by partial methane oxidation. In *Studies in Surface Science and Catalysis*, Vol. 36, ed. D.M. Bibby, C.D. Chang, R.F. Howe, and S. Yurchak, pp. 421–5. Amsterdam: Elsevier Science Publishers.

Edwards, J.H., K.T. Do, and R.J. Tyler. 1990. Reaction engineering studies of methane coupling in fluidized-bed reactors. *Catal. Today* 6:435–44.

Edwards, J.H., and R.J. Tyler. 1988. The production of liquid fuels via the catalytic oxidative coupling of methane. In *Studies in Surface Science and Catalysts*, Vol. 36, ed. D.M. Bibby, C.D. Chang, R.F. Howe, and S. Yurchak, pp. 395–401. Amsterdam: Elsevier Science Publishers.

Edwards, J.H., and R.J. Tyler. 1989. The oxidative coupling of methane in a fluidized-bed reactor. *Catal. Today* 4:345–54.

Edwards, J.H., R.J. Tyler, and S.D. White. 1990. Oxidative coupling of methane over lithium-promoted magnesium oxide catalysts in fixed-bed and fluidized-bed reactors. *Energy and Fuels* 4:85–93.

Ekstrom, A. 1990. The oxidative coupling of methane: Reaction pathways and their process implications. In *Direct Methane Conversion by Oxidative Processes: Fundamental and Engineering Aspects*, ed. E.E. Wolf, Chapter 4. New York: Van Nostrand Reinhold.

Follmer, G., L. Lehmann, and M. Baerns. 1988. The application of laboratory-scale catalytic fixed- and fluidized-bed reactors in the oxidative coupling of methane. American Chemical Society Preprints, Div. Pet. Chem., Vol. 33 (3), pp. 452–9.

Ito, T., J.-X. Wang, C.H. Lin, and J.H. Lunsford. 1985. Oxidative dimerization of methane over lithium-promoted magnesium oxide catalyst. *J. Am. Chem. Soc.* 197:5062–8.

Jackson, P.J., S.C. Thompson, and I.G. Whitehead. 1989. Technical and economic evaluation of the production of liquid fuels via the direct partial oxidation of natural gas. In *Proceedings of the Seventh Australasian Chemical Engineering Conference CHEMECA 89*, pp. 719–25. Conberra: The Institution of Engineers, Australia.

Kunii, D. and O. Levenspiel. 1977. *Fluidization Engineering*. New York: Robert E. Krieger Publishing Co.

Mackie, J.C., J.G. Smith, P.F. Nelson, and R.J. Tyler. 1990. Inhibition of C_2 hydrocarbon oxidation by methane under oxidative coupling conditions. *Energy and Fuels* 4:277–85.

Maitra, A.M. 1990a. Determination of solid state basicity of rare earth oxides by thermal analysis of their carbonates. *J. Thermal Anal.* 36:657.

Maitra, A.M. 1990b. Private communication.

Masters, A.F., and S.J. Brown. 1990. Oligomerization of low concentrations of ethene. *Erdöl und Kohle* 43:15–18.

Mimoun, H., A. Robine, S. Bonnaudet, and C.J. Cameron. 1990. Oxypyrolysis of natural gas. *App. Catal.* 58:269–80.

Nelson, P.F., and N.W. Cant. 1990. Oxidation of C_2 hydrocarbon products during the oxidative coupling of methane over a Li/MgO catalyst. *J. Phys. Chem.* 94:3756–61.

Nelson, P.F., C.A. Lukey, and N.W. Cant. 1988. Isotopic evidence for direct methyl coupling and ethane to ethylene conversion during partial oxidation of methane over Li/MgO. *J. Phys. Chem.* 92:6176–9.

Sofranko, J.A., and J.C. Jubin. 1989. Natural gas to gasoline: The ARCO GTG process. In *Symposium on Methane Activation, Conversion and Utilisation, 1989 International Chemical Congress of Pacific Basin Societies*, pp. 152–5. Washington, DC: American Chemical Society.

Tyler, R.J. and C.A. Lukey. 1987. Catalytic partial oxidation of methane to higher hydrocarbons. American Chemical Society Preprints, Div. Fuel. Chem., Vol. 32, pp. 249–54.

14

Direct Conversion of Methane to C_2's and Liquid Fuels: Process Economics

J.L. Matherne and G.L. Culp

Union Carbide Chemicals and Plastics Company, Inc., South Charleston, West Virginia

TABLE OF CONTENTS

INTRODUCTION

This chapter describes the most significant results of the engineering portion of a two-year project that Union Carbide conducted under contract for the United States Department of Energy (US DOE). That contract (number DE-AC22-87PC79817, administered through the US DOE's Pittsburgh Energy Technology Center) covered both research and engineering programs carried out from September 1987 through December 1989. The objective of the research program was to discover and evaluate novel catalytic systems for the conversion of methane either directly or indirectly to liquid hydrocarbon fuels.

The objective of the engineering program discussed herein was to evaluate, from an engineering perspective, the economics of different conceptualized schemes for conversion of methane to liquid hydrocarbon fuels via ethylene and other light olefins, and to provide economic direction to the chemistry program through sensitivity studies of key process variables. The following topics will be addressed in this chapter:

1. Sensitivity of the methane coupling economics to (a) reactor diluent, (b) cofeed versus sequential feed, and (c) methane-to-oxygen feed ratio, for making ethylene on the current catalysts
2. Economics of producing gasoline and distillate fuels via (a) methane coupling and (b) synthesis gas and methanol
3. Sensitivity of the methane coupling economics to (a) methane conversion and (b) selectivity to C_2s, for making ethylene on a hypothetical catalyst.

DISCUSSION

Methane Coupling Process Simplified Block Diagram

Figure 14-1 summarizes the process arrangement that has been used for the methane coupling sensitivity studies. It depicts not only the integration of this process with the Mobil olefins–to–gasoline and distillates (MOGD) process for producing gasoline and distillate fuels, but also the optional refining system that would be required to produce high-purity ethylene.

In the *catalytic reactor system*, fresh and recycled methane and any diluent are contacted with either air or oxygen supplied externally to conduct the oxidative coupling reaction. Both fixed-bed and fluidized-bed reactor systems have been evaluated.

The fixed-bed reactor system consists of a series of shell-and-tube heat exchangers, the tubes packed with a supported catalyst. It has been assumed that stainless steel tubes would provide sufficient corrosion resistance. The large exothermic heat of reaction dominates the reactor design by requiring that a large heat-transfer area be provided. In the fixed-bed reactor system the reaction heat is removed as superheated 600-psig steam, which is generated on the shell sides of the reactors. This is accomplished by spraying in a mist of hot 600-psig condensate, which vaporizes instantly and superheats rapidly to near the process-side temperature. This superheated steam is used to drive condensing turbines in the compression and refrigeration portions of the process.

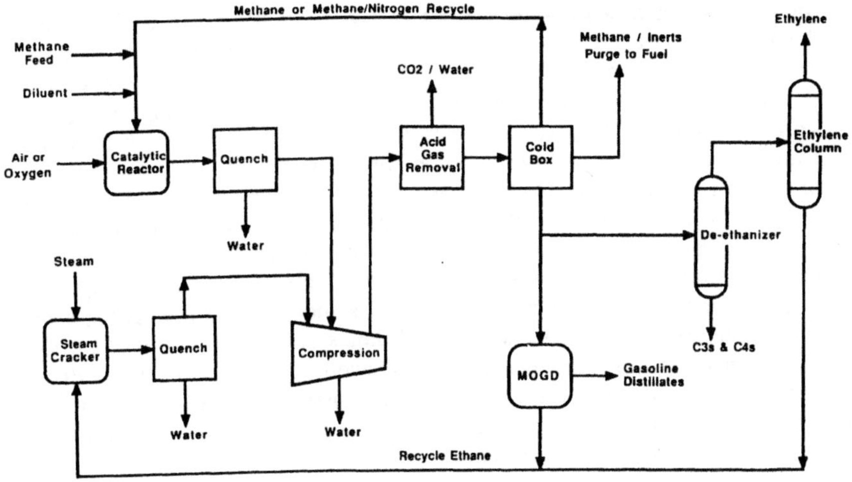

FIGURE 14-1. Methane coupling process simplified block diagram.

Alternatively, a dilute-phase cocurrent downflow reactor design has been used as the model for the fluidized-bed reactor system in this evaluation. The advantage of this system over more conventional fluidized-bed designs is the large catalyst–to–process-flow ratios and the short contact times that are made possible. All of the reactor system vessels are constructed of refractory-lined steel, which confers both temperature and corrosion–erosion resistance. In this system, the large exothermic heat of reaction is removed by spraying hot condensate directly onto the circulating catalyst solids, generating low-pressure superheated steam. This superheated steam is used to drive condensing turbines in the *compression* and *refrigeration* portions of the process.

Whether fixed-bed or fluidized-bed, the reactor feed is brought up to the reactor operating temperature by first cross-exchanging with the hot reactor product stream, followed by additional heating with a direct gas-fired heater. Only a portion of the methane that is fed is converted on each pass, forming mostly ethylene and ethane. The balance of the methane that is converted forms carbon monoxide, carbon dioxide, and C_3 and heavier hydrocarbons.

The reactor product gas is cooled first in the feed cross-exchanger and then is further cooled in a second exchanger to produce hot condensate. The cooled reactor product gas then enters the *quench* portion of the process. This is a large water-cooled partial condenser that cools the gas down to ambient temperature and removes most of the water of reaction as a liquid. From there, the product gas enters the compression system.

The *compression system* consists of four stages of centrifugal compression. The first stage operates at a suction pressure of about 16 psig and handles only the *steam cracker system* product gas. The catalytic reactor product mixes with the steam cracker product at the second stage, which operates at a suction pressure of 40 psig. Water is condensed out of the process stream at each of the compressor intercoolers. The discharge pressure from the fourth stage of the compression system is 300 psig.

The compressed product gas then enters the *acid gas removal system*, which consists of a standard amine-scrubbing system for removal of carbon dioxide from the product gas. This is followed by a molecular sieve dryer system, which removes the last traces of moisture from the product gas. The compressed, scrubbed, and dried product gas then is ready for processing in the cold box system.

The *cold box* feed first enters a series of refrigerated partial condensers, which remove successively larger portions of the stream as liquid. The last of these partial condensers operates at a temperature of less than −95°C (−140°F). These liquids fractions are richer in C_2 and heavier hydrocarbons than the remaining vapor, and are fed to a *methane column* for removal of the light components. The vapor that is left after the final partial condenser then is combined with the vapor off the top of the methane column for further processing in a series of centrifugal expanders. Separators following each of these expanders collect liquids that are rich in C_2s and heavier hydrocarbons. These liquids are recycled to the methane column. A portion of the remaining vapor is removed as a purge from the process, while the rest is recycled back to the catalytic reactor system. The purge gas is then expanded down to 5 psig before it is released from the process. Both the purge and recycle gas streams are heated back up to ambient temperature by cross-exchanging with the gas entering the cold box system. This cross-exchanging reduces the load on the refrigeration system (not shown in Fig. 14-1).

The bottoms product from the methane column in the cold box system consists of C_2 and heavier hydrocarbons. This can be fed directly to the *MOGD system*, where the ethylene, propylene, and a substantial portion of the C_4s are converted to gasoline and distillate fuels. Ethane passes through the MOGD system unconverted.

Alternatively, the bottoms product from the methane column in the cold box system can be fed to a two-column *refining system* for making high-purity ethylene. The *de-ethanizer* separates the C_3 and heavier hydrocarbons from the ethylene and ethane. The overhead make stream from the de-ethanizer is fed to the *ethylene column*. This column employs a vapor recompression system to reduce the refrigeration load on the column condenser. High-purity ethylene (99.85% by volume) is produced as the overhead product, whereas the bottoms product is ethane containing less than 1% ethylene.

Ethane rejected by either the MOGD system or the ethylene column is returned to a conventional *steam cracker system*. About 55% of the ethane that enters the cracking furnaces is converted, with an 85% selectively to ethylene. The rest of the ethane that is converted forms hydrogen, methane, carbon monoxide, carbon dioxide, propylene, and heavier hydrocarbons. The cracked gas exits the furnaces directly into transfer line exchangers, which cool the gas and produce dilution steam for the furnaces.

The furnace gas then enters a *quench* section, where it is further cooled by direct contact with cold water in a large column. The quenched gas then enters the compression system, where it is combined with the product from the catalytic reaction system.

Methane Coupling Case Descriptions

Table 14-1 summarizes the four methane coupling reactor diluent cases that have been evaluated. It was assumed that the maximum amount of oxygen that can be present in the reactor feed is about 10–11% by volume, determined by the explosive limits for the mixture. However, the oxidative coupling reaction consumes only two moles of methane per mole of oxygen. The methane conversion per pass is thereby limited to less than 25% if only methane and oxygen are reacted with an initial oxygen concentration of 10–11%. Therefore, the incentive for adding a diluent is to allow the methane conversion per pass to be increased above what would be possible by feeding methane and oxygen alone.

The *nitrogen diluent* case is the logical starting point for these sensitivity studies because most of the laboratory studies on methane coupling are conducted in nitrogen. The catalyst performance that is reflected in Table 14-1 for the nitrogen diluent case is based on actual Union Carbide laboratory results with a $BaCO_3$ on alumina catalyst with ECl added as a promoter at atmospheric pressure. For this evaluation, the actual observed performance has been extrapolated to what a fully optimized catalyst of this type would ultimately be expected to exhibit. It has been assumed that the reactor pressure can be increased to 50 psig without major impact on the catalyst selectivity. The ratio of methane to oxygen in the feed for this case is 2, which was the midpoint of the range studied. The selectivity to

Table 14-1 Methane Coupling Case Descriptions (Using Nitrogen or Steam as a Diluent Gives High Yield-to-C_2s)

	DILUENT			
	Nitrogen	Methane	Steam	None
Feed System		Cofeed		Sequential
Catalyst		UCC $BaCO_3/Al_2O_3/ECl$		ARCO $Mn/Na/P/SiO_2$
Reactor				
Type		Fixed-bed		Fluidized-bed
Temperature (°C)		750		850
Pressure (psig)		50		50
GHSV[a] (h^{-1})		2133		1200
CH_4 conversion (%)	45	18	45	24
Selectivity to				
C_2H_4 (%)	39	58	39	48
C_2H_6 (%)	11	19	11	14
C_3H_6 (%)	2.2	4.3	2.2	4.6
C_3H_8 (%)	0.7	1.7	0.7	1.5
C_4s + (%)	0.2	—	0.2	10
CO (%)	15	3	15	11
CO_2 (%)	32	14	32	11
C_2 yield (%)	22	14	22	15

[a]Gas hourly space velocity.

C_2s is only 50% for this case, but the methane conversion is 45%, giving a yield to C_2s of 22%. This performance is competitive with any other oxidative coupling catalyst described in the literature. The reactor type is the fixed-bed system described in the previous discussion of the simplified block diagram. An advantage of the nitrogen diluent case is that the oxygen can be added to the process in the form of air. This eliminates the need for an *air separation* system, which is both a high-investment and a high-operating-cost item.

The *methane diluent* case described in Table 14-1 covers the case wherein only methane and oxygen are fed to the catalytic reactor at an initial oxygen concentration of 10–11%. Since the methane is in vast excess to its requirements in the reaction, this excess methane essentially becomes the diluent. The catalyst performance for this case is based on actual Union Carbide laboratory results from the same $BaCO_3$ catalyst that was used for the nitrogen diluent case. Again, the actual observed performance has been extrapolated to what a fully optimized catalyst of this type would ultimately be expected to exhibit. It has been assumed that the atmospheric pressure laboratory results could be translated to 50 psig. The reactor type is fixed-bed. The methane-to-oxygen feed ratio is 7. An air separation plant is needed so that high-purity oxygen can be fed. Note that although the conversion per pass for this case is lower than that for the nitrogen diluent case, the selectivity to C_2s is higher. However, the overall effect is a yield-to-C_2s of only 14%.

The *steam diluent* case considers an alternative to nitrogen as a diluent in the methane coupling process. The catalyst performance for this case has been assumed to be identical to that for the nitrogen diluent case for lack of better information. The primary incentive for the use of steam versus nitrogen as a diluent is that almost all of the steam will condense in the quench step following the catalytic reactor and will not have to be handled in the downstream compression and separation systems. There is some precedent for using steam as a diluent also, because most hydrocarbon cracking operations use steam as a diluent to limit coke formation on the furnace tubes.

For the nitrogen, methane, and steam dilution cases, all of the reactants are mixed at the inlet to the fixed-bed catalytic reactor and enter all at the same time, which is referred to as the *cofeed system*.

In the *sequential feed* system, the methane and oxygen are introduced to the reactor separately, rather than together. This allows the methane conversion per pass to be increased without a diluent, avoiding the explosive mixture limitations of the cofeed cases. To do this, the catalyst is oxidized with air in one step and then reduced in a separate step via the coupling reaction by contacting it with methane. The conversion per pass in this system is limited only by the relative proportions of methane and catalyst present and by the methane residence time in this second reaction zone. The catalyst performance data depicted in Table 14-1 for the sequential feed case is based on literature results (Jones, Leonard, and Sofranko 1987; Sofranko et al. 1987) for an ARCO catalyst system. A 24% methane conversion per pass and 62% selectivity to C_2s have been assumed, giving a 15% yield. Note that this case makes much more heavy C_{4+} hydrocarbons than the cofeed cases. The fluidized-bed reactor design has been assumed for this case, allowing the sequential reaction to be carried out by feeding the methane and air separately but continuously to different points in the catalyst cycle. A fixed-bed sequential feed process is felt to be impractical.

The oxygen conversion in the catalytic reactor has been assumed to be 100% for all cases, versus actual laboratory oxygen conversions of 75–95%. It has been assumed that half of the hydrogen and carbon monoxide that enter the reactor via the methane recycle stream are converted to water and carbon dioxide, respectively. All of the ethane and ethylene that is recycled to the reactor is assumed to be burned to carbon dioxide. Therefore, C_2 recycling is minimized.

All of these cases assume that the ratio of ethylene to ethane in the catalytic reactor product is about 3 : 1. Laboratory studies indicate that using ECl as a promoter can produce this relatively high ratio versus the more typical ethylene to ethane ratio of 1 : 1 that has been reported in the literature without a promoter. Sensitivity studies indicate that achieving this relatively high ratio is important economically, but that still higher ratios become less so. The economic significance of the ethylene to ethane ratio should not be ignored, even though assuming about the same ratio for each of these cases eliminates this variable from this sensitivity analysis.

Methane Coupling Investment Costs

Table 14-2 summarizes the results of the investment estimates that have been prepared for the four methane coupling cases. In each case, facilities are included to produce refined ethylene, but not gasoline and distillate fuel via the MOGD process. The inherent assumption is that the process that is best for producing ethylene will also be the best for making liquid fuels.

All the (*inside battery limits*) (ISBL) investment figures are of $\pm 25\%$ quality. They have been built up from discrete, individual equipment sizes and costs and have only been grouped into process systems to facilitate comparisons among the cases. They are based on 1989 US Gulf Coast nonunion construction costs and include a 15% contingency. The *outside battery limits* (OSBL) investment has been estimated as 40% of the combined ISBL plus air separation investment.

The nitrogen diluent case has by far the highest investment cost of the group. Compared to the other cases, this case is more costly in the compression, acid gas removal, cold box, and refrigeration systems. These systems are all involved in maintaining the large nitrogen diluent cycle, and in separating the excess nitrogen from the unreacted methane. Therefore, although using nitrogen as a diluent may be logical for laboratory studies, a nitrogen diluent full-scale facility is impractical due to the high investment cost.

The methane diluent, steam diluent, and the sequential feed cases all have about the same investment cost, well within the $\pm 25\%$ investment estimate accuracy.

The cost of the catalytic reactor system is most closely related to the amount of heat released in the reaction step. The heat released by the oxidative coupling reaction is much less than that for the nonselective

Table 14-2 Methane Coupling Investment Costs[a] (Nitrogen Diluent Has the Highest Investment Cost)

	DILUENT			
	Nitrogen	Methane	Steam	None
Feed system	Cofeed			Sequential
Catalyst	UCC $BaCO_3/Al_2O_3/ECl$			ARCO $Mn/Na/P/SiO_2$
System				
Catalytic reactor	116	72	113	95
Steam Cracker	25	28	25	26
Compression	149	81	48	77
Acid gas removal	31	14	13	14
Cold box	49	45	26	51
Refining	31	31	31	31
Refrigeration	39	45	38	49
Acetylene hydrogenation	1	1	1	1
Air Compression	66	—	—	36
Total ISBL	**510**	**315**	**295**	**380**
Air separation	—	62	105	—
OSBL	200	153	160	150
Total fixed investment	**710**	**530**	**560**	**530**

[a]All figures are in millions of dollars ($\pm 25\%$) for a unit producing one billion pounds per year (10^9 lb yr^{-1}).

burning reaction. Therefore, the methane diluent case has the lowest catalytic reactor investment, because it has the highest selectivity to C_2s and hence the lowest amount of carbon monoxide and carbon dioxide formed via the burning reaction.

Note that whether a case allows feeding air versus oxygen is not a significant distinguishing feature among these cases. This is because the air compression systems for delivering air to the catalytic reactor are almost equal in cost to the air separation systems.

Methane Coupling Economics

Table 14-3 summarizes the economics of the four methane coupling cases. Consistent with the investment estimates, these economics have been cast on the basis of producing ethylene for sale.

The *total fixed investment* for each case is from Table 14-2. The *working capital* is an estimate of the value of the inventory and accounts receivable

Table 14-3 Methane Coupling Economics[a] (Yield-to-C_2s Is Not a Good General Predictor of Economics)

	DILUTE			
	Nitrogen	Methane	Steam	None
Feed system	Cofeed			Sequential
Catalyst	UCC $BaCO_3/Al_2O_3/ECl$			ARCO $Mn/Na/P/SiO_2$
Total fixed investment	710	530	560	530
Working capital	65	60	70	60
Total utilized investment	**775**	**590**	**630**	**590**
Operating costs				
Methane	146	112	132	124
By-product credits	(12)	(44)	(22)	(50)
Utilities	24	29	56	14
Variable cost	**159**	**97**	**166**	**90**
Fixed cost	53	40	42	41
Total cash cost	**212**	**137**	**209**	**130**
ROIAT[b] (%)	3	14	6	15

[a]All Costs are in millions of dollars per year for a unit producing one billion pounds per year (10^9 lb yr^{-1}).
[b]ROIAT is based on ethylene sales at $0.32 per pound.

that would be involved in a methane coupling operation. The *total utilized investment* is the sum of the total fixed investment and the working capital.

All of the *operating costs* in Table 14-3 have been expressed in terms of millions of dollars for a unit producing one billion (10^9) pounds per year of ethylene. Methane is by far the largest operating cost component. It has been assumed for this analysis that methane is being purchased at the 1989 average US Gulf Coast intrastate delivered price of $2.15 per million (Btu) ($10^6$ Btu), or 5.1¢ per pound. The methane diluent case has the lowest methane usage due to its high selectivity to C_2s.

The *by-product credits* are the sum of the values of the C_3 and heavier coproducts, plus the fuel value of the purge stream of light hydrocarbons. The nitrogen diluent case has a lower by-product credit than the others because the fuel value of its lights purge stream is extremely low. On the other hand, the economics of the sequential feed case is most sensitive to the by-product credits, because of the large amount of C_4 and heavier materials produced in this process.

The *utilities* requirements for the methane coupling cases are mostly the fuel to fire the catalytic reactor preheaters, plus the electrical power to drive the compression, refrigeration, and air separation of air compression systems. However, it has been found that the fuel requirements usually can be satisfied by the inerts purge stream from the cold box. Likewise, credit has been taken for the steam that is produced in the catalytic reactor systems, which is used in turbine drivers and offsets the electrical requirements.

The steam diluent case is distinguished by its high utility cost, because most of the steam generated in the catalytic reactor system must be used as diluent, rather than as motive steam in turbines elsewhere in the process. On the other hand, the sequential feed case is distinguished by its low utility cost. The absence of a diluent that characterizes this case causes the compression and refrigeration system operating costs to be low. Also, air is used to oxidize the catalyst, which eliminates the high operating cost for the air separation system.

The *fixed costs* are mostly the cost of maintenance, which is assumed to be an annual percentage of the total fixed investment. Operating labor is also included.

The *total cash cost* is the sum of the *variable* and *fixed* operating costs. This represents the net cost of producing the product and does not include any ex-plant costs such as distribution, or any marketing, sales, or technical staff support costs.

The *return on investment after tax* (ROIAT) for each case is based on a 1989 average sales price of $0.32 per pound for ethylene. ROIAT is the ratio of the net income after tax (NIAT) to the *total utilized investment*. NIAT is sales income minus cash cost, depreciation, and taxes.

ROIAT can be thought of as the first year's "interest rates" that the facility investment would return to the owners, assuming capacity operation. In this light, the returns from the nitrogen and steam diluent cases are clearly unacceptable. The methane diluent case and the sequential feed cases are approximately equal on an ROIAT basis, both generating an acceptable return.

These returns are all likely to decrease when the price of ethylene returns to more historical levels. As this happens, however, the cases should retain their current relative economic ranking, with the methane diluent and sequential feed cases being clearly superior to either the nitrogen or steam diluent cases.

Referring back to Table 14-1, note that the yield-to-C_2s for both the methane diluent and the sequential feed cases are *lower* than the yields for the nitrogen and steam diluent cases. But Table 14-3 shows that the methane diluent and the sequential feed cases have *better* economics.

The key to understanding this apparently contradictory result is to note that the cases with the lower cash cost have the better economics, and that methane cost is the largest single component of the cash cost. The low methane cost in the methane diluent case is due to its high selectivity to C_2s. The sequential feed case has not only high selectivity to C_2s, but also high by-product credits and low utility costs. The benefits of higher methane conversion and yield-to-C_2s that characterize the nitrogen and steam diluent cases are more than offset by the costs of their lower selectivity to C_2s. Therefore *yield-to-C_2s alone* is *not* a good general predictor of economics.

Although the methane diluent and sequential feed cases are economically equivalent, other factors tip the balance of this overall analysis in favor of the methane diluent case. These noneconomic factors in favor of the methane diluent case are as follows:

- The methane diluent case is based on actual laboratory catalyst results, whereas the sequential feed case is based solely on literature catalyst results.
- The sequential feed case requires a fluidized-bed reactor design, which is inherently more risky from an engineering standpoint than the fixed-bed reactor design that the methane diluent case allows.
- The low cash cost for the sequential feed case is due, in part, to its high coproduct credits, based on the presumption of a ready customer for the heavy C_{4+} stream. No such presumption is necessary for the methane diluent case.

Therefore, the methane diluent case will be used as the preferred model for the methane coupling technology for the balance of this analysis.

Results of Methane Coupling Sensitivity Studies

The sensitivity of the methane coupling economics to the methane-to-oxygen ratio in the feed to the catalytic reactor system is shown in Fig. 14-2. For each of the three graphs in Fig. 14-2, the methane-to-oxygen ratio is the horizontal axis. Moving from left to right, from a low methane-to-oxygen ratio to a high one causes the methane conversion per pass to decrease but the selectivity to C_2s to increase. This effect for the cofeed cases has been quantified on the same $BaCO_3$ catalyst system as has been used elsewhere in this study. The effect of increasing methane conversion on the sequential feed catalyst has been estimated via the literature sources. For this case the methane-to-oxygen feed ratio is the ratio of these components in the two feed streams and is not indicative of the composition of any particular stream in the process.

The graph in the upper left corner of Fig. 14-2 indicates that the total fixed investment (TFI) for the cases varies substantially with methane-to-oxygen ratio. Starting at a ratio of 1, the TFIs for both the steam and the

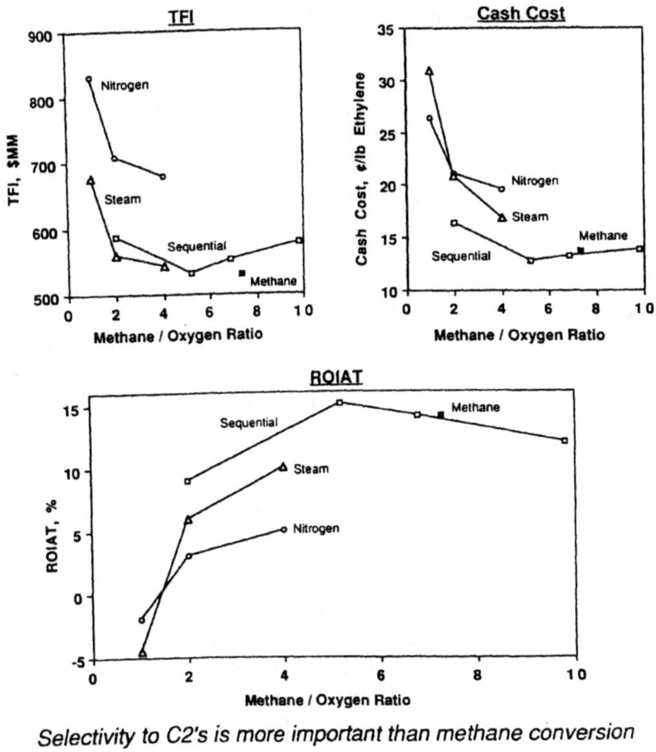

FIGURE 14-2. Results of methane coupling sensitivity studies.

nitrogen diluent systems decrease dramatically with increasing methane-to-oxygen ratio. This is due to the high cost of supplying or recycling the large amounts of diluent needed to produce such a low methane-to-oxygen ratio. The nitrogen diluent case cannot be extended beyond a methane-to-oxygen ratio of about 4 and still have air fed directly to the process. As the methane-to-oxygen ratio is increased beyond 4 in the steam diluent case, both the process itself and its TFI approach that of the methane diluent case. The sequential feed case TFI reaches a minimum at a methane-to-oxygen feed ratio of about 5. Beyond this point, the benefits of higher selectivity to C_2s are overwhelmed by the accompanying cost of lower methane conversion.

The graph in the upper right corner of Fig. 14-2 shows how the *cash cost* of ethylene varies with changes in methane-to-oxygen ratio. The strong sensitivity of the cash cost to selectivity to C_2s discourages operation of any of the cases at a low methane-to-oxygen feed ratio, in spite of the high methane conversion that would result.

The large graph at the bottom of Fig. 14-2 shows how the overall ROIAT for the cases vary with methane-to-oxygen ratio. This graph shows that the economics for the methane diluent case is essentially equal to the best of the sequential feed cases. This graph also underscores the earlier conclusion that across much of the range of conditions that have been studied, selectivity to C_2s is more important economically than methane conversion.

Note that each diluent system has been presented at its midpoint methane-to-oxygen ratio in Tables 14-1–14-3. That is, the results indicated for the nitrogen and steam diluent cases in Tables 14-1–14-3 correspond to a methane-to-oxygen ratio of 2, the methane diluent case is at a methane to oxygen ratio of 7, and the sequential feed case is at a ratio of 5.

Synthesis-Gas–Methanol Process Simplified Block Diagram

The existing technology for producing liquid fuels from methane goes through synthesis gas and methanol as intermediates. The economics of producing liquid fuel by this route have been determined for direct comparison with the methane coupling technology. Figure 14-3 is the simplified block diagram for this synthesis-gas–methanol process.

Methane is first converted to synthesis gas in the *steam reforming* section of the process. The process information for this portion of the process is per SRI International's formulation of an integrated methanol process (SRI International 1983). A large excess of water is used in this system, which is recovered for recycle by condensing it away from the synthesis gas in a heat-integrated quench step.

The synthesis gas is then compressed for delivery to the *methanol synthesis* section of the process. This section is patterned after Imperial Chemical Industries (ICI) low-pressure process by SRI (SRI International 1983). The methanol is condensed away from the unreacted synthesis gas in the second quench step. The methanol goes on to the methanol-to-olefins (MTO) section, while the unreacted synthesis gas is recycled to the methanol synthesis section.

Universal Oil Products' (UOP) *MTO* process (Kaiser 1985) makes both dimethyl ether (DME) and mixed olefins. The unreacted methanol is recovered in a third quench step. The DME is removed in a water absorber and is recycled back to the methanol column. The water of reaction is removed by the methanol column as the bottoms product, and the make is a mixture of methanol, DME, and water, which is recycled to the MTO reactor. The DME is recycled to extinction by converting it to

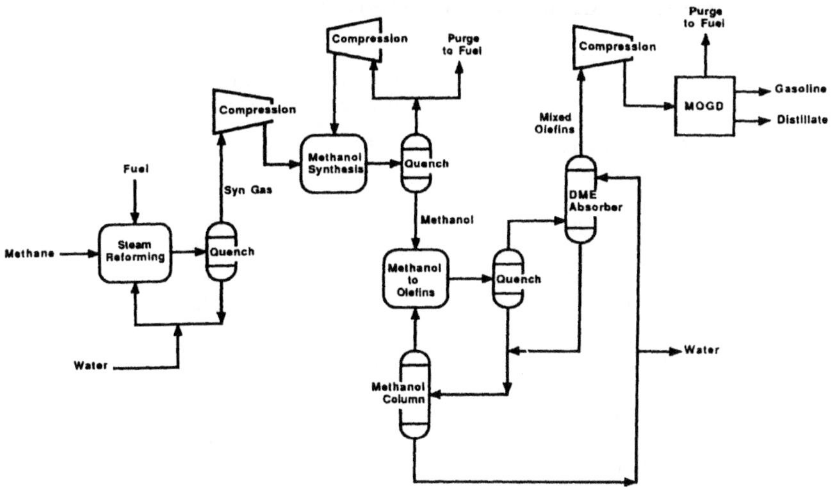

FIGURE 14-3. Synthesis-gas–methanol process simplified block diagram.

additional product. The DME absorber overhead stream contains the mixed olefins product.

The *MOGD* system (Owen et al. 1984) produces gasoline and distillate fuels at a ratio of six to one. A small amount of light hydrocarbons are removed as a purge to fuel.

Gasoline and Distillate Economics

Table 14-4 summarizes the economics of making gasoline and distillate fuels via methane coupling versus the synthesis gas and methanol route. The methane coupling facility does not include the refining system equipment that would be needed to produce high-purity ethylene, but rather is integrated with the MOGD process per the methane coupling simplified block diagram. Both cases are cast on the basis of 1989 US Gulf Coast construction and operating costs, with methane purchased at the intrastate delivered price of $2.15 per million Btu (5.1¢ per pound).

The two routes are similar in that most of the *total fixed investment* is in the methane conversion portion of the processes. The MTO investment is a relatively small component of the synthesis gas and methanol route ISBL investment, and the MOGD investment is a relatively small component of the total fixed investment for both routes. Overall, the investment costs for the two routes are about equal: The difference between them is well within the ±25% investment estimate accuracy.

Table 14-4 Gasoline and Distillate Economics[a] (The Two Routes are Equivalent)

	Methane Coupling	Synthesis Gas and Methanol
OSBL Investment	285	314
Air separation unit	61	—
MOGD	40	35
ISBL investment	154	141
Total fixed investment	**540**	**490**
Working capital	50	46
Total utilized investment	**590**	**536**
Operating Costs		
Methane	101	60
By-product credits	(26)	(2)
Utilities	23	28
Variable cost	**98**	**88**
Fixed cost	38	40
Total cash cost	**136**	**128**
Required sales price[b] (per gallon of product)	$2.09	$1.94

[a]All costs are in millions of dollars per year for a 10,600-bpd unit (10^9 lb of olefins per year).
[b]Required sales price generates a 10% ROIAT.

However the *operating cost* advantage that is shown for the synthesis gas and methanol route is thought to be real. The selectivity to C_2s for the methane coupling route is only about 77%, versus nearly 100% for the competition. This difference in methane efficiency translates directly into a slight but real cash cost advantage for the synthesis gas and methanol route.

Therefore, this route has a slightly lower *required sales price* to produce a 10% ROIAT: $1.94 per gallon versus $2.09 per gallon via methane coupling. This difference is slight when compared to the current wholesale price of liquid fuels of less than $0.60 per gallon. But a low-cost source of methane (such as at a remote location) would make the cost of producing liquid fuels via either of these processes more competitive overall. This would also diminish the impact of the lower methane usage for the synthesis gas and methanol route. Therefore, these two technologies are essentially equivalent economically.

This result is viewed as encouraging, given the early state of development of oxidative coupling catalysis. The synthesis gas and methanol route has had the benefit of many years of development and commercial practice.

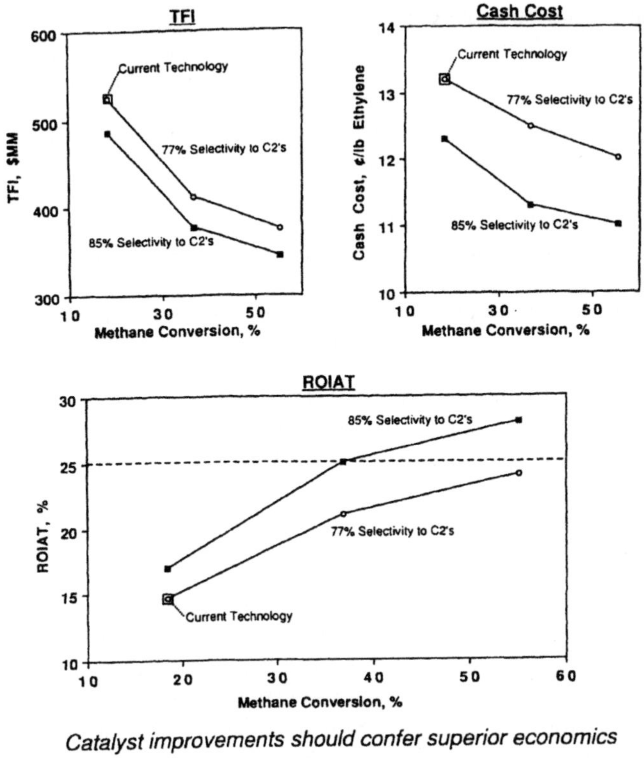

FIGURE 14-4. Hypothetical catalysts performance economics.

The economics of this route are not likely to be improved significantly by further research. In contrast, methane coupling is a comparatively young technology with no current commercial applications. Therefore, it is reasonable to expect that further research might incrementally improve the performance of the current catalyst systems or lead to the discovery of novel catalyst systems with greatly improved performance.

Hypothetical Catalyst Performance Economics

The methane coupling cases considered so far have all been based on current catalyst performance. Figure 14-4 summarizes the economics of hypothetical or future catalyst systems as a function of the methane conversion and selectivity to C_2s using the current technology as the starting point. The current technology is represented by the methane diluent case, which has been shown earlier to be the best of the methane coupling systems that have been considered.

The graph in the upper left corner of Fig. 14-4 shows how the TFI could be reduced from that required for the current technology. If catalyst improvements could be made that would allow the methane conversion to be increased while retaining the current or higher selectivity to C_2s, then the TFI could be reduced from the current $530 million to less than $400 million.

The graph in the upper right corner of Fig. 14-4 shows how the *cash cost* to ethylene could be reduced below that for the current technology. The strong effect of selectivity to C_2s on the cash cost is evident. Even at today's conversion per pass of only 18%, increasing the selectivity to C_2s to 85% would have almost as great an effect on the cash cost to ethylene as would increasing the conversion to 55% at 77% selectivity. Overall, catalyst improvements have the potential for reducing the case cost from the current 13.7¢ per pound to less than 11.5¢ per pound.

The large graph at the bottom of Fig. 14-4 shows how the ROIAT for the methane coupling facility could be improved over that for current technology. The horizontal dashed line at an ROIAT of 25% represents the minimum return that methane coupling would have to offer to attract investment. Any new process must have compelling economics versus the alternatives to justify taking on the associated technical and financial risk. Note that even a methane conversion of 55% does not bring the return up to the economic threshold if the selectivity is 77%. On the other hand, if the selectivity to C_2s could be improved to 85%, then a conversion per pass of 37% or more would generate sufficient return versus the alternatives to attract attention.

The goal is regarded as attainable in the long-term. Therefore, given sufficient time and resources, future catalyst improvements should confer *superior economics* on the methane coupling process.

CONCLUSIONS

1. The preferred methane coupling configuration is as follows:
   a. Methane diluent (versus nitrogen or steam)
   b. Cofeed mode (versus sequential feed mode)
   c. High methane to oxygen ratio (7 : 1), which gives low conversion (18%) but high selectivity (77% to C_2s), (which is the key to success)

These results are for making ethylene the current catalysts. This conclusion is reached in spite of the fact that the other cases that were considered offered a higher yield-to-C_2s. Therefore the yield-to-C_2s *alone* is *not* a good general predictor of economics.

2. The economics are equivalent for producing gasoline and distillate fuels via
   a. Current methane coupling
   b. Synthesis gas and methanol

   which is viewed as an encouraging result given the early stage of development of methane coupling catalysis. However, neither technology produces liquid fuels at anywhere near their current US wholesale price using methane purchased at the current US Gulf Coast price.
3. For a *hypothetical* methane coupling catalyst operating at current conditions, high selectivity to C_2s is more important than high methane conversion. Furthermore, future catalyst improvements should confer *superior economics* on the methane coupling process.

ASSUMPTIONS

Investment Estimate Basis: 1989 US Gulf Coast
Outside battery limits (OSBL) investment: 40% of [inside battery limits (ISBL) + oxygen plant investments]
Working capital: 10% of sales + 20% of [(raw materials) + (utilities costs)]

Ethylene value: 32¢ per pound (used to compare among oxidative coupling cases)
Propylene value: 21¢ per pound (used to compare among oxidative coupling cases)
Propane value: 21.6¢ per gallon (5.11¢ per pound)
Crude C_4s value: 25¢ per gallon (5.13¢ per pound)

Methane (fuel gas) price: $2.15 per million Btu (5.13¢ per pound)
Electricity cost: $43.00/Mkhw

Period costs:

Operating labor: $325M per operating position
Maintenance: 4% of total fixed investment
Laboratory: 20% of operating labor
Location overhead: 50% of [maintenance + (operating labor)]

Depreciation: 10% of total fixed investment
Taxes: 37% of [(operating income) − depreciation]

Listed major equipment: Total major equipment cost based on heat and material balance quantities
Unlisted major equipment: 12.5% of listed major equipment
Direct material: 89% of (listed + unlisted major equipment)
Indirect material: 22% of (listed + unlisted major equipment)
Direct labor: 64% of (listed + unlisted major equipment) (based on 1989 US Gulf Coast nonunion construction)
Indirect labor: 50% of direct labor
Subcontracts: 15% of (listed + unlisted major equipment)
Site administration: 16% of direct labor
On-site cost: sum of listed major equipment and all the above
  Engineering: 28% of on-site cost
  Contingencies: 15% of [(on-site cost) + engineering]
Total installed cost: [(On-site cost) + engineering + contingencies] = 5.5 × (listed major equipment)

The methane raw material is assumed to be of the following composition:

Methane—99.0 mol%
Ethane—0.5 mol%
Nitrogen—0.4 mol%
Carbon dioxide—0.1 mol%

Oxygen produced by an air separation unit is assumed to be of the following composition:

Oxygen—97.5 mol%
Nitrogen—2.5 mol%

The oxygen conversion per pass in the oxidative coupling reaction has been assumed to be 100% in all cases.

References

Jones, C.A., J.J. Leonard, and J.A. Sofranko. 1987. *J. Catal.* 103:311.
Kaiser, S.W. 1985. US Patent No. 4,499,327, to Union Carbide Corporation.
Owen, H., S. K. Marsh, and B. F. Wright 1984. US Patent No. 4,456,779, to Mobil Oil Corporation.
Sofranko, J.A., J.J. Leonard, C.A. Jones, A.M. Gaffney, and H.P. Withers. 1987. American Chemical Society Preprints, Petrol. Div. Vol. 32, p. 770.
SRI International. 1983. Costs of synthesis gases and methanol. PEP Report No. 148.

15

Engineering Evaluation of Direct Methane Conversion Processes

James C.W. Kuo

Mobil Research and Development Corporation, Paulsboro, NJ

TABLE OF CONTENTS

INTRODUCTION

Methane is the most stable hydrocarbon. Conventional conversions of methane into other useful chemicals, such as methanol and higher hydrocarbons, require an initial formation of synthesis gases ($H_2 + CO + CO_2$) at very high temperatures (800–1400°C). The most widely used process for such a conversion is the *steam reforming* process:

$$CH_4 + H_2O \Leftrightarrow 3H_2 + CO \quad \Delta H_f(870°C) = 54 \text{ kcal} \quad \Delta G_f = -15 \text{ kcal} \tag{15-1}$$

$$CO + H_2O \Leftrightarrow H_2 + CO_2 \quad \Delta H_f(870°C) = -8 \text{ kcal} \quad \Delta G_f = 0 \text{ kcal} \tag{15-2}$$

Equation (15-1) is the main reaction and is highly endothermic. The further conversion of synthesis gas is illustrated by the formation of methanol:

$$2H_2 + CO \Leftrightarrow CH_3OH \quad \Delta H_f(270°C) = -24 \text{ kcal} \quad \Delta G_f = 8 \text{ kcal} \tag{15-3}$$

$$3H_2 + CO_2 \Leftrightarrow CH_3OH + H_2O \quad \Delta H_f(270°C) = -14 \text{ kcal} \quad \Delta G_f = 12 \text{ kcal} \tag{15-4}$$

We classify those processes that convert methane into oxygenates or higher hydrocarbons via synthesis gases as *indirect methane conversion*.

Recently, substantial research activity has been conducted in the area of methane conversion without the use of synthesis gases. We classify such processes as *direct methane conversion*. They have the potential of being more energy-efficient because they bypass the energy-intensive step of synthesis gas formation.

The number of direct conversion processes is very large. A logical classification of those processes will help us to appreciate the scope of these processes. Kuo (Kuo 1987) proposed the following classification based on the chemistries of the processes:

1. Direct partial oxidation (to methanol, methyl halides, or others)
   a. Using oxygen
   b. Using nitrous oxide
   c. Using halogens
   d. Using superacids as catalysts
2. Oxidative coupling (to ethylene)
3. Direct conversion to acetylene and other higher hydrocarbons
   a. To acetylene via thermal pyrolysis
   b. To acetylene using electric arc or plasma
   c. To acetylene via partial oxidation
   d. To other higher hydrocarbons
4. Other direct conversion processes
   a. Alkylation with hydrocarbons
   b. Activation using metals, metal oxides, or metal complexes
   c. Conversion using coal or shale

Readers who are interested in literature on these processes are referred to recent surveys (Kuo 1987; Baerns, van der Wiele, and Ross 1989).

The total number of direct conversion processes is very large even if they are restricted to oxidative ones. Due to limitations of time and space, only the following two were selected for engineering evaluation after preliminary screening:

- Direct partial oxidation to methanol (DPOM);
- Oxidative coupling to ethylene (OCE).

The factors used in this screening were process economics, availability of process data, and status of process development.

In DPOM, methane is oxidized to methanol in an oxygen-deficient atmosphere at moderate temperature (350–500°C) and high pressure according to the following main reaction:

$$CH_4 + \tfrac{1}{2}O_2 = CH_3OH \qquad \Delta H_f(400°C) = -30\text{ kcal} \qquad \Delta G_f = -22\text{ kcal} \tag{15-5}$$

which is moderately exothermic. Some formaldehyde also can be formed and both methanol and formaldehyde are further oxidized to form CO_2, H_2O, and CO, which are called the deep oxidation products. Generally, no elementary H_2 or carbon is produced and the product gas may or may not contain any unconverted oxygen. Moderate temperatures and high pressures are essential in promoting methanol formation.

The OCE reactions take place using various reducible metal oxides as oxygen carriers as well as catalysts at moderate to high temperatures (500–900°C) and relatively low pressure (1–30 atm) and are represented by the following two main reactions:

$$2CH_4 + (1/y)MO_{x+y} = C_2H_6 + H_2O + (1/y)MO_x \qquad (15\text{-}6)$$

$$C_2H_6 + (1/y)MO_{x+y} = C_2H_4 + H_2O + (1/y)MO_x \qquad (15\text{-}7)$$

Deep oxidation also occurred, resulting in CO, CO_2, and H_2O formation. The reducible oxides are reoxidized with oxygen according to the following reaction:

$$MO_x + (y/2)O_2 = MO_{x+y} \qquad (15\text{-}8)$$

Combination of reactions (15-6) and (15-7) with reaction (15-8) gives

$$CH_4 + \tfrac{1}{4}O_2 = \tfrac{1}{2}C_2H_6 + \tfrac{1}{2}H_2O \qquad \Delta H_f(800°C) = -21 \text{ kcal}$$

$$\Delta G_f = -14 \text{ kcal} \qquad (15\text{-}9)$$

$$C_2H_6 + \tfrac{1}{2}O_2 = C_2H_4 + H_2O \qquad \Delta H_f(800°C) = -25 \text{ kcal}$$

$$\Delta G_f = -45 \text{ kcal} \qquad (15\text{-}10)$$

These two reactions also represent the performance of a reactor into which oxygen is cofed with methane. They are moderately exothermic.

For both processes, the deep oxidation is thermodynamically favorable (Kuo 1987). The heats of reaction for oxidizing methane to methanol, formaldehyde, ethane, ethylene, CO, and CO_2 are 61, 66, 83, 67, 82, and 95 kcal per O_2, respectively.

Since the 1980s, very extensive research activities have been dedicated to the direct conversion processes, involving hundreds of professionals throughout the world. Among more than a thousand publications, only very few deal with any engineering aspect of the processes (Edwards and Foster 1986; Dautzenberg, Garten and Klingman 1986; Fox, Chen, and Degen 1990; Geerts, Hoebink, and van der Wiele 1990). This author strongly feels that after devoting so much research activities to study the chemistry of these processes it is now time to study some engineering aspects of some of these processes. Results of such a study would provide practical targets for achieving commercial application and would identify potential high-cost areas. This author also feels that it is premature to conduct detailed process and economic studies at the present time because there are insufficient data to define the related reactors and processes properly.

The following engineering parameters were examined:

1. Reactor conditions
   a. Selectivities and conversions from methane and other compounds
   b. Reactor type
   c. Temperature and partial pressure of reactants
2. Recycling of unconverted methane
   a. To achieve acceptably high total conversion
   b. To achieve proper reactor temperature control
3. Separation of products
   a. To concentrate the desirable products
   b. To remove undesirable products

The selectivity and conversion from methane are probably two of the most important issues in process analysis and are discussed separately in the next section. No attempt was made to define the reactor space velocity except that data with unusually low space velocities were excluded from further consideration.

In addition to comparing the two direct conversion processes, it was very important to compare them against indirect conversion processes that have been in commercial practice for many decades. The following were emphasized in performing process comparisons:

- Net thermal efficiency
- Heat-transfer duty [per-product low heating value (LHV)]
- Gas compression duty (per-product LHV)
- Gas expansion duty (per-product LHV).

Definitions of net and chemical thermal efficiencies as well as other guidelines needed for calculating process duties are summarized in Appendix 15-A. The net C-efficiency is the other efficiency often used in comparing methane conversion processes. Although methane has a H–C ratio of 4, all desirable products, such as methanol, olefins, gasoline, or diesel fuel, have substantially lower H–C ratios (1.9 to 2). Such a large imbalance in H–C ratios points to the importance of achieving high C-efficiencies for all methane conversion processes. However, it is often sufficient to consider only the thermal efficiency because the C-efficiencies are usually closely related to thermal efficiencies. Definitions of net and chemical C-efficiencies are also given in the Appendix.

Although the net thermal efficiency is a useful criterion for process comparison, the three process duties are also needed. The major cost of an industrial plant is usually associated with the facilities for moving heat and mass. In this sense, the use of the heat-transfer duties is straightforward. The major cost of mass transfer is the moving of the gaseous streams, that is, gas compression and expansion. The moving of the liquid streams consumes little energy, and there is negligible movement of solids in the current studies.

It is important to understand the ultimate thermal and C-efficiencies of methane conversion reactions. These limits are defined with respect to specific feeds and the final product or products. Table 15-1 summarizes those limits, as well as the minimum heat-transfer duties, of making methanol, ethylene, and 1-decene. In making methanol, methane and air at room temperature and atmospheric pressure react in a black-box factory to produce methanol at the same conditions. The ultimate C-

Table 15-1 Ultimate Limits of Methane Conversion Reactions

Reactions	Max. Carbon Eff. (% Feed)	Max. Thermal Eff. (% Feed LHV)	Min. Heat-Transfer Duty (% Prod. LHV)
Formation of CH_3OH			
$CH_4 + \frac{1}{2}O_2(air) = CH_3OH(l)$	100	80	26
$CH_4 + 0.225CO_2 + 0.451H_2O(l) + 0.162O_2(air) = 1.225CH_3OH(l)$	123	98	0
Formation of C_2H_4			
$CH_4 + \frac{1}{2}O_2(air) = \frac{1}{2}C_2H_4 + H_2O(l)$	100	82	28
$CH_4 + 0.262CO_2 + 0.107O_2(air) = 0.631C_2H_4 + 0.738H_2O(l)$	126	104	0
Formation of $C_{10}H_{20}$			
$CH_4 + \frac{1}{2}O_2(air) = 0.1C_{10}H_{20}(l) + H_2O(l)$	100	77	37
$CH_4 + \frac{1}{3}CO_2 = \frac{4}{30}C_{10}H_{20}(l) + \frac{2}{3}H_2O(l)$	133	103	1

efficiency is 100% whereas the ultimate thermal efficiency, given in LHVs, is 80%. A minimum amount of heat, equivalent to 26% of the methanol LHV, must be transferred out of the system. When there is CO_2 available, the C-efficiency is improved to 123% by maintaining an exact energy (HHV) balance on both sides of the second reaction in Table 15-1. The ultimate thermal efficiency is improved to 98% and the minimum heat-transfer duty becomes nil. These improvements clearly demonstrate the importance of CO_2 availability on methane conversions. This benefit of CO_2 availability can be claimed by indirect conversion processes via CO_2 addition during synthesis gas formation. Based on thermodynamics, the two direct conversion processes cannot claim this benefit.

The other four reactions in Table 15-1 represent the production of hydrocarbons instead of methanol. Except for the case of producing liquid hydrocarbons (exemplified by $C_{10}H_{20}$), production of ethylene and liquid hydrocarbons slightly increases the ultimate efficiencies. Note that the last reaction consumes no O_2.

SELECTIVITY, CONVERSION, AND YIELD

The selectivities and conversions mentioned in the majority of the literature on the direct conversion were single-pass selectivities and conversions. The selectivities were often defined based on the C atoms in the converted methane whereas the conversions were defined based on the C atoms in the feed methane. The yields, which usually implied single-pass yields of useful C atoms, were defined as follows:

$$\text{yield (of useful C atoms)} = \text{selectivity (of useful C atoms)} \times CH_4 \text{ conversion} \qquad (15\text{-}11)$$

One, however, must establish an understanding between the overall and the single-pass selectivities and between the total and the single-pass conversions.

For direct conversion processes, the single-pass conversion is usually too low to be practical. High total conversions are usually achieved during recycling of streams containing unconverted methane. On the contrary, a recycle operation always results in decreasing overall selectivity due to a loss of useful C atoms in product separation.

Unfortunately, the single-pass yield defined by equation (15-11) was often used inappropriately as a criterion for catalyst and/or process selections. This important point is best illustrated by DPOM cases of 10%

yield but of different selectivities:

	Single-Pass Selectivity (C Atom%)	Single-Pass Conversion (C Atom%)	Minimum Heat Rejected (%Prod. LHV)	Molar O_2/CH_3OH
Case I	100	10	26	0.5
Case II	80	12.5	61	1.0
Case III	40	25	235	3.5

In these cases the undesirable products are CO_2 and H_2O. The results show that the 40% selectivity case rejects excess heat as 2.35 times the product heating value versus 0.26 for the 100% selectivity case. Similarly, the oxygen consumption per methanol for the 40% selectivity case is 7 times that of the 100% selectivity case. Furthermore, the reactor temperature control for the low-selectivity case is expected to be much more difficult than that for the high-selectivity case. Because both heat removal and the supply of oxygen can be very capital-intensive, the high-selectivity case is much more preferable than the low-selectivity case. Of course, some of the rejected heat may be recovered as surplus energy, as in steam or electricity, for export. However, the plant would then become a costly combustor. In the case of plants in remote areas for utilizing remote natural gas, export of surplus energy would be impractical.

In conclusion, a single-pass yield is an incorrect criterion for both catalyst and process selection and an oxidative direct conversion process becomes practical only when it has a selectivity higher than 80–85%.

A similar analysis for total conversions is not as clear-cut because it depends very much on the degree of difficulty in separating the useful products from the reactor effluents. In this aspect, the DPOM is better than OCE because methanol can be separated easily from the DPOM reactor effluent by cooling and condensation whereas the separation of C_{2+} from CO, CO_2, and unconverted methane is difficult and expensive. Consequently, the competitive single-pass conversion of a DPOM process should be lower than that of an OCE process.

ENGINEERING EVALUATION OF DIRECT PARTIAL OXIDATION TO METHANOL

Process Description

After examining the available data of methane DPOM, it was concluded that the extensive work done by Professors Gesser and Hunter and their colleagues at the University of Manitoba (Hunter et al. 1984; Gesser,

Table 15-2 Highlights of Direct Partial Oxidation Data (From the University of Manitoba)[a]

1. Noncatalytic system (cold-flame combustion) or catalytic system (SnO_2)
2. 300–500°C and 50–65 atm
3. 10.5-mm inside diameter glass-lined reactor
4. Reaction temperatures decreasing by 100°C with higher methanol selectivity due to the presence of C_{2+} paraffins in feed
5. Methanol selectivities up to 90% and methane conversion up to 15%
6. Increasing selectivity with decreasing conversion
7. Only patent showing negligible formaldehyde formation under noncatalytic system

[a]Hunter et al. 1984; Gesser, Hunter, and Prakash 1985; Hunter et al. 1985; Gesser, Hunter, and Morton 1986; Hunter et al. 1986; Yarlagadda et al. 1987; Gesser et al. 1987; Hunter et al. 1987; Yarlagadda et al. 1988; and Gesser, Hunter, and Morton 1989.

Hunter, and Prakash 1985; Hunter et al. 1985; Gesser, Hunter, and Morton 1986; Hunter et al. 1986; Yarlagadda et al. 1987; Gesser et al. 1987; Hunter et al. 1987; Yarlagadda et al. 1988; Gesser, Hunter, and Morton 1989) gave the most optimistic selectivity and conversion. Highlights of their work are summarized in Table 15-2. All data were obtained using microreactors. Furthermore, the only published results of duplicating their work showed that the best selectivity was less than 70% (Burch, Squire, and Tsang 1989). More work will be needed to confirm their best data.

Figure 15-1 shows a schematic flow diagram of a natural-gas DPOM process. A computer program describing the process operation was constructed. It included the recycle with a specified CO_2 removal, the DPOM reactor mass and heat balances, and product separation. A natural gas

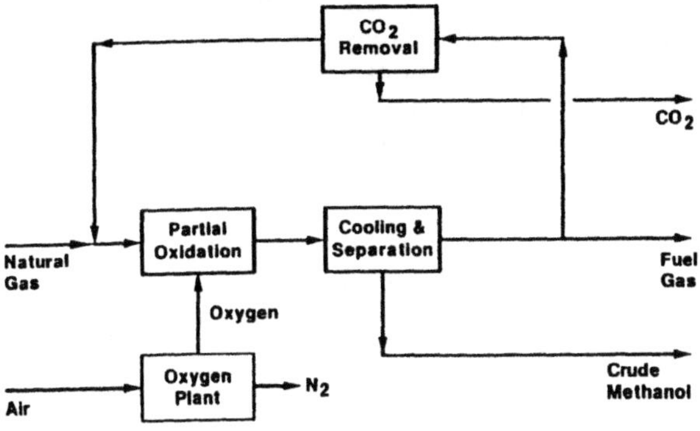

FIGURE 15-1. Schematic flow diagram of a natural gas direct partial oxidation process.

containing 97.2 mol% CH_4, 1.6% C_2H_6, 0.7% CO_2, and 0.5%N_2 was used throughout this and later studies. This composition was similar to the one used by Hunter and co-workers (Hunter et al. 1985) after removing water and C_{3+}.

The natural gas was first compressed and then combined with the recycle gas and oxygen. The oxygen at 98% purity was produced in a typical oxygen plant and was compressed to the same pressure as that of the combined gas. The total gas, which had O_{2+} hydrocarbon partial pressure of 65 atm, was fed into a DPOM reactor. According to data reported by Gesser and co-workers (Hunter et al. 1985; Gesser et al. 1987; Yarlagadda et al. 1988), most runs with high methanol selectivities seemed to occur at 50–65 atm. The high-purity oxygen was needed to avoid very high total pressure.

The design of the DPOM reactor could not be defined at the present time. Both noncatalytic and catalytic reactors were mentioned. For either reactor, a common essential feature was an appropriate reactor temperature control. Although a reactor of adiabatic type could not be completely ruled out, its likelihood was judged to be very small because of large reaction heat. To judge properly the feasibility of using an adiabatic reactor, one needed to determine the minimum reaction initiation temperature and the maximum operating temperature in a scale-up reactor. With reactors of nonadiabatic design, various means of removing reaction heat could be employed.

Throughout this study an operating temperature of 350°C for the DPOM reactor was used. This relatively low temperature was operable for natural gases containing some ethane (Gesser, Hunter, and Morton 1986). Unfortunately, no quantitative relationship between the ethane contents and the operating temperature was ever given. However, using a somewhat higher reactor temperature, such as 450°C, was not expected to alter significantly the results of this study as long as the product selectivity remained unchanged. Use of lower temperature with ethane-containing feed resulted in a higher methanol selectivity.

We assumed that no formaldehyde was formed in the DPOM reactor. This was a simplification but was expected to affect little our conclusions. Formaldehyde could be adopted as a valuable by-product or could be fed together with methanol into a ZSM-5 reactor to be converted into gasoline (MTG). However, the effect of mixing formaldehyde with methanol as feeds to an MTG-type reactor is yet to be investigated. Several sets of methanol selectivity were used in this study. In all cases, a 50/50 split of CO/CO_2 in products was assumed. Furthermore, the CO and ethane conversion rates were assumed to be 5 times that of the methane conversion according to first-order kinetics. In reality, accurate CO and ethane

Table 15-3 Natural Gas Direct Partial Oxidation To Methanol

Major Features	
Natural–gas feed containing 97.2 mol% CH_4, 1.6% C_2H_6, 0.7% CO_2, and 0.5% N_2	
65 atm $CH_4 + O_2$ partial pressure	
Single-pass 90% CO_2 removal in the recycle loop	
98% purity oxygen	
Higher reaction rates for CO and C_2H_6 oxidation	
Base Case (DPOM-1)	
7.5% single-pass conversion and 90% methanol selectivity	
Isothermal reactor at 350°C	
Sensitivity Cases	
Case DPOM-2:	15% single-pass conversion
Case DPOM-3:	15% single-pass conversion and 80% methanol selectivity
Case DPOM-4:	Adiabatic reactor with 300°C inlet and 450°C exit temperatures and without CO_2 removal in the recycle loop

conversion rates was unnecessary as long as they are higher than the methane conversion rate because CO and ethane were essentially completely converted at high total methane conversion.

The DPOM reactor effluent was cooled to 37.8°C and its liquid and vapor streams were separated in a separator. A large portion of the gaseous stream must be recycled to the DPOM reactor in order to achieve a high CH_4 conversion. In later process scoping studies, CO_2 removal in the recycle loop was found to be essential in avoiding extremely high total system pressure and very high recycle ratio in achieving high CH_4 conversion. Conventional CO_2 removal processes, such as the hot carbonate process, could be used here. The small amount of methanol carried in the recycle was expected to be recovered together with CO_2. It could be separated from the CO_2 removal unit as a separate product stream. Recycle-loop pressure drops of 4.4–5.9% of the reactor inlet pressure were used in the calculations. Table 15-3 summarizes the major features of the DPOM process model.

Process Scoping Studies

Process scoping studies were conducted to understand better the effect of process parameter variations on process performance. As demonstrated in Table 15-4 for a 95% total CH_4 conversion, CO_2 removal in the recycle loop was essential for reducing the recycle ratio and the total system pressure. The effect became smaller when the single-pass removal rate was larger than 20%. However, a 90% single-pass CO_2 removal in all process

Table 15-4 Natural-Gas Direct Partial Oxidation to Methanol—Process Scoping (95% Total CH_4 Conversion)

Single-Pass CO_2 Removal[a] (%)	Molar Recycle Ratio	Reactor Inlet Pressure (atm)	Single-Pass CH_4 Conversion[b] (%)	Molar Recycle Ratio	Adiabatic Rise[c] (°C)
0	9.2	103	20.0	5.0	584
20	7.4	87	15.0	7.2	463
90	7.2	85	7.5	15.6	261
100	7.1	84			

[a]Increasing single-pass CO_2 removal reduces recycle ratio and inlet pressure.
[b]Increasing single-pass CH_4 conversion results in reducing recycle and increasing reaction heat removal.
[c]Assuming 65.6°C inlet temperature.

calculations was used because the cost difference between 20 and 90% removal was rather small. Another important process parameter was the single-pass CH_4 conversion. Scoping studies showed that increasing single-pass conversion greatly reduced the recycle ratio required to achieve the same total conversion while increasing the difficulty in reactor temperature control as reflected in increasing adiabatic temperature rises (see also Table 15-4).

A more complete picture of the effect of the single-pass conversion and the molar recycle ratio on the total CH_4 conversion, the reactor inlet pressure, and the adiabatic ΔT is given in Figs. 15-2 and 15-3. The major conclusions were as follows:

- High total CH_4 conversions could be achieved at reasonable single-pass conversions and recycle ratios.
- Higher recycle ratios could greatly reduce the reactor adiabatic ΔT but also increase the reactor pressure due to greater accumulation of inert compounds in the recycle.
- Very large recycle ratios were needed to maintain an adiabatic ΔT below 200°C.

Results of Process Calculations

Table 15-3 also lists the conditions defining a *base case*, designated as DPOM-1. In this base case, process conditions were given to provide reasonable process benefits. The 90% methanol selectivity at 7.5% conversion used here was within the upper range of data reported by Hunter and co-workers (Hunter et al. 1985). Later, this base case was found to be potentially better than the indirect conversion cases. Table 15-3 also lists

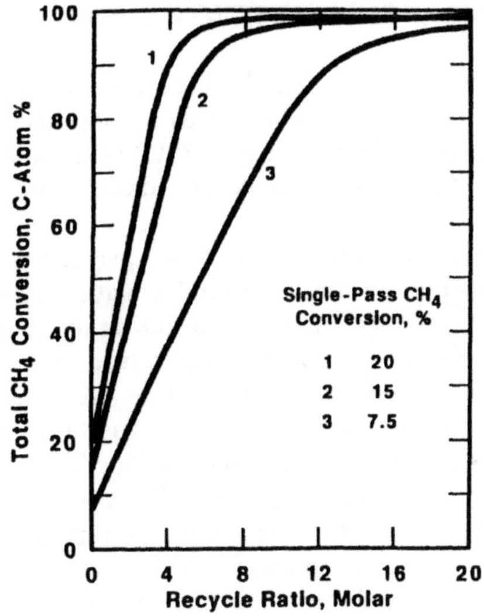

FIGURE 15-2. Natural gas direct partial oxidation: Effect of recycle on total conversion.

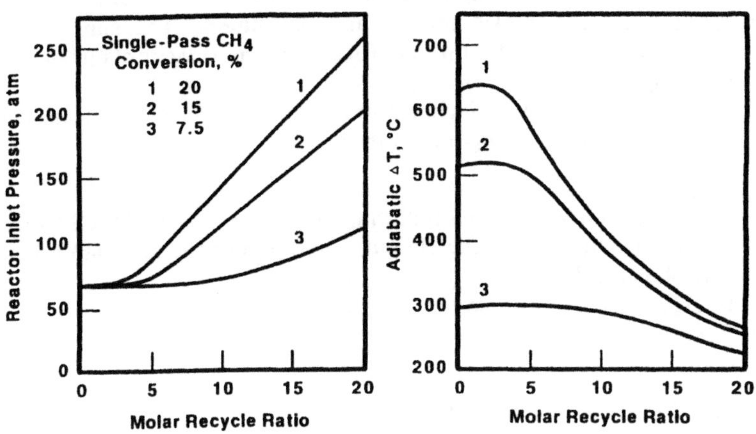

FIGURE 15-3. Natural gas direct partial oxidation: Effect of recycle on reactor pressure and ΔT.

three sensitivity cases, DPOM-2–DPOM-4, containing variations of conditions.

Case DPOM-2 was varied from the base case by increasing the single-pass conversion from 7.5 to 15%. This was an optimistic case and showed the effect of increased conversion. Case DPOM-3 used 80% selectivity and 15% conversion, which were close to the data reported by Gesser and co-workers.

Case DPOM-4 was the same as the base case except that an adiabatic reactor was used instead of an isothermal reactor and there was no CO_2 removal in the recycle loop. In this case, the reaction heat was removed solely using the recycle. An inlet temperature of 300°C and an exit temperature of 450°C were assumed. There were no data supporting operation of such a temperature range. Because CO_2 has a high molar heat capacity, no CO_2 removal in the recycle loop resulted in a lowering of the recycle ratio for the same adiabatic ΔT. A water scrubbing column, however, was added after the separator to minimize the amount of methanol being recycled back to the DPOM reactor. Estimated methanol loss in the recycle gas was about 1% of the total methanol produced after using a 5 molar water–methanol in the scrubber. Further reduction of this loss could be made if desired.

For all cases, the purge gas from the DPOM reactor recycle loop was used as fuel to generate superheated steam using the saturated steam generated from the DPOM reactor and its effluent. The superheated steam was then used in the turbine-drivers for the oxygen plant and for the gas compressors.

The following items were included in the thermal efficiency calculations (in approximately descending importance):

- Methanol
- Steam generated from the following heat sources—the DPOM reactor and its effluent; the recycle purge gas; part of the feed natural gas (case DPOM-4 only)
- Energy consumed for oxygen production
- Energy consumed for gas compression except that of the oxygen plant.

A thermal energy loss was estimated based on a difference between the feed natural gas LHV and the sum of the preceding items. The estimated heat-transfer duties included the following:

- DPOM reactor effluent cooling
- The steam-driven turbine condensate coolers
- The oxygen plant

- Heat removal from the DPOM reactor (except for cases DPOM-1 and DPOM-4)
- The steam superheater
- The compressor precoolers

The estimated gas-compression duties included the following:

- The oxygen plant
- Oxygen
- The process natural-gas feed
- The recycle gas
- Air to the steam superheater

There was no gas-expansion duty. Purge gas of the DPOM reactor recycle loop was the only source from which any expansion work might be recovered. However, its duty was too small to warrant an installation of a gas expander.

Table 15-5 Natural-Gas-to-Methanol via Direct Partial Oxidation: Process Efficiencies and Duties

	CASES			
	DPOM-1	DPOM-2	DPOM-3	DPOM-4
Partial oxidation reactor				
Reactor type	Isothermal	Isothermal	Isothermal	Adiabatic
Inlet temperature (°C)	52.4	52.4	52.4	300.0
Outlet temperature (°C)	350.0	350.0	350.0	450.0
Pressure (atm)	96.7	99.4	108.0	150.0
Single-pass CH_4 conversion (%)	7.5	15.0	15.0	7.5
CH_3OH selectivity (C-atom%)	90.0	90.0	80.0	90.0
Molar recycle ratio	18.0	9.0	10.0	29.5
Total C conversion (%)	96.7	97.3	97.5	98.8
Thermal products (%feed LHV)				
Crude methanol	69 (73)[a]	70 (73)[a]	63 (68)[a]	70 (70)[a]
Steam	19 (—)	18 (—)	24 (—)	15 (—)
O_2 production	−6 (—)	−6 (—)	−8 (—)	−6 (—)
Compression work	−8 (—)	−8 (—)	−9 (—)	−10 (—)
Loss	26 (27)	26 (27)	30 (32)	31 (30)
Total	100 (100)	100 (100)	100 (100)	100 (100)
C efficiencies	87 (91)	88 (91)	78 (85)	88 (87)
Major process duties (%methanol LHV)				
Heat transfer	83 (76)	72 (66)	100 (87)	124 (125)
Compression	21	20	26	23

[a]Numbers in parentheses are values after credit of surplus steam.

Results of process calculations of the base case are given in Table 15-5. A 96.7% total C conversion was obtained using a recycle ratio of 18. With such a high recycle flow, no reaction heat removal from the reactor was needed. In fact, the total reactor feed must be preheated to 119°C by the reactor effluent in order to maintain a reactor temperature of 350°C. The estimated crude methanol thermal efficiency was 69%. Crediting a small surplus of steam increased the efficiency to 73% and reduced the heat-transfer duty from 83% of methanol LHV to 76%. The adjustment of the gas-compression duty was negligible.

Results of process calculations of case DPOM-2 are also given in Table 15-5. With an increase of the single-pass CH_4 conversion from 7.5 to 15%, the recycle ratio could now be reduced to 9.0 while maintaining about the same total C conversion as that of the base case. With reducing recycle flow, no feed preheating was needed and a significant amount of reaction heat was recovered from the DPOM reactor. The heat recovery (steam generation) was slightly less efficient than that of the base case. However, this decrease in steam generation was almost exactly compensated by a decrease in energy required for the recycle compression. Consequently, there was little difference in the thermal efficiencies. There was a small reduction in the heat-transfer duty from that of the base case due to less recycle flow. There was little difference in the gas-compression duties.

A large decrease in the thermal efficiency was estimated when the methanol selectivity was reduced from 90% of case DPOM-2 to 80% of case DPOM-3 (see also Table 15-5). Higher selectivity of carbon oxides resulted in a 23% increase in oxygen consumption and a 35% increase in steam generation. The energy for gas compression was also increased due to higher oxygen consumption. There were substantial increases in both the heat-transfer and the gas-compression duties.

Although the operability of an adiabatic reactor design for the DPOM reactor was highly uncertain, case DPOM-4 was studied in order to understand the magnitude of the potential difficulty. A 29.5 recycle ratio was needed to maintain an adiabatic temperature rise from 300 to 450°C. Such a large recycle flow resulted in (see also Table 15-5) the following:

- Higher reactor pressure
- Less steam generation
- Higher gas-compression work
- Large increase in heat-transfer duty
- Small increase in gas-compression duty

Large recycle flow resulted in large heat degradation and low heat content in the purge gas, both of which contributed to less steam generation. In fact, case DPOM-4 was the only case that required diversion of a portion of natural-gas feed for steam generation.

In summary, the DPOM case studies led to the following conclusions:

- The base case (7.5% single-pass conversion and 90% selectivity using an isothermal reactor) gave 73% thermal efficiency.
- Increasing the single-pass conversion from 7.5 to 15% resulted in little change in thermal efficiency and small decreases in the heat-transfer and compression duties.
- Decreasing the selectivity from 90 to 80% reduced the efficiency to 68% and resulted in about 30% increase in both the heat-transfer and the compression duties.
- An adiabatic reactor required a large recycle for reactor temperature control, resulting in a moderate drop in the efficiency and a large increase in the heat-transfer duty.

ENGINEERING EVALUATION OF OXIDATIVE COUPLING TO ETHYLENE

Process Description—A Base Case (Case OCE-1)

After examining all the data of methane OCE, we concluded that the work done by ARCO Chemical (Jones, Leonard, and Sofranko 1984a–1984g; Withers 1984; Breder, Jaecker, and Johnson 1985; Gaffney 1985a, 1985b; Johnson 1985; Jones, Leonard, and Sofranko 1985a–1985c; Jones et al. 1985; Jones and Sofranko 1985a–1985d; Maffia 1985; Sofranko 1985; Sofranko and Jones 1985; Sofranko and Withers 1985; Withers and Sofranko 1985; Jaecker 1986; Jaecker and Johnson 1986; Jones, Leonard, and Sofranko 1986a, 1986b; Jones et al. 1986; Withers 1986; Jones, Leonard, and Sofranko 1987a–1987c; Jones and Sofranko 1987; Josefowicz 1987; Labinger and Ott 1987; Mazurek 1987; Sofranko 1987; Sofranko, Leonard, and Jones 1987; Withers et al. 1987a, 1987b; Sofranko et al. 1988; Leyshon and Bader 1989) was most extensive and gave one of the most optimistic selectivity and conversion values. Highlights of the work are summarized in Table 15-6. Because there were considerable uncertainties in experimental details, independent experiments to confirm the best data of this work are essential.

Process schemes for OCE were not as straightforward as those for DPOM. The OCE reactor produced a product stream containing a low concentration of ethylene. This ethylene must be further converted to liquid fuels by a "condensation reaction." In the meantime, ethane and unconverted methane must be separated and then recycled back to the

Table 15-6 Highlights of Oxidative Coupling Data From ARCO Chemical[a]

1. More than 38 patents
2. Reducible metal oxides as catalysts and oxygen carriers ($15Mn/5Na_4P_2O_7/SiO_2$ is the best)
3. 750–900°C and 1–30 atm claimed although majority of experiments at 1 atm
4. 14-mm inside diameter fixed-bed quartz reactors used in majority of experiments
5. Less olefins produced as a result of higher pressure operation
6. C_{2+}, especially olefins, in the feed converted at substantially higher rate than methane and yielded more coke and carbon oxides
7. Five modes of operations claimed:
   Fixed-bed cyclic operation
   Fixed-bed O_2 cofeeding operation
   Two fluid beds with one for reoxidation
   Moving-bed operation
   Molten-salt reactor operation
8. Few experiments of air cofeeding operation conducted and demonstrated about 5% less in C_{2+} selectivities than those of cyclic operation
9. Few data on catalyst aging

[a]Jones, Leonard, and Sofranko 1984a–1984g; Withers 1984; Breder, Jaecker, and Johnson 1985; Gaffney 1985a, 1985b; Johnson 1985; Jones, Leonard, and Sofranko 1985a–1985c; Jones et al. 1985; Jones and Sofranko 1985a–1985d; Maffia 1985; Sofranko 1985; Sofranko and Jones 1985; Sofranko and Withers 1985; Withers and Sofranko 1985; Jaecker 1986; Jaecker and Johnson 1986; Jones, Leonard, and Sofranko 1986a, 1986b; Jones et al. 1986; Withers 1986; Jones, Leonard, and Sofranko 1987a–1987c; Jones and Sofranko 1987; Josefowicz 1987; Labinger and Ott 1987; Mazurek 1987; Sofranko 1987; Sofranko, Leonard, and Jones 1987; Withers et al. 1987a, 1987b; Sofranko et al. 1988; and Leyshon and Bader 1989.

OCE reactor. Scoping studies were done on two process schemes, one with the C_{2+} fractionation before the condensation reaction and the other with the C_{3+} fractionation after the condensation reaction. The latter scheme was judged to be better than the former because of easier separation. Figure 15-4 shows a schematic flow diagram of the latter scheme. A computer program describing the process operation was con-

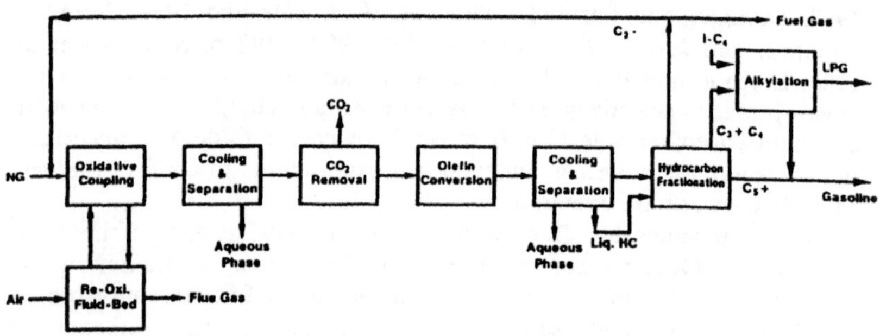

FIGURE 15-4. Schematic flow diagram of an oxidative coupling process.

structed. It included the recycle, the mass and heat balances of OCE, reoxidation, olefin conversion and alkylation reactors, the CO_2 removal, and hydrocarbon separation. The CO_2 removal and the hydrocarbon fractionation were handled using specified separation factors for various components.

ARCO Chemical quoted five modes of OCE reactor operation (see Table 15-6). Except for one study case, the one with two fluid beds was chosen because it did not need an oxygen plant and its operation, which was similar to a fluid catalytic cracking (FCC) operation, was more practical than the other. Recently, OCE process data based on fluid-bed reactors began to appear (Edwards and Tyler 1989; Edwards, Do, and Tyler 1990; Andorf and Baerns 1990). Unfortunately, none of them was based on the two-fluid-beds mode. For this study, the conversion and selectivity of a cyclic fixed-bed mode were adopted for this mode of operation. Furthermore, in one sensitivity case, a mode of using a single fluid-bed with oxygen cofeeding was also adopted. The C_{2+} selectivity of the cofeeding mode was assumed to be about 5% lower than that of the cyclic fixed-bed mode operation at the same conversion level (Jones, Leonard, and Sofranko 1987a). This sensitivity case was designed to see if the oxygen cofeeding operation would given significantly different performance from the other nonoxygen cases.

A base case, designated as OCE-1, was defined to consist of fluid-bed OCE and reoxidation reactors (Fig. 15-4). The process data used in the process design were highly speculative because many assumptions were used. However, when possible, optimistic assumptions were made to show the best side of the process.

The natural gas was first compressed and then combined with the C_2-recycle gas. The combined gas was fed into an OCE reactor together with the oxidized catalyst from a reoxidation reactor. The oxidized catalyst was assumed to carry with it a small amount of inert gas (0.02 mol N_2 for each mole of O_2 it carried). Both reactors were operated at 850°C and 31 atm. This reactor pressure was determined by an assumed required pressure of 34 atm for a downstream hydrocarbon fractionation unit. At this pressure the partial pressure of $C_1 + C_2$ at the OCE reactor inlet was much higher than the atmospheric pressure used in the majority of ARCO's experiments but lower than the maximum pressure of 30 atm claimed in ARCO's patents. The pressure of the reoxidation reactor must be about the same as that of the OCE reactor to maintain proper catalyst flow between them.

The single-pass CH_4 conversion was set at 20% and the reactivity of CO oxidation was assumed to be 5 times of that of CH_4. Because no hydrocarbon reactivity relative to CH_4 was available at the time of this study, the ethane oxidation reactivity was arbitrarily assumed to be 10 times that of CH_4. Furthermore, complete ethylene and C_3 conversion was also assumed. Recently, Labinger and Ott (Labinger and Ott 1987) reported that the relative oxidation reactivities (relative to methane, which is 1) of ethane, ethylene, and propane are 1.9, 1.4, and 3, respectively. The assumed higher C_{2+} reactivities resulted in less accumulation of these hydrocarbons in the recycle. In actuality, the relative reactivities of CO and higher hydrocarbons had little impact on the overall performance of all cases as long as they were higher than that of methane. In contrast, the selectivities of converted C_{2+} were important.

Selectivity of converted methane was the most important process data. Table 15-7 summarizes this selectivity together with selectivities of converted ethane, ethylene, propane, and propylene. The selectivity of methane conversion was adopted from work by Jones and co-workers (Jones, Leonard, and Sofranko 1987b) based on a 15% Mn/5% $Na_4P_2O_7/SiO_2$ catalyst at 850°C and 1 atm (Table 3 of the reference). The C_{2+} selectivity of 78% was in line with the selectivity versus conversion relationship reported for the same catalyst (Jones, Leonard, and Sofranko 1987a, 1987b). Except for ethane, all hydrocarbon products were either useful or very reactive for further oxidation. The ethane selectivity of 14.2% was indeed one of the lowest ever reported for the OCE process.

Table 15-7 Product Selectivities of Hydrocarbon Oxidative Coupling (Case OCE-1; Basis—Moles)

	CONVERTED HYDROCARBON[a]				
Component	Methane	Ethane	Ethylene	Propane	Propylene
Oxygen	−0.743	−0.585	−0.850	−0.733	−1.575
CO	0.105	0.030	0.000	0.045	0.000
CO_2	0.105	0.030	0.500	0.045	0.900
H_2O	1.171	1.080	0.700	1.330	1.350
Methane	−1.000	0.060	0.000	0.000	0.000
Ethylene	0.240	0.900	−1.000	0.000	0.000
Ethane	0.071	−1.000	0.000	0.000	0.000
Propylene	0.020	0.000	0.433	0.890	−1.000
Propane	0.000	0.000	0.000	−1.000	0.000
Butenes	0.010	0.000	0.000	0.000	0.413
Pentenes	0.002	0.000	0.000	0.000	0.000
Benzene	0.007	0.000	0.000	0.000	0.000
Toluene	0.001	0.000	0.000	0.000	0.000
Coke (as C)	0.010	0.080	0.200	0.240	0.450

[a]Selectivity of CH_4 OCE is from Table 3 of Jones, Leonard, and Sofranko 1987b. All other data are assumed.

Typical numbers were 20–30%. Furthermore, this selectivity was obtained at 1 atm whereas the $C_1 + C_2$ partial pressure of the base case was 17 atm. According to thermodynamics (Kuo 1987) and results of Labinger and Ott (Labinger and Ott 1987), the ethane selectivity was expected to increase substantially with increasing pressure.

Selectivities for other converted hydrocarbons (Table 15-7) were highly speculative. 90% of the ethane was assumed to convert to ethylene with the balance of the C atoms going to coke and carbon oxides. The small amount of ethylene, propane, and propylene in the recycle gas were assumed to convert to propylene and butenes at selectivities ranging from 55 to 89%. These selectivities were also optimistic.

In making heat balance calculations, the reaction heat of the OCE reactor was assumed to be nil. The combined feed was preheated to the reaction temperature by the OCE reactor effluent and the reoxidation reactor. Because of this preheating, the reoxidation reactor must be at a temperature higher than that of the OCE reactor. Air, at 15% excess, was compressed and fed to the reoxidation reactor, resulting in complete combustion. The hot flue gas was partially cooled to recover its sensible heat and then expanded in an expander to recover useful work.

The OCE reactor effluent was eventually cooled to 37.8°C and an aqueous stream was separated. The gaseous phase was first compressed to a pressure so that the pressure before the downstream hydrocarbon fractionation unit was at 34 atm. Similarly to the DPOM cases, a CO_2 removal unit was used to remove 90% of the CO_2 from this stream. This CO_2 removal resulted in reducing the recycle flow and increasing the olefin concentration before an olefin conversion unit to condense light olefins into higher hydrocarbons.

After partial CO_2 removal, the remaining gas was fed to the olefin conversion unit. This unit was not yet well-defined because little work was done for a stream containing such a low light olefin concentration (4.2 mol%). To minimize recycling of ethylene and to maximize the liquid hydrocarbon yield, high light olefin conversions in this reactor were essential. Unfortunately, none of the limited data available gave high liquid hydrocarbon selectivities at high conversion. Voltz and Wise (Voltz and Wise 1976) reported exploratory results on the conversion of ethanol over a two-stage dehydration/ZSM-5 reactor system showing gasoline yield similar to those obtained from Mobil's MTG process. Because ethylene was the first product from ethanol dehydration, these results implied that the gasoline yields from ethylene conversion over ZSM-5 catalysts could be approximated by those of MTG processes. Therefore, the best MTG yield, that is, that of a fluid-bed MTG process (Avidan et al. 1986), was adopted as the yield for the olefin conversion unit. The detailed yield of the fluid-bed MTG process is proprietary and is not reported here. This yield structure showed very high efficiency in making gasoline because less than 5% of the C atoms in the light olefins was lost as C_3 paraffins.

An important issue related to the olefin conversion reactor that could not be addressed was the minimum required olefin partial pressure in its feed. Theoretically, a minimum partial pressure must be met in order to achieve high conversions and high gasoline yields. Unfortunately, no such guideline was available. Consequently, the feed olefin partial pressures of all OCE cases were calculated but not used as a process constraint.

The olefin conversion reactor was assumed to be operated at 400°C. The reactor effluent, after cooling to 37.8°C, was sent to a hydrocarbon fractionation unit to be separated into three fractions (i.e., C_{2-}, $C_3 + C_4$, and C_{5+}). This fractionation was a major processing step and detailed process calculations were very cumbersome. Consequently, a simplified approach that was satisfactory for the current purpose was adopted. The C_{2-} fraction was assumed to contain all N_2, CO, CO_2, and C_{2-} together with 25% of propylene and 15% of propane of the feed to this unit. These separation factors were estimated from the streams of a hydrocarbon fractionation unit of a Fischer–Tropsch (F–T) case in a DOE report by Schreiner (Schreiner 1978) in which a refrigeration unit (to −43°C) was used. The $C_3 + C_4$ fraction was sent to a conventional alkylation unit whose yield was approximated by reacting propylene and butenes with isobutane to form i-C_7s and i-C_8s. Because the given yield of the fluid-bed MTG contained more light olefins than isobutane for alkylation, a small amount of isobutane was imported. The alkylate was blended into the gasoline pool together with the remaining C_{4+}.

A major portion of the C_{2-} fraction was recycled to the OCE reactor to achieve a high total conversion. In this case, a recycle ratio of 8.0 was used to achieve a 97.7% total C conversion. The total pressure drop across the recycle loop was guesstimated to be about 10 atm. The estimated $CH_4 + C_2$ partial pressure at the OCE reactor inlet was 17 atm and the light olefin partial pressure at the inlet of the olefin conversion reactor was 1.7 atm (on a C_2H_4 basis). The feed to the olefin conversion reactor must be preheated to above 325°C in order to maintain a 400°C reaction temperature using the reactor effluent and one other heat source. The olefin concentration in that feed was 4.2 mol%.

Process Description—Sensitivity Cases (Cases OCE-2 to OCE-6)

Five sensitivity cases of natural-gas OCE were evaluated to examine ways of improving the base case. This task was essential because, as shown later, case OCE-1 was substantially inferior to the DPOM cases reported earlier, particularly with respect to the process duties. The other objective was to evaluate the sensitivity of some process parameters.

Case OCE-2 was deviated from the base case by increasing the C_{2+} selectivity from 78.0 to 88.3% while reducing the single-pass conversion from 20 to 10%. These selectivity and conversion numbers were in line with the selectivity versus conversion relationship reported by ARCO (Jones, Leonard, and Sofranko 1987a, 1987b). The total product selectivities were obtained by proportionally adjusting the hydrocarbon selectivities upward and the carbon oxide selectivities downward. Because of the reduction of the single-pass conversion, the recycle ratio was increased from 8.0 to 15 to maintain about the same total conversion. Increasing the recycle flow also resulted in the following changes:

- An increase in the $CH_4 + C_2$ partial pressure at the OCE reactor inlet from 17 to 19 atm.
- A significant reduction in the light olefin concentration in the olefin conversion reactor feed from 4.2 to 2.5 mol% with a corresponding reduction in the light olefin partial pressure from 1.7 to 1.0 atm (on a C_2H_4 basis).

A recycle loop pressure drop the same as that of case OCE-1 was used.

Case OCE-3 was a minor modification of case OCE-2. Some process improvements were expected if one could minimize the cooling of the OCE reactor effluent and the reheating of the feed gas to the olefin conversion reactor. The gas compression and the CO_2 removal unit, which were located before the olefin conversion reactor, must now be moved to a location after the cooling and separation of the olefin conversion reactor effluent. The OCE reactor effluent was now cooled to 360°C and fed directly into the olefin conversion reactor. At 360°C, the feed to the olefin conversion reactor contained all the steam formed in the OCE reactor. This large amount of steam might cause permanent aging of the zeolite catalyst used in the olefin reactor. There was only a slight reduction (to 2.3 mol%) of the olefin concentration in the feed to the olefin conversion reactor. However, the light olefin partial pressure in the olefin conversion reactor was greatly reduced (to 0.73 atm). The total pressure drop across the recycle loop was guesstimated to be 9.2 atm.

Because the optimistic C_{2+} selectivity of methane OCE used in this study was established experimentally at 1 atm, a case of lower OCE reactor pressure was evaluated. In case OCE-4, the OCE and the reoxidation reactor pressures were reduced from 31 atm used in cases OCE-1–OCE-3 to 13 atm. All other downstream operations remained the same as those of cases OCE-1 and OCE-2. There was a large pressure letdown when the recycle gas was combined with the OCE reactor fresh feed and a gas expander was used to recover this expansion work.

In this case, the reduction of the OCE reactor pressure resulted in a $CH_4 + C_2$ partial pressure of 11 atm at the reactor inlet. The recycle ratio was also reduced from 15 in case OCE-2 to 10 in OCE-4, resulting in substantial decrease in the compression and expansion duties. The penalty was a reduction of the total conversion from 97 to 91%. The estimated light olefin concentration at the olefin conversion reactor inlet was 3.5 mol% whereas the corresponding partial pressure was 1.4 atm (on a C_2H_4 basis).

Case OCE-5 followed case OCE-4 with the following variations:

- The pressure at both the OCE and the re-oxidation reactors were reduced to 4.1 atm.
- The single-pass conversion was raised to 35%.
- The pressure requirement after the OCE unit was determined by a 1.7-atm olefin partial pressure (on a C_2H_4 basis) at the olefin conversion reactor inlet.
- The requirement of a minimum 34 atm of the gaseous feed stream to the hydrocarbon fractionation unit was removed.

The objective of this case study was to show the targeted process conditions under which the OCE process could become efficient with these process variations. The high single-pass conversion and C_{2+} selectivity adopted in this case were completely hypothetical. In reality, the products at 35% single-pass conversion were mainly carbon oxides.

In this case, the recycle ratio was drastically reduced to 3.5 at a 97.6% total C-conversion. The low OCE reactor pressure resulted in a low $CH_4 + C_2$ partial pressure (2.6 atm) at the OCE reactor inlet. With lower recycle ratio, the olefin concentration at the olefin conversion reactor inlet increased drastically to 10.0 mol%. The total pressure drop across the recycle loop was guesstimated to be 4.7 atm.

Case OCE-6 was the same as case OCE-5 except for replacing the fluid-bed OCE and reoxidation reactors with an oxygen cofeeding fluid-bed reactor. An oxygen plant must now be added. 5% lower C_{2+} selectivity than that of the base case was used according to ARCO's work (Jones, Leonard, and Sofranko 1987a). Nil coke formation was assumed. Among many of ARCO's patents on cofeeding, none used high-purity oxygen. However, to minimize N_2 accumulation in the recycle and to achieve easier product fractionation, use of high-purity oxygen was essential. Similarly to DPOM cases, 98% purity oxygen was used.

In this case, the same recycle ratio as that of case OCE-5 was adopted, achieving a 97.4% total C-conversion. The total pressure drop across the recycle loop remained the same while the olefin concentration at the olefin conversion reactor inlet was essentially unchanged.

Results of Process Calculations

To simplify process calculations of all OCE cases, energy consumption and process duties required for the hydrocarbon fractionation unit were not accounted for. This simplified approach was allright as long as the OCE cases were inferior to the other cases. In reality, energy consumption and process duties for this unit could be quite significant. In the base case, separation factors based on a similar hydrocarbon fractionation unit of a reference F–T case (Schreiner 1978) were used. In the reference case, a refrigeration unit (to −43°C) was needed to cool the feed gas stream to achieve a 74% C_3 recovery. Similar separations in all OCE cases were judged to be more difficult than that of this reference case. This judgment is based on the fact that the C_{3+} concentrations and partial pressures of all OCE cases were much lower than those of the reference case, as shown in Table 15-8. The table also shows that, in all cases, the gaseous streams before the fractionation contained substantial amounts of C_{3+}, which implied that the fractionation unit is a must.

The following items were included in the thermal efficiency estimation:

- C_{4+} gasoline, including a debit of imported isobutane
- Steam generated from the following heat sources: the reoxidation reactor and its flue gas (except for case OCE-6); the OCE reactor effluent (except for case OCE-3); the olefin conversion reactor effluent

Table 15-8 Difficulty in C_{3+} Hydrocarbon Fractionation in Oxidative Coupling Cases

Case	Single-Pass Conversion (%)	Molar Recycle Ratio	Pressure (atm)	C_{3+} Before Fractionation (mol%)	$P_{C_{3+}}$ (atm)	Wt% C_{3+} in Gaseous Phase[b]
F–T[a]	—	—	17.9	5.47	0.98	—
OCE-1	20	8.0	34.0	0.96	0.33	52
OCE-2	10	15.0	34.0	0.67	0.23	65
OCE-3	10	15.0	34.0	0.67	0.23	65
OCE-4	10	10.0	34.0	0.86	0.29	57
OCE-5	35	3.5	13.9	2.13	0.30	42
OCE-6	35	3.5	14.3	2.08	0.30	43

[a]This reference case is an F–T case given in a DOE report (Schreiner 1978) in which a refrigeration unit (to −43°C) is used to recover C_{3+}.
[b]Before fractionation.

- Energy consumed for gas compression except that of the oxygen plant
- Energy recovered from gas expansion except that of the oxygen plant
- Energy consumed for oxygen production (case OCE-6 only)
- C_3 LPG
- By-product fuel gas

A thermal energy loss was estimated based on a difference between the feed natural-gas LHV and the sum of the preceding items. The estimated heat-transfer duties included the following:

- OCE reactor effluent cooling
- Reoxidation reactor effluent cooling (except for case OCE-6)
- Olefin conversion reactor effluent cooling
- The steam-driven turbine condensate coolers
- The oxygen plant (case OCE-6 only)
- Heat removal from the OCE reactor (case OCE-6 only)
- The compressor precoolers.

The estimated gas-compression duties included the following:

- Air for the reoxidation reactor (except for case OCE-6)
- The recycle gas
- The oxygen plant (case OCE-6 only)
- Oxygen (case OCE-6 only)
- Natural gas (except for cases OCE-4–OCE-6).

The estimated gas-expansion duties included the following:

- The flue gas from the reoxidation reactor (except for case OCE-6)
- The recycle gas (cases OCE-4–OCE-6 only)
- Natural gas (cases OCE-5 and OCE-6 only)
- The fuel gas (case OCE-4 only).

Results of process calculations of the base case (case OCE-1) are given in Table 15-9. The estimated C_{3+} thermal efficiency was 56% with 2% of by-product fuel gas. The steam generated plus the expansion work recovered created a large surplus (14% of the natural-gas LHV) of energy that must be exported or disposed of. Including the fuel gas, the total surplus energy amounted to 16% of natural gas. Credit of this surplus energy increased the thermal efficiency to 63%. Similar adjustment of the heat-transfer duty reduced it from 213% of C_{3+} LHV to 186%. The adjustment of the gas-compression duty was negligible and no adjustment was needed for the gas expansion. These adjustments resulting from crediting the

Table 15-9 Natural-Gas–to–Gasoline via Oxidative Coupling: Process Efficiencies and Duties

	CASES[a]					
	OCE-1	OCE-2	OCE-3	OCE-4	OCE-5	OCE-6
Oxidative coupling reactor (850°C)						
Pressure (atm)	31.6	31.6	31.0	13.3	4.1	4.1
Single-pass C conversion (%)	20.0	10.0	10.0	10.0	35.0	35.0
Molar recycle ratio	8.0	15.0	15.0	10.0	3.5	3.5
Total C-conversion (C-atom%)	97.7	97.2	97.2	91.2	97.6	97.4
Product Selectivities (C-atom%)						
C_{2+}	78.0	88.3	88.3	88.3	88.3	85.0
C_2H_6 in hydrocarbons	18.1	18.1	18.1	18.1	18.1	18.3
Carbon in hydrocarbons	1.3	1.3	1.3	1.3	1.3	0.0
Olefin conversion reaction	—	—	Cascaded	—	—	—
Thermal products (%feed LHV)						
C_{4+} Gasoline	53 (63)	59 (63)	59 (65)	56 (63)	59 (73)	58 (67)
C_3 LPG	3 (4)	3 (4)	3 (4)	3 (3)	3 (4)	3 (4)
Fuel gas	2 (—)	3 (—)	3 (—)	9 (—)	3 (—)	3 (—)
Steam	31 (—)	21 (—)	24 (—)	19 (—)	26 (—)	25 (—)
Expansion work	11 (—)	10 (—)	9 (—)	14 (—)	7 (—)	4 (—)
Compression work	−28 (—)	−27 (—)	−27 (—)	−31 (—)	−17 (—)	−10 (—)
O_2 production	0	0	0	0	0	−8 (—)
Loss	28 (33)	31 (33)	29 (31)	30 (34)	19 (23)	25 (29)
Total	100 (100)	100 (100)	100 (100)	100 (100)	100 (100)	100 (100)
C_{3+} C-efficiencies	74 (88)	83 (88)	83 (91)	78 (88)	83 (102)	81 (94)
Major process duties (%C_{3+} LHV)						
Heat transfer	213 (186)	253 (248)	215 (205)	232 (227)	124 (97)	119 (99)
Compression	51	43	43	53	28	30
Expansion	20	15	15	25	11	8

[a] Numbers in parentheses after credit of surplus steam and fuel gas. Energies for product fractionation are not accounted for. Using fluid-bed MTG yields (97.9 C-atom% C_{3+}) for olefin conversion.

surplus energy were very substantial. One should, however, remember that this large energy export implied that the plant was used as an expensive utility generator and the large adjustments gave an impression of high efficiencies that might not be warranted, especially for a remote plant.

Comparing case OCE-1 with the DPOM base case, case OCE-1 was inferior in all aspects. However, this comparison was inexact because these two cases were not making the same products. To make a proper comparison, a fluid-bed MTG unit must be attached to the end of the DPOM process to convert the crude methanol into C_{3+}. Unfortunately, no precise data on the net thermal efficiency and process duties of this MTG process were available. Approximate rules for facilitating these process calculations were suggested later. The expected corrections of the thermal efficiencies and the process duties of the DPOM cases were minor and would not affect the conclusions of their comparison with OCE cases.

Case OCE-2 was designed to investigate the effect of increasing the C_{2+} selectivity from 78% of case OCE-1 to 88.3% while reducing the single-pass conversion from 20 to 10%, following the selectivity–conversion relationship reported by ARCO. To maintain a high total conversion, the recycle ratio was raised from 8.0 to 15. Results of the process calculations were also given in Table 15-9. As shown, the C_{3+} thermal efficiency increased substantially. However, this increase was negated after crediting a smaller surplus energy due to mainly a smaller steam generation. The higher recycle flow greatly downgraded the reaction heat recovered. The increased recycle flow also contributed to a large increase in the heat-transfer duty. The gas compression and expansion duties were somewhat reduced mainly due to less air to the reoxidation reactor resulting from improved C_{2+} selectivity in the OCE reactor.

In case OCE-3, the OCE reactor effluent cooling was minimized and the olefin conversion reactor feed preheating was completely eliminated. This led to a small improvement in reaction heat recovery, resulting in an increase in steam generation and a decrease in the heat-transfer duty, as shown in Table 15-9. However, all process aspects were still far inferior to those of DPOM cases.

Results of process calculations for case OCE-4 are also summarized in Table 15-9. In case OCE-2, the largest compression duty was from the air to the reoxidation reactor and the only expansion duty was from the reoxidation reactor effluent. Substantial reduction in both duties could be expected by lowering the OCE reactor pressure. However, maintaining high pressure in the olefin conversion reactor and the hydrocarbon fractionation unit resulted in a large increase in the recycle gas compression duty and an addition of a large recycle gas expansion duty. In a way, some of the compression and expansion duties in the reoxidation reactor area were moved to the OCE reactor loop. Furthermore, the recycle ratio was reduced from that of case OCE-2 in order to minimize the compression and expansion duties in the recycle loop. This resulted in a decrease of the C_{3+} LHV and an about equal increase in the fuel-gas LHV. The net C_{3+} thermal efficiency was essentially unchanged.

After postanalysis of the preceding four OCE cases, it was concluded that the only way to have a competitive OCE process was to have single-pass conversion and C_{2+} selectivity concurrently. Large single-pass conversion could be translated into drastically lower recycle, which resulted in large decreases of all process duties. Furthermore, lower OCE and reoxidation reactor pressures together with a lower recycle were an ideal combination to further reduce the compression and expansion duties.

Case OCE-5 was designed based on this train of thought. Results of the process calculations are also given in Table 15-9. In comparing against case OCE-2, the low reoxidation reactor pressure (4.1 atm) greatly reduced the air compression and flue gas expansion duties. The increases in the recycle gas compression and expansion duties were largely moderated by the low 3.5 recycle ratio. The net reduction in the compression duty was quite substantial. The low recycle flow also resulted in an improvement in reaction heat recovery, that is, an increase in steam generation. Accordingly, the net C_{3+} thermal efficiency was raised to an impressive level of 77%. The large decrease in the heat-transfer duty, resulted mainly from the low recycle, was also very significant. Although this case appeared to be quite impressive from various aspects, one must notice two counterpoints. One was that a combination of high selectivity and single-pass conversion was never achieved experimentally. The other was that this case must export large quantity of energy as steam and flue gas.

Case OCE-6 was the last OCE case. The intention was to see if oxygen cofeeding made significant differences in the thermal efficiency and the process duties. Other than the oxygen cofeeding, a small decrease in the C_{2+} selectivity, and an elimination of any carbon formation, the case was exactly the same as case OCE-5. Results of the process calculations are also included in Table 15-9 in comparing against case OCE-5. By eliminating the reoxidation reactor, both the compression (excluding that in an oxygen plant) and expansion duties were significantly reduced. However, the saving in the compression work was essentially negated by the energy required for the oxygen production. The net C_{3+} thermal efficiency was lower and there were few differences in process duties.

All OCE cases exported relatively large quantities of surplus steam and fuel gas. As discussed earlier, these exports might be either undesirable or impractical.

In summary, conclusions from the OCE case studies were as follows:

- Better than 35% single-pass C-conversion and 88% C_{2+} selectivity were essential in making the OCE processes competitive with the DPOM processes of lower conversion and about the same selectivity.
- Moderate OCE reactor pressures were essential in minimizing the compression duties required for recycle and hydrocarbon separation.
- All OCE cases studied exported an undesirably large quantity of surplus energy.
- Low OCE reactor pressures combined with high single-pass conversions could significantly reduce the compression and expansion duties.
- An increase of the C_{2+} selectivity from 78 to 88% and a corresponding reduction of the single-pass conversion from 20 to 10% showed no clear trade-off.
- Limited process improvements were obtained by minimizing the cooling, separation, and pre-heating steps between the OCE reactor and the olefin conversion reactor.
- Oxygen cofeeding did not show significant impact on the thermal efficiency and process duties.

COMPARISON OF DIRECT AND INDIRECT CONVERSION PROCESSES

Introduction

As discussed earlier, indirect natural gas conversion can be divided into two steps: natural-gas conversion to synthesis gas and synthesis-gas conversion. Conversion of natural gas to synthesis gas is commercially practiced using the following two processes: *steam reforming* and *noncatalytic partial oxidation*. Currently, steam reforming dominates the commercial practice in the field of conversion of natural gas to methanol. The reforming reactions [equations (15-1) and (15-2)] are promoted by Ni-based catalysts. Excess steam is usually needed to prevent carbon formation on the catalysts. The large amount of reaction heat required by the reforming reaction [equation (15-1)] is supplied through reformer tubes by burning fuels outside. 80–90% CH_4 conversion, which is favored thermodynamically by high temperature and low pressure, is usually obtained. However, the maximum temperature inside the reactor is limited by the metallurgy of the reformer tubes whereas the reactor pressure is usually determined by the pressure requirement of the subsequent synthesis-gas conversion step.

Noncatalytic partial oxidation takes place at very high temperatures (1000–1500°C) using oxygen. This process is an extension of the technology of generating synthesis gas by noncatalytic partial oxidation of liquid hydrocarbons (Reed and Kuhre 1979). The Hydrocol plant, built in 1950 in Texas, for converting natural gas to gasoline used this process for synthesis-gas generation (Keith 1946). Today, noncatalytic partial oxidation is much less popular than steam reforming because it is very energy-intensive due to its high-temperature operation.

A brief review of routes that convert synthesis gas to liquid fuels was given by Kuo (Kuo 1984). Among the various routes of synthesis-gas conversion, only the following two routes entertain large-scale operation today:

- Methanol synthesis
- Fischer–Tropsch synthesis

There is a long history of commercial practice of methanol synthesis. Today, the process is dominated by the highly active Cu–Zn based catalysts that operate at 50–250 atm and 250–300°C. The chemical reactions are given as equations (15-3) and (15-4). Chemical equilibrium strongly limits the single-pass conversion. However, 94–97% total C-conversions are achieved using recycling. Thermodynamically, high pressure and low temperature are required in achieving high conversion. The H_2-rich gases obtained from steam reformers are highly unbalanced for methanol synthesis. The addition of CO_2, when available, is a common commercial practice used to improve the stoichiometrical balance.

The commercial methanol synthesis processes are identified mostly by their reactor designs, which are tailored for removing the large amount of reaction heat. Today, the dominant processes are those of Imperial Chemical Industries (ICI) and Lurgi. The ICI process uses an adiabatic, multizoned, interzone quenching reactor (Pettman and Humphreys 1975), whereas the Lurgi process uses a tubular reactor design (Supp 1973). Other than in reactor design, there seems to be little difference in performance between these two processes. The Lurgi process recovers reaction heat as medium-pressure steam and requires less recycling but has a higher heat-transfer duty. The sensible heat in the ICI reactor effluent can also be recovered to provide the steam required in the synthesis-gas formation step. For these reasons, either process can be used as the base case for indirect conversion.

A combination of steam reforming plus ICI-methanol was chosen as the base case for current study because it is the most widely used today and its process data are more accurate and abundant.

There is also a long history of commercial practice of F–T synthesis. However, only one application was based on natural-gas feed, for example, the Hydrocol plant (Keith 1946). The F–T reactions can be represented by the following:

$$(1 + n/2)H_2 + CO = (CH_n) + H_2O \quad (15\text{-}12)$$

$$H_2O + CO = H_2 + CO_2 \qquad \Delta H_f(230°C) = -10 \text{ kcal}$$

$$\Delta G_f = -5 \text{ kcal} \quad (15\text{-}13)$$

These reactions take place over Fe catalysts (200–325°C, 10–30 atm) or Co catalysts (170–205°C, 5–30 atm). The reaction heat is about 15 kcal per g-mol of H_2 + CO converted. The F–T products have a wide C-number distribution and require extensive product fractionation and upgrading facilities as practiced at SASOL today (Dry 1983). To minimize light paraffin formation, low temperature and high pressure are preferable. When compared with the methanol synthesis, the F–T synthesis has four potential deficiencies:

- Undesirable methane + ethane formation
- Lower C-efficiencies due to useless CO_2 in the synthesis gas
- Lower C-efficiencies due to the conversion of CO to CO_2 [reaction (15-13)]
- More-expensive product separation and upgrading

A recycle operation is usually required in order to increase CO conversion, to facilitate reaction heat removal, and to suppress CO_2 formation. Because of these potential deficiencies, the F–T synthesis was not adopted as a base case.

Substantial research and development activities are currently being conducted in the area of indirect methane conversion, including the following processes:

- Steam reformer and catalytic autothermal reforming in series (Supp 1985)
- Catalytic partial oxidation (Korchnak 1987)
- Liquid-phase methanol synthesis (Alpert et al. 1986)
- Low-temperature methanol synthesis (Sapienza et al. 1985)
- Methanol synthesis with solid methanol absorber (Westerterp and Kuczynski 1986).

The status of these processes is far more advanced than the two direct conversion processes discussed here. In fact, the autothermal reforming may already have reached commercial operation. Furthermore, potential synergetic benefits may be derived through proper matching of a synthesis-gas formation process with a synthesis-gas conversion process. For example, Korchnak (1987) reported a potential 25% reduction in investment using a "catalytic partial oxidation + methanol synthesis" scheme. The selected base cases did not include any potential benefits that might result from adopting the aforementioned developing technologies.

Natural Gas to Methanol via Steam Reforming

A schematic flow diagram of this scheme is given in Fig. 15-5. According to equations (15-3) and (15-4), H_2, CO, and CO_2 must have the following stoichiometry:

$$(H_2 - CO_2)/(CO + CO_2) = 2 \tag{15-14}$$

The synthesis gases produced by steam reforming almost always contain more H_2 than that of equation (15-14). However, this imbalance can be remedied when there is CO_2 available. The methanol yield is maximized when an amount of CO_2 is added to the natural gas such that a stoichiometric ratio slightly higher than that of equation (15-14) is obtained after steam reforming. To cover this extreme, the following two natural-gas–to–methanol via steam reforming (SRM) cases were evaluated:

- Case SRM-1—No CO_2 addition to the reformer feed
- Case SRM-2—Maximum CO_2 addition to the reformer feed

In case SRM-2, CO_2 was assumed to be available at 25° C and 1 atm. Note that, according to thermodynamics, both direct conversion processes studied here cannot have any benefit from the CO_2 availability.

A computer program simulating the operation of this process was constructed. It included the mass and heat balances of both the steam reformer and the methanol synthesis reactor, the recycle in the later reactor, and product separations.

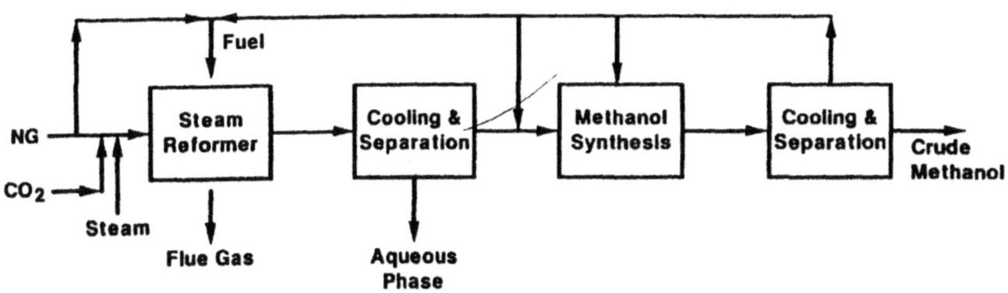

FIGURE 15-5. Schematic flow diagram of natural-gas–to–methanol via steam reforming.

The process natural gas was first compressed and preheated and then was saturated with steam in a saturator using the sensible heat of the methanol reactor effluent. Additional steam was added to make up for a H_2O–(hydrocarbon C atom) ratio of 3. The combined feed was then preheated to 454°C using the reformer effluent. The exit temperature and pressure of the reformer were assumed to be 871°C and 20 atm, respectively. The composition of the effluent was estimated using an 11°C approach to the chemical equilibrium of equations (15-3) and (15-4) using the thermodynamic data from Stull and co-workers (Stull, Westrum, and Sinke 1969). The purge gas from the methanol unit plus a portion of the natural gas were used as fuel in the reformer furnace. The heat duty in the reformer furnace was estimated assuming a 20°C ΔT driving force at the exit of the reformer tubes. The air, preheated to 454°C by the reformer flue gas and in 15% excess, was used in the furnace.

The reformer effluent gas was cooled and compressed further. It was then combined with a recycle gas before charged to an ICI-type methanol synthesis reactor. The methanol reactor exit temperature and pressure were assumed to be 270°C and 96 atm, respectively. The exit compositions were estimated using a 12.5°C approach to the chemical equilibrium. The reactor effluent was then cooled to 37.8°C and a crude methanol stream was withdrawn from the separator as a final product. The purge gas was expanded to 1.7 atm to recover work before being sent to the reformer furnace. A recycle ratio of 5 was used for recycle operation. A pressure drop of 6.8 atm over the recycle loop was assumed.

Table 15-10 summarizes the results of process calculations for both cases SRM-1 and SRM-2. In case SRM-1, 86.5% of the C atoms in the process natural gas was converted to CO and CO_2 and the total CO + CO_2 conversion in the methanol reactor was 96.9%. Both numbers compared favorably with known process data.

Case SRM-2 used the following fresh feed synthesis-gas composition:

$$(H_2 - CO_2)/(CO + CO_2) = 2.03 \tag{15-15}$$

which was slightly above the stoichiometrical ratio of equation (15-14). This ratio was drastically reduced from a ratio of 2.95 for case SRM-1. On a combined-feed basis, this ratio was also reduced from 13.8 for case SRM-1 to 3.8 for case SRM-2. Although there were indications that some ICI methanol reactors were operating using this relatively H_2-poor gas, no process information was available in the public domain. However, a synthesis gas of this ratio was quoted in a case using a Lurgi methanol reactor (Supp 1985) but, again, no process information was made available. With a nearly balanced feed, a recycle ratio about 10% higher might

Table 15-10 Natural-Gas–to–Methanol via Synthesis Gas: Process Efficiencies and Duties

	CASES[a]	
	SRM-1	SRM-2
$H_2 + CO$ formation		
CO_2/C-atoms in feed	0.007	0.274
NG to reactor (%)	96.8	78.3
Useful C atom (%)	86.5	87.3
$(H_2—CO_2)/(CO + CO_2)$	2.95	2.03
Methanol synthesis		
$CO + CO_2$ conversion (%)	96.9	96.4
Thermal products (%NG LHV)		
Crude methanol	65 (66)	68 (68)
Steam	12 (—)	12 (—)
Expansion work	4 (—)	2 (—)
Compression work	−14 (—)	−13 (—)
Loss	33 (34)	31 (32)
Total	100 (100)	100 (100)
C-efficiencies	81 (83)	84 (85)
Major process duties (%methanol LHV)		
Heat transfer	171 (168)	149 (147)
Compression	21	19
Expansion	5	2

[a]Numbers in parentheses are values after credit of surplus steam.

be needed for the same degree of catalyst temperature control. This small difference, however, was not taken into account. An addition of 0.267 mol of CO_2 per hydrocarbon C atom was required for this case.

CO_2 addition to the reformer feed had a large impact on the indirect conversion of natural gas to methanol (Table 15-10). Between case SRM-1 and case SRM-2, the amount of natural gas to the reformer tubes was reduced by 22% for producing the same amount of methanol. There was also a 19% reduction of the moles of fresh synthesis gas fed to the methanol synthesis unit. However, the purge gas from the methanol unit now contained much less energy and a larger fraction of natural gas must not be diverted to the reformer furnace. Nevertheless, about 2–3% increase in thermal efficiency could be expected. The major reduction, however, was in the heat-transfer duty—a 13% reduction.

The following items were included in the thermal efficiency estimation:

- Methanol
- Steam generated from the following heat sources—the steam reformer effluent; the steam reformer furnace flue gas

- Energy consumed for gas compression
- Energy recovered from gas expansion

A thermal energy loss was estimated based on a difference between the total natural-gas LHV and the sum of the preceding items. The estimated heat-transfer duties included the following:

- Heat transfer through steam reformer tubes
- Steam reformer effluent cooling
- Methanol reactor effluent cooling
- Steam reformer flue gas cooling
- The steam-driven turbine condensate coolers
- The compressor precoolers

The estimated gas-compression duties included the following:

- The methanol reactor fresh feed
- Air to the steam reformer furnace
- The methanol reactor recycle gas
- Carbon dioxide (case SRM-2 only)
- Process natural gas

The estimated gas-expansion duties included the following:

- The methanol unit purge gas
- Natural gas to the steam reformer furnace (case SRM-2 only).

Comparison of Direct and Indirect Conversion Processes

Table 15-11 summarizes a comparison between the indirect conversion process and the two direct conversion processes examined earlier. Note that this comparison is inexact because the OCE cases produced C_{3+} whereas the SRM and SPOM cases produced methanol. The methanol produced from the latter cases could, however, be converted to gasoline by attaching a fluid-bed MTG unit to the end of their process schemes. Because no precise data on the thermal efficiency and the process duties were available, the following approximate rules were used:

- A net thermal efficiency of 95.8%
- A heat-transfer duty of 9.6% methanol LHV
- A small compression duty

The net thermal efficiency was the chemical efficiency estimated from a commercial plant design data (Avidan et al. 1986). This was the same as assuming that the energy in the flue gas and the energy recovered from the

Table 15-11 Comparison of Process Efficiencies and Duties[a]

	DESCRIPTIONS				DUTIES (%Product LHV)			
Cases	Conv. (CH_4%)	Selec. (C-Atom%)	Recycle Ratio	Others	Thermal Eff. (%NG LHV)	Heat Transfer	Compression	Expanion
				Methanol synthesis via steam reforming[b]				
SRM-1 (base)	—	—	—	Standard nat. gas	65 (66)	171 (168)	21	5
SRM-2	—	—	—	Max. CO_2 to ref.	68 (68)	149 (147)	19	2
				Direct partial oxidation to methanol[b]				
DPOM-1 (base)	7.5	90	18	Isothermal reactor	69 (73)	83 (76)	21	0
DPOM-2	15	90	9	Isothermal reactor	70 (73)	72 (66)	20	0
DPOM-3	15	80	10	Isothermal reactor	63 (68)	100 (87)	26	0
DPOM-4	7.5	90	30	Adiabatic reactor	70 (70)	124 (125)	23	0
				Gasoline via oxidative coupling to ethylene[c]				
OCE-1 (base)	20	78	8	Two fluid beds	56 (67)	213 (186)	51	20
OCE-2	10	88	15	Two fluid beds	62 (67)	253 (248)	43	15
OCE-3	10	88	15	Cascade reactor	62 (69)	215 (205)	43	15
OCE-4	10	88	10	Medium pressure	59 (66)	232 (227)	53	25
OCE-5	35	88	3.5	Low pressure	63 (77)	124 (97)	28	11
OCE-6	35	85	3.5	O_2 cofeeding	61 (71)	119 (99)	30	8

[a]Numbers in parentheses are values after crediting fuel gas and surplus steam.

[b]These cases produced methanol whereas the six OCE cases produced C_{3+}. A 95% thermal efficiency may be assumed using fluid-bed MTG for methanol conversion.

[c]Substantial energies required for product fractionation in OCE cases are not accounted for.

MTG reaction were balanced by the utility requirement within the unit. The heat-transfer duty was estimated based on the difference of LHVs between the products and the MTG feed at 25°C. The energy needed to provide the compression duty in the fluid-bed MTG unit was judged to be at most a few percent of the feed methanol LHV and would have little impact on the final results.

Using these simple rules, the thermal efficiencies of the SRM and DPOM cases would decrease by less than 5% (in a relative term) for producing C_{3+} instead of methanol. The corresponding heat-transfer duties would increase by less than 14%. These increases were minor and did not have any significant impact on results of our comparison.

With these minor adjustments in mind, the conclusions of the process comparison were as follows:

- A DPOM process could have lower heat-transfer duty and moderately higher thermal efficiency than those of the SRM processes if it could achieve higher than 90% selectivity at 7.5% single-pass conversion or 80% selectivity at 15% single-pass conversion.
- An OCE process could be a potential contender if it could achieve larger than 88% C_{2+} selectivity and 35% single-pass conversion concurrently.

NOMENCLATURE

DOE	Department of Energy
DPOM	Direct partial oxidation to methanol
ERDA	Energy Research and Development Administration
F–T	Fischer–Tropsch
ΔG_f	Gibbs enthalpy of formation (in kilocalories)
ΔH_f	Enthalpy of formation (in kilocalories)
HHV	High heating value (in kilocalories)
ICI	Imperial Chemical Industries
LHV	Low heating value (in kilocalories)
LPG	Liquefied petroleum gas
MTG	Methanol-to-gasoline process
NG	Natural gas
OCE	Oxidative coupling to ethylene
SASOL	South Africa Coal, Oil and Gas Corp., Ltd.
SRM	Steam reforming to make methanol

APPENDIX 15-A GUIDELINES FOR THERMAL AND C-EFFICIENCIES, AND PROCESS DUTIES

Efficiencies are defined as follows:

$$\text{chemical thermal efficiency} = \frac{\text{product LHV}}{\text{process feed LHV}}$$

$$\text{Net Thermal Efficiency} = \frac{\text{product LHV}}{\text{LHV in process feed + fuel}}$$

$$\text{Chemical C-Efficiency} = \frac{\text{C atoms in product}}{\text{C atoms in process feed}}$$

$$\text{Net C-Efficiency} = \frac{\text{C atoms in products}}{\text{C atoms in process feed + fuel}}$$

The following guidelines were used for process schemes:

1. Natural gas was available at 25°C and 13.6 atm.
2. Oxygen was available from an oxygen plant at 98% purity, 25°C and 1 atm.
3. Air and CO_2 were available at 25°C and 1 atm.
4. Fuel gases had a pressure of 1.70 atm.

The following guidelines were used for energy equivalency.

1. Chemical energies: LHV at 25°C, 1 atm.
2. Oxygen: Required 19.6 kcal LHV fuel to produce 1 g-mol of oxygen (Drnevich 1981; Drnevich, Ecelbarger, and Portzer 1981).
3. Thermal energies:
   a. All thermal energies below 176.7°C were discounted.
   b. All excess thermal energies were converted saturated steam at 61.2 atm or superheated steam at 482.2°C and 61.2 atm, whichever was applicable. The boiler feedwater was available at 121.1°C.
4. Steam:
   a. Calculated theoretical work from isentropic expansion to 0.136 atm.
   b. 100 LHV of fuel was equivalent to 42.3 units of theoretical work in steam.

5. Compression work:
   a. Calculated isentropic compression work by limiting compression ratio to less than 3 at each stage.
   b. 100 LHV of fuel was equivalent to 27.0 units of isentropic compression work.
6. Expansion work:
   a. Gases were preheated to a minimum temperature of 148.9°C.
   b. Calculated isentropic work by limiting the expansion ratio so that the gas temperature after expansion was more than 26.7°C.
   c. 100 LHV of fuel is equivalent to 42.3 units of isentropic expansion work.
7. Electricity: 100 LHV of fuel was needed to generate 34.0 units of electricity.

References

Alpert, S.B., R. Wolk, R.W. Kirkland, and R.W. Weatherington. 1986. Coproduction of methanol and electric power in an IGCC plant. Paper presented at the American Society of Mechanical Engineering Region V Technical Conference, June 26–27, at Columbus, OH.

Andorf, R., and M. Baerns. 1990. Oxidative coupling of methane in fluidized- and packed-fluidized-bed reactors. *Catal. Today* 6:445–52.

Avidan, A.A., M. Edwards, W. Loeffler, H.-H. Gierlich, N. Thiagarajan, and E. Nitschke. 1986. The fluid-bed MTG process. Paper presented at the 21st American Chemical Society State-of-the-Art Symposium on Methanol as a Raw Material for Fuels and Chemicals, June 15–18, at Marco Island, FL.

Baerns, M., K. van der Wiele, and J.R.H. Ross. 1989. Methane activation—A bibliography. *Catal Today* 4:471–94.

Breder, E.W., Jr., J.A. Jaecker, and M.F.L. Johnson. 1985. Methane conversion using a magnesia/silica support. US Patent 4,544,786 (Oct. 1).

Burch, R., G.D. Squire, and S.C. Tsang. 1989. Direct conversion of methane into methanol. *J. Chem. Soc., Faraday Trans. 1* 85:3561–8.

Dautzenberg, F.M., R.L. Garten, and G. Klingman. 1986. Fuels from remote natural gas: Defining the research and development challenge. Paper presented at the 21st American Chemical Society State-of-the-Art Symposium on Methanol as a Raw Material for Fuels and Chemicals, June 15–18, at Marco Island, FL.

Drnevich, R.F. 1981. Preliminary oxygen plant assessment for Texaco gasifier-based gasification-combined-cycle systems. Union Carbide Corp., EPRI report AP-1942 (Project 239-5) (July).

Drnevich, R.F., E.J. Ecelbarger, and J.W. Portzer. 1981. Industrial oxygen plants—A technology overview for users of coal gasification-combined-cycle systems. Union Carbide Corp., EPRI report AP-1674 (Project 239-5) (Jan.).

Dry, M.E. 1983. The Sasol Fischer–Tropsch processes. In *Applied Industrial Catalysis*, Vol. 2, ed. B.E. Leach, Chapter 5. New York: Academic Press.

Edwards, J.H., K.T. Do, and R.J. Tyler. 1990. Reaction engineering studies of methane coupling in fluidized-bed reactors. *Catal. Today* 6:435–44.

Edwards, J.H., and N.R. Foster. 1986. The potential for methanol production from natural gas by direct catalytic partial oxidation. *Fuel Sci. Tech. Int.* 4:365–90.

Edwards, J.H., and R.J. Tyler. 1989. The oxidative coupling of methane in a fluidized-bed reactor. *Catal. Today* 4:345–54.

Fox, J.M., III, T.-P. Chen, and B.D. Degen. 1990. An evaluation of direct methane conversion processes. *Chem. Eng. Prog.* 86(4):42–50.

Gaffney, A.M. 1985a. Methane conversion. US Patent 4,499,323 (Feb. 12).

Gaffney, A.M. 1985b. Methane conversion. US Patent 4,499,324 (Feb. 12).

Geerts, J.W.M.H., J.H.B.J. Hoebink, and K. van der Wiele. 1990. Methanol from natural gas. Proven and new technologies. *Catal. Today* 6:613–20.

Gesser, H.D., N.R. Hunter, and L. Morton. 1986. Direct conversion of natural gas to methanol by controlled oxidation. US Patent 4,618,732 (Oct. 21).

Gesser, H.D., N.R. Hunter, and L.A. Morton. 1989. The partial oxidation of methane to methanol at high pressures: A status report. Paper presented at the 1989 International Chemical Congress of Pacific Basin Societies, Dec. 17–20, at Honolulu, HI.

Gesser, H.D., N.R. Hunter, L. Morton, P.S. Yarlagadda, and D.P.C. Fung. 1987. The direct conversion of methane to methanol by a high pressure partial oxidation. Paper presented at the 194th American Chemical Society National Meeting (FUEL), Aug. 31–Sept. 4, at New Orleans, LA.

Gesser, H.D., N.R. Hunter, and C.B. Prakash. 1985. The direct conversion of methane to methanol by controlled oxidation. *Chem. Rev.* 85:235–44.

Hunter, N.R., H.D. Gesser, L.A. Morton, and D.P.C. Fung. 1985. The direct conversion of natural gas to methanol by the controlled oxidation at high pressure. Paper presented at the 35th Canadian Chemical Engineering Conference, Oct. 6–9, at Calgary, Canada.

Hunter, N.R., H.D. Gesser, L.A. Morton, and C.B. Prakash. 1984. The direct conversion of methane to methanol. In *Proceedings of the VI International Symposium on Alcohol Fuels Technology, May 21–25, at Ottawa, Canada*, Vol. 2, pp. 2/147–53.

Hunter, N.R., H.D. Gesser, L.A. Morton, P.S. Yarlagadda, and D.P.C. Fung. 1986. The direct conversion of natural gas to alcohols. Paper presented at the VII International Symposium on Alcohol Fuels, Oct. 20–23, at Paris, France.

Hunter, N.R., H.D. Gesser, L.A. Morton, P.S. Yarlagadda, and D.P.C. Fung. 1987. The catalyzed conversion of methane to methanol at high pressure. Paper presented at the 194th American Chemical Society National Meeting (PETR), Aug. 31–Sept. 4, at New Orleans, LA.

Jaecker, J.A. 1986. Preparative process for methane conversion agents. US Patent 4,568,785 (Feb. 4).

Jaecker, J.A., and M.F.L. Johnson. 1986. Hydroxylated magnesia support. US Patent 4,613,718 (Sep. 23).

Johnson, M.F.L. 1985. Methane conversion using a silica/Magnesia support. US Patent 4,547,608 (Oct. 15).

Jones, C.A., J.J. Leonard, and J.A. Sofranko, 1984a. Methane conversion. US Patent 4,443,644 (Apr. 17).

Jones, C.A., J.J. Leonard, and J.A. Sofranko. 1984b. Methane conversion. US Patent 4,443,645 (Apr. 17).

Jones, C.A., J.J. Leonard, and J.A. Sofranko. 1984c. Methane conversion. US Patent 4,443,646 (Apr. 17).

Jones, C.A., J.J. Leonard, and J.A. Sofranko. 1984d. Methane conversion. US Patent 4,443,647 (Apr. 17).

Jones, C.A., J.J. Leonard, and J.A. Sofranko. 1984e. Methane conversion. US Patent 4,443,648 (Apr. 17).

Jones, C.A., J.J. Leonard, and J.A. Sofranko. 1984f. Methane conversion. US Patent 4,443,649 (Apr. 17).

Jones, C.A., J.J. Leonard, and J.A. Sofranko. 1984g. Methane conversion. US Patent 4,444,984 (Apr. 24).

Jones, C.A., J.J. Leonard, and J.A. Sofranko. 1985a. Methane conversion process. US Patent 4,547,607 (Oct. 15).

Jones, C.A., J.J. Leonard, and J.A. Sofranko. 1985b. Methane conversion. US Patent 4,554,395 (Nov. 19).

Jones, C.A., J.J. Leonard, and J.A. Sofranko. 1985c. Methane conversion. US Patent 4,560,821 (Dec. 24).

Jones, C.A., J.J. Leonard, and J.A. Sofranko. 1986a. Two-step methane conversion. US Patent 4,567,307 (Jan. 28).

Jones, C.A., J.J. Leonard, and J.A. Sofranko. 1986b. The oxidative conversion of methane to higher hydrocarbons. Paper presented at the 1986 Spring Symposium of the Catalysis Club of Philadelphia, May 14, at University of Delaware, Newark, DE.

Jones, C.A., J.J. Leonard, and J.A. Sofranko. 1987a. Fuels for the Future: Remote Gas Conversion. *J. Energy and Fuels* 1:12–16.

Jones, C.A., J.J. Leonard, and J.A. Sofranko. 1987b. The oxidative conversion of methane to higher hydrocarbons over alkali-promoted Mn/SiO_2. *J. Catal.* 103:311–19.

Jones, C.A., J.J. Leonard, and J.A. Sofranko. 1987c. Methane conversion. US Patent 4,665,260 (May 12).

Jones, C.A., J.J. Leonard, and J.A. Sofranko, and H.P. Withers, Jr. 1985. Methane conversion process. US Patent 4,523,049 (Jun. 11).

Jones, C.A., J.J. Leonard, J.A. Sofranko, H.P. Withers, E.W. Breder, Jr., M.F.L. Johnson, and J.A. Jaecker. 1986. Alkali promoted manganese oxide compositions containing silica and/or alkaline earth oxides. US Patent 4,629,718 (Dec. 16).

Jones, C.A., and J.A. Sofranko. 1985a. Methane conversion. US patent 4,495,374 (Jan. 22).

Jones, C.A., and J.A. Sofranko. 1985b. Methane conversion. International Patent Application WO 85/00804 (Feb. 28).

Jones, C.A., and J.A. Sofranko. 1985c. Methane conversion process. US Patent 4,523,050 (June 11).

Jones, C.A., and J.A. Sofranko. 1985d. Methane conversion. US Patent 4,547,611 (Oct. 15).

Jones, C.A., and J.A. Sofranko. 1987. Alkali promoted manganese oxide compositions containing zirconium. US Patent 4,650,781 (Mar. 15).

Josefowicz, J.Y. 1987. Reducible metal oxide compositions containing zirconium oxide, yttrium oxide and an alkali metal component. US Patent 4,656,155 (Apr. 7).

Keith, P.C. 1946. Gasoline from natural gas. *Oil and Gas J.* 104–12 (June 15).

Korchnak, J. 1987. Advances in methanol technology. Paper presented at the Auckland Methane Conversion Symposium, Apr. 26–30, at Auckland, New Zealand.

Kuo, J.C.W. 1984. Gasification and indirect liquefaction. In *The Science and Technology of Coal and Coal Utilization*, ed. B.R. Cooper and W.A. Ellingson, Chapter 5. New York: Plenum Publishing Corp.

Kuo, J.C.W. 1987. Evaluation of direct methane conversion to higher hydrocarbons and oxygenates, Mobil Research & Development Corp., DOE final report DOE/PC/90009-3 (DOE Contract No. DE-AC22-86PC90009, Aug.).

Labinger, J.A., and K.C. Ott. 1987. Mechanistic studies on the oxidative coupling of methane. *J. Phys. Chem.* 91:2682–4.

Leyshon, D.W., and R.A. Bader. 1989. Thin bed cofeed reactors for methane conversion. US Patent 4,876,409 (Oct. 24).

Maffia, G.J. 1985. Natural gas conversion. US Patent 4,533,780 (Aug. 6).

Mazurek, H. 1987. Hydrocarbon conversion process using a molten salt. US patent 4,665,261 (May 12).

Pettman, M.J., and G.C. Humphreys. 1975. Improved designs to save energy. *Hydrocarbon Processing* 54(1):77–81.

Reed, C.L., and C.J. Kuhre. 1979. Make syn gas by partial oxidation. *Hydrocarbon Processing* 58(9):191–4.

Sapienza, R.A., W.A. Slegeir, D. Mahajan, M. Foran, T.E. O'Hare, and G.T. Skaperdas. 1985. Low temperature methanol synthesis catalyst and process development. Paper presented at the Tenth Annual EPRI Contractors' Conference on Clean Liquid and Solid Fuels, Apr. 23–25, at Palo Alto, CA.

Schreiner, M. 1978. Research guidance studies to assess gasoline from coal by methanol-to-gasoline and SASOL-type Fischer–Tropsch technologies. Mobil Research & Development Corp., DOE final report FE-2447-13 (DOE Contract No. EF-76-C-01-2447, Aug.).

Sofranko, J.A. 1985. Hydroxylated magnesia support. US Patent 4,517,398 (May 14).

Sofranko, J.A. 1987. Alkali promoted manganese oxide compositions containing barium carbonate. US Patent 4,654,459 (Mar. 31).

Sofranko, J.A., and C.A. Jones. 1985. Methane conversion. US patent 4,544,784 (Oct. 1).

Sofranko, J.A., J.J. Leonard, and C.A. Jones. 1987. The oxidative conversion of methane to higher hydrocarbons. *J. Catal.* 103:302–10.

Sofranko, J.A., J.J. Leonard, C.A. Jones, A.M. Gaffney, and H.P. Withers. 1988. Catalytic oxidative coupling of methane over sodium-promoted Mn/SiO_2 and Mn/MgO. *Catal. Today* 3:127–35.

Sofranko, J.A., and H.P. Withers, Jr. 1985. Methane conversion. US Patent 4,547,610 (Oct. 15).

Stull, D.R., E.F. Westrum, and G.C. Sinke. 1969. *The Chemical Thermodynamics of Organic Compounds*. New York: John Wiley & Sons.

Supp, E. 1973. Technology of Lurgi's low pressure methanol process. *Chem. Tech.* 3:430–5.

Supp, E. 1985. Improved methanol production and conversion technologies. *Energy Prog.* 5:127–30.

Voltz, S.E., and J.J. Wise. 1976. Development studies on conversion of methanol and related oxygenates to gasoline. Mobil Research & Development Corp., ERDA final report FE-1773-25 (ERDA Contract No. EX-76-C-01-1773, Nov.).

Westerterp, K.R., and M. Kuczynski. 1986. Retrofit methanol plants with this converter system. *Hydrocarbon Processing* 65(11):80–3.

Withers, H.P. 1984. Methane conversion. US Patent 4,489,215 (Dec. 18).

Withers, H.P. 1986. Methane conversion. US Patent 4,593,139 (Jun. 3).

Withers, H.P., Jr., C.A. Jones, J.J. Leonard, and J.A. Sofranko. 1987a. Methane conversion process. US Patent 4,634,800 (Jan. 6).

Withers, H.P., Jr., C.A. Jones, J.J. Leonard, J.A. Sofranko, and A.M. Gaffney. 1987b. Methane conversion process. US Patent 4,670,619 (Jun. 6).

Withers, H.P., and J.A. Sofranko. 1985. Methane conversion. US Patent 4,544,785 (Oct. 1).

Yarlagadda, P.S., L.A. Morton, N.R. Hunter, and H.D. Gesser. 1987. Direct catalytic conversion of methane to higher hydrocarbons. *Fuel Sci. Tech. Int.* 5(2):169–83.

Yarlagadda, P.S., L.A. Morton, N.R. Hunter, and H.D. Gesser. 1988. Direct conversion of methane to methanol in a flow reactor. *I&EC Res.* 27:252–6.

16

Summary

Eduardo E. Wolf

University of Notre Dame, Notre Dame

When I decided to write an overall summary of the book, I had the hope that from the sum of all the parts a grand generic mechanism, as well as engineering considerations different than those that have been presented in each chapter, would emerge. I am not sure that such a synthesis has emerged but I will attempt to present a summary nonetheless. After some pondering I decided to summarize first the key points made by each author and then to incorporate all the common elements in some generic reaction pathway. The current table of contents is not the original one and, with the exception of the chapter by Lunsford, it is not based on the authors' importance but is in a sequence that deals first with the mechanistic studies, followed by kinetic studies, then catalyst characterization, and finally with engineering and economic considerations. This chapter follows the sequence of the table of contents; afterward I propose some generic reaction pathway, followed by recommendations for further research.

Chapter 1, by Lunsford, focuses on the formation of methyl radicals as the source of C_2 products, a theme that permeates this book, so it is the appropriate starting point of the book. The nature of Lunsford's work is fundamental and aims at establishing a relation between methane conversion and the generation of methyl radicals. The results for Li/MgO and related catalysts are presented first, followed by those on lanthanide oxides and transition-metal oxides.

It is clearly demonstrated that methyl radicals are produced on alkali-promoted Li/MgO and Na/CaO catalysts, as the rate increases with alkali promotion, and that the radicals couple in the gas phase. Once the radicals are generated, they can diffuse into the bulk gas phase and be detected as such or they can couple within the boundary layer surrounding the catalyst. Several factors are involved in methyl radical production and in the coupling reaction: (1) OH^- ions must be formed on the surface; (2) the presence of CO_2 inhibits both methane conversion and methyl radical production; (3) the presence of a third body, seldom mentioned, is needed for the coupling reaction. Regarding the reactions of methyl radicals on the surface, it is shown that MgO does not adsorb, or oxidize such radicals at 750° C.

In the case of lanthanide oxide catalysts, it was found that a good correlation existed between the basicity of the oxide and methyl radical generation except for those oxides with multiple oxidation states wherein methyl radicals undergo a reductive addition through which CO_2 formation occurs. Addition of Na_2CO_3 decreases the ability of CeO_2 to oxidize methyl radicals due to its coverage of the surface with sodium compounds. In the case of metal oxides, such as $LiNiO_2$ and $NaMnO_4/MgO$, the C_2 yield correlates well with the production of methyl radicals present in the bulk gas phase without a measurable contribution of surface-generated C_2 species. The sticking coefficients of ZnO, and CeO_2, the most active catalysts for the oxidation of methyl radicals, were severalfold higher than on MgO. The main theme of Lunsford's work is that those catalysts that are active in producing methyl radicals are active and selective for the coupling reaction. Oxidation of the methyl radicals can occur on surfaces or in the gas phase. Although O^- species are mentioned throughout the chapter as the sites for hydrogen abstraction, the nature of the active sites is not emphasized.

The second chapter of this book was placed this early because the theme of gas-phase reactions is mentioned in several other chapters, and it was more appropriate to discuss this issue at the beginning. Chapter 2 basically focuses on studies of activity and selectivity both without and with a catalyst. The main theme is that, under a specific set of operating conditions, significant C_2 selectivities are observed in the absence of a catalyst. It is also shown that C_2 hydrocarbons can be oxidized in the gas phase at temperatures lower than those used in oxidative coupling. However, the fact that these species survive during the coupling reaction indicates that such reactions could be inhibited under reaction conditions or that the surviving products are generated only near the reactor exit. In the gas phase, selectivity decreases with conversion. Catalysts modify the reaction pathway by increasing both the rate of methane conversion and the selectivity, which does not necessarily decrease with conversion in the presence of catalysts. The catalyst also modifies the pathway for oxidation of the C_2 products and leads to the formation of CO_2 instead of CO as observed in the gas phase. The gas-phase results are consistent with the reaction proceeding via methyl radicals that couple in the gas phase. A reaction network based only on methane combustion kinetics does not fit experimental results without introducing additional parameters.

The contributions of the prolific group of Professor Otsuka and collaborators are summarized in Chapter 3. The results and conclusions are mainly based on kinetic and isotopic exchange experiments focusing on Sm_2O_3 and $LiNiO_2$ catalysts, although other catalysts also were studied. Otsuka and Hatano find that the kinetics of all the catalysts that they have studied, except FeNbB–O, follow a generic expression containing five constants and that this expression can be derived from a Langmuir–Hinshelwood type of mechanism involving adsorbed CH_4 and molecular or dissociatively adsorbed oxygen. Isotopic CH_4 and CD_4 experiments show that hydrogen abstraction is the rate-determining step and that methyl radicals couple to form C_2H_6. The role of oxygen species was investigated for Sm_2O_3 and $LiNiO_2$ and it was found to be quite different depending on the catalyst. On Sm_2O_3 the main pathway involves adsorbed oxygen and the active sites are assumed to be associated with O_2^{2-} superoxide species. On $LiNiO_2$ these authors found a different kinetic expression and concluded that lattice oxygen associated with $Ni^{3+}O^{2-}$ species was involved.

Chapter 4 summarizes work by Ekstrom, which focuses upon mechanistic studies and their process implications. To establish the mechanism of CO_x formation, addition of ^{13}C-labeled ethane and ethylene were used, resulting in a decrease of unlabeled C_2 products and in an increase in the contribution of total CO_x formation. The decrease in methane conversion is attributed to the competition between methane and the added hydrocarbons for the same sites. A reaction pathway involving these features is proposed, emphasizing the formation of CO_x from C_2 species and the competition of the latter with methane for the same sites. Oxygen is assumed to form two types of oxidant, one responsible for the formation of carbon oxides and the other as the precursor of free radicals. On the basis of $^{18}O_2$ exchange experiments, a mechanism incorporating oxygen exchange to regenerate O^- species (presumably for the second type of oxygen) is proposed. The effect of pressure was also investigated and these authors found agreement with previous results with blank runs on gas-phase reactions. The effect of pressure depended on the catalyst used.

The contribution by Peil, Marcelin, and Goodwin (Chapter 5) summarizes their work utilizing a steady-state transient isotopic technique. Using this method, the authors found that the diffusivity of lattice oxygen increased by an order of magnitude when Li was added to MgO and it was about a hundredfold higher for monoclinic Sm_2O_3 and thousandsfold higher for the cubic form. The number of equivalent surface layers of lattice oxygen exchange also increased in the same sequence. The diffusivity of lattice oxygen on SnO and SnO/MgO was found to be of the same order of magnitude as Li/MgO, but in NiO was of the same order as on the cubic form of Sm_2O_3. Similar studies made in the presence of methane lead these authors to the conclusion that lattice oxygen participated in the reaction in all the catalyst studied. In the case of Sn/MgO the presence of methane increased oxygen exchange. From these results the authors concluded that the catalysts with the lowest lattice oxygen diffusivity were the least selective and as the diffusivity increased so did the C_2 selectivity. In the case of NiO this correlation did not hold even though this catalyst was very active for total methane oxidation. This discrepancy in the trend between oxygen mobility and product selectivity was attributed to a nonselective oxygen species, the form of which is not specified. These authors also concluded that during reaction, lattice oxygen also participated in the formation of CO_2 whereas in the absence of gas-phase oxygen, lattice oxygen is preferentially involved in the formation of C_2H_6.

The contribution from J. Ross' group, Chapter 6, summarizes activity–selectivity results on Li/MgO and in the development of Li/X/MgO low-temperature catalysts. The stability of Li/MgO is addressed first, and it is reported that catalyst deactivation occurs due to Li losses; addition of CO_2 decreases the activity but the effect is less severe when quartz is present. Likewise, addition of water resulted in Li losses due to evaporation of volatile LiOH. A reaction model for the Li/MgO catalyst incorporates several Li species, including $LiCO_3$, LiO_2, Li_2SiO_3, and LiOH as well as Li^+, which is assumed to be associated with the active site. Reactor considerations are presented and it is demonstrated that back-mixing is detrimental to methane conversion. Using a continuous stirred tank reactor (CSTR) reactor, these authors corroborated the sequential reaction pathway $CH_4 \rightarrow C_2 \rightarrow C_2^{=} \rightarrow CO_x$.

A number of oxides supported on Li/MgO were tested in the search for a low temperature catalyst. Among several oxides tested, it was found that SnO reduced the temperature for maximum C_2 yield, with a corresponding increase in the stability of Li. The decrease in temperature and the increase in stability was attributed to the higher capacity for oxygen adsorption of the added oxide. The reaction pathway was not altered by addition of the dopant.

Chapter 7, by Hutchings and Scurrell, focuses on the effects of various oxidants in the elucidation of the reaction mechanism on Li/MgO. The effect of N_2O, ozone, and mixed feeds is reviewed, and a good deal of references are presented. At similar conversions N_2O was found to be more selective than O_2, whereas the use of ozone did not enhance activity or selectivity. In studies with mixtures of methane–oxidant–NO, addition of NO decreased significantly the C_2 selectivity with a concomitant increase in CO_2 production when NO was added. This effect was not observed on MgO. In the case of mixtures containing CO_2 it was found that addition of 10% CO_2 inhibited C_2 production, but this effect was not found at low CO_2 feed concentrations. These results are explained by a competitive adsorption of NO and CO_2 with the oxidant.

These authors also report that on MgO the activity depended on the crystal morphology, and they propose that the mechanism in the unpromoted oxide might be different than in the alkali-promoted oxide. This proposal is based on their previous studies with N_2O, in which the authors speculate the existence of two types of oxygen, a selective and a nonselective species. The issue about the primary product of oxidative coupling is discussed at length but it is concluded that it is mainly ethane. Some discussion is also devoted to the active structural form of samarium. The mechanistic picture proposed is similar to what has been discussed in previous chapters.

Chapter 8 by van der Wiele and collaborators focuses on activity and kinetic studies and on modeling the reaction pathway. It presents results corroborating that methane conversion to C_2 hydrocarbons occurs in the gas phase in the absence of a catalyst and that selectivity decreases with increasing conversion. These authors found that the gas-phase reactions are not affected by addition of CO_2, whereas addition of water and hydrogen accelerates slightly the reaction. Addition of ethane increases methane conversion but decreases the selectivity. A complex kinetic model based solely on gas-phase results for methane combustion is presented; however, it could be reduced to a pathway consisting of reactions in series involving methyl radical coupling, ethane dehydrogenation to ethylene, and further oxidation to CO. Studies with Li/MgO result in the same type of pathway as for the gas-phase reaction, except that the surface also catalyzes the dehydrogenation of ethane to ethylene and the oxidation of ethylene to CO and to CO_2. These authors concluded that little oxidation of ethane to CO and CO_2 occurs on the catalyst. The conversion of CO to CO_2 is seen mainly as a catalytic step. Detailed kinetic models based only on methane combustion kinetics were not satisfactory to explain all the experimental trends observed unless arbitrary reaction parameters are used to fit the results.

Chapter 9, by McCarty, dwells on efforts to model the reaction pathway utilizing a combination of homogeneous and heterogeneous steps in a similar way as done in the previous chapters. The model incorporates explicitly the concentration of O_2 sites and the rate constants of some of the heterogeneous steps are fitted to experimental data obtained on a Na/CaO catalyst. The model overpredicts the concentration of formaldehyde and underpredicts the concentration of CO_2, whereas it predicts qualitatively other trends such as the effect of residence time and oxygen partial pressure. The simulation also predicts that the C_2 selectivity reaches a maximum as the concentration of active sites increases. The effect of diffusion is also analyzed using a unidimensional geometry and selective homogeneous–heterogeneous reactions. It is found that for a 25-μm particle most methyl radicals couple within the pores, so intraphase diffusional mass transport should affect C_2 selectivity.

The analysis also involves thermodynamic calculations of alkali and alkaline-earth concentrations of carbonate, oxides, and hydroxides to estimate the concentration of the most stable oxygen anion. The calculations indicate that superoxide and peroxide anions are the most likely species to populate the catalyst surface. A calculation based on the oxygen concentration predicted by thermodynamics follows qualitatively the trends observed by others about the effect of gas composition and temperature on methane conversion.

Chapter 10, a contribution by Mirodatos and Martin, summarizes results on the loss of matter in PbO- and Li_2O-supported catalysts. Studies of the effect of the support on Li volatilization indicates that the stability varies according to $Cr_2O_3 > Al_2O_3 > SiO_2 \gg MgO$ and that this effect increases from Li to K. The stability of Li deposited on basic Mg silicates was similar to that of Li deposited on silica. The concentration of CO_2 and H_2O in the reaction environment also affects the losses of Li. The effect of surface area on the methane oxidative coupling reaction is analyzed, but no firm conclusions are reached because in such studies changes in area are also accompanied by changes in surface composition. Studies of the effect of sintering indicate that it depends on the alkali and the oxide used according to $MgO > SiO_2 = Cr_2O_3 > Al_2O_3$ and K, Na > Li. Sintering also occurs during reaction. The structure sensitivity of the reaction in different oxide catalysts is also reviewed, along with a discussion of the phase organization of various Li-supported catalysts.

Chapter 11 deals with the effect of catalyst acidity and basicity, and summarizes studies by M. Baerns' group on PbO, alkali–alkaline-earth compounds, and La_2O_3/CaO catalysts. This group reports that surface acidity is detrimental to C_2 selectivity whereas basicity (the less-acid materials such as MgO) is a requirement for the formation of C_2 hydrocarbons. The selectivity of alkaline-earth metal oxides also increases with basicity but the volatility of some of these materials prevents its application to methane oxidative coupling. The basicity concept when applied to a La_2O_3/CaO catalyst did not confirm the predictions of the correlation.

The chapter on partial oxidation of methane to methanol is the lone contribution in this area because not too much information is available in the literature. A review of the various methods of activation of the C—H bond, excluding oxidative coupling, is presented, as well as the recent work on this conversion route. The engineering aspects of the conversion of methane to methanol are discussed later in the chapter by Kuo.

Although only three contributions are presented about the engineering aspects of oxidative conversion processes, some are very detailed. The chapter by Edwards, Do, and Tyler summarizes engineering work done on a process combining the high exothermicity of the oxidative coupling reaction with alkane pyrolysis—the OXCO process. The key of the process is the demonstration that the coupling and pyrolysis reactions can occur at acceptable conversion levels in a single fluidized-bed reactor. The pyrolysis reactions are conducted by injecting the alkanes separated from natural gas and those produced in the process into the oxygen-free zone of the fluid-bed reactor to yield mainly olefins. The olefins produced are oligomerized in a separate process to yield gasoline and diesel fuels. The carbon efficiency can be improved by adding a methanation unit. The effect of the various operating variables in a small-scale fluidized-bed reactor is presented for proprietary catalysts and compared with the performance of Li/MgO catalysts. A computer simulation of the integrated OXCO process based on the laboratory results indicates that the performance of the integrated plant is equivalent to a conversion per pass of the carbon input of about 29% with an overall selectivity to unsaturates of 86%.

One of the few examples including economic evaluations is Chapter 14 by Matherne and Culp, and it is a good illustration of the engineering complexities associated with a full chemical plant, which goes well beyond mechanistic details. The purpose was to provide economic direction to the proper process configuration and operating conditions in terms of methane conversion and selectivity. The process consists of a configuration that includes methane coupling with a plant to convert olefins to gasoline (MOGD) and an optional refining system to produce high-purity ethylene. A fixed-bed and a fluidized-bed reactor are considered, both producing superheated steam from heat removal. The reaction products are cooled and then compressed before entering a CO_2 removal unit and a drying system. The dried products are then cooled to separate methane, which is recycled to the reactor, and C_2 and heavier hydrocarbons, which are sent to the MOGD process where they are converted to gasoline, with ethane passing through this process unconverted. Alternatively, the liquids can be sent to a refining system to produce high-purity ethylene. The ethane rejected is sent to a steam cracker where it is selectively converted to ethylene, which is recycled back to the MOGD process. A maximum oxygen concentration of about 10%, dictated by explosion limits, indicates the desirability of using a diluent to increase conversion per pass. Catalyst performance was based on a Union Carbide $BaCO_3/Al_2O_3$ catalyst and on results published by ARCO for a $Mn/Na/P/SiO_2$ catalyst.

Four basic cases are analyzed: three on a cofed fixed-bed reactor using nitrogen, methane (i.e., excess methane), or steam as diluent, and a fourth case in a sequential fluidized-bed reactor (ARCO process). Conversion and selectivities vary in each case. The calculations are presented for the production of ethylene assuming that the same economics indicators will apply for gasoline production. The results for these basic cases indicate that the cofed process with methane diluent and the sequential feed have the best economics, the main components of the cost being the fixed costs and the cost of methane as operating costs (assumed to be 5.1¢ per pound). A sensitivity study indicates that a high methane to oxygen ratio of 7 : 1 results in low conversions (18%) but also in higher selectivities (77%), which yield the lowest costs. A comparison of the economics of producing gasoline via oxidative coupling versus the alternative route of producing synthesis gas, then methanol, followed by its conversion to gasoline (i.e., the Mobil plant in New Zealand) indicates that the economics are comparable, although the price of the liquid product is about $2 per gallon, which is much higher than current prices in the United States. A sensitivity study of the effect of an hypothetical catalyst capable of yielding about 37% conversion with an 85% selectivity (31.4% yield) meets the 25% return of investment after taxes that is required to attract investment.

The chapter by Kuo analyzes the engineering aspects of methane conversion to methanol and to ethylene via oxidative coupling. Process comparisons are given in terms of thermal efficiencies, heat-transfer duties, and gas-compression and gas-expansion duties, which are related to the facilities for heat transfer and separations. The direct conversion of methane to methanol at high pressure, based on the data of Gesser and Hunter of 7.5% conversion and 90% selectivity, resulted in a thermal efficiency of 73%. A process based on the conversion of methane to produce gasoline using the ARCO sequential feed oxidative coupling process indicates a lower thermal efficiency of 56% and larger heat-transfer duties than the direct oxidation, although this process yielded a different product than the direct oxidation of methane to methanol. In the configurations analyzed, a large quantity of undesirable surplus energy is produced. Increasing the selectivity from the base case of 78% to 88% with the corresponding increase of conversion from 20 to 35%, resulted in comparable thermal efficiencies and heat-transfer duties than the direct methane conversion to methanol. Finally, calculations for conversion of natural gas to methanol via conventional steam reforming plus methanol synthesis resulted in thermal efficiencies of 66% and heat-transfer duties higher than the direct conversion to methanol but lower than base oxidative coupling case. No economic comparisons are presented and for this reason it is difficult to reach a final verdict among all the various processes considered because there are significant differences in process costs and configurations among the various processes analyzed.

Summarizing the economic and engineering calculations, it can be said that even though there are orders of magnitude difference in the complexity between the bench-scale laboratory studies and an integrated chemical plant yielding gasoline or other chemical product, the engineering performance and economic estimates depend critically on the conversion and selectivity attained in the methane conversion step. The two studies presented in Chapters 14 and 15 point out the need of attaining single-pass conversions of the order of 35–37% and selectivities of the order of 85–88% to achieve commercial feasibility. This justifies all the efforts devoted in the previous chapters to improving our understanding of the catalytic results as well as the reaction mechanism, which is required to increase both conversion and selectivity to the desirable levels, because by now the periodic table has been visited extensively in catalytic screening studies. Only by furthering our knowledge of the reaction mechanism and the role of the catalyst will it be possible to achieve this difficult but hopefully not impossible goal.

Next, I will attempt to synthesize schematically our current basic knowledge of the reaction mechanism based on the results presented in previous chapters as well as on experience in our laboratory. There are several scales (in size or time) to consider in order to understand the process in an integral manner. Only in this manner can all the interacting factors be properly ascertained.

The first and most significant scale is the surface and the gas-phase volume in its vicinity, which is of the order of a micrometer. The interaction of the various species with the surface and its various sites is the subject of many of the studies reported in Chapters 1–12. In most of the mechanisms, the surface is usually considered to be homogeneous with little detail as to the interaction of the promoter with the support. Even in the simplest case of the ubiquitous Li/MgO catalyst there are several phases present: Li can form a solid solution with the oxide support, or it can exist as an oxide or a carbonate. The coverage and morphology of these species depend on the level of loading and the reaction mixture and operating conditions used, which establish the thermodynamically stable phase concentration. Each of these phases will exhibit a different type of oxygen anion, which is the potential site for hydrogen abstraction. The

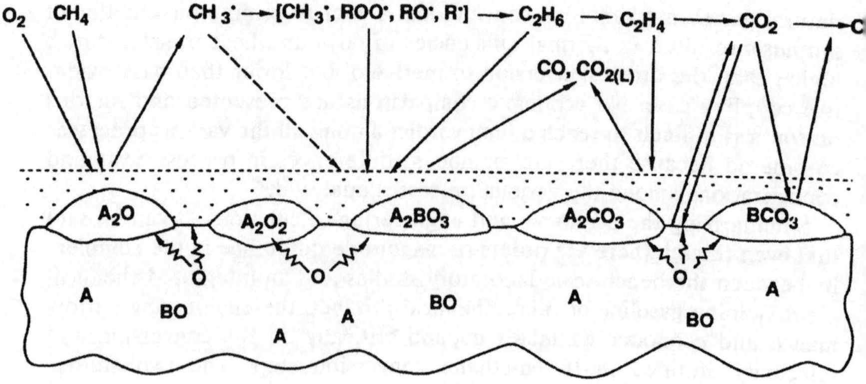

FIGURE 16.1. Schematic representation of the reaction pathway for methane oxidative dimerization on alkali (A) promoted oxide catalyst (BO). The pathway represents the most probable routes, including the coexistence of various phases in equilibrium with the reaction environment.

mobility of oxygen in each phase is also a factor in conversion and selectivity determining the extent of participation of lattice oxygen. If one considers the usual situation of a promoter A_2O supported in an oxide BO, the possible phases present on the surface are those shown in Fig. 16-1, which includes oxides, hydroxides, and carbonates, with their corresponding concentrations of oxygen anions, including oxides, superoxide, and peroxides, each with a population ultimately determined by the gas-phase concentration of oxygen, water, and CO_2, and with a mobility determined by the prevailing crystallographic form. The existence of binary oxides of the type A_2BO_2 is also possible; as well as the various oxides and carbonate of the support. Depending on the temperature, it is possible that some of these phases exist in a quasiliquid form rather than as a crystalline state. The composition of these phases changes with the environment and is quite different at the reaction conditions than at ambient conditions (where most characterization studies are conducted), which makes the evaluation of their contributions rather difficult. The composition of the surface also changes during reaction due to the volatilization of some of the oxides that have a high vapor pressure, resulting in the loss of the active species or in the loss of surface area due to sintering caused by the formation of solid phases with low surface area, a process that is often catalyzed by the active component.

The reacting gases, radicals and products formed interact in a complex manner in a network of reactions that includes unselective gas-phase homogeneous reactions and more-selective surface reactions. The relative importance between the homogeneous and heterogeneous reactions is the key to the overall selectivity. Methane interacts with oxygen anions of the surfaces that abstract hydrogen in an Eley–Rideal (E–R) homolytic-type interaction because methane is not adsorbed by most of the active coupling catalysts at the conditions where the reaction occurs. Furthermore, the fact that the reaction can also occur in the gas phase under the proper reaction conditions also suggests that a homolytic E–R mechanism prevails. It is now clearly established that methyl radicals are produced from the interaction of methane with the surface. These methyl radicals can dimerize in the gas phase and can also produce a myriad of other radicals with relatively short life times of which hydroperoxy and methylperoxy radicals are often quoted as being important. The interaction of these radicals with the other reactants and oxygen species on the surface has not been unequivocally determined except for the case of methyl radicals, which are relatively unreactive on good coupling catalysts whereas they are oxidized to CO_2 on the surface of catalysts with multiple oxidation states.

The reaction products can generate other radicals that can react in the gas phase or interact with the surface to yield additional radicals, or they can be further oxidized to CO_2. This process is seen as a limiting factor to the C_2 yield but the fact of the matter is that all the kinetic models depend on assumptions about the relative rate of the homogeneous unselective gas-phase reactions versus the selective heterogeneous surface reactions. The crucial factor is the selectivity of the surface because no model based exclusively on gas-phase methane combustion kinetics can account for the experimental results except at high pressures, where the relative concentration of gas-phase species surpasses that of the surface. Although the trend of gas-phase results is decreasing selectivities with increasing conversions, the surface clearly alters this pattern to yield higher selectivities at high conversion levels.

The second scale that affects the selectivity and conversion is diffusion of the reactants through the catalyst pores. Model calculations (Chapter 9) indicate that coupling occurs preferentially inside the pores in a particle of 25 μm in diameter. The effect of this time scale is shown in Fig. 16-2 in terms of the intraphase and interphase profiles for methane and ethane inside a catalyst pore. Clearly, higher C_2 selectivities are obtained on catalysts with an open pore structure and low surface area. Most results reported in the literature have been obtained using powders, in which diffusional effects are not important; however, such effects will be relevant in fixed-bed reactors requiring the use of catalysts in a pelletized form.

The third scale (Fig. 16-2) is the reactor configuration, with the two main types of reactor considered being a multiple-tube fixed-bed heat

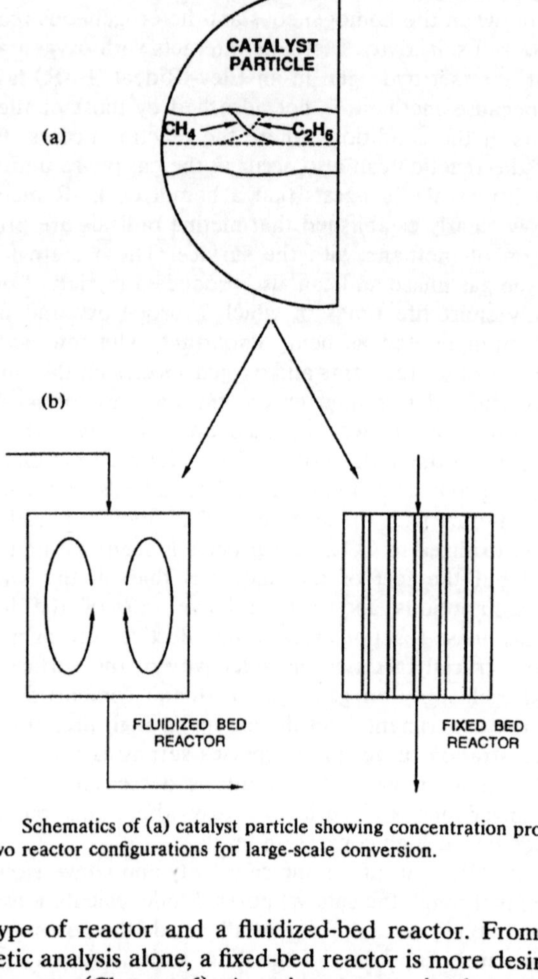

FIGURE 16.2. Schematics of (a) catalyst particle showing concentration profiles inside a pore and (b) two reactor configurations for large-scale conversion.

exchanger type of reactor and a fluidized-bed reactor. From the standpoint of kinetic analysis alone, a fixed-bed reactor is more desirable than a well-mixed reactor (Chapter 6). A major concern in the reactor design, however, is the rejection of large amounts of heat of reaction at a relatively high temperature. This is achieved more efficiently in fluidized-bed reactors than in a fixed-bed reactor. The majority of the results presented in the literature, however, have been obtained in laboratory microreactors in which small beds are used, except for the results reported in Chapter 13, which were obtained in a small pilot-scale reactor. Although mixing is detrimental to selectivity, the results in a fluidized-bed

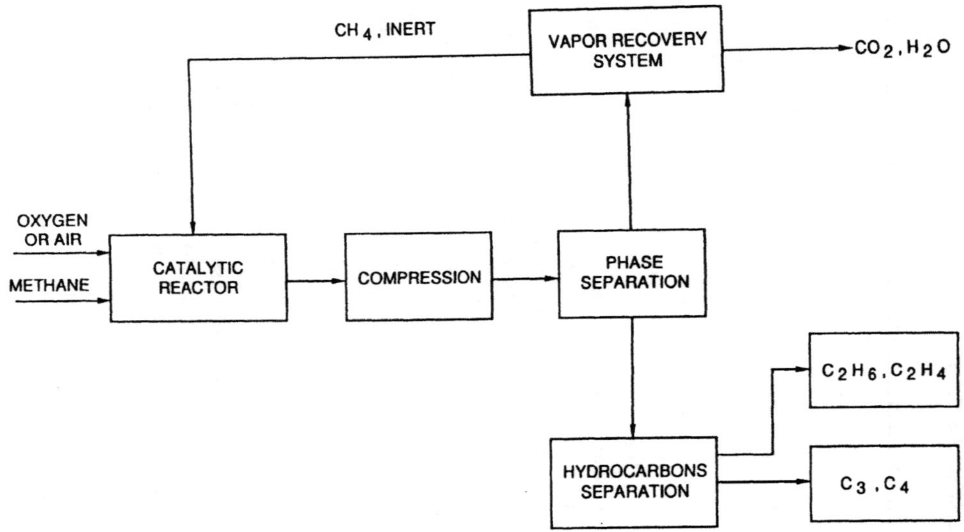

FIGURE 16.3. Schematic of a flow diagram of an integrated plant, showing the main process involved in methane conversion by oxidative coupling process.

reactor suggest a complex mixing mode that does not fit the fully mixed situation. An important aspect of the operation of these reactors is the mechanical resistance of the catalyst to attrition, a condition that is not met by the most-studied catalyst, Li/MgO. Although there is little information in the literature regarding the effect of scaling up on the basis of the effect of oxygen partial pressure on the reaction, it is clear that further reaction engineering work is needed to improve conversion and selectivity. Integration of the oxidative coupling reactor with a pyrolysis unit, which is the basis of the OXCO process, is an example of a promising reactor design that can help to achieve high overall conversion.

The last and largest scale (Fig. 16-3) is the integrated plant that not only includes the catalytic reactor but all the other processes involved in compressing and separating the reaction products, recycling the unconverted product, and transferring heat among the various streams. The equipment and facilities involved translate into fixed and operating costs. The configuration of these processes is crucial to the best thermal efficiency and lowest heat-transfer area and compression and expansion efficiencies. At this moment, the configurations analyzed are assumed to follow established practices common to other processes; however, higher levels of integration can significantly affect the resulting economics. The few economic evaluations available in the literature for relatively nonintegrated processes indicate that the feasibility of oxidative coupling largely depends on the catalyst performance; a conversion in the 35% range and C_2 selectivity of 85% can render this process economically attractive. Of course, the economics are highly sensitive to the price of fuel, which affects not only the products but also the raw materials. Thus the calculations done so far must be constantly revised in light of the changing situation of the price of liquid fuels and the improvements in processes and catalysts affecting conversion and selectivity. In summary, the conversion of methane by oxidative processes is still a promising technology and its economic success rests heavily on catalyst and process improvements.

Index

GPSR Compliance

The European Union's (EU) General Product Safety Regulation (GPSR) is a set of rules that requires consumer products to be safe and our obligations to ensure this.

If you have any concerns about our products, you can contact us on ProductSafety@springernature.com

In case Publisher is established outside the EU, the EU authorized representative is:

Springer Nature Customer Service Center GmbH
Europaplatz 3
69115 Heidelberg, Germany

Batch number: 10584512

Printed by Printforce, the Netherlands